CALCULUS
PROBLEMS AND SOLUTIONS

CALCULUS
PROBLEMS AND SOLUTIONS

A. Ginzburg

DOVER PUBLICATIONS, INC.
Mineola, New York

Bibliographical Note

This Dover edition, first published in 2003, is an unabridged reprint of the edition published by Holden-Day, Inc., San Francisco, 1963.

Library of Congress Cataloging-in-Publication Data

Ginzburg, Abraham
Calculus : problems and solutions / A. Ginzburg.
p. cm.
Originally published: San Francisco : Holden-Day, 1963
Includes bibliographical references and index.
ISBN-13: 978-0-486-43277-9 (pbk.)
ISBN-10: 0-486-43277-7 (pbk.)
1. Calculus—Problems, exercises, etc. I. Title.

QA303.2.G5 2003
515'.076—dc22

2003055518

Manufactured in the United States by LSC Communications
4500056911
www.doverpublications.com

PREFACE

This is the first of a series of problem books in analysis, analytic geometry, and higher algebra. The main purpose of this series is to provide the student with a rich collection of carefully selected material designed to increase his understanding and skill in handling problems in the above fields. To this end a large number of problems are presented and their solutions are given in full detail. To improve understanding, some problems of more difficult character are included, the solution of which requires deeper insight in the topics treated.

More than 1200 problems are presented in this book. Some of these were taken from examination papers of the Technion, Israel Institute of Technology. Others were drawn from various sources and there seems to be no point in endeavoring to trace their origins.

The order of exposition adopted here seems quite natural; however, care has been taken so that the book may be used in a course in which the topics are arranged differently.

Every section begins with a brief explanation of the basic notions and theorems to be used. In general, the theorems are given without proof. The main part of every section is devoted to problems, a large number of which are immediately followed by solutions; others have solutions to be found only at the end of the book. It is believed that this will encourage the reader to solve the problem by himself and only afterwards to look for the printed solution. The comparison of solutions will often be beneficial, as it will afford a check on the work and, occasionally, encounters with new methods.

Some problems, of course, have a variety of solutions and it may easily happen that the one given here is not the simplest possible. Remarks in this and other regards will be welcomed by the author.

My thanks are due to Dr. M. Edelstein, who read the manuscript and made many valuable suggestions, and to Dr. Emory P. Starke who supervised the production of the book.

Haifa, March 1963

A. Ginzburg
Lecturer in Mathematics
Technion, Israel Institute of Technology, Haifa

CONTENTS

I SEQUENCES

1.1 Basic definitions and theorems 1
1.2 Examples and exercises on general notions 3
1.3 Representation of a number by sequences 5
1.4 Evaluation of $N(\epsilon)$ 7
1.5 Sequences given in the form $n_{n+1} = f(u_n)$ 8
1.6 Methods for the evaluation of limits 9

II FUNCTIONS OF A SINGLE VARIABLE

2.1 Definition and notation 12
2.2 The elementary functions 13
2.3 Domain of definition 16
2.4 Even and odd functions 16
2.5 Rational functions 17
2.6 Logarithmic functions 18
2.7 Trigonometric functions 18
2.8 Hyperbolic functions 19
2.9 Inverse functions 19
2.10 The inverse trigonometric functions 20
2.11 The inverse hyperbolic functions 21
2.12 Composite functions 21
2.13 Periodic functions 22

III LIMIT OF A FUNCTION

3.1 Definitions and general exercises 23
3.2 Evaluation of limits 26
3.3 Continuity 30

IV DIFFERENTIAL CALCULUS FOR FUNCTIONS OF A SINGLE VARIABLE

4.1 The notion of derivative and its physical and geometric interpretation 36
4.2 Evaluating derivatives 38
4.3 Evaluating derivatives of explicit functions 42
4.4 Differentiation of implicit functions 45
4.5 Parametric differentiation 46
4.6 Special cases in calculating derivatives 46
4.7 Higher derivatives 49
4.8 Calculation of $y^{(n)}$ 51
4.9 Graphical differentiation 53
4.10 Various examples 55

V FUNDAMENTAL THEOREMS OF THE DIFFERENTIAL CALCULUS

5.1 The theorems of Rolle, Lagrange, and Cauchy 57
5.2 Taylor's and Maclaurin's formulas 60
5.3 Indeterminate forms: L'Hôpital's rule 65

VI APPLICATIONS OF DIFFERENTIAL CALCULUS

6.1 Rate of change 68
6.2 Locating intervals in which a function increases or decreases 69
6.3 Minima and maxima 71
6.4 Concavity: points of inflection 85
6.5 Asymptotes 88
6.6 Curve tracing 93
6.7 Graphs in polar coordinates 101
6.8 Parametric equations 105
6.9 Tangent and normal 107
6.10 The order of contact 113
6.11 Osculating circle, radius of curvature 114
6.12 Evolute and involute 118
6.13 Solution of equations by Newton's approximation method 120

VII THE DIFFERENTIAL

7.1 Definition of the differential 125
7.2 The invariance of the form of the differential 125
7.3 The differential as the principal part of the increment of the function: application to approximate calculations 129
7.4 Higher order differentials 131

VIII THE INDEFINITE INTEGRAL

8.1 Definition and basic properties 134
8.2 Immediate integrals 136
8.3 The method of substitution 140
8.4 Integration by parts 144
8.5 Integrals of rational functions 151
8.6 Irrational integrals 157
8.7 Trigonometric integrals 165
8.8 Integrals of exponential and hyperbolic functions 169
8.9 Miscellaneous integrals 170

IX THE DEFINITE INTEGRAL

9.1 Definition 172
9.2 Basic properties of the definite integral 174
9.3 Evaluation of the definite integral from its definition 176
9.4 Estimation of definite integrals 179
9.5 The mean value theorem of integral calculus 180
9.6 Integrals with variable limits 181
9.7 Evaluation of definite integrals 182
9.8 Changing the variable of integration 184
9.9 Approximate integration 187
9.10 Improper integrals 191
9.11 Miscellaneous problems 195

X APPLICATIONS OF THE DEFINITE INTEGRAL

10.1 Computation of plane areas 200
10.2 Computation of arc length 204
10.3 Computation of volumes 207
10.4 Area of a surface of revolution 214
10.5 Moment of mass: centroids 215
10.6 Pappus' theorems 219
10.7 Moment of inertia 220
10.8 Physics problems 222

XI INFINITE SERIES

11.1 The general notion of a number series 227
11.2 Convergence of series with positive terms 228
11.3 Convergence of series with positive and negative terms 235
11.4 Arithmetic operations on series 237

11.5 Series of functions 240
11.6 Power series: radius of convergence 248
11.7 Taylor's and Maclaurin's series: operations on power series 250
11.8 Applications of Taylor's and Maclaurin's expansions 259

XII VARIOUS PROBLEMS 265

PART TWO: SOLUTIONS, HINTS, ANSWERS 273

INDEX 451

LIST OF GREEK LETTERS 457

CALCULUS
PROBLEMS AND SOLUTIONS

I

SEQUENCES

1.1 BASIC DEFINITIONS AND THEOREMS

If corresponding to each positive integer n there is assigned a number u_n, then the numbers $u_1, u_2, \ldots, u_n, \ldots$ are said to form an infinite sequence.

The elements u_n $(n = 1, 2, \ldots)$ can, thus, be considered as values of a function (see Chap. II) whose argument varies over the set of all positive integers: $u_n = f(n)$.

A sequence can be represented graphically by marking off points on a number axis as in Figure 1.

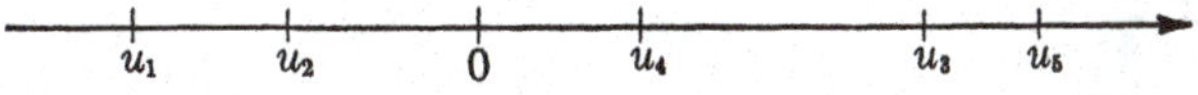

FIGURE 1

A sequence is said to be monotone increasing if $u_{n+1} > u_n$ and monotone decreasing if $u_{n+1} < u_n$ for every n.

If $u_{n+1} \geqslant u_n (u_{n+1} \leqslant u_n)$ for every n, the sequence is nondecreasing (or, correspondingly, nonincreasing).

We shall often call a sequence monotone also when the above inequalities hold only beyond a certain finite n.

A point a on the number axis is said to be an accumulation point of a sequence if for every $\epsilon > 0$, no matter how small, there are infinitely many distinct indices n such that $a - \epsilon < u_n < a + \epsilon$.

A sequence can possess many points of accumulation. If a point of accumulation exists such that no point to its right (left) is an accumulation point, then the corresponding number is called the upper (lower) limit of the sequence.

The above definition shows that it is possible for a sequence to have elements greater than its upper limit and smaller than its lower limit.

A sequence is said to be bounded from above if there exists a number B such that for every n: $u_n \leqslant B$. B is termed an upper bound of the sequence. M is the least upper bound if M is an upper bound and for every $\epsilon > 0$, no matter how small, there are elements of the sequence greater than $M - \epsilon$.

In the same way we define a lower bound and the greatest lower bound.

A sequence bounded both from above and from below is called a bounded sequence.

We now introduce a concept which is central in the theory of sequences:

Definition of a limit. A number a is said to be the limit of the sequence $u_1, u_2, \ldots, u_n, \ldots$ if for every positive ϵ, however small, there exists an integer N (which may depend on ϵ) such that $|u_n - a| < \epsilon$ for every $n > N$.

The graphical interpretation of this concept is given below. We surround the point a by an interval of length 2ϵ. Then every element u_n of the sequence with $n > N$ will lie in this interval or, in other words, outside of this interval only finitely many elements of the sequence are to be found (Fig. 2). (We remark that the possibility is not excluded that some elements with $n < N$ also lie in the interval.)

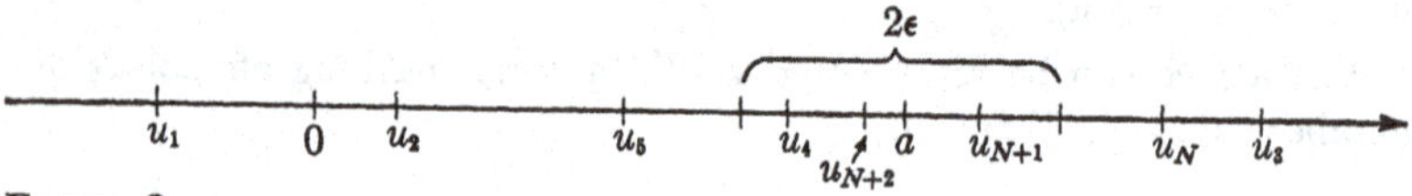

Figure 2

The fact that a is the limit of the sequence $u_1, u_2, \ldots, u_n, \ldots$ will be denoted by

$$\lim_{n \to \infty} u_n = a \quad \text{or} \quad u_n \underset{n \to \infty}{\longrightarrow} a.$$

Read: limit of u_n when n tends to infinity equals a. The notation $n \to \infty$ is to be understood as saying that, however large N is taken, n ultimately exceeds it.

We remark that it follows from the definition of a limit that if a sequence has more than one point of accumulation, it has no limit.

A sequence that has a limit is called convergent.

We now quote a number of basic facts concerning limits:

(1) If $\lim_{n \to \infty} u_n = a$ and $\lim_{n \to \infty} v_n = b$, then

$$\lim_{n \to \infty} (u_n + v_n) = a + b, \; \lim_{n \to \infty} (u_n - v_n) = a - b, \text{ and } \lim_{n \to \infty} u_n v_n = ab.$$

(2) If $\lim_{n\to\infty} u_n = a$ and $\lim_{n\to\infty} v_n = b \neq 0$, then

$$\lim_{n\to\infty} \frac{u_n}{v_n} = \frac{a}{b}.$$

We remark that the sequence u_n/v_n is defined only when $v_n \neq 0$ for every n.

(3) Every monotone increasing sequence bounded from above has a limit, and every monotone decreasing sequence bounded from below has a limit.

(4) A general principle of convergence: A sequence u_n has a limit if and only if for every $\epsilon > 0$, no matter how small, there can be found a natural number N such that $|u_n - u_{n'}| < \epsilon$ for every $n > N$ and $n' > N$.

1.2 EXAMPLES AND EXERCISES ON GENERAL NOTIONS

1. Prove that the sequence $u_n = \dfrac{2n-3}{n+2}$ is (a) monotone, (b) bounded.

Solution. (a) Let us find the difference $u_{n+1} - u_n$:

$$\begin{aligned} u_{n+1} - u_n &= \frac{2(n+1)-3}{n+1+2} - \frac{2n-3}{n+2} \\ &= \frac{2n^2+4n-n-2-2n^2-6n+3n+9}{(n+3)(n+2)} \\ &= \frac{7}{(n+3)(n+2)}. \end{aligned}$$

This difference is positive for every n, that is, $u_{n+1} > u_n$ for every n and hence the sequence is monotone increasing.

(b) We find

$$u_1 = \frac{2-3}{1+2} = -\frac{1}{3}.$$

The sequence increases and consequently $u_n \geqslant -\frac{1}{3}$ for every n. We now show that the sequence is bounded from above by evaluating the difference:

$$A - u_n = A - \frac{2n-3}{n+2} = \frac{An+2A-2n+3}{n+2} = \frac{n(A-2)+2A+3}{n+2}.$$

For $A = 2$ this difference is positive, that is, $2 > u_n$ for every n.

Remark. The boundedness of u_n can be shown also in the following way:

$$u_n = \frac{2n-3}{n+2} < \frac{2n}{n+2} < \frac{2n}{n} = 2.$$

2. Discuss the sequence $u_n = \dfrac{1}{n+2} + \cos\dfrac{n\pi}{3}$.

Solution. We begin by evaluating a few terms of the sequence:

$$u_1 = \tfrac{1}{3} + \cos\tfrac{1}{3}\pi = \tfrac{1}{3} + \tfrac{1}{2};\quad u_2 = \tfrac{1}{4} + \cos\tfrac{2}{3}\pi = \tfrac{1}{4} - \tfrac{1}{2};$$
$$u_3 = \tfrac{1}{5} + \cos\pi = \tfrac{1}{5} - 1;\quad u_4 = \tfrac{1}{6} + \cos\tfrac{4}{3}\pi = \tfrac{1}{6} - \tfrac{1}{2};$$
$$u_5 = \tfrac{1}{7} + \cos\tfrac{5}{3}\pi = \tfrac{1}{7} + \tfrac{1}{2};\quad u_6 = \tfrac{1}{8} + \cos 2\pi = \tfrac{1}{8} + 1.$$
$$u_7 = \tfrac{1}{9} + \cos\tfrac{7}{3}\pi = \tfrac{1}{9} + \cos\tfrac{1}{3}\pi = \tfrac{1}{9} + \tfrac{1}{2};$$
$$u_8 = \tfrac{1}{10} + \cos\tfrac{8}{3}\pi = \tfrac{1}{10} + \cos\tfrac{2}{3}\pi = \tfrac{1}{10} - \tfrac{1}{2}.$$

We observe that the second part of every element changes periodically and that the first part tends to zero (as $n \to \infty$). Beginning with n large enough, all elements of the sequence lie in intervals of length 2ϵ (for arbitrary $\epsilon > 0$) about the points $-1, -\frac{1}{2}, \frac{1}{2}, 1$; these are the points of accumulation of the sequence (Fig. 3). The sequence is clearly bounded. It is

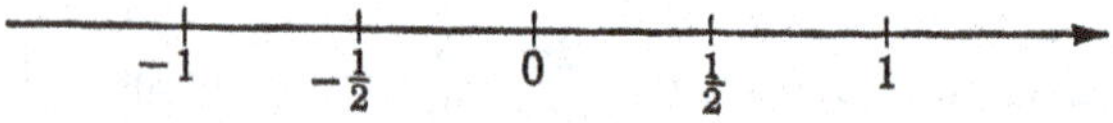

FIGURE 3

not monotone. The least upper bound is $1\frac{1}{8}$ and the greatest lower bound -1. The upper limit is 1 (the largest point of accumulation) and the lower limit is -1. The sequence has no limit.

Remark. We can describe the sequence also in the following way:

$$u_{6k+1} = \frac{1}{6k+3} + \frac{1}{2};\qquad u_{6k+2} = \frac{1}{6k+4} - \frac{1}{2};\qquad u_{6k+3} = \frac{1}{6k+5} - 1;$$
$$u_{6k+4} = \frac{1}{6k+6} - \frac{1}{2};\qquad u_{6k+5} = \frac{1}{6k+7} + \frac{1}{2};\qquad u_{6k+6} = \frac{1}{6k+8} + 1;$$
$$(k = 0, 1, 2, \ldots).$$

The elements u_{6k+1} establish a monotone decreasing subsequence (i.e., a sequence built by a part of the elements of the given sequence) with the limit $\frac{1}{2}$. In the same way we can choose from the given sequence five other subsequences with limits corresponding to the points of accumulation of the given sequence.

3. Find the limit of $u_n = 1/n$.

Solution. We shall prove that this limit is 0. According to the definition of a limit we consider the expression

$$\left|\frac{1}{n} - 0\right| = \frac{1}{n},$$

and shall prove that it is less than an arbitrary ϵ if n is greater than a suitable $N = N(\epsilon)$. Indeed, if $n > 1/\epsilon$ then $1/n < \epsilon$, and consequently every $N > 1/\epsilon$ satisfies the required conditions. Hence, to sum up, for every ϵ we found N (viz. $N = 1/\epsilon$) such that if $n > N$ then $|1/n - 0| < \epsilon$. The limit of $u_n = 1/n$ is indeed 0.

4. Find the limit of the sequence $u_n = \dfrac{n^2 - n + 2}{3n^2 + 2n - 4}$.

5. Evaluate the first six elements of each of the following five sequences. Discuss: (a) monotonicity, (b) values of the least upper and greatest lower bounds (if any), (c) points of accumulation, (d) upper and lower limits, (e) convergence.

$$(1)\quad u_n = \frac{2n+1}{2n} \qquad\qquad (2)\quad u_n = \frac{(-1)^n}{-n} + 1$$

$$(3)\quad u_n = n + (-1)^n n^2 \qquad\qquad (4)\quad u_n = \frac{2}{n} + \sin\frac{n\pi}{3} + \cos\frac{n\pi}{3}$$

$$(5)\quad u_n = \frac{n + (-1)^n n}{n - \dfrac{(-1)^n n}{2}}$$

6. Find a formula for u_n in the most obvious sequence beginning with

$$\tfrac{1}{2}, \tfrac{1}{3}, \tfrac{2}{3}, \tfrac{1}{4}, \tfrac{2}{4}, \tfrac{3}{4}, \tfrac{1}{5}, \tfrac{2}{5}, \tfrac{3}{5}, \tfrac{4}{5}, \tfrac{1}{6}, \tfrac{2}{6}, \tfrac{3}{6}, \tfrac{4}{6}, \tfrac{5}{6}, \tfrac{1}{7}, \ldots.$$

Does this sequence converge? Answer parts (b), (c), and (d) of Problem 5 for this sequence.

1.3 REPRESENTATION OF A NUMBER BY SEQUENCES

It is well known that $\sqrt{2}$ is an irrational number. Approximate values of it with increasing accuracy are

$$\begin{array}{lcl}
1 & < \sqrt{2} < & 2 \\
1.4 & < \sqrt{2} < & 1.5 \\
1.41 & < \sqrt{2} < & 1.42 \\
1.414 & < \sqrt{2} < & 1.415 \\
\cdot & \cdot & \cdot \\
\cdot & \cdot & \cdot \\
\cdot & \cdot & \cdot \\
x_n & < \sqrt{2} < & y_n \\
\cdot & \cdot & \cdot
\end{array}$$

The two sequences thus determined define together the irrational number $\sqrt{2}$. They have the following properties:

(1) $x_{n+1} \geqslant x_n$ (this sequence increases monotonically);
(2) $y_{n+1} < y_n$ (this sequence decreases monotonically);
(3) $y_n > x_n$ (each element of the decreasing sequence is greater than the corresponding element of the increasing one);
(4) $\lim_{n\to\infty} (y_n - x_n) = 0$.

Remark. It follows from (1,2,3) that $y_m > x_n$ for every m and n.

The above four conditions, and the natural postulate that the number axis has no gaps, ensure that the two sequences define one number, in this case their common limit.

7. $u_n = \dfrac{n^2 - 1}{3n^2 + 2}$; $v_n = \dfrac{n + 3}{3n - 2}$.

Show that these two sequences define a number, and find it.

8. Given two positive numbers a, b with $a > b$. Denote by a_1 their arithmetic mean $(a + b)/2$ and by b_1 their geometric mean $\sqrt{ab}$. Repeat the same for a_1, b_1 and so on; in general,

$$a_{n+1} = \frac{a_n + b_n}{2},\ b_{n+1} = \sqrt{a_n b_n}.$$

Prove that the sequences a_n and b_n define a number.

Solution. First we shall prove that $a_1 > b_1$.

$$a_1 - b_1 = \frac{a + b}{2} - \sqrt{ab} = \tfrac{1}{2}(a - 2\sqrt{ab} + b) = \tfrac{1}{2}(\sqrt{a} - \sqrt{b})^2 > 0.$$

Now let us prove that $a > a_1$ and $b < b_1$:

$$a - a_1 = a - \frac{a + b}{2} = \frac{a - b}{2} > 0,$$

$$b - b_1 = b - \sqrt{ab} = \sqrt{b}(\sqrt{b} - \sqrt{a}) < 0.$$

In both cases we used the given inequality $a > b$.

We collect the results: $a > a_1 > b_1 > b$. In the same way we obtain $a_1 > a_2 > b_2 > b_1$, and by mathematical induction,

$$a_n > a_{n+1} > b_{n+1} > b_n.$$

This last inequality shows that both sequences are monotone. The sequence a_n decreases and b_n increases. Moreover, for every n, $a_n > b$ and $b_n < a$, i.e., both sequences are bounded. Now a monotone bounded sequence has a limit; denote $\lim_{n\to\infty} a_n = \alpha$, $\lim_{n\to\infty} b_n = \beta$; but,

$$\lim_{n\to\infty} a_{n+1} = \lim_{n\to\infty} \frac{a_n + b_n}{2}.$$

Using the simple fact that $\lim_{n\to\infty} a_{n+1} = \lim_{n\to\infty} a_n = \alpha$

and the theorem about the limit of a sum, we get

$$\alpha = \frac{\alpha + \beta}{2}, \quad \text{i.e.,} \quad \alpha = \beta.$$

The limits are equal. Gauss called this limit the arithmetic-geometric mean of the numbers a and b and denoted it $\mu(a, b)$.

9. Using the data of the previous example, construct the arithmetic mean $a_{n+1} = (a_n + b_n)/2$ and the harmonic mean $b_{n+1} = 2a_nb_n/(a_n + b_n)$. Find the arithmetic-harmonic mean of the two numbers (i.e., the common limit of the two sequences).

1.4 EVALUATION OF $N(\epsilon)$

10. Given the sequence $u_n = (2n - 3)/(n + 1)$. Show that beginning with a certain n each element of the sequence differs from 2 by less than 0.001. Does this prove that $\lim_{n\to\infty} u_n = 2$?

Solution. We solve the inequality

$$|2 - u_n| = \left|2 - \frac{2n - 3}{n + 1}\right| = \left|\frac{5}{n + 1}\right| = \frac{5}{n + 1} < 0.001;$$

$$0.001\,n + 0.001 > 5; \quad 0.001\,n > 4.999; \quad n > 4999.$$

Each u_n with $n > 4999$ differs from 2 by less than 0.001. This result is certainly insufficient to prove that $\lim_{n\to\infty} u_n = 2$. We have showed only that if we surround 2 by an interval of length 0.002, the elements of the sequence beginning with u_{5000} will lie in this interval. Several points of accumulation might be situated inside this interval. Even if there is a limit it could have the value 1.9995 or any other value in the said interval.

To show that 2 is the limit of u_n we must prove that for *every* ϵ there can be found an N such that, beginning with the smallest $n > N$, each u_n differs from 2 by less than this ϵ. In this way we shall be able to exclude 1.9995 (or any number different from 2) from an interval about 2 in which all elements of the sequence beginning with a certain n are contained. From the inequality

$$|2 - u_n| = \frac{5}{n + 1} < \epsilon$$

we obtain

$$n > \frac{5}{\epsilon} - 1 = N,$$

and 2 is shown to be the limit of the given sequence.

11. Given the sequence $u_n = \dfrac{n^2 + n - 1}{3n^2 + 1}$. Find a number N such that for every $n > N$, $|u_n - \frac{1}{3}| < \epsilon$.

12. The following sequences converge to 0. Find in every case a number N such that $|u_n| < \epsilon$ if $n > N$. It is not required to find the least possible N.

(a) $u_n = \dfrac{n}{n^3 + n^2 + 1}$ (b) $u_n = \dfrac{1 + \sqrt{n}}{n^3}$

(c) $u_n = \dfrac{\sin n + 2\cos^2 n}{\sqrt{n}}$ (d) $u_n = \sqrt{n+1} - \sqrt{n}$

1.5 SEQUENCES GIVEN IN THE FORM $u_{n+1} = f(u_n)$

13. A sequence is given by the formula

$$u_{n+1} = \frac{6(1 + u_n)}{7 + u_n}, \quad u_1 = c > 0.$$

Prove that the sequence is monotone. Find when it increases or decreases. Find the limit if any.

Solution:

$$u_{n+1} - u_n = \frac{6(1 + u_n)}{7 + u_n} - u_n = \frac{6 - u_n - u_n^2}{7 + u_n} = \frac{(3 + u_n)(2 - u_n)}{7 + u_n}.$$

Clearly, $u_n > 0 \Rightarrow u_{n+1} > 0$.* But $u_1 = c > 0$; consequently, all elements are positive. The sign of $u_{n+1} - u_n$ depends on the sign of $2 - u_n$. If $u_n > 2$, we can write $u_n = 2 + \alpha$, $\alpha > 0$. Then

$$u_{n+1} = \frac{6(1 + u_n)}{7 + u_n} = \frac{6(1 + 2 + \alpha)}{7 + 2 + \alpha} = \frac{18 + 6\alpha}{9 + \alpha} > \frac{18 + 2\alpha}{9 + \alpha} = 2.$$

If $0 < u_n < 2$, assume $u_n = 2 - \alpha$, $0 < \alpha < 2$. Then

$$u_{n+1} = \frac{6(1 + 2 - \alpha)}{7 + 2 - \alpha} = \frac{18 - 6\alpha}{9 - \alpha} < \frac{18 - 2\alpha}{9 - \alpha} = 2.$$

Finally, if $u_n = 2$,

$$u_{n+1} = \frac{6(1 + 2)}{7 + 2} = \frac{18}{9} = 2.$$

We have proved

$$u_n < 2 \Rightarrow u_{n+1} < 2; \; u_n = 2 \Rightarrow u_{n+1} = 2;$$
$$u_n > 2 \Rightarrow u_{n+1} > 2.$$

* The sign $\Rightarrow$ is used with the meaning "implies."

Now we conclude:

(1) $u_1 = c > 2 \Rightarrow u_n > 2 \Rightarrow u_{n+1} < u_n$;
the sequence decreases monotonically and is bounded from below ($u_n > 0$).

(2) $u_1 = c < 2 \Rightarrow u_n < 2 \Rightarrow u_{n+1} > u_n$;
the sequence increases monotonically and is bounded from above by 2.

(3) $u_1 = c = 2 \Rightarrow u_n = 2$ for every n.

In every case there exists a limit. Denote it for the case (1) by k:

$$\lim_{n\to\infty} u_{n+1} = \lim_{n\to\infty} \frac{6(u_n+1)}{7+u_n} \Rightarrow k = \frac{6(k+1)}{7+k} \Rightarrow k^2 + k - 6 = 0;$$

$$k_1 = 2, \quad k_2 = -3.$$

Only the first solution agrees with $u_n > 0$. 2 is also the limit of the sequence in the other cases, because the last calculation does not depend on u_1 and is based only on the fact, common to all cases, that the sequence has a limit.

14. Given $u_{n+1} = 6/(1 + u_n)$ and $u_1 = 1$. Prove that the sequence converges and find its limit. Find the sequence elements by means of the graph $y = 6/(1 + x)$.

15. Show that the following sequences converge and find their limits:

(a) $a_{n+1} = \sqrt{2a_n}$; $a_1 = \sqrt{2}$ (b) $a_{n+1} = \sqrt{a_n + k}$, $a_1 = 0$; $k > 0$

1.6 METHODS FOR THE EVALUATION OF LIMITS

16. $u_n = \dfrac{n^2 - n + 2}{3n^2 + 2n - 4}$; $\lim\limits_{n\to\infty} u_n = ?$

Solution. We start by dividing both the numerator and denominator by n^2:

$$u_n = \frac{\dfrac{n^2 - n + 2}{n^2}}{\dfrac{3n^2 + 2n - 4}{n^2}} = \frac{1 - \dfrac{1}{n} + \dfrac{2}{n^2}}{3 + \dfrac{2}{n} - \dfrac{4}{n^2}}.$$

Using the theorems on limits of a quotient and a sum we get:

$$\lim_{n\to\infty} u_n = \frac{\lim\limits_{n\to\infty}\left(1 - \dfrac{1}{n} + \dfrac{2}{n^2}\right)}{\lim\limits_{n\to\infty}\left(3 + \dfrac{2}{n} - \dfrac{4}{n^2}\right)} = \frac{\lim\limits_{n\to\infty} 1 - \lim\limits_{n\to\infty} \dfrac{1}{n} + \lim\limits_{n\to\infty} \dfrac{2}{n^2}}{\lim\limits_{n\to\infty} 3 + \lim\limits_{n\to\infty} \dfrac{2}{n} - \lim\limits_{n\to\infty} \dfrac{4}{n^2}}$$

$$= \frac{1 - 0 + 0}{3 + 0 - 0} = \frac{1}{3}.$$

We used here also the well-known fact:

$$\text{for } p > 0, \lim_{n\to\infty} 1/n^p = 0.$$

17. $u_n = \dfrac{3n^2 - n + 1}{5n^3 - 4n}$; $\lim u_n = ?$

Solution. Divide the numerator and denominator by n^2:

$$u_n = \frac{3 - \dfrac{1}{n} + \dfrac{1}{n^2}}{5n - \dfrac{4}{n}}.$$

When n increases without bound, the numerator tends to 3 and the denominator tends to ∞. The limit of the fraction is consequently 0:

$$\lim_{n\to\infty} u_n = \lim_{n\to\infty} \frac{3 - \dfrac{1}{n} + \dfrac{1}{n^2}}{5n - \dfrac{4}{n}} = 0.$$

From the last two examples we learn that division by a suitable power of n will often prove helpful.

Evaluate the limits of the following sequences.

18. $u_n = \dfrac{n^4 - 5n}{n^2 - 3n + 1}$

19. $u_n = \dfrac{n^3}{2n^2 - 1} - \dfrac{n^2}{2n + 1}$

20. $u_n = \dfrac{1 + 2 + \ldots + n}{n(n + 2)}$

21. $u_n = \dfrac{\sqrt{n^2 + 1} + \sqrt{n}}{\sqrt[4]{n^3 + n} - n}$

22. $u_n = \dfrac{(\sqrt{n^2 + 1} + n)^2}{\sqrt[3]{n^6 + 1}}$

23. $u_n = \sqrt{n + a} - \sqrt{n}$

*Solution.** The following simple device (often useful) enables immediate evaluation:

$$u_n = \frac{(\sqrt{n + a} - \sqrt{n})(\sqrt{n + a} + \sqrt{n})}{\sqrt{n + a} + \sqrt{n}} = \frac{n + a - n}{\sqrt{n + a} + \sqrt{n}}$$

$$= \frac{a}{\sqrt{n + a} + \sqrt{n}};$$

$$\lim_{n\to\infty} u_n = \lim_{n\to\infty} \frac{a}{\sqrt{n + a} + \sqrt{n}} = 0.$$

Evaluate the limits of the following sequences.

24. $u_n = n(\sqrt{n^2 + 1} - n)$

* To Problem 23. Solutions *always* refer to problem immediately preceding, whether problem is in left or right column.

25. $u_n = \sqrt[3]{(n+1)^2} - \sqrt[3]{(n-1)^2}$

26. $u_n = n^3(\sqrt{n^2 + \sqrt{n^4+1}} - n\sqrt{2})$

27. $u_n = \dfrac{1 + 2^2 + 3^2 + \ldots + n^2}{n^3}$

28. $u_n = \dfrac{10^{12}n}{n^2 - 1}$

29. $u_n = \dfrac{1 + 3 + 5 + \ldots + (2n-1)}{1 + 2 + \ldots + n}$

30. $u_n = \dfrac{\sqrt[3]{n^2} \sin n!}{n+1}$

31. $u_n = (\sqrt{n^4 + n + 1} - n^2)(n+3)$

32. $u_n = \sqrt[n]{a^n + a^{-n}}, \qquad a > 0, \quad a \neq 1$

Evaluation of limits will be dealt with once more in the section on limits of functions.

II

FUNCTIONS OF A SINGLE VARIABLE

2.1 DEFINITION AND NOTATION

Consider two sets X and Y and a correspondence which assigns to each element x in X exactly one element $y = f(x)$ in Y. The set f of ordered pairs (x,y) generated by this correspondence is called a function, x is called the argument of the function, and $y = f(x)$ is called the value of the function at x. The set X is called the domain of f and the set of values of f is called the range. Note that the range need not be all of Y. Intuitively a function f is like a machine which produces an element $f(x)$ in Y when an element x in X is inserted.

In this chapter we will consider (real valued) functions of a single variable, i.e., functions whose domain and range are subsets of the set of real numbers. Usually such a function is specified analytically by giving a formula for $f(x)$ in terms of x with the understanding that the domain consists of all numbers x for which the formula makes sense.

When there is no chance for confusion it is customary to abbreviate "the function f defined by $f(x)$ = (a formula in x)" by "the function y = (a formula in x)."

33. Express the radius r of a cylinder of constant volume $V = 1$, as a function of its height h, analytically, graphically, and by means of a table.

Solution. The volume V of a cylinder is given by the formula $V = \pi r^2 h$. In our case $1 = \pi r^2 h$ and the analytic expression of the desired function is $r = 1/\sqrt{\pi}\sqrt{h}$. There follows a table for some values of h, and the graph-

h	1	2	3	4	5	6
r	0.563	0.397	0.324	0.281	0.252	0.23

ical representation is given in Figure 4.

34. Given $f(x) = \dfrac{x + a}{x - a}$, find $f(a + b), f\left(\dfrac{1}{a}\right) + f\left(\dfrac{1}{b}\right)$.

35. Given $f(x) = x^3 - 4x$, compute $\dfrac{f(x+h) - f(x)}{h}$.

36. Given $\varphi(x) = \dfrac{|x - 2|}{x + 1}$, compute $\varphi(4)$, $\varphi(0)$, $\varphi(-2)$.

37. Given $\varphi(x) = 5x$, $F(x) = x^2 + 6$. Solve the equation $F(x) = |\varphi(x)|$.

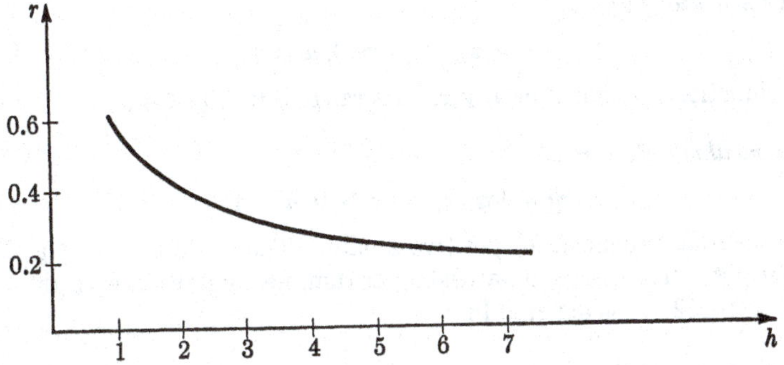

FIGURE 4

2.2 THE ELEMENTARY FUNCTIONS

The Rational Functions. In the polynomial or integral rational function

$$y = a_0x^n + a_1x^{n-1} + \ldots + a_{n-1}x + a_n,$$

n is a positive integer and the a's are any constants. This function is defined for every x.

The rational fraction or rational function is a quotient of two polynomials:

$$y = \frac{a_0x^n + a_1x^{n-1} + \ldots + a_{n-1}x + a_n}{b_0x^m + b_1x^{m-1} + \ldots + b_{m-1}x + b_m}$$

(m and n positive integers). The function is defined for any x except the values for which the denominator is 0.

The Power Function $y = x^\mu$. For integral μ this is a rational function. For rational μ this is a radical, i.e., when $\mu = m/n$, $y = \sqrt[n]{x^m}$.

The function $y = \sqrt[n]{x}$ is defined for any x if n is odd, and for nonnegative x only, when n is even.

For irrational μ the function is defined only for $x \geqslant 0$.

If $\mu \leqslant 0$ the power function is not defined at $x = 0$.

The rational functions and power function with μ rational (i.e., radicals) are examples of algebraic functions. An algebraic function is any continuous (see Sec. 3.3), function f for which x and $f(x)$ satisfy a polynomial equation

$$a_0(x)(f(x))^n + a_1(x)(f(x))^{n-1} + \dots + a_n(x) = 0$$

in which $a_0(x), a_1(x), \dots, a_n(x)$ are polynomials. All other functions (including power functions for irrational μ) are named transcendental.

Exponential Functions

$$y = a^x, \qquad a > 0, a \neq 1.$$

The function is defined for any x. Its range is $0 < y < \infty$.

Logarithmic Functions

$$y = \log_a x, \qquad a > 0, a \neq 1.$$

The function is defined for positive x only. The range is $-\infty < y < \infty$.

Remark. The so-called natural logarithm, i.e., logarithm with the base $e = 2.718282\dots$ is denoted $\ln x$.

Trigonometric Functions

$y = \sin x$ and $y = \cos x$ are defined for any x, and range over the interval $-1 \leqslant y \leqslant 1$. $y = \tan x$, $y = \sec x$ are defined for any x except $x = (\pi/2)(2n + 1)$; $y = \cot x$, $y = \csc x$ for any x except $x = \pi n$ ($n = 0, \pm 1, \pm 2, \dots$).

Hyperbolic Functions. Here we shall merely define these functions. Their properties will be studied later.

hyperbolic sine: $$\sinh x = \frac{e^x - e^{-x}}{2}$$

hyperbolic cosine: $$\cosh x = \frac{e^x + e^{-x}}{2}$$

hyperbolic tangent: $$\tanh x = \frac{e^x - e^{-x}}{e^x + e^{-x}}$$

hyperbolic cotangent: $$\coth x = \frac{e^x + e^{-x}}{e^x - e^{-x}}.$$

e is the base of the natural logarithms.

The hyperbolic functions are defined for any x, except that $\coth x$ is not defined for $x = 0$.

Inverse Functions. Given a function f. Suppose that in a certain part X of the domain of f, the values of f at different x's are different. Denote by Y the set of values of f at numbers in X. If the above condition holds then for each number y in Y there corresponds a unique number $x = g(y)$ in X such that $y = f(x)$. This defines a new function g with domain Y

and range X such that for every y in Y, $f(g(y)) = y$. The two functions f and g are called inverse functions. The analytical expression for $g(y)$ in terms of y can often be found by solving for x the equation $y = f(x)$.

The conditions for existence of an inverse function will be given elsewhere. Examples of inverse functions are $f(x) = x^3$ and $g(y) = \sqrt[3]{y}$, $f(x) = \log_a x$ and $g(y) = a^y$.

Remarks. (a) In the expression of the inverse function $x = g(y)$ we often change the notation so as to have x the argument. We thus write the inverse function in the form $y = g(x)$. For example, the inverse function of $y = x^3$ is denoted by $y = \sqrt[3]{x}$; the inverse function of $y = \log_a x$ by $y = a^x$.

(b) A function $y = f(x)$ can have in different parts of its domain of definition different analytic expressions for its inverse functions. For example, $y = x^2$ has in the interval $x \geqslant 0$ the inverse function $x = \sqrt{y}$, and in the interval $x < 0$ the inverse function $x = -\sqrt{y}$.

The Inverse Trigonometric Functions. The function $y = \arcsin x$ (read: arc sine x) is the inverse function of $y = \sin x$ in the interval $-\pi/2 \leqslant x \leqslant \pi/2$. It follows that $y = \arcsin x$ is defined in the interval $-1 \leqslant x \leqslant 1$ and ranges over the interval $-\pi/2 \leqslant \arcsin x \leqslant \pi/2$. The other inverse trigonometric functions are defined similarly:

$$
\begin{aligned}
&y = \operatorname{arccos} x; \qquad && -1 \leqslant x \leqslant 1,\ 0 \leqslant y \leqslant \pi. \\
&y = \arctan x; && -\infty < x < \infty,\ -\pi/2 < y < \pi/2. \\
&y = \operatorname{arccot} x; && -\infty < x < \infty,\ 0 < y < \pi.
\end{aligned}
$$

The Inverse Hyperbolic Functions. The function $y = \arg\sinh x$ is the inverse function of $y = \sinh x$ in the whole interval $-\infty < x < \infty$. $y = \arg\sinh x$ is defined for any x. $x = \arg\cosh y$ is the inverse function of $y = \cosh x$ in the interval $x \geqslant 0$. $y = \arg\cosh x$ is defined for $x \geqslant 1$ and ranges over the interval $0 \leqslant y < \infty$. In a similar manner we can define functions inverse to $\tanh x$ and $\coth x$.

A Function of a Function (Composite Function). Given two functions f and φ such that X is the domain of f. Let $y = f(x)$ and $z = \varphi(y)$. If the range of f is a part (possibly the whole) of the domain of definition of φ, we can construct the function $z = \varphi[f(x)] = F(x)$ defined on X; e.g., if $y = \sin x$ and $z = y^2$, we have $z = \sin^2 x$, which is defined for any x. This function will be called the composite function (of f and φ).

Periodic Functions. A function f such that there exists a constant T such that for any x, $f(x + T) = f(x)$, is named periodic and T is called a period of f. The most familiar examples of periodic functions are the trigonometric functions: $\sin(x + 2\pi) = \sin x$, $\tan(x + \pi) = \tan x$, and so on.

Using the elementary functions defined above, new elementary functions can be constructed by performing arithmetical operations on them and on composite functions obtained from them.

We proceed now to solve a number of exercises which help to clarify the various concepts introduced in this section.

2.3 DOMAIN OF DEFINITION

Describe the domains of definition of the following functions:

38. $y = \ln(x + 3)$

39. $y = \sqrt{5 - 2x}$

40. $y = \dfrac{1}{x^2 - 1}$

41. $y = \dfrac{x}{\sqrt{x^2 - 3x + 2}}$

42. $y = \sqrt{\ln \dfrac{5x - x^2}{4}}$

43. $y = \sqrt{3 - x} + \arcsin \dfrac{3 - 2x}{5}$

44. $y = \sqrt{x} + \sqrt[3]{\dfrac{1}{x - 2}} - \ln(2x - 3)$

45. $y = \sqrt{\sin x} + \sqrt{16 - x^2}$

46. $y = \ln \dfrac{x - 5}{x^2 - 10x + 24} - \sqrt[3]{x + 5}$

47. $y = \sqrt{\dfrac{x - 2}{x + 2}} + \sqrt{\dfrac{1 - x}{\sqrt{1 + x}}}$

48. $y = \arccos \dfrac{2}{2 + \sin x}$

49. $y = \log_{10}[1 - \log_{10}(x^2 - 5x + 16)]$

2.4 EVEN AND ODD FUNCTIONS

A function f is said to be even if $f(-x) = f(x)$. A function f is said to be odd if $f(-x) = -f(x)$. For example, if $f(x)$ is an even power of x it is an even function.

A product of an even function by an odd one is an odd function. A product of two odd functions is an even function.

Remark. The properties of evenness and oddness of functions have simple graphical interpretations. This topic will be dealt with in Chapter VI.

50. Find which of the following functions are even, which are odd, and which are neither even nor odd:

(a) $y = x^4 - 2x^2 + 1$ (b) $y = x - x^2$
(c) $y = \cos x$ (d) $y = \sin x$
(e) $y = 2^x$ (*f*) $y = -\frac{x^3}{6} + \frac{x^5}{120}$
(g) $y = 2^{-x^2}$ (h) $y = \cosh x$
(k) $y = \sinh x$ (l) $y = \tanh x$
(m) $y = \frac{x \sin x}{(x^2 + 5)\tanh x}$

51. Show that every function can be expressed as a sum of an even function and an odd one.

Solution. The following identity solves the problem:

$$f(x) = \frac{f(x) + f(-x)}{2} + \frac{f(x) - f(-x)}{2} = F_1(x) + F_2(x).$$

Indeed,

$$F_1(-x) = \frac{f(-x) + f(x)}{2} = F_1(x) \qquad \text{(even)};$$

$$F_2(-x) = \frac{f(-x) - f(x)}{2} = -F_2(x) \quad \text{(odd)}.$$

52. Represent each of the following functions as a sum of an even and an odd function.

(a) $y = ax^2 + bx + c$ (b) $y = e^x$
(c) $y = A \sin (kx + t)$ (d) $y = \sqrt{1 - (x - a)^2}$

2.5 RATIONAL FUNCTIONS

53. Are the following functions rational? Draw their graphs.

(a) $y = |x - 3| - 2|x + 1| + 2|x| - x + 1.$
(b) $y = \min [(x - 1)^2, (x + 1)^2, 1] - 1, -3 \leqslant x \leqslant 3.$

54. Given $f(x - 1) = 2x^2 - 3x + 1$, find $f(x + 1)$.

55. Given $y = f(x) = \frac{x + 2}{3x - 1}$, prove $x = f(y)$.

56. Given $f(x) = \left(\frac{x - 1}{x + 1}\right)^2$, prove $f\left(\frac{1}{x}\right) = f(x)$.

57. Solve the inequality

$$|f(x) + g(x)| < |f(x)| + |g(x)|$$

if $f(x) = x - 3$ and $g(x) = 4 - x$.

58. Find two numbers whose sum is a and whose product is maximal.

59. Given $y = \dfrac{x^2 + 2x + c}{x^2 + 4x + 3c}$, $0 < c < 1$; show that y takes any real value.

2.6 LOGARITHMIC FUNCTIONS

60. Prove $\log_a b \cdot \log_b c \cdot \log_c a = 1$.

Solution. First we shall prove the formula

$$\log_a x = \frac{\log_b x}{\log_b a}.$$

Let $\log_a x = u$; that is, $x = a^u$. Then $\log_b x = u \log_b a$. This is the required formula. Now

$$\log_a b \cdot \log_b c \cdot \log_c a = \log_a b \frac{\log_a c}{\log_a b} \cdot \frac{\log_a a}{\log_a c} = \log_a a = 1.$$

Remark. It is important to remember the following two formulas:

$$a^x = e^{x \ln a} \quad \text{and} \quad \log_a x = \frac{\ln x}{\ln a}.$$

2.7 TRIGONOMETRIC FUNCTIONS

61. Given $f(x) = a \cos(bx + c)$. Find a, b, c from the condition

$$f(x + 1) - f(x) = \sin x.$$

62. Given $y = a \sin x + b \cos x$. Express y in the form $y = A \sin(x + \varphi)$.

Solution. For any x we have the condition

$$a \sin x + b \cos x = A \sin x \cos \varphi + A \cos x \sin \varphi.$$

It follows that $A \cos \varphi = a$, $A \sin \varphi = b$,

$$A^2(\sin^2 \varphi + \cos^2 \varphi) = a^2 + b^2, \quad \text{i.e.,} \ A^2 = a^2 + b^2.$$

Take $A = \sqrt{a^2 + b^2}$. The solutions are those φ that satisfy

$$\sin \varphi = \frac{b}{\sqrt{a^2 + b^2}}; \quad \cos \varphi = \frac{a}{\sqrt{a^2 + b^2}}.$$

To compute φ it is easier to use $\tan \varphi = b/a$ and the signs of $\sin \varphi$ and $\cos \varphi$ which correspond to the signs of b and a respectively.

63. Prove

$$\sin x + \sin 2x + \ldots + \sin nx = \frac{\sin \dfrac{nx}{2} \sin \dfrac{(n+1)x}{2}}{\sin \dfrac{x}{2}}; \quad \sin \frac{x}{2} \neq 0.$$

2.8 HYPERBOLIC FUNCTIONS

64. Prove $\cosh(x \pm y) = \cosh x \cosh y \pm \sinh x \sinh y$.

Solution: For the + case we have

$$\cosh x \cosh y + \sinh x \sinh y = \frac{e^x + e^{-x}}{2} \cdot \frac{e^y + e^{-y}}{2} + \frac{e^x - e^{-x}}{2} \cdot \frac{e^y - e^{-y}}{2}$$

$$= \frac{e^{x+y} + e^{-x+y} + e^{x-y} + e^{-x-y} + e^{x+y} - e^{-x+y} - e^{x-y} + e^{-x-y}}{4}$$

$$= \frac{e^{x+y} + e^{-x-y}}{2} = \cosh(x + y).$$

Prove the following hyperbolic identities:

65. $\cosh^2 x - \sinh^2 x = 1$

66. $1 - \tanh^2 x = \dfrac{1}{\cosh^2 x}$

67. $(\cosh x \pm \sinh x)^n = \cosh nx \pm \sinh nx$

68. Given $x = \ln(\sec t + \tan t)$, prove $\sec t = \cosh x$

69. $\dfrac{\sinh 2x}{\cosh 2x + 1} = \tanh x$

70. $\dfrac{\cosh 2x + \cosh 4y}{\sinh 2x + \sinh 4y} = \coth(x + 2y)$

2.9 INVERSE FUNCTIONS

Find the inverse of each of the following functions:

71. $y = 2x$

Solution. We solve this equation for x and find $x = y/2$. This is the inverse function. We shall write it in the form $y = g(x)$, i.e., $y = x/2$.

72. $y = x^2 + 1$

73. $y = \sqrt[3]{x^2 + 1}$

74. $y = \dfrac{ax - b}{cx - a}$

75. $y = \sqrt[3]{x + \sqrt{1 + x^2}} + \sqrt[3]{x - \sqrt{1 + x^2}}$

76. $y = 10^{x+1}$

77. $y = 1 + \ln (x + 2)$

78. $y = \log_x 2$

79. $y = \dfrac{2^x}{1 + 2^x}$

80. Given $y = \dfrac{Ax^2 + Bx + C}{Kx^2 + Lx + M}$. When is also the inverse function rational?

2.10 THE INVERSE TRIGONOMETRIC FUNCTIONS

Find the inverse of each of the following functions.

81. $y = 2 \sin 3x, \quad -\pi/6 \leqslant x \leqslant \pi/6$

Solution:

$$\sin 3x = \frac{y}{2}, \quad 3x = \arcsin \frac{y}{2}, \quad x = \frac{1}{3} \arcsin \frac{y}{2};$$

in the usual notation,

$$y = \frac{1}{3} \arcsin \frac{x}{2}.$$

The domain of definition is $|x/2| \leqslant 1$, i.e., $|x| \leqslant 2$.

82. $y = 1 + 2 \sin \dfrac{x - 1}{x + 1}, \quad x \geqslant \dfrac{2 - \pi}{2 + \pi}$

83. $y = 4 \arcsin \sqrt{1 - x^2}$

Prove the following equalities.

84. $\arcsin x + \arccos x = \pi/2$

Solution. Denote $\arcsin x = \alpha$, $\arccos x = \beta$, i.e.,

$$\sin \alpha = x, \quad \cos \beta = x.$$

Consequently,

$$\sin \alpha = \cos \beta = \sin \left(\frac{\pi}{2} - \beta\right).$$

If $x > 0$, α and β are in the interval $[0,\pi/2]$ and from the above equality follows $\alpha + \beta = \pi/2$. If $x = 0$, then $\alpha = 0$, $\beta = \pi/2$, $\alpha + \beta = \pi/2$. If $x < 0$, $-\pi/2 \leqslant \alpha < 0$ and $\pi/2 < \beta \leqslant \pi$; $(\pi/2) - \beta$ is negative, and

$$\sin \alpha = \sin \left(\frac{\pi}{2} - \beta\right) \Rightarrow \alpha = \frac{\pi}{2} - \beta,$$

i.e., in this case too, $\alpha + \beta = \pi/2$.

85. $\arcsin x = \begin{cases} \operatorname{arccos} \sqrt{1 - x^2}, & 0 \leqslant x \leqslant 1 \\ -\operatorname{arccos} \sqrt{1 - x^2}, & -1 \leqslant x < 0 \end{cases}$

86. $\arctan x = \begin{cases} \operatorname{arccot} \dfrac{1}{x}, & x > 0 \\ \operatorname{arccot} \dfrac{1}{x} - \pi, & x < 0 \end{cases}$

87. $\arctan x + \arctan y = \arctan \dfrac{x + y}{1 - xy} + C,$

where $C = 0$ if $xy < 1$, $C = \pi$ if $xy > 1$ and $x > 0$, and $C = -\pi$ if $xy > 1$ and $x < 0$.

88. $\arccos x + \arccos \left(\dfrac{x}{2} + \dfrac{1}{2}\sqrt{3 - 3x^2}\right) = \dfrac{\pi}{3}, \dfrac{1}{2} \leqslant x \leqslant 1.$

89. $\arccos \dfrac{1 - x^2}{1 + x^2} = 2 \arctan x, 0 \leqslant x < \infty$. Why does not this identity hold for $x < 0$?

2.11 THE INVERSE HYPERBOLIC FUNCTIONS

90. Express $y = \text{arg sinh } x$ and $y = \text{arg cosh } x$ in terms of other elementary functions.

Solution. $y = \text{arg sinh } x$, $\sinh y = x$; $\dfrac{e^y - e^{-y}}{2} = x$;

$$e^y - e^{-y} = 2x; \quad e^{2y} - 2xe^y - 1 = 0; \quad e^y = x + \sqrt{x^2 + 1}.$$

($e^y > 0$ for any y, and the radical must be taken with the positive sign.)

$$y = \text{arg sinh } x = \ln (x + \sqrt{x^2 + 1}).$$

In the same way we get $\text{arg cosh } x = \ln (x + \sqrt{x^2 - 1})$; (we take here the + sign because $\text{arg cosh } x \geqslant 0$ by definition).

91. The same question for $y = \text{arg tanh } x$.

92. Solve again Problem 68 using the inverse hyperbolic functions.

93. Given $x = \ln \tan \left(\dfrac{\pi}{4} + \dfrac{t}{2}\right)$, prove $\sin t = \tanh x$.

2.12 COMPOSITE FUNCTIONS

94. Given $y = \sqrt{z + 1}$, $z = \tan^2 x$. Find y as a function of x.

95. Given $y = \sin x$, $v = \ln y$, $u = \sqrt{1 + v^2}$; find u as a function of x.

In the following exercises introduce new letters so as to exhibit the given functions as composite functions.

96. $y = \sin^3(2x + 1)$ **97.** $y = 5^{(3x+1)^2}$

98. Given $f(x) = x^3 - x$, $g(x) = \sin 2x$, find

(a) $f\left[g\left(\frac{\pi}{12}\right)\right]$, (b) $f\{f[f(0)]\}$.

2.13 PERIODIC FUNCTIONS

99. Is the function $y = \sin^2 x$ periodic?

Solution. $y = \sin^2 x = \frac{1}{2} - \frac{1}{2}\cos 2x.$

The function is periodic and its least period is π. First,

$$\begin{aligned}(fx + \pi) &= \tfrac{1}{2} - \tfrac{1}{2}\cos 2(x + \pi)\\ &= \tfrac{1}{2} - \tfrac{1}{2}\cos(2x + 2\pi)\\ &= \tfrac{1}{2} - \tfrac{1}{2}\cos 2x = f(x).\end{aligned}$$

Secondly, no positive number less than π satisfies the equality

$$f(x + T) = f(x).$$

Suppose that there exists such a number T. Then $\frac{1}{2} - \frac{1}{2}\cos 2(x + T) = \frac{1}{2} - \frac{1}{2}\cos 2x$ for any x. Take $x = 0$. We have $\frac{1}{2}\cos 2(0 + T) = -\frac{1}{2}\cos 0$; i.e., $\cos 2T = 1$; $2T = 2\pi n$, $T = \pi n$, and the least positive number satisfying this equation is π.

100. Is the function $y = \sin(x^2)$ periodic?

101. Find the least period of $y = 2\sin 3x + 3\sin 2x$.

Solution. The least period of $2\sin 3x$ is $2\pi/3$, and of $3\sin 2x$, $2\pi/2 = \pi$. The least period of the sum is equal in this case to the least common multiple of these numbers, i.e., to 2π.

Find the least periods of the following functions.

102. $y = \sin\frac{\pi t}{3} + \sin\frac{\pi t}{4}$ **103.** $y = |\sin x| + |\cos x|$

104. Find one function satisfying the three conditions, (a) for any integer x, $f(x) = x^2$; (b) for any other x, $f(x) \neq x^2$; (c) the function is defined by elementary functions only.

III

LIMIT OF A FUNCTION

3.1 DEFINITIONS AND GENERAL EXERCISES

In the following, when dealing with functions we shall assume that their domains of definition consist of an interval or of several intervals. Exceptions will be explicitly noted.

Consider a function f and a sequence of numbers $x_1, x_2, \ldots, x_n, \ldots$ belonging to its domain of definition X, such that $\lim_{n \to \infty} x_n = a$. The number a itself may or may not belong to X, and x_n is to be different from a for all n.

Consider the sequence $y_1 = f(x_1), y_2 = f(x_2), \ldots, y_n = f(x_n), \ldots$. This sequence may or may not have a limit as $n \to \infty$.

If for any choice of a sequence as above, the limit of y_n exists and is always the same number b, then $f(x)$ is said to have the limit b when x tends to a:

$$\lim_{x \to a} f(x) = b.$$

The concept of a limit of a function is one of the most important in the calculus. We give now another definition of it which does not use the notion of a sequence.

The number b is said to be the limit of $f(x)$ when x tends to a if for every positive ϵ, no matter how small, there exists a δ (which may depend on ϵ) such that for every x different from a and satisfying $|x - a| < \delta$, the inequality $|f(x) - b| < \epsilon$ holds.

This definition has a clear geometrical interpretation: It means that $f(x)$ can be made arbitrarily near to b by taking x sufficiently close to a.

It can be shown that the two definitions of the limit of a function are equivalent. The notion of a limit can be extended to the case when $x \to \infty$:

$\lim_{x\to\infty} f(x) = b$ if for any $\epsilon > 0$ a number M exists such that if $x > M$, $|f(x) - b| < \epsilon$.

All theorems of Chapter I concerning limits of sequences hold also for limits of functions.

It can be shown that for any elementary function, $\lim_{x\to a} f(x) = f(a)$, providing that a belongs to the domain of definition of $f(x)$ (see Prob. 140).

If we allow x to approach a from one side only, the limit is said to be one-sided, from the left or from the right. These limits are denoted by $\lim_{x\to a-} f(x)$ (from the left) and $\lim_{x\to a+} f(x)$ (from the right).

We proceed now to discuss a few problems concerning limits of functions.

105. Given $a > 1$, find $\lim_{x\to\infty} a^x/x$.

Solution. Assume that x takes only integral values $1, 2, \ldots, n, \ldots,$ and look for $\lim_{n\to\infty} a^n/n$. $a > 1$, i.e., $a = 1 + \alpha$, $\alpha > 0$. Now

$$a^n = (1 + \alpha)^n = 1 + n\alpha + \frac{n(n-1)}{1\cdot 2}\alpha^2 + \ldots > \frac{n(n-1)}{2}\alpha^2.$$

It follows that $\frac{a^n}{n} > \frac{n-1}{2}\alpha^2$. Clearly, $\frac{n-1}{2}\alpha^2$ tends to infinity together with n; consequently $\lim_{n\to\infty} a^n/n = \infty$. It is clear also that

$$\lim_{n\to\infty}\frac{a^n}{n+1} = \infty \quad \text{and} \quad \lim_{n\to\infty}\frac{a^{n+1}}{n} = \infty.$$

We now return to the original problem. For any x two integers n and $n + 1$ can be found such that $n \leqslant x < n + 1$. But then

$$\frac{a^n}{n+1} < \frac{a^x}{x} < \frac{a^{n+1}}{n},$$

and this inequality together with the previous results shows that $\lim_{x\to\infty} a^x/x = \infty$.

To sum up: for any positive M no matter how large, a positive N can be found such that $x > N$ ensures $a^x/x > M$.

106. Given $a > 1$, $k > 0$, find $\lim_{x\to\infty} a^x/x^k$.

Solution:

$$\frac{a^x}{x^k} = \left[\frac{(a^{1/k})^x}{x}\right]^k.$$

If $a > 1$, also $a^{1/k} > 1$, and by the previous problem $(a^{1/k})^x/x$ tends to infinity as $x \to \infty$. $k > 0$ is constant and it is clear that $\lim_{y\to\infty} y^k = \infty$. To conclude, $\lim_{x\to\infty} a^x/x^k = \infty$.

We now mention two basic well-known limits:

$$\lim_{x \to 0} \frac{\sin x}{x} = 1; \qquad \lim_{x \to \infty} \left(1 + \frac{1}{x}\right)^x = e.$$

107. Find $\lim_{x \to 0} \frac{(1+x)^n - 1}{x}$; n is a positive integer.

Solution:

$$\lim_{x \to 0} \frac{(1+x)^n - 1}{x} = \lim_{x \to 0} \frac{1 + nx + \frac{n(n-1)}{1 \cdot 2} x^2 + \ldots + x^n - 1}{x}$$

$$= \lim_{x \to 0} \left[n + \frac{n(n-1)}{1 \cdot 2} x + \ldots + x^{n-1} \right] = n.$$

We may clearly divide by x, because it is assumed that $x \neq 0$ when it tends to 0. (At $x = 0$ the function is not defined.) Every term in the brackets except the first tends to 0 together with x.

108. Given $y = \frac{x - 1}{2(x + 1)}$; denote $y_0 = \lim_{x \to 3} y$. Find a δ such that if $|x - 3| < \delta$, then $|y - y_0| < 0.01$.

Solution. y is defined at $x = 3$:

$$y_0 = \lim_{x \to 3} y = y(3) = \frac{3 - 1}{2(3 + 1)} = \frac{1}{4}.$$

$$\left| y - \frac{1}{4} \right| = \left| \frac{x - 1}{2(x + 1)} - \frac{1}{4} \right| = \left| \frac{x - 3}{4(x + 1)} \right|.$$

When x is near 3, $x + 1 > 0$ and we can write

$$\left| y - \frac{1}{4} \right| = \frac{|x - 3|}{4(x + 1)} < 0.01; \qquad |x - 3| < 0.04(x + 1).$$

x is near 3 and we can assume, e.g., $x + 1 > 3.5$. Then

$$|x - 3| < 0.04 \cdot 3.5 = 0.14$$

is surely less than $0.04(x + 1)$. We have found $\delta = 0.14$. The assumption $x > 2.5$ was satisfactory because it is compatible with $|x - 3| < 0.14$.

The δ found is not the best possible (i.e., the greatest possible). To find the best value of δ we have to solve the inequality $|x - 3| < 0.04(x + 1)$. This can be illustrated graphically as in Figure 5. $x_1 \approx 2.85$ is the solution of $3 - x = 0.04(x + 1)$, and $x_2 \approx 3.17$ the solution of $x - 3 = 0.04(x + 1)$. The δ obtained now is the smaller of the two differences $x - x_1$ and $x_2 - x$, i.e., $\delta \approx 0.15$.

In the same way we can for any ϵ find a δ such that if $|x - 3| < \delta$ then $|y - \frac{1}{4}| < \epsilon$; and this proves, according to the definition, that $\lim_{x \to 3} y = \frac{1}{4}$.

109. Given $y = \dfrac{x^2 - 1}{x^2 + 3}$, find $y_0 = \lim\limits_{x\to\infty} y$ and a number N such that $x > N$ ensures $|y - y_0| < \epsilon$ for a given ϵ.

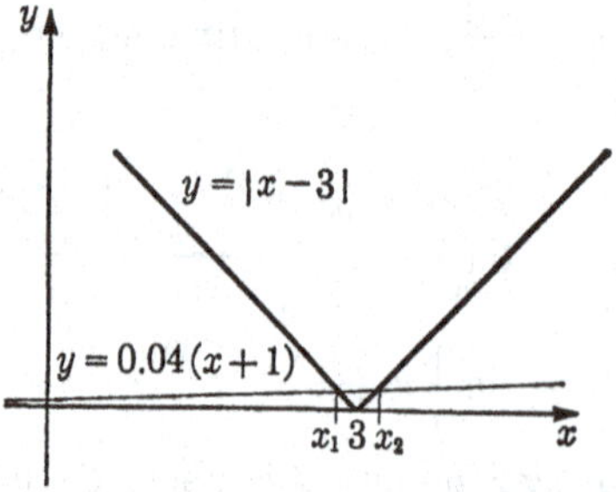

FIGURE 5

Solution:

$$\lim_{x\to\infty} \frac{x^2 - 1}{x^2 + 3} = \lim_{x\to\infty} \frac{1 - \dfrac{1}{x^2}}{1 + \dfrac{3}{x^2}} = 1.$$

$$|y - 1| = \left|\frac{x^2 - 1}{x^2 + 3} - 1\right| = \left|\frac{-4}{x^2 + 3}\right| = \frac{4}{x^2 + 3} < \epsilon;$$

$$x^2 > \frac{4}{\epsilon} - 3; \quad x > \sqrt{\frac{4}{\epsilon} - 3}.$$

The required $N = \sqrt{(4/\epsilon) - 3}$. (If $\epsilon > \frac{4}{3}$, $(4/\epsilon) - 3$ is negative; show that $|y - 1| \leqslant \frac{4}{3}$ for any x.)

110. Given $f(x) = 6x$. Find, for a given positive ϵ, a δ that depends on ϵ and x_0, $\delta = \delta(x_0, \epsilon)$, such that if $|x - x_0| < \delta$ then $|f(x) - f(x_0)| < \epsilon$. What is δ if $x_0 = 2$ and $\epsilon = \frac{1}{10}$? Solve the same problem for the function $f(x) = x^2 - 2x$.

3.2 EVALUATION OF LIMITS

Evaluate the following limits.

111. $\lim\limits_{x\to 2} \dfrac{x^2 - 5x + 6}{x^2 - 7x + 10}$

Solution. At $x = 2$ the numerator and denominator are equal to 0. We have here an indeterminate form which is usually denoted by the symbol 0/0. This symbol is used to indicate the situation

$$\lim_{x\to a}\frac{f(x)}{g(x)};\ \lim_{x\to a} f(x) = 0,\ \lim_{x\to a} g(x) = 0.$$

To find the given limit we make use of a fact known from algebra: If $f(x)$ is a polynomial and $f(a) = 0$, then $x - a$ is a factor of $f(x)$, i.e., $f(x) = \varphi(x)(x - a)$ where $\varphi(x)$ is a polynomial.

In our case both the numerator and denominator are polynomials equal to zero at $x = 2$, hence $x - 2$ can be factored out of both. (*Remark.* $x - 2 \neq 0$, because we explicitly require $x \neq 2$ when $x \to 2$.)

$$\lim_{x\to 2}\frac{x^2 - 5x + 6}{x^2 - 7x + 10} = \lim_{x\to 2}\frac{(x-2)(x-3)}{(x-2)(x-5)} = \lim_{x\to 2}\frac{x-3}{x-5}.$$

At this point it is permissible to substitute 2 for x (why?), resulting in

$$\lim_{x\to 2}\frac{x-3}{x-5} = \frac{1}{3}.$$

112. $\lim_{x\to 1}\frac{3x^4 - 4x^3 + 1}{(x-1)^2}$

113. $\lim_{x\to 1}\frac{x^m - 1}{x^n - 1}$, m and n positive integers

114. $\lim_{x\to 1}\left(\frac{1}{1-x} - \frac{3}{1-x^3}\right)$

Solution. This is an example of another indeterminate form, denoted by $\infty - \infty$ (both fractions tend to infinity when $x \to 1$):

$$\lim_{x\to 1}\left(\frac{1}{1-x} - \frac{3}{1-x^3}\right) = \lim_{x\to 1}\frac{1 + x + x^2 - 3}{1 - x^3}$$

$$= \lim_{x\to 1}\frac{(x-1)(x+2)}{(1-x)(1+x+x^2)} = -1.$$

115. $\lim_{x\to 0}\frac{\sqrt[3]{1+x} - 1}{x}$

116. $\lim_{x\to -1}\frac{1 + \sqrt[3]{x}}{1 + \sqrt[5]{x}}$

117. $\lim_{x\to 2}\frac{\sqrt{2+x} - \sqrt{3x-2}}{\sqrt{4x+1} - \sqrt{5x-1}}$

118. $\lim_{x\to 0}\frac{\sqrt{1+x+x^2} - 1}{x}$

119. $\lim_{x\to 0}\frac{\sqrt[3]{1+3x} - \sqrt{1-2x}}{x + x^2}$

120. $\lim_{x\to 0}\frac{\sqrt[5]{1+3x^4} - \sqrt{1-2x}}{\sqrt[3]{1+x} - \sqrt{1+x}}$

Solution. We use here a special device:

$$\lim_{x\to 0}\frac{\sqrt[5]{1+3x^4} - \sqrt{1-2x}}{\sqrt[3]{1+x} - \sqrt{1+x}} = \lim_{x\to 0}\frac{(\sqrt[5]{1+3x^4} - 1) - (\sqrt{1-2x} - 1)}{(\sqrt[3]{1+x} - 1) - (\sqrt{1+x} - 1)}$$

$$= \lim_{x\to 0}\frac{\dfrac{\sqrt[5]{1+3x^4} - 1}{x} - \dfrac{\sqrt{1-2x} - 1}{x}}{\dfrac{\sqrt[3]{1+x} - 1}{x} - \dfrac{\sqrt{1+x} - 1}{x}}$$

$$= \frac{0 - (-1)}{\frac{1}{3} - \frac{1}{2}} = \frac{1}{-\frac{1}{6}} = -6.$$

Every one of the limits in the numerator and denominator was found separately, for example:

$$\lim_{x\to 0} \frac{\sqrt[5]{1+3x^4}-1}{x}$$

$$= \lim_{x\to 0} \frac{1+3x^4-1}{x(\sqrt[5]{(1+3x^4)^4}+\sqrt[5]{(1+3x^4)^3}+\sqrt[5]{(1+3x^4)^2}+\sqrt[5]{1+3x^4}+1)}$$

$$= \frac{0}{5} = 0.$$

121. $\lim_{x\to 0} \frac{\sin 4x}{x}$

Solution. We know already that $\lim_{u\to 0} (\sin u)/u = 1$. To bring the given expression to this form we multiply the numerator and denominator by 4:

$$\lim_{x\to 0} \frac{\sin 4x}{x} = \lim_{4x\to 0} \frac{4 \sin 4x}{4x} = 4 \lim_{4x\to 0} \frac{\sin 4x}{4x} = 4.$$

122. $\lim_{x\to \pi} \frac{\sin mx}{\sin nx}$, m, n integers

123. $\lim_{x\to 0} \frac{1-\cos x}{x^2}$

124. $\lim_{x\to 0} \frac{\tan x - \sin x}{x^3}$

125. $\lim_{x\to 0} \frac{\sin (a+x) + \sin (a-x) - 2 \sin a}{x^2}$

126. $\lim_{x\to \infty} (\sin \sqrt{x+1} - \sin \sqrt{x})$

127. $\lim_{x\to 0} \frac{\cos x - \sqrt[3]{\cos x}}{\sin^2 x}$

128. $\lim_{x\to 0} \frac{\sqrt{1+x \sin x} - \cos x}{\sin^2 \frac{x}{2}}$

129. $\lim_{x\to 0} \frac{1-\cos 2x + \tan^2 x}{x \sin x}$

130. $\lim_{x\to 0} \frac{1-\cos (1-\cos x)}{x^4}$

The indeterminate form 1^∞. Although $\lim_{v\to \infty} 1^v = 1$ the form $\lim_{\substack{u\to 1\\ v\to\infty}} u^v$ is indeterminate. This situation is covered by the following theorem.

THEOREM. Suppose $\lim_{x\to a} f(x) = 1$ and $\lim_{x\to a} g(x) = \infty$. Then $\lim_{x\to a} f(x)^{g(x)}$ exists if and only if $\lim_{x\to a} (f(x)-1)g(x)$ exists and in that case

$$\lim_{x\to a} f(x)^{g(x)} = e^{\lim(f(x)-1)g(x)}$$

A heuristic argument in favor of this theorem (but not a proof) is obtained if we let $u = f(x)$, $v = g(x)$, and $t = \dfrac{1}{u-1}$ (assuming $u \to 1$ but $u \neq 1$).

If u is close to 1 then t will be large and $\left\{1 + \dfrac{1}{t}\right\}^t$ will be close to e. Hence if $v(u-1)$ is close to b it is plausible that

$$u^v = \left\{1 + \frac{1}{t}\right\}^v = \left[\left\{1 + \frac{1}{t}\right\}^t\right]^{v/t} = \left[\left\{1 + \frac{1}{t}\right\}^t\right]^{v(u-1)}$$

will be close to e^b.

131. $\lim\limits_{x\to\infty} \left(\dfrac{x+1}{x-1}\right)^x$

Solution:

$$\lim_{x\to\infty} \frac{x+1}{x-1} = \lim_{x\to\infty} \frac{1 + \frac{1}{x}}{1 - \frac{1}{x}} = 1.$$

This is the case of 1^∞. Let us use the above formula:

$$\lim_{\substack{u\to 1\\ v\to\infty}} (u-1)v = \lim_{x\to\infty} \left(\frac{x+1}{x-1} - 1\right) x = \lim_{x\to\infty} \frac{2}{x-1}\, x = 2.$$

Consequently,

$$\lim_{x\to\infty} \left(\frac{x+1}{x-1}\right)^x = e^2.$$

132. $\lim\limits_{x\to 0} (1 + \tan x)^{\cot x}$

133. $\lim\limits_{m\to\infty} \left(\cos \dfrac{x}{m}\right)^m$

134. $\lim\limits_{x\to\infty} \left(1 + \dfrac{1}{x^2}\right)^x$

135. $\lim\limits_{x\to\frac{\pi}{4}} (\tan x)^{\tan 2x}$

There follow a few examples of limits of logarithmic and exponential functions.

136. $\lim\limits_{x\to\infty} \left(\dfrac{x^2+x+1}{2x^2-x+1}\right)^{x^2}$

Solution:

$$\lim_{x\to\infty} \frac{x^2+x+1}{2x^2-x+1} = \frac{1}{2}.$$

The given limit is not indeterminate and is equal to $\lim_{x\to\infty} (\frac{1}{2})^{x^2} = 0$.

137. $\lim_{x\to 0} \dfrac{\ln(1+x)}{x}$

Solution:

$$\lim_{x\to 0} \frac{\ln(1+x)}{x} = \lim_{x\to 0} \ln(1+x)^{1/x} = \lim_{y\to\infty} \ln\left(1+\frac{1}{y}\right)^y = \ln e = 1.$$

We have put $x = 1/y$ and $\lim_{x\to 0} y = \infty$.

138. $\lim_{x\to 0} \dfrac{a^x - 1}{x}$, $a > 0$

Solution. Put $a^x - 1 = \alpha$, $\lim_{x\to 0} \alpha = 0$, and $a^x = \alpha + 1$; i.e.,

$$x = \frac{\ln(\alpha+1)}{\ln a}.$$

Now,

$$\lim_{x\to 0} \frac{a^x - 1}{x} = \lim_{\alpha\to 0} \frac{\alpha \ln a}{\ln(1+\alpha)} = 1 \cdot \ln a = \ln a.$$

We used here the result of the previous exercise.

139. Prove $\lim_{x\to 0} x^{\sin x} = 1$.

Solution. The function $x^{\sin x}$ is defined only for positive x; for these x, $x = e^{\ln x}$, hence $\lim_{x\to 0} x^{\sin x} = \lim_{x\to 0} e^{\sin x \ln x}$; we now find

$$\lim_{x\to 0} \sin x \ln x = \lim_{x\to 0} \frac{\sin x}{x} \cdot x \ln x = \lim_{x\to 0} x \ln x.$$

Put $x = e^{-z} = 1/e^z$; then $\lim_{x\to 0} z = \infty$ and $\ln x = -z$; $\lim_{x\to 0} x \ln x = \lim_{z\to\infty} -z/e^z = 0$. (Compare with Problem 105.) To conclude:

$$\lim_{x\to 0} x^{\sin x} = \lim_{x\to 0} e^{\sin x \ln x} = e^0 = 1.$$

Note the intermediary result: $\lim_{x\to 0} x \ln |x| = 0$.

3.3 CONTINUITY

A function f is said to be continuous at a point $x = a$ if it is defined at this point, and if $\lim_{x\to a} f(x)$ exists and is equal to $f(a)$; i.e., $\lim_{x\to a} f(x) = f(a)$. Two other (equivalent) definitions of this basic concept will now be given.

A function f is said to be continuous at a point $x = a$ if for any $\epsilon > 0$, no matter how small, a $\delta = \delta(\epsilon)$ can be found such that $|x - a| < \delta$ implies $|f(x) - f(a)| < \epsilon$.

The third definition will be stated in terms of increments of the argument and the function. Given f with the value $f(a)$ at a. Let us change x by an increment Δx (read delta x), i.e., consider the point $a + \Delta x$ (Δx may, of course, be positive or negative). The value of f at this point is $f(a + \Delta x)$. The difference $f(a + \Delta x) - f(a)$ is denoted by Δf and called the increment of f corresponding to the increment Δx of the argument. Using the above notions we now give the third definition of continuity:

A function $f(x)$ defined at a is continuous there if $\Delta x \to 0$ implies that there $\Delta f \to 0$.

A somewhat more general notion of continuity, namely that of continuity from one side, can be defined: e.g., $f(x)$ is said to be continuous at $x = a$ from the right if $\lim_{x \to a+} f(x) = f(a)$.

A function is said to be continuous in an interval if it is continuous at every point of the interval. If the interval is closed, we only require continuity at an endpoint from inside the interval.

We now mention some theorems concerning continuity.

(1) If $f(x)$ and $g(x)$ are continuous at $x = a$, then $f(x) \pm g(x)$, $f(x) \cdot g(x)$, and $f(x)/g(x)$ are also continuous at $x = a$, ($f(x)/g(x)$ only if $g(a) \neq 0$).

(2) Suppose that $y = f(x)$ is continuous at $x = a$, and $f(a) = b$. Suppose also that $z = g(y)$ is continuous at $y = b$. Then the composite function $z = g[f(x)]$ is also continuous at $x = a$.

(3) If $y = f(x)$ is defined, strictly monotone increasing (or decreasing), and continuous in the interval X, then in the corresponding range Y of this function a function inverse to $f(x)$ exists, and it is also monotone increasing (or decreasing) and continuous.

(4) Let $f(x)$ be defined and continuous in an interval X, and $f(a) = A$ and $f(b) = B$ where a and b are two points of X. Then for every C between A and B there exists at least one point c between a and b such that $f(c) = C$. It follows directly that if A and B have different signs, then there exists at least one c between a and b such that $f(c) = 0$.

(5) A function $f(x)$ continuous in a closed interval $[a,b]$ must assume in this interval a largest and a smallest value.

If $f(a)$ is not defined, the function f is clearly not continuous at a, but if $\lim_{x \to a} f(x)$ exists and is equal to l we can extend the definition of f by postulating $f(a) = l$. The function thus obtained is clearly continuous at $x = a$.

It is obvious that if $\lim_{x \to a} f(x)$ does not exist it is impossible to define $f(x)$ at $x = a$ so as to make it continuous there.

We conclude with a definition of the concept of uniform continuity: If for every $\epsilon > 0$ a $\delta > 0$ can be found such that $|x_1 - x_2| < \delta$ implies $|f(x_1) - f(x_2)| < \epsilon$ for any position of x_1 and x_2 in the given interval, then f is said to be uniformly continuous in the interval.

It can be shown that every continuous function in a closed and bounded interval is uniformly continuous there.

140. Prove the continuity of the elementary functions.

Solution. The function $f(x) = x$ is continuous in the whole interval $-\infty < x < \infty$, because $\lim_{x \to a} f(x) = \lim_{x \to a} x = a = f(a)$.

By the above theorems, a polynomial is continuous everywhere and a rational function is continuous at any point except those for which the denominator equals zero.

We now prove the continuity of $y = a^x$. Suppose $a > 1$. Let n be a positive integer and put $a = b^n$; then $b = a^{1/n}$ and $b > 1$. We have

$$\begin{aligned} b^n &= (1 + b - 1)^n \\ &= 1 + n(b - 1) + \frac{n(n-1)}{1 \cdot 2}(b - 1)^2 + \ldots > 1 + n(b - 1). \end{aligned}$$

$$a > 1 + n(a^{1/n} - 1), \quad \text{i.e.,} \quad 0 < a^{1/n} - 1 < \frac{a - 1}{n};$$

but $\lim_{n \to \infty} \frac{a-1}{n} = 0$, consequently $\lim_{n \to \infty} (a^{1/n} - 1) = 0$, $\lim_{n \to \infty} a^{1/n} = 1$.

It is easy to see that even when x tends to infinity in any manner we have also $\lim_{x \to \infty} a^{1/x} = 1$. Now

$$\lim_{x \to -\infty} a^{1/x} = \lim_{y \to \infty} \frac{1}{a^{1/y}} = \frac{1}{1} = 1.$$

As a consequence of the above limits we obtain: $\lim_{x \to 0} a^x = 1$. This proves that any exponential function with base $a > 1$ is continuous at $x = 0$.

At any x_0 we have, for $a > 1$

$$\lim_{x \to x_0} a^x = \lim_{x \to x_0} a^{x_0}(a^{x - x_0}) = a^{x_0} \lim_{x - x_0 \to 0} a^{x - x_0} = a^{x_0} \cdot 1 = a^{x_0}.$$

Thus, if $a > 1$, a^x is continuous for all x. Continuity of a^x for $0 < a < 1$ follows by noting that $a^x = (1/a)^{-x}$. As a direct consequence we have that all hyperbolic functions are continuous (except $\coth x$ at $x = 0$).

To prove the continuity of the trigonometric functions let us first observe that for $-\pi/2 \leqslant \alpha \leqslant \pi/2$, $|\sin \alpha| < |\alpha|$ (the proof can be based on the fact that a chord is shorter than the arc spanned by it). For $|\alpha| > \pi/2$ we have $|\sin \alpha| \leqslant 1 < \pi/2 < |\alpha|$. The inequality $|\sin \alpha| < |\alpha|$ is thus true for any α.

Now we use the second definition of continuity. For any ϵ we shall find a δ such that $|x - a| < \delta$ implies $|\sin x - \sin a| < \epsilon$. But

$$|\sin x - \sin a| = \left|2 \sin \frac{x-a}{2} \cos \frac{x+a}{2}\right| = 2\left|\sin \frac{x-a}{2}\right|\left|\cos \frac{x+a}{2}\right|$$

$$\leqslant 2\left|\sin \frac{x-a}{2}\right| \leqslant 2\left|\frac{x-a}{2}\right| = |x-a|.$$

We can choose $\delta = \epsilon$ and $|\sin x - \sin a| < |x - a| < \epsilon$. The continuity of $\sin x$ is proved. In the same way we can prove continuity of $\cos x$, but we can also use the identity $\cos x = \sin\left(\frac{\pi}{2} - x\right)$ and the theorem about the continuity of a composite function.

The functions $\tan x = \dfrac{\sin x}{\cos x}$ and $\cot x = \dfrac{\cos x}{\sin x}$ are continuous by the theorems of continuity of the quotient of two continuous functions, except at the points $x = (\pi/2)(2n + 1)$ (for $\tan x$) and $x = \pi n$ (for $\cot x$), where these functions are not defined.

By the theorems about the continuity of inverse and composite functions, the continuity of $y = \ln x$, $y = x^\mu$ (use the identity $x^\mu = e^{\mu \ln x}$), and the inverse trigonometric functions can be proved. Using now the theorems of continuity of sum, product, quotient, and composite functions, we conclude finally that all elementary functions are continuous at every point at which they are defined.

This proves now the statement that for every elementary $f(x)$, $\lim_{x\to a} f(x) = f(a)$, assuming that $f(a)$ exists, which we used earlier to evaluate limits.

141. $f(x) = \dfrac{x^2 - 1}{x - 1}$ is not defined at $x = 1$. How can the definition of $f(x)$ be extended so as to obtain continuity at $x = 1$?

Solution. Let us find $\lim_{x\to 1} f(x)$ (if any).

$$\lim_{x\to 1} \frac{x^2 - 1}{x - 1} = \lim_{x\to 1} (x + 1) = 2.$$

We define $f(1) = 2$, and the function defined by $f(x) = \dfrac{x^2 - 1}{x - 1}$ and $f(1) = 2$ is continuous at $x = 1$. Compare the graphs of the original and extended function (Figs. 6, 7).

142. Prove that the equation $x^5 - 3x = 1$ has at least one solution between $x = 1$ and $x = 2$.

143. Consider the continuity of $y = f(x) = \lim_{n\to\infty} \dfrac{x^{2n} - 1}{x^{2n} + 1}$.

Examine the following functions for continuity.

144. $f(x) = \sin(1/x)$

145. $f(x) = x \sin(1/x)$

146. $f(x) = e^{1/x}$

147. A function is defined as follows:

$$f(x) = 2x, \qquad 0 \leqslant x < 1$$
$$f(x) = 3 - x, \qquad 1 \leqslant x \leqslant 2.$$

Is the function continuous in the whole interval $0 \leqslant x \leqslant 2$?

148. A function is defined as follows:

$$f(x) = x \ln x^2, \quad f(0) = a.$$

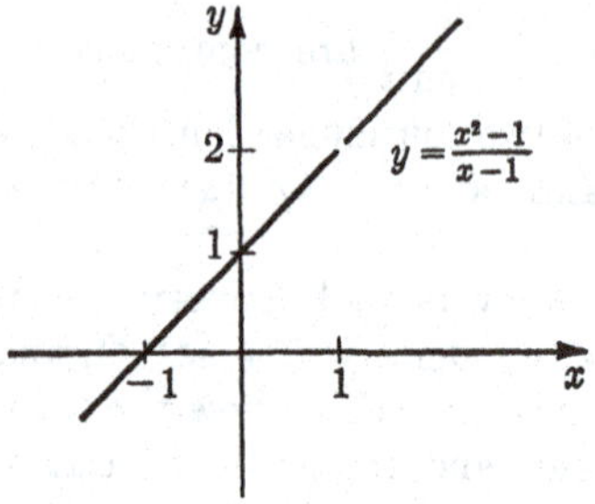

FIGURE 6

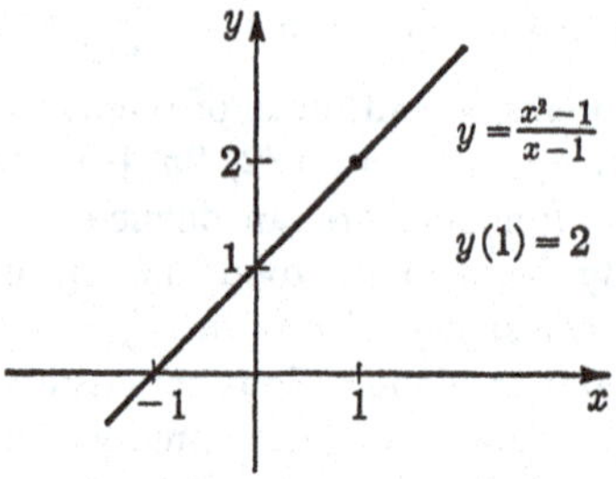

FIGURE 7

For what a is the function continuous on the whole x axis?

Locate any points of discontinuity in the following functions:

149. $y = \dfrac{x}{x^2 - 4}$

150. $y = \dfrac{x^2 + 3}{x^2 - 3}$

151. $y = \dfrac{1}{x - x^3}$

152. $y = \dfrac{x^3 + 3x + 7}{x^2 - 6x + 8}$

153. $y = \dfrac{x^2 + 3x + 7}{x^2 - 6x + 10}$

154. $y = \dfrac{\tan 3x}{\tan 2x}$

155. $y = \dfrac{x}{\sin x}$

156. $y = \ln \ln (1 + x^2)$

157. $y = \dfrac{\cos (\pi x/2)}{x^2(x - 1)}$

158. Prove that $f(x) = x^2 - 5x$ is continuous at $x_0 = -1$. Find a δ such that if $|x - x_0| < \delta$, then $|f(x) - f(x_0)| < 0.01$.

DEFINITION: $[x]$ denotes the maximal integer not greater than x; e.g., $[5.2] = 5$, $[7] = 7$, $[-\frac{1}{2}] = -1$.

159. Given $f(x) = x - [x]$, $-\infty < x < \infty$. Prove that $f(x)$ is continuous for every x which is not an integer and discontinuous for integer values of x. At what points is the function continuous from one side, and which side is it?

160. Prove that $f(x) = [x] + \sqrt{x - [x]}$ is defined, continuous, and monotone increasing for any x.

IV

DIFFERENTIAL CALCULUS FOR FUNCTIONS OF A SINGLE VARIABLE

4.1 THE NOTION OF A DERIVATIVE AND ITS PHYSICAL AND GEOMETRIC INTERPRETATIONS

The derivative with respect to x of the function f at the point x_0 is the limit (if it exists) of the following expression:

$$\lim_{\Delta x \to 0} \frac{f(x_0 + \Delta x) - f(x_0)}{\Delta x}.$$

The denominator is an increment of the argument and the numerator, which can also be denoted by Δf, is the corresponding increment of the function. The derivative is denoted by

$$y'(x_0), \quad f'(x_0), \quad \text{or} \quad \frac{dy}{dx}_{(x = x_0)}.$$

Assume now that the derivative of f exists at any point of a set X. By assigning to each x_0 of X the value of $f'(x_0)$ we obtain a new function defined on X called the derived function or the derivative of f and denoted by f' or, if $y = f(x)$, by y' or dy/dx.

The limit of $\Delta f/\Delta x$ when $\Delta x \to 0$ can only exist if Δf tends to zero together with Δx, i.e., when f is continuous. Continuity is thus necessary (but as we shall see later, not sufficient) for the existence of a derivative.

To find an analytical formula for f' we take the following steps:

(1) Give x an increment Δx.

(2) Find the value of the function at x, i.e., $f(x)$, and at $x + \Delta x$, i.e., $f(x + \Delta x)$.

(3) Find the increment of f by subtracting: $\Delta f = f(x + \Delta x) - f(x)$.

(4) Find the ratio $\dfrac{\Delta f}{\Delta x} = \dfrac{f(x + \Delta x) - f(x)}{\Delta x}$.

(5) Evaluate the limit of this ratio, when Δx tends to zero:

$$f'(x) = \lim_{\Delta x \to 0} \frac{f(x + \Delta x) - f(x)}{\Delta x}.$$

To obtain the value of the derivative at a certain point $x = x_0$ we can evaluate the limit

$$\lim_{\Delta x \to 0} \frac{f(x_0 + \Delta x) - f(x_0)}{\Delta x},$$

or we can put x_0 in the derived function f'.

The calculation of derivatives is also called differentiation.

The notion of the derivative has many applications in physics and in geometry. We list a few examples:

(a) The velocity of a moving particle is the derivative of the displacement with respect to time (i.e. ds/dt, where s is the displacement given as a function of the time t).
(b) The acceleration is the derivative of the velocity with respect to time.
(c) The slope of the tangent of a curve $y = f(x)$ at a point $(x_0, f(x_0))$ is the derivative $f'(x_0)$.
(d) Denote by Q the electrical charge, by J the intensity of the electrical current, and by t the time. Then $J = dQ/dt$.

Find the derivatives of the following functions, from the definition:

161. $y = 1/x$ at $x = 5$ and at x (i.e., find the derived function).

Solution:

$$y'(5) = \lim_{\Delta x \to 0} \frac{\dfrac{1}{5 + \Delta x} - \dfrac{1}{5}}{\Delta x} = \lim_{\Delta x \to 0} \frac{5 - 5 - \Delta x}{\Delta x(25 + 5\Delta x)}$$

$$= \lim_{\Delta x \to 0} \frac{-1}{25 + 5\Delta x} = -\frac{1}{25}.$$

$$y'(x) = \lim_{\Delta x \to 0} \frac{\dfrac{1}{x + \Delta x} - \dfrac{1}{x}}{\Delta x} = \lim_{\Delta x \to 0} \frac{x - x - \Delta x}{\Delta x(x + \Delta x)x}$$

$$= \lim_{\Delta x \to 0} \frac{-1}{x(x + \Delta x)} = -\frac{1}{x^2}.$$

162. $y = a^x$

163. $y = \tan x$

164. $y = x \sin x$

4.2 EVALUATING DERIVATIVES

First let us write down the basic theorems and formulas for differentiation:

(1) The derivative of a sum of a finite number of functions is equal to the sum of their derivatives:

$$(u + v + \ldots + w)' = u' + v' + \ldots + w'.$$

(2) The derivative of a product is $(uv)' = u'v + uv'$.
In a special case when one of the factors is constant, $(Cu)' = Cu'$.

(3) The derivative of a quotient is

$$\left(\frac{u}{v}\right)' = \frac{u'v - uv'}{v^2}.$$

(4) The derivative of an inverse function: If $y = f(x)$, $x = g(y)$, and f and g are inverse functions, and if $f'(x_0)$ exists and is different from zero, then also $g'(y_0)$ exists and its value is $g'(y_0) = 1/f'(x_0)$. Here $y_0 = f(x_0)$. The formula for the derivative of the inverse function (assuming that it exists) is therefore

$$\frac{dx}{dy} = \frac{1}{\frac{dy}{dx}}.$$

(5) The derivative of a composite function:
If $y = f(u)$ and $u = g(x)$, then

$$\frac{dy}{dx} = \frac{dy}{du} \cdot \frac{du}{dx},$$

or in another form,

$$\{f[g(x)]\}' = f'(u) \cdot g'(x).$$

Remark. We assume here that $u = g(x)$ and $y = f(u)$ are differentiable at corresponding points.

The accompanying table lists the most common derivatives.

Function	*Derivative*
(1) $y = C$	$y' = 0$
(2) $y = x$	$y' = 1$
(3) $y = x^\alpha$	$y' = \alpha x^{\alpha-1}$

Two important cases of this formula are

$$\left(\frac{1}{x}\right)' = -\frac{1}{x^2}; \quad (\sqrt{x})' = \frac{1}{2\sqrt{x}}.$$

(4) $y = a^x, a > 0$ $\qquad y' = a^x \ln a$

Note especially: $(e^x)' = e^x$

(5) $y = \log_a x,$ $\qquad y' = \frac{1}{x} \log_a e = \frac{1}{x \ln a}$

$x > 0, a > 0, a \neq 1$

Note especially: $(\ln x)' = 1/x$.

(6) $y = \sin x$ $\qquad y' = \cos x$

(7) $y = \cos x$ $\qquad y' = -\sin x$

(8) $y = \tan x$ $\qquad y' = \frac{1}{\cos^2 x} = \sec^2 x$

(9) $y = \cot x$ $\qquad y' = -\frac{1}{\sin^2 x} = -\csc^2 x$

(10) $y = \arcsin x$ $\qquad y' = \frac{1}{\sqrt{1 - x^2}}$

(11) $y = \arccos x$ $\qquad y' = -\frac{1}{\sqrt{1 - x^2}}$

(12) $y = \arctan x$ $\qquad y' = \frac{1}{1 + x^2}$

(13) $y = \text{arccot}\, x$ $\qquad y' = -\frac{1}{1 + x^2}$

(14) $y = \sinh x$ $\qquad y' = \cosh x$

(15) $y = \cosh x$ $\qquad y' = \sinh x$

(16) $y = \tanh x$ $\qquad y' = \frac{1}{\cosh^2 x}$

(17) $y = \coth x$ $\qquad y' = -\frac{1}{\sinh^2 x}$

(18) $y = \text{arg sinh}\, x$ $\qquad y' = \frac{1}{\sqrt{1 + x^2}}$

(19) $y = \text{arg cosh}\, x$ $\qquad y' = \frac{1}{\sqrt{x^2 - 1}}$

(20) $y = \text{arg tanh}\, x$ $\qquad y' = \frac{1}{1 - x^2}$

(21) $y = \text{arg coth}\, x$ $\qquad y' = -\frac{1}{1 - x^2}$

The following examples will serve to illustrate ways of calculating derivatives. Find the derivatives of the following functions.

165. $y = \arcsin x$

Solution. We shall use the law for differentiation of an inverse function:

$$x = \sin y, \qquad \frac{dx}{dy} = \cos y = \sqrt{1 - x^2}$$

(by definition, $-\pi/2 \leqslant y \leqslant \pi/2$ and $\cos y \geqslant 0$);

$$\frac{dy}{dx} = \frac{1}{\dfrac{dx}{dy}} = \frac{1}{\sqrt{1 - x^2}}.$$

This proves formula (*10*).

166. $y = \arctan x$

Solution. $x = \tan y$. $\dfrac{dx}{dy} = \sec^2 y = 1 + \tan^2 y = 1 + x^2$;

$$\frac{dy}{dx} = \frac{1}{\dfrac{dx}{dy}} = \frac{1}{1 + x^2}.$$

This proves formula (*12*).

167. $y = \log_a x$

Solution. $x = a^y$, $\dfrac{dx}{dy} = a^y \ln a = x \ln a$;

$$\frac{dy}{dx} = \frac{1}{\dfrac{dx}{dy}} = \frac{1}{x \ln a}.$$

This proves formula (*5*).

168. $y = (x^4 + 2x^3 + 3) \sin x$

Solution. Here we shall use the law for differentiating a product:

$$\begin{aligned} y' &= (x^4 + 2x^3 + 3)' \sin x + (x^4 + 2x^3 + 3)(\sin x)' \\ &= (4x^3 + 6x^2) \sin x + (x^4 + 2x^3 + 3) \cos x. \end{aligned}$$

169. $y = \dfrac{3x + 2}{x^2 + 1}$

Solution. We find the derivative by the law for differentiating a quotient:

$$\begin{aligned} y' &= \frac{(3x + 2)'(x^2 + 1) - (3x + 2)(x^2 + 1)'}{(x^2 + 1)^2} \\ &= \frac{3(x^2 + 1) - (3x + 2)2x}{(x^2 + 1)^2} = \frac{-3x^2 - 4x + 3}{(x^2 + 1)^2}. \end{aligned}$$

170. $y = \dfrac{xe^x - x^2}{x + e^x}$

Solution. Let us find the derivatives of the numerator and denominator and then substitute in the formula for the derivative of a quotient:

$$(xe^x - x^2)' = (xe^x)' - (x^2)' = (x)'e^x + x(e^x)' - 2x = e^x + xe^x - 2x.$$
$$(x + e^x)' = (x)' + (e^x)' = 1 + e^x.$$

So then

$$y' = \frac{(e^x + xe^x - 2x)(x + e^x) - (xe^x - x^2)(1 + e^x)}{(x + e^x)^2}$$
$$= \frac{xe^x + x^2e^x - 2x^2 + e^{2x} + xe^{2x} - 2xe^x - xe^x + x^2 - xe^{2x} + x^2e^x}{(x + e^x)^2}$$
$$= \frac{2x^2e^x - 2xe^x + e^{2x} - x^2}{(x + e^x)^2}.$$

171. $y = \sqrt{x^3 + 2x + 1}$

Solution. We shall differentiate this as a composite function:

$$y = \sqrt{u}, \qquad u = x^3 + 2x + 1, \qquad \frac{dy}{dx} = \frac{dy}{du}\frac{du}{dx};$$

$$\frac{dy}{du} = (\sqrt{u})' = \frac{1}{2\sqrt{u}}; \qquad \frac{du}{dx} = (x^3 + 2x + 1)' = 3x^2 + 2;$$

$$\frac{dy}{dx} = \frac{1}{2\sqrt{u}} \cdot (3x^2 + 2) = \frac{3x^2 + 2}{2\sqrt{x^3 + 2x + 1}}.$$

172. $y = \ln \tan x$

Solution. $y = \ln u,\ u = \tan x;$

$$\frac{dy}{du} = \frac{1}{u}; \quad \frac{du}{dx} = \frac{1}{\cos^2 x}; \quad \frac{dy}{dx} = \frac{dy}{du}\frac{du}{dx} = \frac{1}{u} \cdot \frac{1}{\cos^2 x} = \frac{1}{\tan x \cos^2 x} = \frac{2}{\sin 2x}.$$

173. $y = \arctan x^3$

Solution. $y = \arctan u,\ u = x^3;$

$$\frac{dy}{dx} = \frac{dy}{du}\frac{du}{dx} = \frac{1}{1 + u^2} \cdot 3x^2 = \frac{3x^2}{1 + x^6}.$$

174. $y = \sin\tan^2 \sqrt{x}$

Solution. $y = \sin u,\ u = z^2,\ z = \tan t,\ t = \sqrt{x};$

$$\frac{dy}{du} = \cos u, \qquad \frac{du}{dz} = 2z, \qquad \frac{dz}{dt} = \frac{1}{\cos^2 t}, \qquad \frac{dt}{dx} = \frac{1}{2\sqrt{x}};$$

$$\frac{dy}{dx} = \frac{dy}{du}\frac{du}{dz}\frac{dz}{dt}\frac{dt}{dx} = \cos u \cdot 2z \cdot \frac{1}{\cos^2 t} \cdot \frac{1}{2\sqrt{x}}$$

$$= \operatorname{costan}^2 \sqrt{x} \cdot 2 \tan \sqrt{x} \cdot \frac{1}{\cos^2 \sqrt{x}} \cdot \frac{1}{2\sqrt{x}} = \operatorname{costan}^2 \sqrt{x} \frac{\sin \sqrt{x}}{\cos^3 \sqrt{x}} \frac{1}{\sqrt{x}}.$$

175. Given $f'(x)$, the derivative of $f(x)$. Find the derivatives with respect to x of the following functions:

$$\text{(a)}\ \ [f(x)]^3, \qquad \text{(b)}\ \ \tan f(x), \qquad \text{(c)}\ \ e^{f(x)}.$$

Solution:

(a) $[f^3(x)]' = 3f^2(x) \cdot f'(x)$.

(b) $[\tan f(x)]' = \dfrac{1}{\cos^2 f(x)} \cdot f'(x)$.

(c) $[e^{f(x)}]' = e^{f(x)} \cdot f'(x)$.

176. Find the derivatives with respect to t of the functions

$$\text{(a)}\ \ f(t^3), \qquad \text{(b)}\ \ f(\tan t), \qquad \text{(c)}\ \ f(e^t).$$

Solution:

(a) $[f(t^3)]' = f'(t^3) \cdot 3t^2$.

(b) $[f(\tan t)]' = f'(\tan t) \cdot \dfrac{1}{\cos^2 t}$.

(c) $[f(e^t)]' = f'(e^t) \cdot e^t$.

4.3 EVALUATING DERIVATIVES OF EXPLICIT FUNCTIONS

The above examples show that differentiation of elementary functions can in practice be performed with the aid of a number of formulas and laws. To find the derivative of a composite function we usually avoid the intermediate variables. We shall illustrate this by solving again Problem 174:

$$y = \operatorname{sintan}^2 \sqrt{x}.$$

$$\frac{dy}{dx} = \operatorname{costan}^2 \sqrt{x}$$

$$\cdot 2 \tan \sqrt{x}$$

$$\cdot \frac{1}{\cos^2 \sqrt{x}}$$

$$\cdot \frac{1}{2\sqrt{x}}$$

$$= \operatorname{costan}^2 \sqrt{x} \frac{\sin \sqrt{x}}{\cos^3 \sqrt{x}} \cdot \frac{1}{\sqrt{x}}.$$

First appears the function sin. Its derivative is cos of the same argument. Then appears the square of a function. Its derivative is twice this function. Then appears the function tan. Its derivative is one divided by the square of cos of the same argument.

The last function is $\sqrt{x}$. Its derivative is $1/2\sqrt{x}$. The product is equal to the derivative of the given function and is the same as the earlier result.

Find the derivatives of the following functions.

177. $y = \frac{4}{x^3} + 5x^4 - 7x^{-5} + \frac{a}{x^8}$

178. $y = \frac{1}{\sqrt[3]{2x-1}} + \frac{5}{\sqrt[4]{(x^2+2)^3}}$

179. $y = \sqrt[3]{\frac{1}{1+x^2}}$

180. $y = \sinh x$

181. $y = (\sin x + \cos 2x)^3$

182. $y = \arcsin \sqrt{x}$

183. $y = \cosh \frac{1}{x}$

184. $y = \sqrt{x+1} - \ln (x + \sqrt{x+1})$

185. $y = \sqrt{\arctan \left(\sinh \frac{x}{3}\right)}$

186. $y = \frac{1}{x} + \frac{2}{\sqrt{x}} + \frac{3}{\sqrt[3]{x}}$

187. $y = \frac{1}{2}\left(a^2 \arcsin \frac{x}{a} + x\sqrt{a^2 - x^2}\right)$

188. $y = \frac{1+x^2}{x\sqrt{1-x^2}}$

189. $y = \frac{1}{2} \arctan x + \frac{1}{4} \ln \frac{(x+1)^2}{x^2+1}$

190. $y = \frac{\sqrt{x^2+x-1}}{x} + \frac{1}{2} \arcsin \frac{x-2}{\sqrt{5}x}$

191. $y = \frac{1}{\sqrt{a^2-b^2}} \arcsin \frac{a \sin x + b}{a + b \sin x}, \quad |b| < |a|, \quad -\frac{\pi}{2} < x < \frac{\pi}{2}$

192. $y = -x + \frac{1}{2} \ln (1 + e^{2x}) - e^{-x} \operatorname{arccot} (e^x)$

193. $y = \sinh^3 x$

194. $y = \cosh (\sinh x)$

195. $y = \tanh (\ln x)$

196. $y = \sqrt[4]{(1 + \tanh^2 x)^3}$

197. $y = \frac{1}{2} \tanh x + \frac{\sqrt{2}}{8} \ln \frac{1 + \sqrt{2} \tanh x}{1 - \sqrt{2} \tanh x}$

Logarithmic Differentiation. Find derivatives of the following functions.

198. $y = u^v$ (u and v are functions of x).

Solution:

$$(u^v)' = (e^{v \ln u})' = e^{v \ln u}\left(v' \ln u + v\frac{u'}{u}\right) = u^v\left(\frac{u'}{u}v + v' \ln u\right).$$

The same derivative can be obtained in a different way:

$$y = u^v; \qquad \ln y = v \ln u; \qquad (\ln y)' = (v \ln u)';$$

$$\frac{y'}{y} = \frac{v}{u}u' + v' \ln u; \qquad y' = y\left(\frac{u'}{u}v + v' \ln u\right);$$

$$(u^v)' = u^v\left(\frac{u'}{u}v + v' \ln u\right).$$

199. $y = x^x$

200. $y = x^{\sin x}$

201. $y = x^{x^x}$

202. $y = \sqrt[x]{(x+1)^2}$

203. $y = \dfrac{(x+1)^3 \sqrt[4]{x-2}}{\sqrt[5]{(x-3)^2}}$

204. $y = \sqrt{x \sin x \sqrt{1 - e^x}}$

205. Prove that $y = \dfrac{x^2}{2} + \dfrac{1}{2}x\sqrt{x^2+1} + \ln\sqrt{x + \sqrt{x^2+1}}$ is a solution of the differential equation

$$2y = xy' + \ln y'.$$

(A differential equation is an equation in which derivatives of the function in question appear.)

206. Given $u = \dfrac{1}{2}\ln\dfrac{1+v}{1-v}$, check the identity $\dfrac{du}{dv} \cdot \dfrac{dv}{du} = 1.$

207. Given the identity $1 + x + x^2 + \ldots + x^n = \dfrac{x^{n+1} - 1}{x - 1}$; find a formula for the sum: $1 + 2x + 3x^2 + \ldots + nx^{n-1}$.

208. Given $\cos x + \cos 3x + \ldots + \cos (2n-1)x = \dfrac{\sin 2nx}{2 \sin x}$; find a formula for the sum

$$\sin x + 3 \sin 3x + \ldots + (2n-1) \sin (2n-1)x.$$

209. Given $f(x) = \sqrt{\dfrac{x+1}{x-1}}$, find $f'(2)$.

4.4 DIFFERENTIATION OF IMPLICIT FUNCTIONS

An expression $f(x,y) = 0$ may define a function g in the sense that $f[x,g(x)] \equiv 0$ ($\equiv$ means identically equal) for all x's of some interval. For example, $\dfrac{2x + y}{x - y} - 1 = 0$ gives $y = -\frac{1}{2}x$ $(x \neq 0)$. Indeed,

$$\frac{2x - \frac{1}{2}x}{x + \frac{1}{2}x} - 1 = \frac{3x}{3x} - 1 = 1 - 1 = 0.$$

Sometimes the formula for $y = g(x)$ in terms of x can be obtained as above by solving $f(x,y) = 0$ with respect to y. When this is inconvenient or impossible, we shall learn to deal directly with the function in its implicit form $f(x,y) = 0$.

Remark. Not every expression of the above form defines y as a function of x. For example, $x^2 + y^2 + 1 = 0$ cannot be satisfied by any pair of real values (x,y). Conditions ensuring that an expression of the form $f(x,y) = 0$ defines y as a function of x will not be stated here.

If the equation can be solved for y, the derivative y' can be found as usual.

y' can also be obtained directly from the given expression in the following way:

We consider $f(x,y)$ as a composite function of x (by considering y as a function of x) that equals 0 for any x (in a certain domain). We now use the rule for differentiation of composite functions. The derivative of $f(x,y)$ so obtained is equal to 0 because $f(x,y)$ is a constant ($\equiv 0$). Ey solving for y' we express the derivative explicitly (generally in terms of x and y).

Find dy/dx of the following expressions.

210. $\dfrac{x^2}{a^2} + \dfrac{y^2}{b^2} = 1$

211. $x^{1/2} + y^{1/2} = a^{1/2}$

212. $x^3 + y^3 - 3axy = 0$

213. $y^2 \cos x = a^2 \sin 3x$

214. $x^y = y^x$

215. $y = 1 + xe^y$

216. $y = x + \arctan y$

217. $x \sin y - \cos y + \cos 2y = 0$

218. Prove that the function defined by $xy - \ln y = 1$ satisfies the equation

$$y^2 + (xy - 1)\frac{dy}{dx} = 0.$$

219. $\ln \sqrt{x^2 + y^2} = \arctan (y/x)$

220. Given the equation of a circle $(x - 1)^2 + (y + 3)^2 = 17$; find y' at the point (2,1).

4.5 PARAMETRIC DIFFERENTIATION

Two equations $y = f(t)$, $x = \varphi(t)$ usually define y as a function of x. To find dy/dx, let us consider y as a composite function. Then:

$$\frac{dy}{dx} = \frac{dy}{dt} \cdot \frac{dt}{dx} = \frac{dy}{dt} \cdot \frac{1}{\dfrac{dx}{dt}} = \frac{\dfrac{dy}{dt}}{\dfrac{dx}{dt}}.$$

Find dy/dx in the following.

221. $x = a \cos t,\ y = b \sin t$

222. $x = a(\varphi - \sin \varphi),\ y = a(1 - \cos \varphi)$

223. $x = \dfrac{t+1}{t},\ y = \dfrac{t-1}{t}$

224. $x = \ln(1 + t^2),\ y = t - \arctan t$

225. $x = e^t \sin t,\ y = e^t \cos t$

226. Given $x = \dfrac{1+t}{t^3},\ y = \dfrac{3}{2t^2} + \dfrac{2}{t}$, prove that

$$x(y')^3 = 1 + y' \qquad (y' = dy/dx).$$

227. Given

$$x = \frac{1}{\sqrt{1+t^2}} - \ln \frac{1 + \sqrt{1+t^2}}{t}, \qquad y = \frac{t}{\sqrt{1+t^2}},$$

prove that

$$y\sqrt{1 + y'^2} = y' \qquad (y' = dy/dx).$$

228. Calculate y' at the point $(1, -\sqrt{3}/2)$ of the curve

$$x = 2 \cos t, \qquad y = \sin t.$$

4.6 SPECIAL CASES IN CALCULATING DERIVATIVES

(a) *Onesided Derivatives.* If

$$\lim_{\Delta x \to 0+} \frac{f(x_0 + \Delta x) - f(x_0)}{\Delta x}$$

exists, it is called the derivative from the right of f at x_0. The derivative from the left is defined in the same way.

$f'(x_0)$ exists only if the above two derivatives exist and are equal. An example in which they exist and are not equal is given in Figure 8, at the

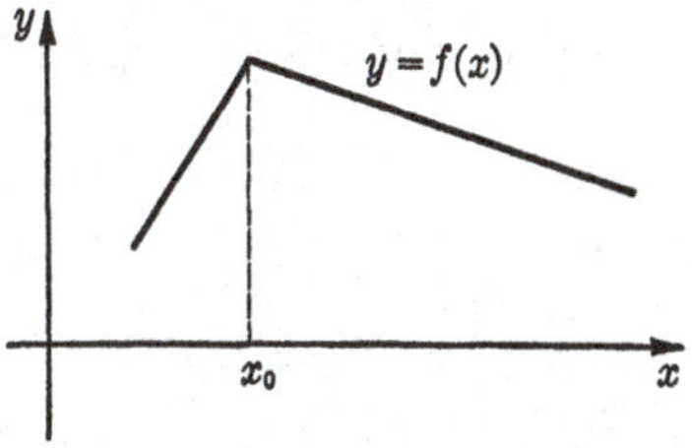

FIGURE 8

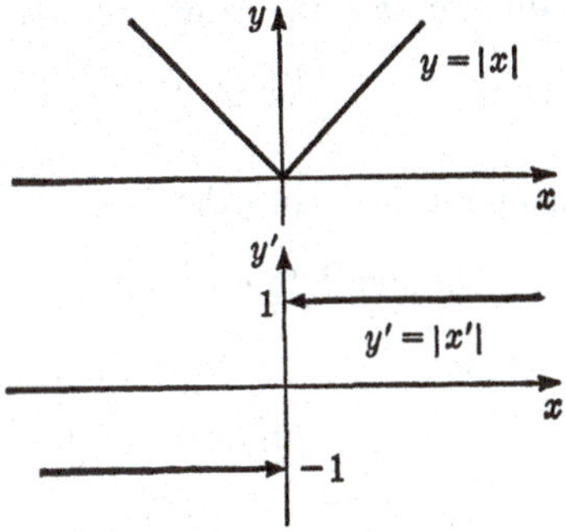

FIGURE 9

point x_0. The function f represented in this figure has no derivative at x_0. This simple example shows that continuity is not sufficient to ensure the existence of the derivative.

(b) If the tangent to the curve representing $y = f(x)$ at $x = x_0$ is perpendicular to the x axis, then $\Delta y/\Delta x \to \infty$ as $\Delta x \to 0$. The derivative does not exist (is also often said to be "infinite") at this point.

(c) A function can be continuous at a certain point and not have any derivative there (not even a onesided derivative).

(d) The derived function may be discontinuous at certain points.

The following examples serve to clarify the above statements.

229. Differentiate the function $y = f(x) = |x|$.

Solution. The function can be divided into three parts:

$$\begin{aligned} &y = x, && x > 0, && y' = 1; \\ &y = -x, && x < 0, && y' = -1; \\ &y = 0, && x = 0. \end{aligned}$$

The derivative does not exist at $x = 0$. But clearly,

$$\lim_{\Delta x \to 0+} \frac{f(0 + \Delta x) - f(0)}{\Delta x} = 1 \quad \text{and} \quad \lim_{\Delta x \to 0-} \frac{f(0 + \Delta x) - f(0)}{\Delta x} = -1,$$

i.e., there exist derivatives from both right and left. Figure 9 illustrates the situation.

230. Differentiate $y = x^{1/3}$.

Solution. $y' = \frac{1}{3}x^{-2/3}$. This derivative does not exist at $x = 0$. We try to differentiate the given function at $x = 0$ by definition:

$$\lim_{\Delta x \to 0} \frac{f(0 + \Delta x) - f(0)}{\Delta x} = \lim_{\Delta x \to 0} \frac{(\Delta x)^{1/3}}{\Delta x} = \lim_{\Delta x \to 0} \frac{1}{(\Delta x)^{2/3}}.$$

This limit does not exist. (It is the case of an "infinite" derivative.)

231. Differentiate $y = \sqrt{x}$.

Solution. $y' = 1/2\sqrt{x}$ is not defined at $x = 0$.

$$\lim_{\Delta x \to 0} \frac{f(0 + \Delta x) - f(0)}{\Delta x} = \lim_{\Delta x \to 0} \frac{(\Delta x)^{1/2}}{\Delta x} = \lim_{\Delta x \to 0} \frac{1}{(\Delta x)^{1/2}}.$$

The derivative does not exist at $x = 0$. (Here we have also an "infinite" derivative but only from the right. The function is not defined at all to the left of $x = 0$.)

232. Differentiate $y = x^{2/3}$.

Solution. $y' = \frac{2}{3}x^{-1/3}$. At $x = 0$ we have

$$\lim_{\Delta x \to 0} \frac{f(0 + \Delta x) - f(0)}{\Delta x} = \lim_{\Delta x \to 0} \frac{(\Delta x)^{2/3}}{\Delta x} = \lim_{\Delta x \to 0} \frac{1}{(\Delta x)^{1/3}}.$$

For $\Delta x \to 0+$ this tends to $+\infty$ and for $\Delta x \to 0-$, to $-\infty$. The derivative does not exist at $x = 0$. (We sometimes say we have here two different onesided "infinite" derivatives.)

233. Given $f(x) = x \sin \frac{1}{x}$, $f(0) = 0$; differentiate this function.

Solution:

$$f'(x) = \sin \frac{1}{x} + x \cos \frac{1}{x} \cdot \left(-\frac{1}{x^2}\right) = \sin \frac{1}{x} - \frac{1}{x} \cos \frac{1}{x}.$$

This is not defined at $x = 0$. By the definition:

$$\lim_{\Delta x \to 0} \frac{f(0 + \Delta x) - f(0)}{\Delta x} = \lim_{\Delta x \to 0} \frac{\Delta x \sin \dfrac{1}{\Delta x}}{\Delta x} = \lim_{\Delta x \to 0} \sin \frac{1}{\Delta x}.$$

This limit does not exist (when $1/\Delta x \to 0$, $\sin(1/\Delta x)$ oscillates between -1 and $+1$). The given function is continuous at $x = 0$ (Prob. 145) and does not have any derivative there.

234. Differentiate the function $f(x) = x^2 \sin \frac{1}{x}$, $f(0) = 0$.

Solution. At $x \neq 0$,

$$f'(x) = 2x \sin \frac{1}{x} - \cos \frac{1}{x}.$$

At $x = 0$,

$$\lim_{\Delta x \to 0} \frac{f(0 + \Delta x) - f(0)}{\Delta x} = \lim_{\Delta x \to 0} \frac{(\Delta x)^2 \sin (1/\Delta x)}{\Delta x}$$
$$= \lim_{\Delta x \to 0} \Delta x \sin \frac{1}{\Delta x} = 0,$$

i.e., $f'(0) = 0$. Let us consider

$$\lim_{\Delta x \to 0} f'(x) = \lim_{x \to 0} \left(2x \sin \frac{1}{x} - \cos \frac{1}{x}\right).$$

This limit does not exist and the derivative, which exists at $x = 0$, is not continuous there.

4.7 HIGHER DERIVATIVES

The derived function can itself be differentiated to obtain the so-called second derivative, denoted by y'' or d^2y/dx^2. By definition, $y'' = (y')'$, and using induction derivatives of any order n can be defined: $y^{(n)} = [y^{(n-1)}]'$ or, in a different notation,

$$\frac{d^n y}{dx^n} = \frac{d\left(\dfrac{d^{n-1}y}{dx^{n-1}}\right)}{dx}.$$

Find the higher derivatives as indicated in the following examples.

235. $y = x^3 - x^2 + x - 1$, $\quad y^{iv} = ?$

236. $y = \arctan x$, $\quad y'' = ?$

237. $y = \sqrt[5]{x^3}$, $\quad y''' = ?$

238. $y = x^5 \ln x$, $\quad y^{vi} = ?$

239. $y = \dfrac{x^3}{x - 1}$, $\quad y''' = ?$

240. $y = x^2 e^{2x}$, $\quad y^{iv} = ?$

241. $y = (1 + x^2) \arctan x$, $\quad y'' = ?$

242. Prove that $y = e^x \sin x$ is a solution of the differential equation

$$y'' - 2y' + 2y = 0.$$

243. Prove that $y = A \sin(\omega t + \omega_0) + B \cos(\omega t + \omega_0)$ is a solution of the equation

$$\frac{d^2y}{dx^2} + \omega^2 y = 0.$$

When y is an implicit function of x, we find y' as before, and then differentiate the whole expression once again, remembering that also y' is a function of x. In this way y'' is expressed in terms of x, y and y'. y' can be replaced by its expression in x and y and then y'' is given in terms of x and y only. In the same way higher derivatives of y can also be obtained.

244. $b^2x^2 + a^2y^2 = a^2b^2,$ $\quad y'' = ?$

245. $y = \tan(x + y),$ $\quad y''' = ?$

246. $e^{x+y} = xy,$ $\quad y'' = ?$

247. $e^y + xy = e,$ $\quad y''(x = 0) = ?$

248. $x^2 + 2xy + y^2 - 4x + 2y - 2 = 0,$ $\quad y'''(x = 1, y = 1) = ?$

Higher derivatives of functions may be given by parametric equations. Suppose $y = f(t)$ and $x = g(t)$. We saw, in Section 4.5,

$$\frac{dy}{dx} = \frac{f'(t)}{g'(t)} = F(t).$$

In the same way we obtain:

$$\frac{d^2y}{dx^2} = \frac{dF(t)}{dx} = \frac{\frac{dF(t)}{dt}}{\frac{dx}{dt}} = \frac{F'(t)}{g'(t)} = \frac{f''(t)g'(t) - f'(t)g''(t)}{g'(t)[g'(t)]^2}$$

$$= \frac{f''(t)g'(t) - f'(t)g''(t)}{[g'(t)]^3}.$$

Similarly we can find d^3y/dx^3, etc.

To solve related exercises it is not always preferable to use the above formulas. It is better to perform the calculations by the steps which were exhibited above, especially for the third, fourth, and higher derivatives, for which the general form becomes involved.

249. $x = at^2,\ y = bt^3;$ $\quad \frac{d^2y}{dx^2} = ?$

250. $x = a \cos t,\ y = a \sin t;\qquad \dfrac{d^3y}{dx^3} = ?$

251. $x = at \cos t,\ y = at \sin t;\qquad \dfrac{d^2y}{dx^2} = ?$

252. Prove that the function given by $x = e^t \sin t$, $y = e^t \cos t$ is a solution of the equation

$$\frac{d^2y}{dx^2}(x + y)^2 = 2\left(x\frac{dy}{dx} - y\right).$$

4.8 CALCULATION OF $y^{(n)}$

In many cases we can give an explicit expression for $y^{(n)}$. The corresponding formulas can be proved by induction.

Leibniz derived the following formula for the nth derivative of a product:

$$(uv)^{(n)} = u^{(n)}v + nu^{(n-1)}v' + \frac{n(n-1)}{1\cdot 2}u^{(n-2)}v'' + \dots$$
$$+ \frac{n(n-1)\cdots(n-k+1)}{1\cdot 2\cdots k}u^{(n-k)}v^{(k)} + \dots + uv^{(n)}.$$

The analogy to Newton's binomial expansion is clear.

Leibniz's formula is especially suitable when one of the factors is a polynomial and its derivatives vanish beginning with a certain k. In this case the number of terms in the formula is the same for all values of n greater than k.

In finding the nth derivative we also use the following two obvious formulas:

$$(u \pm v)^{(n)} = u^{(n)} \pm v^{(n)}, \qquad (Cu)^{(n)} = Cu^{(n)}.$$

253. Find $y^{(n)}$ if $y = x^p$ (p is any number).

Solution:

$$y' = px^{p-1}, \qquad y'' = p(p-1)x^{p-2}.$$

The rule is clear and we can write:

$$y^{(n)} = p(p-1)(p-2)\cdots(p-n+1)x^{p-n}.$$

We shall prove this formula by induction. For $n = 1$ it is true, as was shown above. Suppose it is true for $n = k$, i.e.,

$$y^{(k)} = p(p-1)(p-2)\cdots(p-k+1)x^{p-k}.$$

Let us differentiate this equation:

$$y^{(k+1)} = p(p-1)(p-2)\cdots(p-k+1)(p-k)x^{p-k-1}.$$

This is exactly the formula for $n = k + 1$. By induction, therefore, the formula is true for any n. If $p = m$ (m a positive integer), we have:

$$y^{(m)} = m(m-1)(m-2)\cdots(m-m+1) = m!;$$
$$y^{(m+1)} = 0.$$

The $(m + 1)$th derivative and all higher derivatives of x^m are equal to 0.

For a polynomial $y = a_0x^m + a_1x^{m-1} + \ldots + a_m$ we have:

if $k < m$: $y^{(k)} = a_0m(m-1)(m-2)\cdots(m-k+1)x^{m-k}$
$+ a_1(m-1)(m-2)\cdots(m-k)x^{m-k-1} + \ldots + a_{m-k}k!$;

if $k = m$: $y^{(k)} = y^{(m)} = a_0m!$;

if $k > m$: $y^{(k)} = 0$.

Find $y^{(n)}$ in the following exercises.

254. $y = \sqrt{x}$

255. $y = (ax + b)^p$

256. $y = \dfrac{1}{(2x+3)^4}$

257. $y = \dfrac{1}{1-x}$

258. $y = \ln(ax + b)$

259. $y = \ln\dfrac{x+1}{3-2x}$

260. $y = \dfrac{1}{x^2-1}$

261. $y = \dfrac{1}{x^2-3x+2}$

262. $y = \dfrac{1+x}{1-x}$

263. $y = a^{px}$

264. $y = \sin x$

265. $y = \cos 3x$

266. $y = \sin^2 x$

267. $y = x^3 \ln x$

268. Prove $(\sin^4 x + \cos^4 x)^{(n)} = 4^{n-1}\cos\left(4x + n\dfrac{\pi}{2}\right)$.

269. $y = \dfrac{x^4+1}{x^3-x}$

270. $y = e^{ax}\sin bx$

The use of Leibniz's formula is illustrated in the following exercises.

271. $y = x^2e^{2x}$, $\quad y^{iv} = ?$

Solution. Put $u = e^{2x}$, $v = x^2$. By Leibniz's formula:

$$y^{iv} = (e^{2x}x^2)^{iv} = (e^{2x})^{iv}x^2 + 4(e^{2x})'''(x^2)' + \frac{4\cdot 3}{1\cdot 2}(e^{2x})''(x^2)''$$
$$+ \frac{4\cdot 3\cdot 2}{1\cdot 2\cdot 3}(e^{2x})'(x^2)''' + e^{2x}(x^2)^{iv}$$
$$= 16e^{2x}x^2 + 4\cdot 8e^{2x}\cdot 2x + 6\cdot 4e^{2x}\cdot 2 = 16e^{2x}(x^2 + 4x + 3).$$

It could be seen beforehand that $(x^2)''' = 0$, $(x^2)^{iv} = 0$. Compare this solution with the one given in Problem 240.

272. $y = x^2 \cos ax, \qquad y^{(30)} = ?$

273. $y = e^x \sin x, \qquad y^{iv} = ?$

274. $y = x^3 \ln x, \qquad y^{(n)} = ?$

4.9 GRAPHICAL DIFFERENTIATION

A function may sometimes be given only graphically (for example, when it results from an experiment). Its derivative can be found by a method of graphical differentiation.

Suppose we have to find the value of the derivative at M (Fig. 10). We

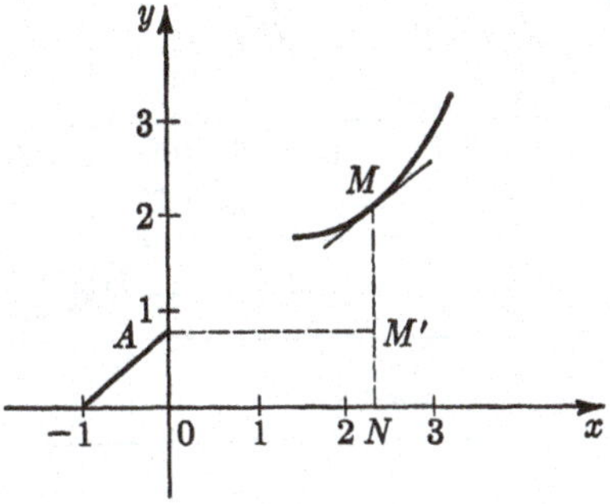

FIGURE 10

draw, as accurately as possible, the tangent line to the graph at this point. Through the point $(-1,0)$ we draw a line parallel to the tangent. A is the point of intersection of this parallel with y axis. Through A we draw a parallel to Ox, which intersects at M' the ordinate of M. NM' then gives the value of the derivative corresponding to M. Now we divide the domain of definition of the function in intervals and repeat the above construction for a point in each interval. By joining the points M' we obtain an approximate graph of the derived function.

If it is not possible to work with the same scale on both axes, the value of the derivative will differ from NM' by a constant factor.

To obtain the tangent more easily we can draw chords between the ends of the arc in each interval, and the tangent at the midpoint of this arc (if it is small enough) will be nearly parallel to the chord.

Now we shall solve one problem of graphical differentiation.

275. A graphical representation of the displacement vs. time of an output valve in a steam engine is given in Figure 11. Time is given in seconds and displacement in millimeters. Plot the curve of velocity vs. time.

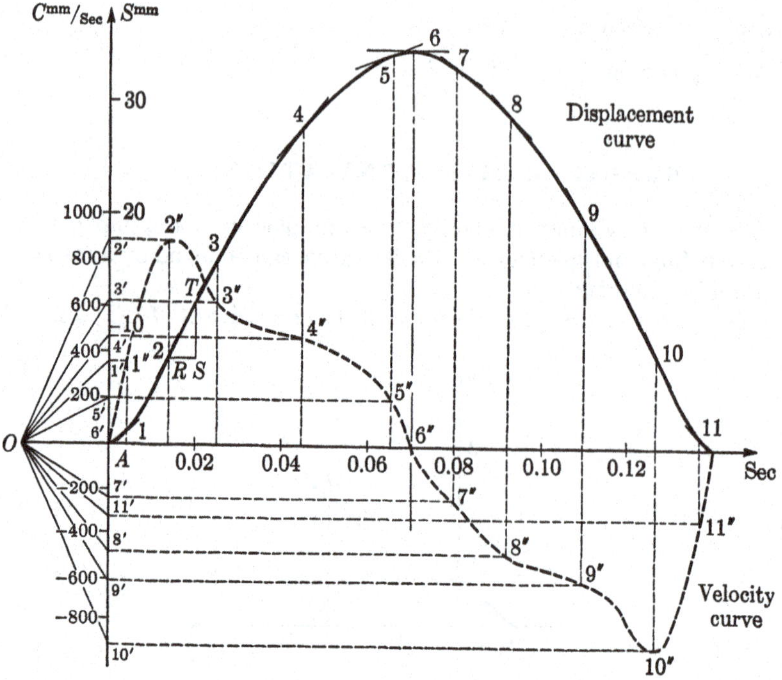

FIGURE 11

Solution. We choose the point O as in Figure 11 and mark on the curve a number of points: 1, 2, . . . , 11. We draw at these points tangents to the curve and then draw through O parallels to these tangents ($O1'$, $O2'$, . . . , $O11'$). On the ordinates of 1, 2, . . . , 11 we mark off segments $A1'$, $A2'$, . . . , $A11'$. We join the resulting points $1''$, $2''$, $3''$, . . . , $11''$, and thus obtain the velocity curve approximately.

Now we shall find by the following calculation the scale of the velocity. Suppose that the scale on the x axis is m_1, i.e., m_1 is the length of the segment we choose as unit of this axis.

In our case $m_1 = \dfrac{25 \text{ mm}}{0.04 \text{ sec}} = 625 \text{ mm/sec}.$

In the same way the scale on the y axis, denoted by m_2, will be

$$m_2 = \frac{16 \text{ mm}}{10 \text{ mm}} = 1.6 \text{ mm/mm}.$$

We take one of the segments Ai' ($i = 1, 2, \ldots, 11$), e.g., $A2'$, and we find what value it expresses:

$$A2' = OA \cdot \tan \measuredangle\, 2'OA = a \cdot \tan \measuredangle\, 2'OA.$$

In our case $a = 12.5$ mm. On the other hand:

$$y'(2) = \frac{\Delta y}{\Delta x} = \frac{\dfrac{TS}{m_2}}{\dfrac{RS}{m_1}} = \frac{m_1}{m_2} \cdot \frac{TS}{RS} = \frac{m_1}{m_2} \tan \measuredangle\, TRS = \frac{m_1}{m_2} \tan \measuredangle\, 2'OA.$$

This gives:
$$y'(2) = \frac{m_1}{m_2} \cdot \frac{A2'}{a} = \frac{A2'}{am_2/m_1}.$$

Hence, if we want to obtain $y'(2)$ directly from $A2'$, we choose for the derivative a scale equal to

$$a \frac{m_2}{m_1} \frac{\text{cm}}{\text{units of the derivative}}.$$

In the present case the scale is

$$\frac{12.5\text{ mm} \cdot 1.6\text{ mm/mm}}{625\text{ mm/sec}} = 0.032 \frac{\text{mm}}{\text{mm/sec}}.$$

4.10 VARIOUS EXAMPLES

276. Prove that

$$\arccos \frac{a \cos x + b}{a + b \cos x} - 2 \arctan\left(\sqrt{\frac{a-b}{a+b}} \tan \frac{x}{2}\right), \quad 0 < b \leqslant a,\ 0 \leqslant x < \pi,$$

is a constant.

277. Prove that the derivative of the function

$$f(x) = \frac{\sin x}{x}, \qquad f(0) = 1$$

exists at every point and equals

$$f'(x) = \frac{x \cos x - \sin x}{x^2}, \qquad f'(0) = 0.$$

Show that $f'(x)$ is continuous.

278. Draw the graphs of $f(x) = x/[x]$ and of $f'(x)$ and $f''(x)$.

279. Prove: If $y = y_1 \cdot y_2 \cdots y_n$ and $y_i = f_i(x) \neq 0$, then

$$\frac{dy}{dx} = \sum_{k=1}^{n} \frac{y}{y_k} \frac{dy_k}{dx}.$$

280. (a) Prove that if $(x - a)^n$ (n an integer greater than 1) divides a polynomial $f(x)$, then $(x - a)^{n-1}$ divides $f'(x)$.
(b) Prove that if $x - a$ divides $f(x)$ and $(x - a)^n$ divides $f'(x)$, then $(x - a)^{n+1}$ divides $f(x)$.

Is the condition that $x - a$ divide $f(x)$ necessary?

281. $x^2 + y^2 = e^{2 \arctan (y/x)}$ defines y as an implicit function of x. Find y' and y''.

282. Prove that $y = \cos (m \ln x)$ is a solution of the equation

$$x^2y^{(n+2)} + (2n + 1)xy^{(n+1)} + (m^2 + n^2)y^{(n)} = 0.$$

283. Prove by induction

$$(e^{x^2/2})^{(n)} = u_n(x)e^{x^2/2}$$

where $u_n(x)$ is a polynomial of degree n. Prove the reduction formula

$$u_{n+1} = xu_n + u_n'.$$

284. Given

$$x = f'(t), \qquad y = tf'(t) - f(t),$$

find dy/dx and d^2y/dx^2.

285. Prove that

$$x = \frac{1 + \ln t}{t^2}, \qquad y = \frac{3 + 2 \ln t}{t}$$

is a solution of

$$y \frac{dy}{dx} = 2x \left(\frac{dy}{dx}\right)^2 + 1.$$

V

FUNDAMENTAL THEOREMS OF THE DIFFERENTIAL CALCULUS

5.1 THE THEOREMS OF ROLLE, LAGRANGE, AND CAUCHY

ROLLE'S THEOREM. If $f(x)$ is a continuous function in a closed interval $[a,b]$, and $f'(x)$ exists in the open interval (a,b), and $f(a) = f(b)$, then there is at least one point c between a and b such that $f'(c) = 0$ (see Fig. 12).

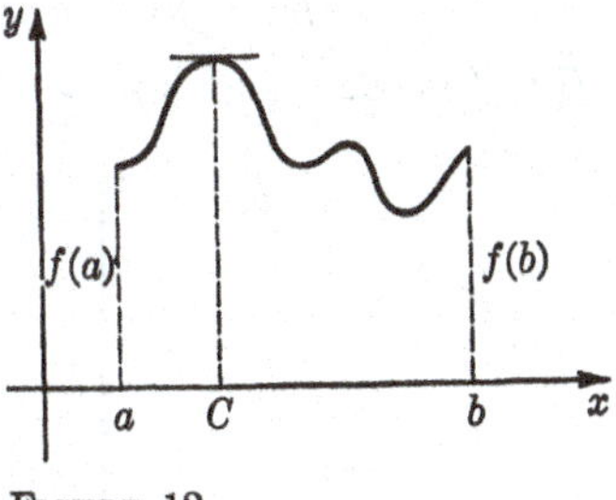

FIGURE 12

LAGRANGE'S THEOREM. If $f(x)$ is continuous in the closed interval $[a,b]$ and $f'(x)$ exists in the open interval (a,b), then there is at least one point c between a and b such that

$$f(b) - f(a) = (b - a)f'(c).$$

This theorem is also called the theorem of the mean. Rolle's theorem is a particular case of this theorem when $f(a) = f(b)$.

The theorem of Lagrange can also be expressed in the following form:

$$\Delta f(x) = f(x + \Delta x) - f(x) = f'(x + \theta\Delta x)\,\Delta x,$$

where θ is a number satisfying $0 < \theta < 1$.

An approximation (often a very rough one) we obtain by taking $\theta = 0$. Then

$$\Delta f(x) \approx f'(x)\,\Delta x.$$

CAUCHY'S THEOREM. If $f(x)$ and $g(x)$ are continuous functions in $[a,b]$ and have in (a,b) derivatives $f'(x)$ and $g'(x)$ and if $g'(x) \neq 0$ in this interval, then there is a point c between a and b such that

$$\frac{f(b) - f(a)}{g(b) - g(a)} = \frac{f'(c)}{g'(c)}.$$

Lagrange's theorem can be obtained from this by setting $g(x) = x$.

A number of problems concerning the above theorems follow.

286. Verify Rolle's theorem for the function $y = x^3 + 4x^2 - 7x - 10$ in the interval $[-1,2]$.

287. The same question for $y = (2 - x^2)/x^4$ in the interval $[-1,1]$.

288. Prove that if the equation $f(x) = a_0x^n + a_1x^{n-1} + \ldots + a_{n-1}x = 0$ has a positive solution $x = x_0$, then also the equation

$$na_0x^{n-1} + (n-1)a_1x^{n-2} + \ldots + a_{n-1} = 0$$

has a positive solution smaller than x_0.

289. Given $f(x) = x(x-1)(x-2)(x-3)(x-4)$. Determine how many solutions the equation $f'(x) = 0$ has, and find intervals including every one of them, without calculating $f'(x)$.

290. Prove that the equation $f(x) = x^n + px + q = 0$ cannot have more than two real solutions for an even n nor more than 3 real solutions for an odd n.

291. (a) Give the geometrical interpretation of Lagrange's formula

$$f(x+h) = f(x) + hf'(x + \theta h), \qquad 0 < \theta < 1.$$

(b) Calculate θ for $f(x) = x^2$ and explain the meaning of the result.

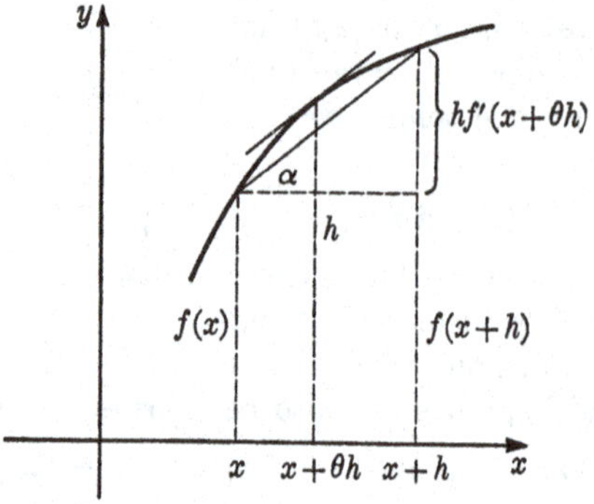

FIGURE 13

Solution. (a) From Figure 13 we see that

$$\tan \alpha = \frac{hf'(x + \theta h)}{h} = f'(x + \theta h).$$

$f'(x + \theta h)$ is the slope of the tangent at $x + \theta h$. This clarifies the geometrical meaning of the formula: there is always a point c between a and b

such that the tangent at $(c,f(c))$ to the curve $y = f(x)$ is parallel to the chord joining the points $(a,f(a))$ and $(b,f(b))$.

(b) $f(x) = x^2$; $f'(x) = 2x$; $f'(x + \theta h) = 2(x + \theta h)$;
$(x + h)^2 = x^2 + h2(x + \theta h)$;
$x^2 + 2hx + h^2 = x^2 + 2hx + 2\theta h^2$; $h^2 = 2\theta h^2$; $\theta = \frac{1}{2}$.

The result shows that in the parabola $y = x^2$ the line parallel to the above chord is tangent at the point on the arc whose abscissa is $x = (a + b)/2$, i.e., at the center of the interval $[a,b]$.

292. Write down Lagrange's formula for the function $f(x) = \sqrt{x}$ in the interval $[1,4]$, and find c.

Answer the preceding question for the following functions in the given intervals.

293. $f(x) = x^3$, $[a,b]$

294. $f(x) = \arctan x$, $[0,1]$

295. $f(x) = \arcsin x$, $[0,1]$

296. $f(x) = \ln x$, $[1,2]$

297. Write down Cauchy's formula and find c for the two functions $f(x) = \sin x$, $g(x) = \cos x$ in the interval $[0,\pi/2]$.

298. The preceding question for $f(x) = x^2$, $g(x) = \sqrt{x}$ in the interval $[1,4]$.

299. Using Lagrange's formula, prove the inequalities

$$nb^{n-1}(a - b) < a^n - b^n < na^{n-1}(a - b), \qquad a > b > 0,\ n > 1.$$

300. Prove

$$\frac{a - b}{a} \leqslant \ln \frac{a}{b} \leqslant \frac{a - b}{b}, \qquad 0 < b \leqslant a.$$

301. An approximate form obtained from Lagrange's formula is

$$f(x + \Delta x) \approx f(x) + f'\left(x + \frac{\Delta x}{2}\right) \Delta x.$$

Use this to compute approximately $\log_{10} 11$.

302. Prove the inequality $e^x > 1 + x$.

303. Show that the theorem of the mean does not hold for $f(x) = \tan x$ in the interval $(1.5,1.6)$. Why?

304. Two functions u and v and their derivatives u', v' are continuous in the interval $a \leqslant x \leqslant b$, and the expression $uv' - u'v$ does not vanish at any point of this interval. Prove that between each two solutions of the equation $u = 0$ lies one solution of the equation $v = 0$. Illustrate this theorem for $u = \cos x$, $v = \sin x$.

5.2 TAYLOR'S AND MACLAURIN'S FORMULAS

We first show Taylor's formula for a polynomial. Given a polynomial

$$p(x) = a_0 + a_1x + a_2x^2 + \ldots + a_nx^n,$$

its expansion around a, i.e., in powers of $x - a$, is given by

$$p(x) = p(a) + p'(a)(x - a) + \frac{p''(a)}{2!}(x - a)^2 + \ldots + \frac{p^{(n)}(a)}{n!}(x - a)^n.$$

Taylor's formula for any function is:

$$f(x) = f(a) + \frac{f'(a)}{1!}(x - a) + \frac{f''(a)}{2!}(x - a)^2 + \ldots + \frac{f^{(n-1)}(a)}{(n - 1)!}(x - a)^{n-1} + R_n.$$

R_n, the so-called remainder term (i.e., the one that balances the two sides), can be represented in several forms. Lagrange's form is:

$$R_n = \frac{(x - a)^n}{n!} f^{(n)}(c),$$

where c is a point between a and x. Cauchy's form of the remainder is

$$R_n = \frac{f^{(n)}[a + \theta(x - a)]}{(n - 1)!}(1 - \theta)^{n-1}(x - a)^n, \qquad 0 < \theta < 1.$$

For $a = 0$ Taylor's formula reduces to the so-called Maclaurin's formula:

$$f(x) = f(0) + \frac{f'(0)}{1!}x + \frac{f''(0)}{2!}x^2 + \ldots + \frac{f^{(n-1)}(0)}{(n - 1)!}x^{n-1} + R_n,$$

and
$$R_n = \frac{x^n}{n!} f^{(n)}(\theta x), \qquad 0 < \theta < 1;$$

or (in Cauchy's form)

$$R_n = \frac{f^{(n)}(\theta x)}{(n - 1)!}(1 - \theta)^{n-1}x^n.$$

By fixing n, i.e., the number of terms in the formula, we obtain in general an approximate value of $f(x)$. An estimate of the remainder may serve here to evaluate the closeness of the approximation.

By setting $n = 1$ in Taylor's formula we obtain:

$$f(x) = f(a) + R_1 = f(a) + (x - a)f'(c),$$

which evidently is Lagrange's formula.

It is important to emphasize that all these "formulas of the mean" (Lagrange's, Cauchy's, Taylor's) are exact equalities. However, they all include a term with an indeterminate c (known only to lie in a certain interval). This makes it impossible, in general, to find $f(x)$ exactly by the above formulas. On the other hand, this unknown term can often be estimated.

In Taylor's formula R_n tends in many cases to 0 when $n \to \infty$, and this enables us to find $f(x)$ with any desired accuracy by taking n large enough.

305. Expand the polynomial $f(x) = x^3 - 6x + 2$ in powers of $x - 2$, i.e., around the point $x = 2$. Use this expansion to calculate $f(2.003)$ with accuracy to 0.0001.

Solution. The expansion is given by Taylor's formula:

$$f(x) = f(2) + (x-2)f'(2) + \frac{(x-2)^2}{2!}f''(2) + \frac{(x-2)^3}{3!}f'''(2) + \dots.$$

$$\begin{aligned} f(x) &= x^3 - 6x + 2, & f(2) &= -2, \\ f'(x) &= 3x^2 - 6, & f'(2) &= 6, \\ f''(x) &= 6x, & f''(2) &= 12, \\ f'''(x) &= 6, & f'''(2) &= 6. \end{aligned}$$

Substitution leads to

$$x^3 - 6x + 2 = -2 + 6(x-2) + 6(x-2)^2 + (x-2)^3.$$

Remark. This equality can be easily checked by opening the parentheses and summing the right-hand side.

To calculate $f(2.003)$ we substitute 2.003 for x in the above expansion:

$$f(2.003) = -2 + 6 \cdot 0.003 + 6 \cdot 0.003^2 + 0.003^3.$$

It is easy to see that to obtain the required accuracy it is enough to calculate only the first three terms:

$$f(2.003) \approx -2 + 0.018 + 0.000054 \approx -1.9815.$$

306. Find the expansion of the polynomial $f(x) = x^4 - 5x^3 + x^2 - 3x + 4$ in powers of $x - 4$, i.e., around the point $x = 4$.

307. Find the expansion of $f(x) = (x^2 - 3x + 1)^3$ in powers of x (i.e., around $x = 0$).

308. Given $f(x) = x^{80} - x^{40} + x^{20}$. Find $f(1.005)$ by evaluating the first three terms of the expansion around 1.

309. Find the Maclaurin formula for $f(x) = e^x$ and estimate the remainder.

Solution. We know $f^{(n)}(x) = e^x$, $f^{(n)}(0) = e^0 = 1$. By Maclaurin's formula we have:

$$e^x = 1 + \frac{x}{1!} + \frac{x^2}{2!} + \dots + \frac{x^n}{n!} + R_{n+1},$$

$$R_{n+1} = \frac{f^{(n+1)}(\theta x)}{(n+1)!} x^{n+1}, \quad 0 < \theta < 1; \quad \text{i.e.,} \quad R_{n+1} = \frac{e^{\theta x}}{(n+1)!} x^{n+1}.$$

Estimating R_{n+1},

$$\text{for } x > 0: \quad e^{\theta x} < e^x, \quad \text{i.e.,} \quad R_{n+1} < \frac{e^x}{(n+1)!} x^{n+1}.$$

$$\text{for } x < 0: \quad e^{\theta x} < e, \quad \text{i.e.,} \quad R_{n+1} < \frac{e}{(n+1)!} x^{n+1}.$$

310. Prove that if we calculate e^x $(0 \leqslant x \leqslant \frac{1}{2})$ by the approximate formula

$$e^x \approx 1 + x + \frac{x^2}{2} + \frac{x^3}{6},$$

then the error will not exceed 0.01.

311. Find the Maclaurin formula for $f(x) = \sin x$ and estimate the remainder.

312. Calculate $\sin 17°21'$ with accuracy to four places of decimals $(\pi = 3.14159)$.

Solution. We have seen in Problem 311,

$$\sin x = x - \frac{x^3}{3!} + \frac{x^5}{5!} - \ldots + (-1)^{k-1} \frac{x^{2k-1}}{(2k-1)!} + R_{2k+1},$$

$$|R_{2k+1}| \leqslant \frac{|x|^{2k+1}}{(2k+1)!}.$$

In our case x is equal to $\dfrac{(17 \cdot 60 + 21)\pi}{180 \cdot 60} = \dfrac{347\pi}{3600}.$

We need $|R_{2k+1}| < \frac{1}{2} \cdot 0.0001$. We try $k = 2$;

$$\left(\frac{347\pi}{3600}\right)^5 \cdot \frac{1}{5!} < \left(\frac{1}{3}\right)^5 \cdot \frac{1}{5!} < \frac{1}{243 \cdot 120} < \frac{1}{2} \cdot 10^{-2} \cdot 10^{-2} = \frac{1}{2} 10^{-4}.$$

It can be shown that $k = 1$ does not suffice. Now

$$\sin 17°21' \approx \frac{347\pi}{3600} - \left(\frac{347\pi}{3600}\right)^3 \cdot \frac{1}{3!} \approx 0.2982.$$

It should be remarked that in evaluating the terms of the expansion we must work with accuracy to five decimal places and of course use the value of π with at least five decimal places.

313. Find the Maclaurin expansion of $f(x) = \cos bx$ and estimate the remainder.

314. Calculate $\cos^2 (0.1)$ by evaluating the first three nonzero terms of Maclaurin's expansion. Estimate the error.

Solution:

$$\cos^2 x = \frac{1}{2}(1 + \cos 2x) = \frac{1}{2}\left[1 + 1 - \frac{(2x)^2}{2!} + \frac{(2x)^4}{4!} - \ldots + R_{2k}\right],$$

$$|R_{2k}| < \frac{2^{2k}x^{2k}}{(2k)!}.$$

The first three nonvanishing terms give $\cos^2 x = 1 - x^2 + \frac{x^4}{3} - \ldots;$

$$\cos^2(0.1) \approx 1 - 0.01 + \frac{0.0001}{3} \approx 0.9900333.$$

The error is smaller than

$$\frac{1}{2}\cdot\frac{2^6(0.1)^6}{6!} = \frac{32\cdot 10^{-6}}{720} < (0.5)10^{-7}.$$

315. Find the first three terms of the Maclaurin expansion of $y = \sin^2 x$.

316. Find the Maclaurin expansion of $f(x) = \ln(1 + x)$.

317. Find the Maclaurin expansion of $f(x) = (1 + x)^m$, $m \neq 0$ and not a positive integer.

318. Expand $y = \dfrac{x}{x-1}$ in powers of $x - 2$ up to $(x - 2)^3$.

319. Evaluate sin 41° with accuracy to three decimal places.

Solution. We expand $y = \sin x$ around $x = \pi/4$. By Taylor's formula:

$$f(x) = f\left(\frac{\pi}{4}\right) + \left(x - \frac{\pi}{4}\right)f'\left(\frac{\pi}{4}\right) + \frac{1}{2!}\left(x - \frac{\pi}{4}\right)^2 f''\left(\frac{\pi}{4}\right) + \ldots$$

$$+ \frac{1}{n!}\left(x - \frac{\pi}{4}\right)^n f^{(n)}\left(\frac{\pi}{4}\right) + R_{n+1}.$$

$$f(x) = \sin x; \qquad f^{(n)}(x) = \sin\left(x + \frac{n\pi}{2}\right).$$

$$R_{n+1} = \frac{1}{(n+1)!}\left(x - \frac{\pi}{4}\right)^{n+1}\sin\left[\frac{\pi}{4} + \theta\left(x - \frac{\pi}{4}\right) + \frac{(n+1)\pi}{2}\right], \quad 0 < \theta < 1,$$

$$|R_{n+1}| \leqslant \left|x - \frac{\pi}{4}\right|^{n+1}\frac{1}{(n+1)!}.$$

In the present case $\quad x - \dfrac{\pi}{4} = -\dfrac{4\pi}{180} = -\dfrac{\pi}{45}.$

We need $\quad \left(\dfrac{\pi}{45}\right)^{n+1}\cdot\dfrac{1}{(n+1)!} < (0.5)\cdot 10^{-3}.$

For $n = 2$ we obtain

$$\left(\frac{\pi}{45}\right)^3 \cdot \frac{1}{3!} < \left(\frac{1}{10}\right)^3 \cdot \frac{1}{6} < \frac{1}{6} \cdot 10^{-3}.$$

This is enough for the required accuracy. We obtain

$$\sin 41° \approx \frac{\sqrt{2}}{2} - \frac{\pi}{45} \cdot \frac{\sqrt{2}}{2} - \left(\frac{\pi}{45}\right)^2 \cdot \frac{\sqrt{2}}{2} \cdot \frac{1}{2} \approx 0.656.$$

320. Find the Maclaurin expansions of $y = \cosh x$ and $y = \sinh x$.

321. Find the Maclaurin expansion of $f(x) = x^2e^x$.

322. Find the Maclaurin expansion of $f(x) = \tan x$.

323. Find the Maclaurin expansion of $f(x) = \cos \sin x$ (up to x^4).

324. Find the Maclaurin expansion of $f(x) = \arctan x$.

Solution:

$$y' = \frac{1}{1 + x^2}; \qquad y'(1 + x^2) = 1.$$

Now we differentiate this product n times by Leibniz's formula. We suppose $u = y'$, $v = 1 + x^2$. Then:

$$(1 + x^2)y^{(n+1)} + n \cdot 2x \cdot y^{(n)} + \frac{n(n-1)}{1 \cdot 2} \cdot 2 \cdot y^{(n-1)} = 0.$$

We are interested in $y^{(n)}(0)$. Substituting $x = 0$ in the above equation we obtain:

$$y^{(n+1)}(0) + n(n-1)y^{(n-1)}(0) = 0.$$

By direct evaluation,

$$y'' = \frac{-2x}{(1 + x^2)^2}; \qquad y''(0) = 0.$$

It follows that all even derivatives equal 0 at $x = 0$; the odd derivatives are:

$$y'(0) = 1, \qquad y'''(0) = -2(2-1) \cdot 1 = -2,$$
$$y^{v}(0) = -4 \cdot 3(-2) = 4!, \quad y^{vii} = -6 \cdot 5 \cdot 4! = -6!.$$

In general, $$y^{(2n+1)} = (-1)^n(2n)!.$$

Hence

$$\arctan x = x - \frac{x^3}{3} + \frac{x^5}{5} - \ldots + (-1)^{m-1} \frac{x^{2m-1}}{(2m-1)!} + R_{2m+1}.$$

We remark that here the nth derivative could have been found more easily by using complex numbers and partial fraction decomposition.

Questions similar to those of this section will be considered again in Chapter XI.

5.3 INDETERMINATE FORMS: L'HÔPITAL'S RULE

This section deals with l'Hôpital's rule and its application to the evaluation of limits.

L'HÔPITAL'S RULE. Given

$$\lim_{x\to a} f(x) = 0, \quad \lim_{x\to a} g(x) = 0, \quad \text{and} \quad \lim_{x\to a} \frac{f'(x)}{g'(x)} = m,$$

then also

$$\lim_{x\to a} \frac{f(x)}{g(x)} = m.$$

The rule remains true when $x \to \infty$ (instead of approaching a).

If $\lim_{x\to a} f'(x) = 0$ and $\lim_{x\to a} g'(x) = 0$, but $\lim_{x\to a} f''(x)/g''(x) = m$, then a repeated application of the above rule gives $\lim_{x\to a} f(x)/g(x) = m$. In an analogous way we may, if necessary, repeat the above process more than twice.

L'Hôpital's rule is true also if $\lim_{x\to a} f(x) = \infty$ and $\lim_{x\to a} g(x) = \infty$. If in this case $\lim_{x\to a} f'(x)/g'(x) = m$, then also $\lim_{x\to a} f(x)/g(x) = m$. The same holds for the case when $x \to \infty$.

L'Hôpital's rule enables us to evaluate limits in the case of the indeterminate forms (*1*) $0/0$ and (*2*) ∞/∞. We shall now show that it can be used also in other cases of indeterminate forms.

(*3*) $0 \cdot \infty$, that is,

$$\lim_{x\to a} f(x)g(x) \quad \text{where} \quad \lim_{x\to a} f(x) = 0 \quad \text{and} \quad \lim_{x\to a} g(x) = \infty.$$

In this case we write:

$$f(x)g(x) = \frac{f(x)}{1/g(x)} = \frac{g(x)}{1/f(x)},$$

and obtain in this way either $0/0$ or ∞/∞. In both cases l'Hôpital's rule applies.

(*4*) $\infty - \infty$, that is,

$$\lim_{x\to a} [f(x) - g(x)] \quad \text{where} \quad \lim_{x\to a} f(x) = \infty,$$

$$\lim_{x\to a} g(x) = \infty.$$

We proceed as follows:

$$f(x) - g(x) = \frac{1}{\frac{1}{f(x)}} - \frac{1}{\frac{1}{g(x)}} = \frac{\frac{1}{g(x)} - \frac{1}{f(x)}}{\frac{1}{f(x)g(x)}},$$

and this is an indeterminate form 0/0. In many practical cases simpler transformations lead to the desired form.

(5) 1^∞; (6) 0^0; (7) ∞^0. Any of these indeterminate forms can be transformed into the form $0 \cdot \infty$ by use of logarithms. After that we proceed as in case (3), obtaining thereby the limit m of the logarithm of the given expression. Hence the limit of the expression itself will equal e^m.

Remarks. (a) L'Hôpital's rule applies only in the cases 0/0 and ∞/∞. Consider, for instance,

$$\lim_{x\to 1} \frac{x^2 + 2}{x - 3} = -\frac{3}{2}.$$

L'Hôpital's procedure would lead to

$$\lim_{x\to 1} \frac{(x^2 + 2)'}{(x - 3)'} = \lim_{x\to 1} \frac{2x}{1} = 2,$$

which is absurd. We must, then, satisfy ourselves that we are dealing with an indeterminate form 0/0 or ∞/∞, before using l'Hôpital's rule. Should repeated application be necessary, it is advisable that whatever simplifications possible be performed before proceeding to a next differentiation.

(b) It may happen that $\lim_{x\to a} f(x)/g(x)$ exists even when $\lim_{x\to a} f'(x)/g'(x)$ does not. For example, $\lim_{x\to\infty} (2x + \cos x)/x = 2$, but

$$\lim_{x\to\infty} \frac{(2x + \cos x)'}{(x)'} = \lim_{x\to\infty} (2 - \sin x)$$

does not exist.

Find the following limits.

325. $\lim_{x\to 0} \frac{\sin x}{x}$

Solution. $$\lim_{x\to 0} \frac{\sin x}{x} = \lim_{x\to 0} \frac{(\sin x)'}{(x)'} = \lim_{x\to 0} \frac{\cos x}{1} = 1.$$

We thus obtain easily this well-known limit. However, the formula $(\sin x)' = \cos x$ was derived on the basis of this limit, thus its truth was assumed before the necessary differentiation was performed. A limit that can be obtained by l'Hôpital's rule can also be evaluated by elementary methods, but the use of l'Hôpital's rule makes the work in general much easier.

326. $\lim_{x\to 0} \frac{\sin x - x}{\sinh x}$

327. $\lim_{x\to 0} \frac{\sqrt{1+x} - \sqrt[3]{1-x}}{2x}$

328. $\lim_{x\to 0} \frac{e^{\sin x} - e^x}{\sin x - x}$

329. $\lim_{x\to \pi} \frac{\sin 3x}{\tan 5x}$

330. $\lim_{x\to 0} \frac{\ln(1-x) + x^2}{(1+x)^m - 1 + x^2}$

331. $\lim_{x\to \infty} \frac{x^\alpha}{a^x}, a > 1, \alpha > 0$

332. $\lim_{x\to \infty} \frac{\ln x}{x^\alpha}, \alpha > 0$

333. $\lim_{x\to \pi/4} \frac{\sin x - \cos x}{1 - \tan^2 x}$

334. $\lim_{h\to 0} \frac{f(x+h) - 2f(x) + f(x-h)}{h^2}$

335. $\lim_{x\to 0} \frac{\tan ax - a \sin x}{x(1 - \cos ax)}$

336. $\lim_{x\to 0} \frac{\pi/x}{\cot(\pi x/2)}$

337. $\lim_{x\to 0} \frac{\ln(\sin ax)}{\ln \sin x}$

338. $\lim_{x\to 0} x^\alpha \ln x, \alpha > 0$

339. $\lim_{x\to 1-} \ln x \ln(1-x)$

340. $\lim_{x\to 1} \left(\frac{x}{x-1} - \frac{1}{\ln x}\right)$

341. $\lim_{x\to 0} \left(\frac{1}{x} - \frac{1}{\cosh x - 1}\right)$

342. $\lim_{x\to \infty} \left[x - x^2 \ln\left(1 + \frac{1}{x}\right)\right]$

343. $\lim_{x\to 1} x^{\frac{1}{x-1}}$

344. $\lim_{x\to \pi/2} (\tan x)^{2x-\pi}$

345. $\lim_{x\to 0} x^{\frac{1}{\ln(e^x - 1)}}$

346. $\lim_{x\to 0} \left(\frac{2}{\pi} \arccos x\right)^{1/x}$

347. $\lim_{x\to 0} \frac{\ln(1 + x + x^2) + \ln(1 - x + x^2)}{x \sin x}$

348. $\lim_{x\to \infty} \left(1 + \frac{1}{x^2}\right)^x$

349. $\lim_{x\to a} \left(2 - \frac{x}{a}\right)^{\tan \frac{\pi x}{2a}}$

350. $\lim_{x\to \pi/4} (1 + \ln \tan x)^{\frac{1}{1 - \cot x}}$

351. In a given circle, arc AB subtends a central angle α. Consider the chord AB and the tangent lines to the circle at A and B. Letting S_1 be the area between the chord and the arc, and S_2 the area between the tangents and the arc, find $\lim S_1/S_2$.

VI

APPLICATIONS OF DIFFERENTIAL CALCULUS

6.1 RATE OF CHANGE

352. Given the equation of a rectilinear motion $S = t^3 + (3/t)$, find the average velocity between $t = 4$ and $t = 6$ and the velocity at $t = 4$.

353. Given a rod AB of length $L = 20$ cm. The mass of a section AM of the rod is proportional to the square of the length AM. The mass of the section $AN = 2$ cm is 8 g. Find (a) the average density of the section AN; (b) the average density of the whole rod; (c) the density of the rod at the point M.

354. The amount of heat Q needed to increase the temperature of a unit of mass of water from 0°C. to t°C. is given by the formula

$$Q = t + 0.00002\, t^2 + 0.0000003\, t^3 \text{ cal/kg.}$$

Find the specific heat of water at $t = 100$°C.

355. Given $y^2 = 12x$. Let the rate of increase of x be 2 units per second. What is the rate of change of y when $x = 3$?

356. At what point of the ellipse $16x^2 + 9y^2 = 400$ does the ordinate decrease at the same rate as the abscissa increases?

357. The radius of a sphere changes at the rate v. What are the rates of change of the volume and the surface area of this sphere?

358. At what values of x does $\sin x$ change at a rate the absolute value of which is one half the rate of change of its argument?

359. A ladder stands on a horizontal floor and leans against a vertical wall. Its lower end moves away from the wall by a velocity of 2 m/sec.

What is the velocity of its top at the moment when the angle between the ladder and the floor is 60°?

360. An airplane A flies horizontally with constant velocity v in the direction MA (see Fig. 14). What should be the angular velocity of a gun positioned at O, in order that it be pointed always towards the plane?

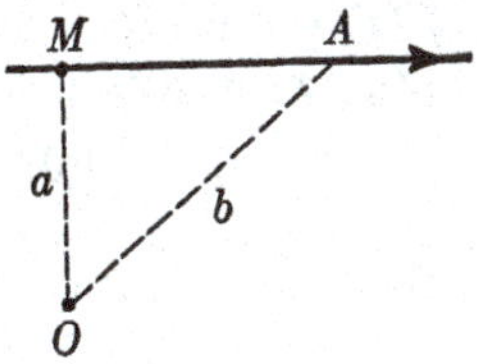

FIGURE 14

6.2 LOCATING INTERVALS IN WHICH A FUNCTION INCREASES OR DECREASES

The following theorems have important applications in connection with a discussion of the behavior of a function.

(1) If $f'(x) = 0$ for all x's satisfying $a < x < b$, then $f(x) =$ const. there.

(2) If $f'(x) = g'(x)$ for every $a < x < b$, then in the interval (a,b), $f(x) = g(x) +$ const.

(3) A differentiable function $f(x)$ does not decrease (increase) in the interval $a < x < b$ if and only if for every x in this interval

$$f'(x) \geqslant 0 \qquad (\text{or } f'(x) \leqslant 0, \text{ correspondingly}).$$

(4) $f(x)$ increases (decreases) in the interval $a < x < b$ if and only if

$$f'(x) \geqslant 0 \ (\leqslant 0) \quad \text{for} \quad a < x < b$$

and the derivative is not identically zero in any partial interval of $a < x < b$.

(5) If $f(x)$ and $g(x)$ are defined in $a \leqslant x < b$, and $f(a) = g(a)$ and $f'(x) > g'(x)$ for every $a < x < b$, then $f(x) > g(x)$ in $a < x < b$.

361. Prove that $f(x) = \arcsin x$ and $g(x) = \arctan (x/\sqrt{1 - x^2})$, $|x| < 1$, are identical functions.

Solution:

$$f'(x) = \frac{1}{\sqrt{1 - x^2}};$$

$$g'(x) = \frac{1}{1 + \dfrac{x^2}{1 - x^2}} \cdot \frac{\sqrt{1 - x^2} + \dfrac{x^2}{\sqrt{1 - x^2}}}{1 - x^2} = \frac{1}{\sqrt{1 - x^2}}.$$

$f'(x) = g'(x)$ for every x in the given interval, and according to theorem (2), $f(x) = g(x) + c$. To find c, write $x = 0$. We obtain

$$f(0) = 0; \qquad g(0) = 0; \qquad \text{i.e.,} \qquad c = 0.$$

Hence
$$\arcsin x \equiv \arctan \frac{x}{\sqrt{1 - x^2}}.$$

This identity can also be proved directly from the definition of inverse trigonometric functions.

362. The same question for

$$f(x) = 2 \arctan x, \qquad g(x) = \arctan \frac{2x}{1 - x^2}.$$

363. Where does the function $f(x) = 2x^5 - 5$ increase?

Solution. $f'(x) = 10x^4 \geqslant 0$. The function increases in the whole interval $(-\infty, \infty)$. The derivative vanishes only at $x = 0$. This being a single point, theorem (4) applies.

364. The same question for $f(x) = ax - \sin ax$, $a > 0$.

365. The same question for $f(x) = \dfrac{x}{1 + x^2}$.

366. Where does the function $f(x) = x^3(1 - x)$ decrease?

367. Prove that $y = x^3 - 5x + 7$ increases at $x_1 = 2$ and decreases at $x_2 = 0$.

Solution. We shall first discuss the problem in general. Let x_0 be arbitrary:

$$y'(x_0) = \lim_{\Delta x \to 0} \frac{\Delta y}{\Delta x}.$$

If $y'(x_0) > 0$ there must be an interval around x_0 where Δy and Δx have the same signs. This is the meaning of the statement that $f(x)$ increases at $x = x_0$. In the same manner, $f'(x_0) < 0$ implies that $f(x)$ decreases at $x = x_0$. Consider now the given function, $y' = 3x^2 - 5$:

$$y'(2) = 3 \cdot 4 - 5 = 7; \qquad \text{the function increases at 2;}$$
$$y'(0) = 3 \cdot 0 - 5 = -5; \qquad \text{the function decreases at 0.}$$

368. Determine whether $f(x) = \sin x + \cos x$ decreases or increases at the points $x = 0$ and $x = 1$.

369. Prove that $y = 5 + \sqrt{2x - x^2}$ increases in the interval (0,1) and decreases in the interval (1,2).

Determine the intervals of monotonicity of the following functions.

370. $y = (x - 2)^5(2x + 1)^4$ **371.** $y = x - e^x$

372. $y = x\sqrt{ax - x^2}, \quad a > 0$ **373.** $y = x^2e^{-x}$

374. $y = \dfrac{x}{\ln x}$

375. Show that $f(x) = \cos x > g(x) = 1 - \frac{1}{2}x^2, \quad x \neq 0$.

Solution:

$$f(0) = 1 = g(0).$$

Now $$f'(x) = -\sin x, \qquad g'(x) = -x.$$

It is well known that for $x > 0$, $x > \sin x$. (This follows, e.g., from Problem 364.) It follows that $-\sin x > -x$, i.e., $f'(x) > g'(x)$, $(x > 0)$, and by theorem (5) of this section, $f(x) > g(x)$ for $x > 0$. But both functions are even, and consequently $f(x) > g(x)$ also for $x < 0$. (At $x = 0$, $f(0) = g(0)$ as we have seen above.)

376. Show that for $x > 0$, $x - \dfrac{x^2}{2} < \ln(1 + x) < x$.

377. Show that for $0 < x < \pi/2$, $\tan x > x + \frac{1}{3}x^3$.

378. Prove that $2\sqrt{x} > 3 - \dfrac{1}{x}$ for $x > 1$.

379. Prove that $1 + \ln(x + \sqrt{1 + x^2}) \geqslant \sqrt{1 + x^2}$ for $0 \leqslant x \leqslant 1$.

380. Show that the perimeter of a regular polygon inscribed in a circle increases (and that of a circumscribed polygon decreases) when the number of sides increases.

6.3 MINIMA AND MAXIMA

The function $f(x)$ is said to have a maximum (minimum) at the point x_0 if there exists around this point an interval $(x_0 - \delta, x_0 + \delta)$ such that for every point x of this interval, except $x = x_0$, the inequality $f(x_0) > f(x)$ $(f(x_0) < f(x))$ is fulfilled (see Fig. 15). It is clear that the maxima and minima defined here are of a local nature, e.g., there may be values of f in a larger interval which are greater than $f(x_0)$.

A necessary condition for $f(x)$ to have an extremum (i.e., a maximum or minimum) at $x = x_0$ is that $f'(x_0) = 0$ or that $f'(x)$ does not exist at x_0. This condition is, however, not a sufficient one and therefore all points satisfying it are to be regarded only as "candidates" for extrema. The following two sufficient conditions will serve to discover extrema among these "candidates":

(1) If $f'(x_0 -) > 0$ (i.e., to the left of x_0 an interval exists in which

$f'(x)$ is positive) and $f'(x_0+) < 0$, then $f(x)$ has a maximum at $x = x_0$. In other words, if $f'(x)$ changes sign from + to − as x passes through x_0 and $f'(x_0) = 0$ or $f'(x_0)$ does not exist, then this is a maximum point. If $f'(x_0-) < 0$ and $f'(x_0+) > 0$, $f(x)$ has a minimum at $x = x_0$. If $f'(x)$ does not change sign at $x = x_0$, this point is not an extremum.

(2) Let $x = x_0$ be, as before, a "candidate" point. If $f''(x_0) < 0$, $f(x)$ has a maximum at $x = x_0$, and if $f''(x_0) > 0$, $f(x)$ has a minimum there. In the case when $f''(x_0) = 0$ we can try to use the first condition, or proceed as follows: we compute $f'''(x_0)$, $f^{iv}(x_0)$, and so on. If the first nonvanishing derivative (at x_0) is of odd order, there is no extremum; if it is at even order, then there is a maximum if this derivative is negative at this point, and a minimum if it is positive.

An extremum can often be located by an investigation of the function itself in the neighborhood of a "candidate" point. In general, however, we find it simpler to use the above conditions.

In many problems we are asked to find the greatest (or smallest) value of $f(x)$ in a given interval. To answer such a question we have to compare the values of the function at the extrema and at the ends of the interval. The largest (smallest) of these values is the required one. In Figure 16 the greatest value is at $x = b$ and the smallest at $x = x_3$.

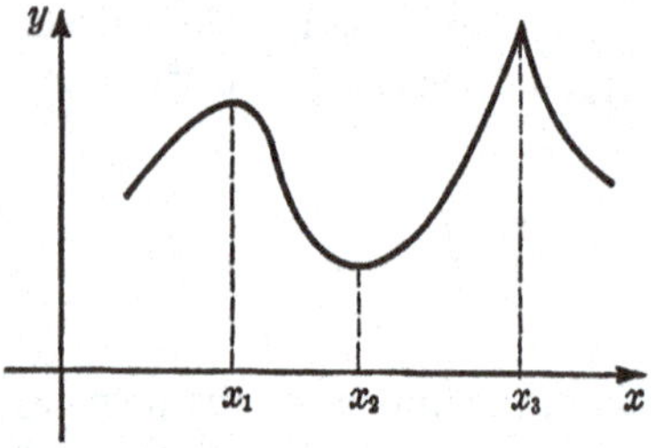

FIGURE 15

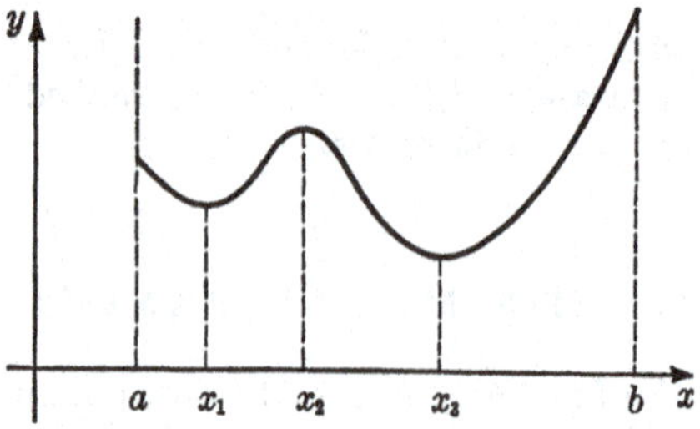

FIGURE 16

The following remark is often useful. If a function is continuous and has only *one* extremum inside a given interval, a maximum (or a minimum), it obtains its greatest (or smallest) value at this extremum.

In the following exercises find extrema of the given functions. (Problems involving applications of extrema will be dealt with later on.)

381. $y = f(x) = x^3 - 12x$

Solution. First we find $y' = 3x^2 - 12$.

$$y' = 0; \qquad 3x^2 - 12 = 0, \qquad x_1 = 2, \qquad x_2 = -2.$$

These are the only "candidate" points (y' exists everywhere).

Let us use the first method:

$$y' = 3(x + 2)(x - 2);$$

$y'(-2\,-) > 0$, because both factors are negative there. (The constant factor 3 clearly does not affect the sign of y'.) $y'(-2\,+)$ is negative, because close to the right of -2 the first factor is positive and the second negative. The given function has consequently a maximum at $x = -2$. Its value is $(-2)^3 - 12(-2) = 16$. Now to the left of $x = 2$ and close to it the first factor is positive and the second still negative, i.e., $y'(2\,-) < 0$. To the right of $x = 2$ both factors are positive, i.e., $y'(2\,+) > 0$. $(2, -16)$ is a minimum point of the function. We can now show the results, in the accompanying table.

		Sign of		
x	y	$y'(x\,-)$	$y'(x\,+)$	*Result*
-2	16	$(-)(-) = +$	$(+)(-) = -$	Max
2	-16	$(+)(-) = -$	$(+)(+) = +$	Min

The same results can be obtained in this case very easily by the second method:

$$y'' = 6x, \qquad y''(-2) = -12 < 0, \qquad \text{Max } (-2, 16);$$
$$y''(2) = 12 > 0, \qquad \text{Min } (2, -16).$$

The use of the second derivative is, of course, preferable here. A sketch of the function is shown in Figure 17.

382. $y = (x - 2)^5(2x + 1)^4$

Solution. In Problem 370 we found

$$y' = (x - 2)^4(2x + 1)^3(18x - 11);$$
$$y' = 0; \qquad x_1 = -\tfrac{1}{2}, \qquad x_2 = \tfrac{11}{18}, \qquad x_3 = 2.$$

Let us make a table as before.

		Sign of		
x	y	$y'(x\,-)$	$y'(x\,+)$	*Result*
$-\frac{1}{2}$	0	$(+)(-)(-) = +$	$(+)(+)(-) = -$	Max
$\frac{11}{18}$	$-\frac{2^3 \cdot 5^{14}}{3^{18}}$	$(+)(+)(-) = -$	$(+)(+)(+) = +$	Min
2	0	$(+)(+)(+) = +$	$(+)(+)(+) = +$	No extremum

A sketch of $f(x)$ is given in Figure 18. In this case the first method seems to be preferable.

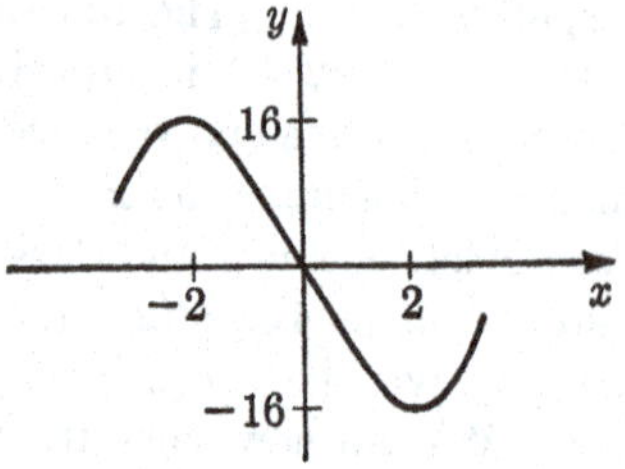

FIGURE 17

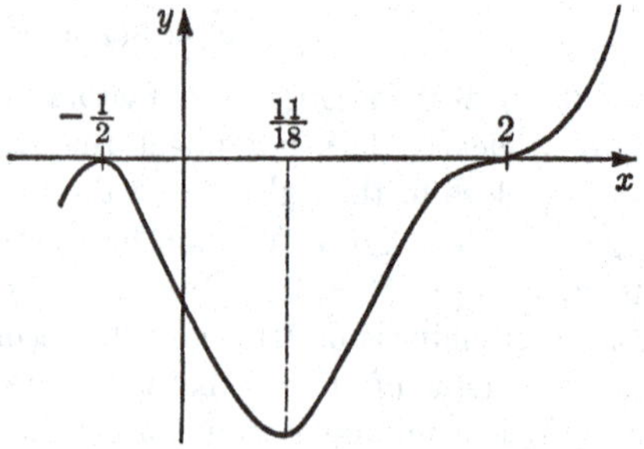

FIGURE 18

383. $y = x - e^x$

Solution:

$$y' = 1 - e^x; \qquad y' = 0; \qquad e^x = 1, \qquad x = 0;$$

$$y'' = -e^x, \qquad y''(0) = -1; \qquad \text{Max } (0,-1).$$

Compare Problem 371.

384. $y = x\sqrt{ax - x^2},\ a > 0$

Solution. We found in Problem 372,

$$y' = \frac{3ax - 4x^2}{2\sqrt{ax - x^2}}.$$

At $y' = 0$, $x = 3a/4$. The derivative does not exist at $x = a$, but $x = 0$ and $x = a$ are the end points of the domain of definition of the function and they cannot give extrema. (According to our definition, an extremum must be at an inner point of the domain of definition of the function.)

To investigate $x = 3a/4$, let us write

$$y' = \frac{x(3a - 4x)}{2\sqrt{ax - x^2}}.$$

It is easy to see that y' changes sign from + to − when x passes through $3a/4$ ($a > 0$). As a result, $(3a/4, 3\sqrt{3}a^2/16)$ is a maximum point of the given function. Use of the second derivative is here inconvenient.

The following device seems to be helpful in this and in similar cases. Suppose that in general $y' = f(x)/g(x)$, and we investigate for an extremum a point x_0 obtained by solving the equation $y' = 0$, i.e., $f(x) = 0$. (Of course $g'(x_0) \neq 0$, otherwise y' is not defined at x_0.) Let us use the second method:

$$y'' = \frac{f'(x)}{g(x)} - f(x)\frac{g'(x)}{g^2(x)};$$

$$y''(x_0) = \frac{f'(x_0)}{g(x_0)} - f(x_0)\frac{g'(x_0)}{g^2(x_0)} = \frac{f'(x_0)}{g(x_0)},$$

because $f(x_0) = 0$. It follows that we can limit our attention to the sign of $f'(x)/g(x)$ at x_0, and if $g(x)$ is positive (negative) for every x, to that of $f'(x)$ only.

In the above example,

$$y' = \frac{3ax - 4x^2}{2\sqrt{ax - x^2}}, \qquad g(x) > 0 \text{ for any } 0 < x < a.$$

$f'(x) = 3a - 8x$. $f'(\frac{3}{4}a) = 3a - 6a < 0$, and we have a maximum.

385. $y = \dfrac{10\sqrt[3]{(x-1)^2}}{x^2+9}$

Solution. Let us use logarithmic differentiation:

$$\ln y = \ln 10 + \tfrac{2}{3}\ln(x-1) - \ln(x^2+9).$$

$$\frac{y'}{y} = \frac{2}{3}\frac{1}{x-1} - \frac{2x}{x^2+9} = \frac{2(x^2+9-3x^2+3x)}{3(x-1)(x^2+9)} = \frac{2(-2x^2+3x+9)}{3(x-1)(x^2+9)};$$

$$y' = \frac{10\sqrt[3]{(x-1)^2}}{x^2+9}\cdot\frac{2(-2x^2+3x+9)}{3(x-1)(x^2+9)} = \frac{20(9+3x-2x^2)}{3\sqrt[3]{x-1}\,(x^2+9)^2};$$

$$y' = 0; \qquad 2x^2 - 3x - 9 = 0, \qquad x = \frac{3 \pm \sqrt{9+72}}{4} = \frac{3 \pm 9}{4},$$

$$x_1 = 3; \qquad x_2 = -1.5.$$

An additional "candidate" point is $x = 1$ where y' does not exist (although y does exist there!).

Let us investigate this point first. Close to the left of it, the numerator is positive and the denominator negative (the cubic root of a negative number), i.e., $y'(1-) < 0$. In the same manner we obtain $y'(1+) > 0$, and consequently the point (1,0) is a minimum of the function. For the other two points we shall use the device described in the previous problem. Put

$$f(x) = 9 + 3x - 2x^2, \qquad f'(x) = 3 - 4x.$$

$$f'(-1.5) = 3 + 6 = 9 > 0, \qquad f'(3) = 3 - 12 = -9 < 0.$$

At -1.5 the denominator of y' is negative and at 3 it is positive. In both cases y'' is negative, and the points $(-1.5, \sim 1.6)$, $(3, \sim 0.8)$ are maximum points. A rough sketch of the function is given in Figure 19. We use

$$\lim_{x \to 1-} y' = -\infty, \qquad \lim_{x \to 1+} y' = \infty.$$

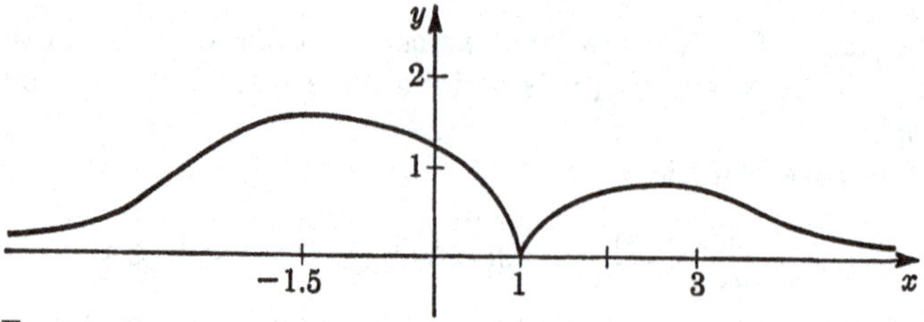

FIGURE 19

386. $f(x) = \frac{1}{6}x^6 - x^4 + 5$

Solution:

$f'(x) = x^5 - 4x^3; \qquad x^5 - 4x^3 = 0; \qquad x_1 = -2; \qquad x_2 = 2; \qquad x_3 = 0.$

$f''(x) = 5x^4 - 12x^2;$

$f''(-2) = f''(2) = 5 \cdot 16 - 12 \cdot 4 = 80 - 48 > 0;$

Min $(-2, -\frac{1}{3})$, Min $(2, -\frac{1}{3})$.

$f''(0) = 0.$

To decide on the nature of this last point we differentiate repeatedly.

$$f'''(x) = 20x^3 - 24x, \qquad f'''(0) = 0,$$

$$f^{iv}(x) = 60x^2 - 24, \qquad f^{iv}(0) = -24 < 0.$$

The first nonvanishing derivative at this point is of even order, and its value is negative. The point $(0,5)$ is a maximum of the function.

We remark that after finding the two minimum points it was very easy to see that $x = 0$ is a maximum point of the function, without the above computation.

387. $f(x) = \dfrac{x^3}{3} - 2x^2 + 4x$

Solution:

$$f'(x) = x^2 - 4x + 4, \qquad x^2 - 4x + 4 = 0, \qquad x = 2;$$

$$f''(x) = 2x - 4, \qquad f''(2) = 0;$$

$$f'''(x) = 2 \neq 0.$$

The first nonvanishing derivative is of odd order and $x = 2$ does not give an extremum.

388. $f(x) = x^2\left(2 + \sin\frac{1}{x}\right)$, $\quad f(0) = 0$

Solution. $\lim\limits_{x\to 0} f(x) = 0$ and the given function is continuous at 0.

$$f'(x) = 2x\left(2 + \sin\frac{1}{x}\right) - \cos\frac{1}{x}.$$

The derivative at $x = 0$ is not defined by this formula. We shall evaluate it by definition:

$$f'(0) = \lim_{\Delta x\to 0}\frac{f(0 + \Delta x) - f(0)}{\Delta x} = \lim_{\Delta x\to 0}\frac{(\Delta x)^2\left(2 + \sin\frac{1}{\Delta x}\right)}{\Delta x} = 0.$$

It is easy to see that $f'(x)$ changes its sign infinitely many times in every neighborhood of $x = 0$. The first method cannot be used. Now $\lim\limits_{x\to 0} f'(x)$ does not exist, i.e., $f'(x)$ is not continuous at this point. It follows that $f''(x)$ does not exist at $x = 0$, i.e., the second method cannot be used either. But it is clear that $f(x) > 0$ for $x \neq 0$, and as $f(0) = 0$ the point (0,0) is a minimum of the function.

389. $y = 6x - x^2$

390. $y = a + (x - b)^4$

391. $y = a + (x - b)^3$

392. $y = x + \dfrac{1}{x}$

393. $y = \dfrac{x^2}{x^4 + 4}$

394. $y = x \ln x$

395. $y = x^x$

396. $y = x^2 e^{-x}$

397. $y = \cosh x$

398. $y = \cos x + \cosh x$

399. $y = \ln x - \arctan x$

400. $y = e^x \cos x$

401. $y = \sin 3x - 3 \sin x$

402. $y = x^3\sqrt[3]{x - 1}$

403. $y = e^{-1/x^2}$, $y(0) = 0$

Solution. $\lim\limits_{x\to 0} y = 0$ and the given function is continuous at 0. It can be seen at once that at $x = 0$ there is a minimum, because for $x \neq 0$, $y > 0$. We shall see that no use can be made of the method involving the higher derivatives in this case.

$$y' = \frac{2}{x^3}e^{-1/x^2}; \qquad y'' = \left(-\frac{6}{x^4} + \frac{4}{x^6}\right)e^{-1/x^2}$$

By induction we obtain that $y^{(n)}(x) = P_n(1/x)e^{-1/x^2}$ where $P_n(1/x)$ is a polynomial in $1/x$. $y^{(n)}(0)$ cannot be obtained by substituting $x = 0$ in the above formula. We must evaluate it directly:

$$f'(0) = \lim_{\Delta x \to 0} \frac{f(0 + \Delta x) - f(0)}{\Delta x} = \lim_{\Delta x \to 0} \frac{e^{-1/(\Delta x)^2}}{\Delta x} = \lim_{\Delta x \to 0} \frac{\frac{1}{\Delta x}}{e^{1/(\Delta x)^2}}$$

$$= \lim_{\alpha \to \infty} \frac{\alpha}{e^{\alpha^2}} = \lim_{\alpha \to \infty} \frac{1}{e^{\alpha^2} \cdot 2\alpha} = 0.$$

Assume $f^{(k)}(0) = 0$;

$$f^{(k+1)}(0) = \lim_{\Delta x \to 0} \frac{P_k\left(\frac{1}{\Delta x}\right) e^{-1/(\Delta x)^2}}{\Delta x} = \lim_{\alpha \to \infty} \frac{\alpha P_k(\alpha)}{e^{\alpha^2}} = 0.$$

The last limit follows from Problem 331.

By induction $f^{(n)}(0) = 0$ for any n, and the extremum cannot be found by the use of higher derivatives. We shall return to this example in Chapter XI, and find another important consequence of the fact that $f^{(n)}(0) = 0$ for any n, though $f(x)$ is not identically 0.

404. Find the greatest and smallest value of $f(x) = x^3 - 3x + 2$ in the interval $[-3,1.5]$.

Solution:

$$f'(x) = 3x^2 - 3, \quad f'(x) = 0, \quad x = \pm 1;$$

$$f''(x) = 6x, \quad f''(-1) = -6 < 0, \quad \text{Max } (-1,4),$$

$$f''(1) = 6 > 0, \quad \text{Min } (1,0).$$

We find also the values of the function at the ends of the interval:

$$f(-3) = -27 + 9 + 2 = -16;$$

$$f\left(\frac{3}{2}\right) = \frac{27}{8} - \frac{9}{2} + 2 = \frac{27 - 36 + 16}{8} = \frac{7}{8}.$$

The function attains its greatest value 4 at the point $x = -1$ which is also a (local) maximum. Its least value -16 is reached at the left end of the interval for $x = -3$.

405. The same question for $f(x) = \sqrt{x(6 - x)}$ in the whole domain of definition of this function.

In the following problems find extrema of functions given in implicit or parametric form.

406. $y^2 + 2yx^2 + 4x - 3 = 0$

Solution:

$$2yy' + 2y'x^2 + 4yx + 4 = 0; \qquad y' = -2\frac{xy+1}{x^2+y};$$

$$y' = 0, \qquad xy = -1.$$

It follows that the solutions of the system

$$\begin{cases} y^2 + 2yx^2 + 4x - 3 = 0 \\ xy = -1 \end{cases}$$

are "candidates" for extrema.

$$y = -\frac{1}{x}; \quad \frac{1}{x^2} - \frac{2}{x}x^2 + 4x - 3 = 0, \quad 1 - 2x^3 + 4x^3 - 3x^2 = 0,$$

$$2x^3 - 3x^2 + 1 = 0, \quad 2x^2(x-1) - (x^2-1) = 0,$$

$$(x-1)(2x^2 - x - 1) = 0; \quad x_1 = 1, \quad x_{2,3} = \frac{1 \pm \sqrt{1+8}}{4} = \frac{1 \pm 3}{4};$$

$$x_2 = 1; \quad x_3 = -\tfrac{1}{2}.$$

The corresponding values of y are

$$x_1 = 1, \quad y_1 = -1; \qquad x_3 = -\tfrac{1}{2}, \quad y_3 = 2.$$

Both points have to be checked. We begin with the second:

$$2(y'^2 + yy'' + y''x^2 + 2xy' + 2y + 2xy') = 0,$$

substituting $x = -\frac{1}{2}$, $y = 2$, $y' = 0$;

$$2y'' + \tfrac{1}{4}y'' + 4 = 0, \qquad y'' = -\tfrac{16}{9} < 0; \qquad \text{Max } (-\tfrac{1}{2}, 2).$$

The first derivative does not exist at $(1, -1)$, so that the second also cannot exist there.

The given equation defines two functions:

$$y = -x^2 \pm \sqrt{x^4 - 4x + 3} = -x^2 \pm \sqrt{(x-1)^2(x^2+2x+3)}$$

$$= -x^2 \pm |x-1|\sqrt{x^2+2x+3};$$

$$y' = -2x \pm \frac{2x^3 - 2}{\sqrt{x^4 - 4x + 3}} = -2x \pm \frac{2(x-1)(x^2+x+1)}{|x-1|\sqrt{x^2+2x+3}}.$$

Now for $\quad x < 1, \qquad y' = -2x \mp \dfrac{2(x^2+x+1)}{\sqrt{x^2+2x+3}};$

$$x > 1, \qquad y' = -2x \pm \frac{2(x^2+x+1)}{\sqrt{x^2+2x+3}}.$$

For $y_1 = -x^2 + \sqrt{x^4 - 4x + 3}$, $y'(1-) < 0$ and $y'(1+) > 0$. The first inequality is clear and the second is obtained as follows:

$$g(x) = -2x + \frac{2(x^2 + x + 1)}{\sqrt{x^2 + 2x + 3}} \text{ is continuous;}$$

$$g(1) = -2 + \frac{2\cdot 3}{\sqrt{6}} = -2 + \sqrt{6} > 0;$$

consequently there must be an interval to the right of $x = 1$, where $g(x) > 0$;

$$y_1 = -x^2 + \sqrt{x^4 - 4x + 3} \text{ has a minimum at } (1, -1).$$

In the same way we show that $y_2 = -x^2 - \sqrt{x^4 - 4x + 3}$ has a maximum at $(1, -1)$. A substitution of $y = -x^2$ (to account for the points where the derivative does not exist) into the given equation does not lead to additional results.

407. $x^2 + 2xy + y^2 - 4x + 2y - 2 = 0$

408. $x^3 + y^3 - 3axy = 0,\ x > 0$

409. $x^4 + y^4 = x^2 + y^2$

410. $x = t^3 + 3t + 1,\ y = t^3 - 3t + 1$

Solution:

$$\frac{dy}{dx} = \frac{\dfrac{dy}{dt}}{\dfrac{dx}{dt}} = \frac{3t^2 - 3}{3t^2 + 3} = \frac{t^2 - 1}{t^2 + 1}; \quad y' = 0, \quad t = \pm 1;$$

$$\frac{d^2y}{dx^2} = \frac{\dfrac{d}{dt}\left(\dfrac{dy}{dx}\right)}{\dfrac{dx}{dt}} = \frac{2t(t^2 + 1) - 2t(t^2 - 1)}{3(t^2 + 1)^3} = \frac{4t}{3(t^2 + 1)^3}.$$

$$y''(t = 1) > 0; \quad \text{Min } (5, -1).$$

$$y''(t = -1) < 0; \quad \text{Max } (-3, 3).$$

411. $x = te^t,\ y = te^{-t}$

The following problems involve extrema methods.

412. The sum of two nonnegative numbers is 6. What are these numbers if the product of the square of the first by the second is maximal?

Solution. Let x be one of the numbers, $0 < x < 6$. The other is $6 - x$. The required product is $y = x^2(6 - x) = 6x^2 - x^3$;

$$\frac{dy}{dx} = 12x - 3x^2; \qquad 12x - 3x^2 = 0; \qquad x_1 = 0, \quad x_2 = 4.$$

$x_1 = 0$ is clearly not the solution. The maximum is at $x = 4$ and its value equals $4^2 \cdot 2 = 32$. This conclusion is clear from the conditions of the problem, but can also be verified by a second differentiation:

$$y'' = 12 - 6x, \qquad y''(4) = 12 - 24 < 0.$$

We have here only one extremum inside the domain of definition and this is clearly the greatest value of the product.

413. The area of the printed text on a page is S cm^2. The left and right margins are each a cm wide, and the upper and lower margins each b cm. What are the most economical dimensions of the page if only the amount of paper matters?

Solution. Let the width of the text be x and its height y. The area of the page is $P = (x + 2a)(y + 2b)$. This area is to be made minimal. P is expressed in terms of x and y, but these two unknowns are connected by the given condition that $xy = S$. We substitute $y = S/x$ and obtain

$$P = xy + 2ay + 2bx + 4ab = S + \frac{2aS}{x} + 2bx + 4ab.$$

$$\frac{dP}{dx} = -\frac{2aS}{x^2} + 2b; \qquad \frac{dP}{dx} = 0, \qquad x = \sqrt{\frac{aS}{b}};$$

$$\frac{d^2P}{dx^2} = \frac{4aS}{x^3} > 0.$$

The value of x is the required minimum (there is just one extremum inside the domain of definition). The required dimensions of the page are

$$\text{width: } x + 2a = \sqrt{\frac{aS}{b}} + 2a, \qquad \text{height: } y + 2b = \sqrt{\frac{bS}{a}} + 2b.$$

Consideration of the last two examples enables us to make the following general remarks on procedure in extremum problems. To solve such a problem:

(1) Express the quantity which is to be extremized as a function of some argument, chosen out of the parameters of the problem. If this cannot be done directly, use may be made of several variables, which are then to be expressed in terms of the argument, using the conditions of the problem.

(2) Find the extremum of the function thus obtained.

We remark that in practical problems we are interested in finding the greatest or smallest value of the function, and thus its values at the ends of the interval of definition must be investigated also. Here much use can be made of the remark about the greatest or smallest value of a continuous function which has one extremum in an inner point of its domain of definition (see p. 72). In practical problems we can often decide on the character (maximum or minimum) of the "candidate" point from the physical conditions of the problem without further discussion.

414. Find a point on the altitude of an isosceles triangle (inside the triangle) such that the sum of its distances from the vertices is the smallest possible.

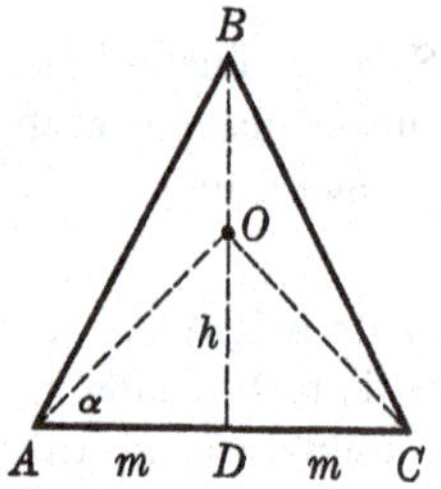

FIGURE 20

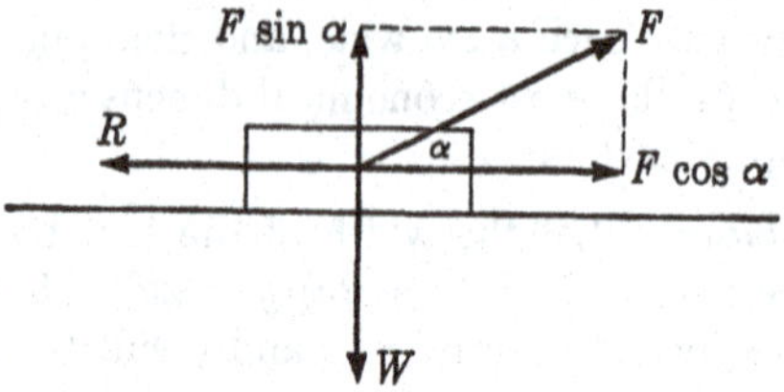

FIGURE 21

Solution. See Figure 20. Let us put $AD = m$, $BD = h$. For the argument we choose the angle α $(0 < \alpha < \sphericalangle BAD)$. The required sum of distances y is equal to

$$y = 2AO + OB = 2\frac{m}{\cos\alpha} + h - m\tan\alpha.$$

$$\frac{dy}{d\alpha} = 2m\frac{\sin\alpha}{\cos^2\alpha} - m\frac{1}{\cos^2\alpha} = \frac{m(2\sin\alpha - 1)}{\cos^2\alpha};$$

$$\frac{dy}{d\alpha} = 0, \qquad \sin\alpha = \frac{1}{2}, \qquad \alpha = \frac{\pi}{6}.$$

By differentiating the numerator of $dy/d\alpha$ we obtain $2m\cos\alpha$; at $\alpha = \pi/6$ this is positive, and because the denominator also is positive we have a minimum at $\alpha = \pi/6$;

$$y_{\min} = \frac{2m}{\sqrt{3}/2} + h - m\frac{\sqrt{3}}{3} = h + m\sqrt{3}.$$

We leave to the reader to show that when $\sphericalangle BAD < \pi/6$ the solution is at the end of the domain of definition of α; i.e., the required point is the vertex B.

415. A weight W lies on a horizontal plane. A force F is applied to it.

What is the angle between F and the plane to make minimal the force needed to displace W? The coefficient of friction, f, is given.

Solution. See Figure 21. To balance R (the force of friction) we need $R = F\cos\alpha$. (Of course a displacement occurs only when $F\cos\alpha$ exceeds R.) Now

$$R = (W - F\sin\alpha)f.$$

We obtain

$$(W - F\sin\alpha)f = F\cos\alpha, \qquad F = \frac{fW}{\cos\alpha + f\sin\alpha}.$$

The product fW is constant. F will be minimal when $g(\alpha) = \cos\alpha + f\sin\alpha$ is maximal.

$$g'(\alpha) = -\sin\alpha + f\cos\alpha = 0, \qquad \tan\alpha = f;$$

$$g''(\alpha) = -\cos\alpha - f\sin\alpha < 0, \qquad 0 < \alpha < \pi/2.$$

So, $\alpha_0 = \arctan f$ gives the minimum. (α_0 is called the friction angle.) The corresponding force is equal to

$$F_{\min} = \frac{fW}{\cos\alpha_0 + f\sin\alpha_0}$$

$$= \frac{fW}{\cos\alpha_0 + \tan\alpha_0\sin\alpha_0} = fW\cos\alpha_0 = \frac{fW}{\sqrt{1+\tan^2\alpha_0}} = \frac{fW}{\sqrt{1+f^2}}.$$

416. Given a circular cone. Inscribe in it another so that its vertex will be on the center of the base of the first cone. Show that the inner cone will have maximum volume when its altitude is equal to 1/3 of the altitude of the given cone.

417. Given a point (x_0, y_0) $(x_0 > 0; y_0 > 0)$. Find a straight line through this point, such that the segment cut from it by the positive half axes Ox and Oy will be of minimal length.

418. The straight line $x = m^2/2p$ cuts a segment from the parabola $y^2 = 2px$. Inscribe in this segment a rectangle of maximal area S.

419. Inscribe in a given sphere a cylinder such that its lateral surface (without the bases) shall be maximal.

420. Two points move along the x, y axes towards the origin with the constant velocities v_1 and v_2. The points start from $(a,0)$ and $(0,b)$ respectively $(a > 0, b > 0)$. What will be the minimal distance between the points during their motion?

421. A rod AB (see Fig. 22) of length $2b$ leans on two planes that make angles α and β with the horizontal line MN. AB can move in the plane POQ. In what position will the middle of AB be at maximal height?

422. A funnel in the form of a cone with height H and radius of the base

R is filled with water, and then a sphere is inserted so as to rest on the wall of the funnel. What is the radius of the sphere that displaces the maximum amount of water?

423. A series of n measurements of some quantity A resulted in n numbers $x_1, x_2, \ldots, x_n$. The size of A is often approximated by A_1 which results from making $S = \sum_{k=1}^{n} (A_1 - x_k)^2$ minimal (this is known as the least squares principle). Find A_1.

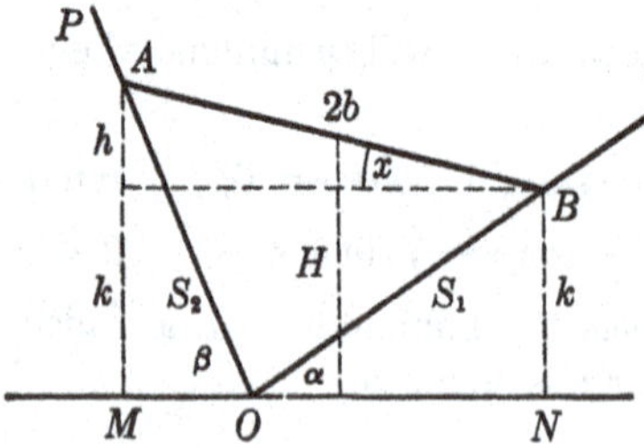

FIGURE 22

424. Among all cones having a generating line of given length L find the one that has the greatest volume.

425. Find the smallest distance between the point $M(p,p)$ and the parabola $y^2 = 2px$.

426. Find the distance from the given point $M(x_0,y_0)$ to the given straight line $ax + by + c = 0$, using extremum methods.

427. A cylindrical box with cover, of inner volume V, is made of iron sheet of thickness d. Find the inside radius r and the inside altitude h of the box, so that the amount of material needed for it shall be minimal.

428. Into a cup that has the form of a hemisphere of radius a, a thin rod of length b $(2a < b < 4a)$ is introduced. Find the position of equilibrium of the rod, i.e., the position in which its centroid is at the lowest level. (The cup should be considered immovable.)

429. The centers of three elastic balls are on one straight line. The first ball, of mass m_1, moves with constant velocity v_1 and hits the second ball. This ball in turn begins to move and hits the third ball, the mass of which is m_3. What is the mass of the second ball, if the velocity obtained by the third ball is the largest possible? (A resting ball of mass c obtains the velocity $2bv/(c + b)$ when hit by a ball of mass b moving with the velocity v.

6.4 CONCAVITY; POINTS OF INFLECTION

Given a curve $y = f(x)$ and a point $A(x_0,f(x_0))$ on it. Let us draw the tangent to the curve at A. If there exists an interval $(x_0 - \delta, x_0 + \delta)$ such that the points on the curve corresponding to points of this interval are above (below) the tangent at A, then the curve is called concave upward (or concave downward, respectively) at A; the curve in Figure 23 is concave upward at A_1 and concave downward at A_2.

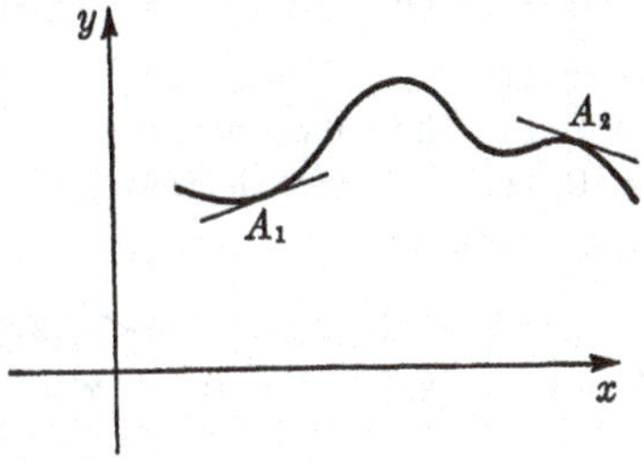

FIGURE 23

A point at which the direction of the concavity of the curve changes is called an inflection point. We can use the following rules:

(1) The curve is concave upward at $(x_0,f(x_0))$ if $f''(x_0) > 0$, and concave downward if $f''(x_0) < 0$.
(2) Necessary conditions for a curve $y = f(x)$ to have an inflection point at $(x_0,f(x_0))$ are: $f''(x_0) = 0$ or $f''(x_0)$ does not exist.

A sufficient condition is that $f''(x_0)$ changes its sign when x passes through x_0.

Another sufficient condition (for the case $f''(x_0) = 0$) is that the first nonvanishing derivative among $f'''(x_0)$, $f^{iv}(x_0)$, $f^{v}(x_0)$, etc., is of odd order.

430. Find the inflection points of the curve $y = x^5 + x$.

Solution:

$$y' = 5x^4 + 1, \qquad y'' = 20x^3;$$
$$y'' = 0, \qquad x = 0.$$

When x passes through $x = 0$, y'' changes sign from $-$ to $+$, i.e., (0,0) is an inflection point. By the second method we find

$$y''' = 60x^2, \qquad y^{iv} = 120x, \qquad y^{v} = 120;$$

y^{v} is the first derivative (higher than y'') different from zero at $x = 0$ and as it is of odd order, (0,0) is an inflection point.

431. The same question for $y = x^4 + x$.

Solution:

$$y' = 4x^3 + 1, \quad y'' = 12x^2, \quad 12x^2 = 0, \quad x = 0.$$
$$y''' = 24x, \quad y'''(0) = 0; \quad y^{iv}(x) = 24, \quad y^{iv}(0) = 24 \neq 0.$$

The first derivative (beyond y'') different from zero is of even order and (0,0) is not an inflection point. The same result follows also from the fact that $y'' = 12x^2$ does not change sign when x passes 0.

432. Investigate the behavior (domains of increasing and decreasing values, concavity, and inflection points) of the curve $y = f(x) = e^{-x^2}$.

Solution. $f'(x) = -2xe^{-x^2}$.
For $x < 0$, $f'(x) > 0$, i.e., the curve increases.
For $x > 0$, $f'(x) < 0$, i.e., the curve decreases.
At (0,1) it has a maximum.

$$f''(x) = -2e^{-x^2} + 4x^2e^{-x^2} = 2e^{-x^2}(2x^2 - 1);$$
$$f''(x) = 0, \quad 2x^2 - 1 = 0, \quad x = \pm\sqrt{2}/2.$$

For $-\infty < x < -\sqrt{2}/2$ and $\sqrt{2}/2 < x < \infty$, $f''(x) > 0$ and the curve is concave upward.

For $-\sqrt{2}/2 < x < \sqrt{2}/2$, $f''(x) < 0$ and the curve is concave downward in this interval.

The points $(\pm\sqrt{2}/2, e^{-1/2})$ are inflection points.

Now $\lim_{x \to \pm\infty} f(x) = 0$. The curve is given in Figure 24.

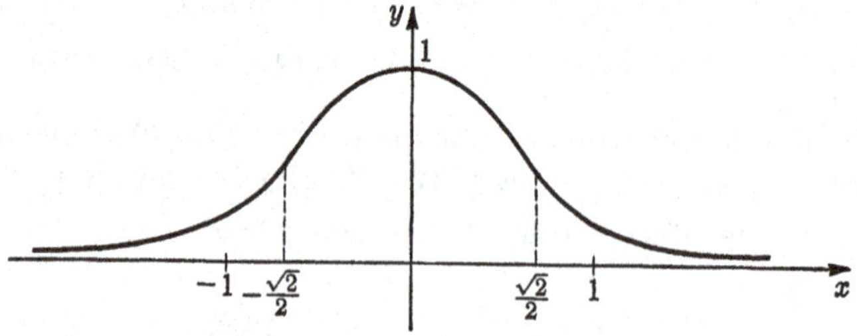

FIGURE 24

433. For what x is $y = x^3 + ax + b$ concave downward?

434. What is the concavity of the curve $x(x^2 - y^2) + y^2 = 0$ at $x = \frac{3}{2}$, $y > 0$?

435. Determine the concavity of $y = x^2 \ln x$ at the points (1,0) and $(1/e^2, -2/e^4)$.

In the following examples find the concavity and inflection points of the given functions.

436. $y = x^3 - 5x^2 + 3x - 5$

437. $y = (x + 1)^4 + e^x$

438. $y = (x + 2)^6 + 2x + 2$

439. $y = 3x^5 - 5x^4 + 3x - 2$

440. $y = \dfrac{x^3}{x^2 + 3a^2}, \ a > 0$

441. $y = a - \sqrt[3]{x - b}$

442. $y = \ln (1 + x^2)$

443. $y = e^{\arctan x}$

444. Prove that the inflection points of $y = x \sin x$ lie on the curve

$$y^2(4 + x^2) = 4x^2.$$

445. For what value of a does $y = e^x + ax^3$ have inflection points?

446. Prove that for the curve given by $x = g(t)$, $y = f(t)$ the points t for which the expression $\dfrac{g'f'' - f'g''}{g'}$ changes its sign $(g' \neq 0)$ are inflection points.

Solution. We have seen in Section 4.7 that

$$\frac{d^2y}{dx^2} = \frac{g'f'' - f'g''}{(g')^3}.$$

But the sign of $(g')^3$ is equal to the sign of g', so that we can limit our attention to the given expression.

Find the inflection points of the following curves.

447. $x = t^2, \ y = 3t + t^3$

Solution. We use the expression of the preceding problem.

$$g' = 2t, \qquad g'' = 2; \qquad f' = 3 + 3t^2, \qquad f'' = 6t.$$

Now $$\frac{g'f'' - f'g''}{g'} = \frac{2t \cdot 6t - (3 + 3t^2) \cdot 2}{2t} = \frac{3(t^2 - 1)}{t}.$$

This expression changes sign when t passes through $t = 1$, $t = -1$. The inflection points are (1,4), (1,−4).

448. $x = e^t, \ y = \sin t$

449. Prove that if $f''(x) \geqslant 0$ for all x in the interval $a \leqslant x \leqslant b$, then

(a) the graph of $y = f(x)$ is above the tangent to it at any point $(c,f(c))$, $a < c < b$;

(b) the graph in any partial interval $(a \leqslant) x_1 \leqslant x \leqslant x_2 (\leqslant b)$ is below the segment joining the two points $(x_1,f(x_1))$ and $(x_2,f(x_2))$;

(c) $f\left(\dfrac{x_1 + x_2}{2}\right) \leqslant \dfrac{f(x_1) + f(x_2)}{2}$.

6.5 ASYMPTOTES

A straight line l is said to be an asymptote of a given curve if the distance to l of a point $P(x,y)$ on a branch of the curve tends to zero when at least one of the coordinates of P tends to $+\infty$ or $-\infty$. For example, the curve in Figure 25 has three asymptotes. We say that an asymptote is horizontal (vertical) if it is parallel to the x axis (or y axis correspondingly). Any other asymptotes are called oblique.

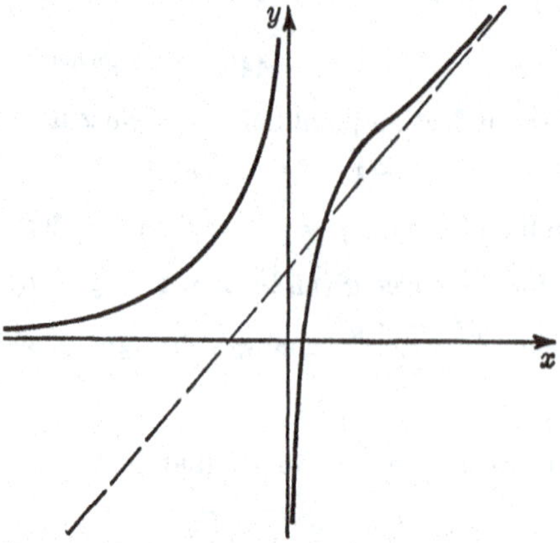

FIGURE 25

To find horizontal asymptotes we compute $\lim_{x\to\infty} y$ and $\lim_{x\to-\infty} y$. If, for example, $\lim_{x\to\infty} y = b$, then the straight line $y = b$ is an asymptote.

By computing $\lim_{y\to\infty} x$ and $\lim_{y\to-\infty} x$ we obtain the vertical asymptotes, if any. For example, if $\lim_{y\to\infty} x = a$, then $x = a$ is an asymptote. In many practical cases we find this limit by investigating what values of x in $y = f(x)$ make y tend to ∞. These are clearly to be found among the points of discontinuity.

To find oblique asymptotes we can use one of the following methods:

(1) If, for (x,y) on one branch of the curve, $\lim_{\substack{x\to\infty \\ (x\to-\infty)}} y/x = k$ and

$$\lim_{\substack{x\to\infty \\ (x\to-\infty)}} [y - kx]$$

exists and equals b, then the straight line $y = kx + b$ is an asymptote.

(2) If it is possible to put the equation of one branch of the curve in the form $y = f(x) = kx + b + g(x)$, where $\lim_{\substack{x\to\infty \\ (x\to-\infty)}} g(x) = 0$, then $y = kx + b$ is an asymptote.

(3) If the equation of the curve is $F(x,y) = 0$, where $F(x,y)$ is a polynomial in x and y (or can be brought to such a form), we substitute for y the expression $kx + b$ and equate to zero the coefficients of the two highest powers of x in the expression so obtained. If from these two equations k and b can be determined, then $y = kx + b$ is an asymptote.

Remark. The above methods also locate the horizontal asymptotes.

Find the asymptotes of the following curves.

450. $y = a_0x^n + a_1x^{n-1} + \ldots + a_{n-1}x + a_n, \qquad n > 1,\ a_0 \neq 0$

Solution. When $x \to \pm\infty$, y tends to infinity (+ or −); hence there are no horizontal asymptotes. The function is defined (and continuous) for every x, i.e., there are no vertical asymptotes. $\lim\limits_{x\to\pm\infty} y/x$ is also infinity (+ or −), i.e., there are no oblique asymptotes.

451. $y = \dfrac{3x^2 + 2x - 1}{x^2 + 2}$

Solution. $\lim\limits_{x\to\pm\infty} y = 3$; i.e., $y = 3$ is a horizontal asymptote. The given function is continuous for every x and consequently has no vertical asymptote.

$$\lim_{x\to\pm\infty} y/x = 0; \qquad \lim_{x\to\pm\infty} (y - 0) = \lim_{x\to\pm\infty} y = 3.$$

We obtain once more the horizontal asymptote.

452. $y = \dfrac{3x^2 - 2x + 3}{x - 1}$

Solution. $\lim\limits_{x\to\pm\infty} y = \pm\infty$; there is no horizontal asymptote.

$$\lim_{y\to\pm\infty} x = 1,$$

because for $x = 1$ the denominator vanishes and the numerator is not zero. (More strictly, $\lim\limits_{y\to\pm\infty} x_2 = 1$, while $\lim\limits_{y\to\pm\infty} x_1 = \pm\infty$. If we solve the given equation for x we obtain two functions,

$$x_1, x_2 = \frac{2 + y \pm \sqrt{y^2 - 8y - 32}}{6}.)$$

Further,

$$\lim_{x\to\infty} \frac{y}{x} = \lim_{x\to\infty} \frac{3x^2 - 2x + 3}{x^2 - x} = 3;$$

$$\lim_{x\to\infty} (y - 3x) = \lim_{x\to\infty} \frac{3x^2 - 2x + 3 - 3x^2 + 3x}{x - 1} = \lim_{x\to\infty} \frac{x + 3}{x - 1} = 1.$$

The straight line $y = 3x + 1$ is an oblique asymptote. (The same is obtained by letting x tend to $-\infty$.) We can also obtain this asymptote by writing

$$y = 3x + 1 + \frac{4}{x - 1}.$$

Now $\lim_{x\to\infty} \frac{4}{x - 1} = 0$, i.e., $y = 3x + 1$ is an asymptote.

453. $y = \sqrt{\frac{x^3 + 1}{x + 3}}$

Solution. $\lim_{x\to\infty} y = \infty$; there is no horizontal asymptote. $\lim_{y\to\infty} x = -3$; $x = -3$ is a vertical asymptote.

$$\lim_{x\to\infty} \frac{y}{x} = \lim_{x\to\infty} \sqrt{\frac{x^3 + 1}{x^2(x + 3)}} = 1;$$

$$\lim_{x\to\infty} (y - x) = \lim_{x\to\infty} \left(\sqrt{\frac{x^3 + 1}{x + 3}} - x\right) = \lim_{x\to\infty} \frac{\frac{x^3 + 1}{x + 3} - x^2}{\sqrt{\frac{x^3 + 1}{x + 3}} + x}$$

$$= \lim_{x\to\infty} \frac{\frac{x^3 + 1 - x^3 - 3x^2}{x + 3}}{\sqrt{\frac{x^3 + 1}{x + 3}} + x}$$

$$= \lim_{x\to\infty} \frac{-3x^2 + 1}{\sqrt{(x + 3)(x^3 + 1)} + x(x + 3)} = -\frac{3}{2};$$

$y = x - \frac{3}{2}$ is an oblique asymptote. For $x \to -\infty$ we obtain:

$$\lim_{x\to-\infty} \frac{y}{x} = \lim_{x\to-\infty} \frac{\sqrt{\frac{x^3 + 1}{x + 3}}}{x} = \lim_{x\to-\infty} -\sqrt{\frac{x^3 + 1}{x^2(x + 3)}} = -1;$$

$$\lim_{x\to-\infty} (y + x) = \lim_{x\to-\infty} \left(\sqrt{\frac{x^3 + 1}{x + 3}} + x\right) = \lim_{x\to-\infty} \frac{\frac{x^3 + 1}{x + 3} - x^2}{\sqrt{\frac{x^3 + 1}{x + 3}} - x}$$

$$= \lim_{x\to-\infty} \frac{x^3 + 1 - x^3 - 3x^2}{(x + 3)\sqrt{\frac{x^3 + 1}{x + 3}} - x(x + 3)}$$

$$= \lim_{x \to -\infty} \frac{-3 + \dfrac{1}{x^2}}{-\left(1 + \dfrac{3}{x}\right)\sqrt{\dfrac{x + \dfrac{1}{x^2}}{x+3}} - 1 - \dfrac{3}{x}} = \frac{3}{2};$$

$y = -x + \frac{3}{2}$ is a second oblique asymptote.

In Figure 26 we sketch the graph of this function. It is defined only for $\dfrac{x^3+1}{x+3} \geqslant 0$, i.e., $\dfrac{x+1}{x+3} \geqslant 0$ (because $x^2 - x + 1 > 0$ always). This inequality holds for $x \geqslant -1$ and $x < -3$. This example shows clearly that

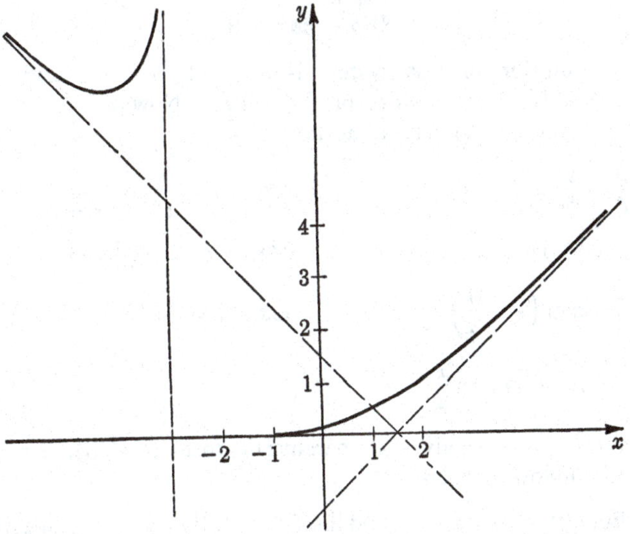

FIGURE 26

we must investigate the behavior of the function for both $x \to +\infty$ and $x \to -\infty$.

454. $y = x + \ln x$

Solution. $\lim\limits_{x\to\infty} y = \infty$; for $x \to 0+$, $y \to -\infty$, i.e., $x = 0$ is a vertical asymptote.

$$\lim_{x\to\infty} \frac{y}{x} = \lim_{x\to\infty} \left(1 + \frac{\ln x}{x}\right) = 1, \quad \text{because} \quad \lim_{x\to\infty} \frac{\ln x}{x} = \lim_{x\to\infty} \frac{1}{x} = 0.$$

Now $$\lim_{x\to\infty} (y - x) = \lim_{x\to\infty} (x + \ln x - x) = \infty.$$

There is no oblique asymptote.

455. $y = x \arctan x$

456. $y = \dfrac{2x^4 + x^3 + 1}{x^3}$

457. $x^3 + y^3 - 3axy = 0$

Solution. We shall use here the third method. We substitute $y = kx + b$:

$$x^3 + k^3x^3 + 3k^2bx^2 + 3kb^2x + b^3 - 3akx^2 - 3abx = 0,$$

or

$$x^3(1 + k^3) + x^2(3k^2b - 3ak) + 3kb^2x - 3abx + b^3 = 0.$$

We equate to zero the coefficients of the two highest powers of x:

$$1 + k^3 = 0,$$
$$3k^2b - 3ak = 0.$$

The only solution of this system is $k = -1$, $b = -a$ and therefore $y = -x - a$ is an asymptote of the curve. Now $y \to \infty$ only when $x \to \infty$, i.e., there is no vertical asymptote.

458. $\dfrac{x^2}{a^2} - \dfrac{y^2}{b^2} = 1$

459. $y = \dfrac{1}{x^2 - 4x - 5}$

460. $2y(x + 1)^2 = x^3$

461. $xy^2 + x^2y = a^3$

462. $y = x \ln\left(e + \dfrac{1}{x}\right)$

463. $y = xe^{1/x^2}$

464. $y = 2x + \arctan \dfrac{x}{2}$

When the curve is given in parametric form $y = f(t)$, $x = g(t)$, we can use the following rules:

(1) If $\lim_{t \to t_0} g(t) = \infty$ (or $-\infty$) and $\lim_{t \to t_0} f(t) = b$, then $y = b$ is an asymptote.

(2) If $\lim_{t \to t_0} g(t) = a$ and $\lim_{t \to t_0} f(t) = \infty$ (or $-\infty$), then $x = a$ is an asymptote.

(3) If $\lim_{t \to t_0} g(t) = \infty$ (or $-\infty$), $\lim_{t \to t_0} f(t) = \infty$ (or $-\infty$), $\lim_{t \to t_0} \dfrac{f(t)}{g(t)} = k$, and $\lim_{t \to t_0} [f(t) - kg(t)] = b$, then $y = kx + b$ is an asymptote.

Note that in any case the limits must be investigated also for $t \to \pm\infty$. These rules correspond clearly to those given at the beginning of this section.

Find the asymptotes of the following curves:

465. $x = \dfrac{1}{t},\ y = \dfrac{t}{t + 1}$

Solution:

$\lim_{t\to 0\pm} x = \pm\infty, \quad \lim_{t\to 0\pm} y = 0;$ i.e., $y = 0$ is an asymptote.

$\lim_{t\to -1\pm} y = \mp\infty, \quad \lim_{t\to -1} x = -1;$ i.e., $x = -1$ is also an asymptote.

There is no t_0 such that x and y both tend to infinity when $t \to t_0$, i.e., there is no oblique asymptote.

466. $x = \dfrac{2e^t}{t-1}, \; y = \dfrac{te^t}{t-1}$

Solution. x and y both tend to infinity when $t \to 1$.

$$\lim_{t\to 1} \frac{y}{x} = \lim_{t\to 1} \frac{t}{2} = \frac{1}{2};$$

$$\lim_{t\to 1} \left(y - \frac{1}{2}x\right) = \lim_{t\to 1} \left(\frac{te^t}{t-1} - \frac{e^t}{t-1}\right) = \lim_{t\to 1} e^t = e;$$

$y = \dfrac{x}{2} + e$ is an oblique asymptote. When $t \to \infty$, x and y but also y/x tend to ∞.

467. $x = \dfrac{2t}{1-t^2}, \; y = \dfrac{t^2}{1-t^2}$

468. $x = \dfrac{3at}{1+t^3}, \; y = \dfrac{3at^2}{1+t^3}$

469. $x = \dfrac{t-8}{t^2-4}, \; y = \dfrac{3}{t(t^2-4)}$

6.6 CURVE TRACING

We shall now describe a general scheme for the investigation of a function $y = f$ with the intention of finding points at which the corresponding curve has special features (extrema, points of inflection, intercepts, etc.), and use these features to make graphs.

We find:

(1) The domain of definition of the function.
(2) Points of discontinuity and intervals of continuity.
(3) Points of intersection with the axes.
(4) Symmetry.
(5) Periodicity.
(6) Asymptotes and the position of the graph with respect to them.
(7) Points of intersection of the graph with its asymptotes.
(8) Intervals of monotonicity.

(9) Extrema.
(10) Concavity and inflection points.

Remarks. (*a*) For the first 7 items no derivatives are needed.
(*b*) The symmetry can be found with the aid of the table shown.

If the given equation, being satisfied by the point (a,b),	*is satisfied also by the point*	*then the graph of this equation is symmetric with respect to the*	*Examples*
	$(-a,b)$	y axis	$y = f(x)$ where $f(x)$ is even, i.e., $f(-x) = f(x)$
	$(a,-b)$	x axis	$y = \pm\sqrt{a^2 - x^2}$, $x^3 + 2xy^2 + y^4 = 6$
	$(-a,-b)$	origin (0,0)	Any $y = f(x)$ where $f(x)$ is odd, i.e., $f(-x) = -f(x)$
	(b,a)	straight line $y = x$	$x^3 + y^3 + 3axy = 0$, $x^2y + xy^2 = 5$
	$(-b,-a)$	straight line $y = -x$	$xy = k$

We remark that the graph can also have other axes or centers of symmetry.

(*c*) It is often preferable, before sketching the graph, to sum up the results of the above investigation in a table, as will be done in Problem 470.

(*d*) In many cases a discussion of some of the listed properties will suffice for the tracing.

(*e*) It is often advisable to find additional points of the graph between the characteristic points mentioned above.

(*f*) In this section we shall sketch graphs of equations given in explicit, implicit, or parametric form. The next section will deal with graphs in polar coordinates.

Rational Functions

Draw the graphs of the curves given by the following equations:

470. $y = \dfrac{x^2}{1 + x}$

Solution. In this first example we shall follow exactly the scheme given above.

(1) The domain of definition consists of the whole x axis with the exception of $x = -1$.

(2) $x = -1$ is a point of discontinuity. The function is continuous for any other x.

(3) Points of intersection with the axes: $x = 0$, $y = 0$.

(4) Symmetry: The function is not even and not odd. The coordinate axes and the straight lines $y = \pm x$ are not axes of symmetry of the graph. The origin is also not a center of symmetry.

(5) The function is not periodic.

(6) $\lim_{x \to \infty} y = \infty$; $\lim_{y \to \infty} x = -1$. The straight line $x = -1$ is a vertical asymptote. Now

$$y = \frac{x^2}{x+1} = x - 1 + \frac{1}{x+1} \quad \text{and} \quad \lim_{x \to \infty} \frac{1}{x+1} = 0,$$

i.e., $y = x - 1$ is also an asymptote of the curve.

For $x > -1$ we have $\frac{1}{x+1} > 0$, i.e., when $x > -1$ the graph is above the asymptote.

For $x < -1$ the graph is below the asymptote.

Now $\lim_{x \to -1+} y = \infty$ and $\lim_{x \to -1-} y = -\infty$, and this determines the behavior of the graph with respect to the vertical asymptote.

(7) The equation $\frac{x^2}{x+1} = x - 1$ or $x^2 = x^2 - 1$ has no solutions, i.e., the graph does not intersect the asymptote $y = x - 1$.

(8) Let us differentiate:

$$y' = \frac{2x(1+x) - x^2}{(1+x)^2} = \frac{x(x+2)}{(1+x)^2}.$$

$$y' = 0; \qquad x_1 = 0, \qquad x_2 = -2$$

For $x < -2$, $y' > 0$ and the function increases.

For $-2 < x < -1$ and $-1 < x < 0$, $y' < 0$ and the function decreases.

For $x > 0$, $y' > 0$ and the function increases.

(9) From the above it follows directly that $(-2, -4)$ is a maximum and $(0,0)$ is a minimum of the function. y' is not defined at $x = -1$, but there the function itself also is not defined.

(10) $$y'' = \frac{(2x+2)(1+x) - (x^2+2x)\cdot 2}{(1+x)^3} = \frac{2}{(1+x)^3}.$$

For $x < -1$, $y'' < 0$ and the curve is concave downward.

For $x > -1$, $y'' > 0$ and the curve is concave upward.

$y'' \neq 0$ and there are no inflection points. ($x = -1$ is not in question.)

Before drawing the curve let us arrange these results in tabular form, as shown.

x	y	*Behavior of the function*
$x \to -\infty$	$y \to -\infty$	Tends from below to the straight line $y = x - 1$.
$x < -2$		Increases, concave downward.
-2	-4	Maximum.
$-2 < x < -1$		Decreases, concave downward.
$x \to -1-$	$y \to -\infty$	$x = -1$ is an asymptote.
$x = -1$		Not defined, point of discontinuity.
$x \to -1+$	$y \to +\infty$	$x = -1$ is an asymptote.
$-1 < x < 0$		Decreases, concave upward.
0	0	Intersection with the axes, minimum.
$x > 0$		Increases, concave upward.
$x \to \infty$	$y \to \infty$	Tends from above to the asymptote $y = x - 1$.

Now we shall draw the curve. We begin with the two asymptotes, mark off the extremum points, the points of intersection with the axes, and all other characteristic points. If we are interested in a more precise graph, we compute additional points, e.g., ($x = 1, y = \frac{1}{2}$), ($x = -3, y = -4.5$), and so on. The graph is given in Figure 27.

471. $y = \dfrac{x^3 - 9x}{10}$ **472.** $y = \dfrac{x(x-1)(x-2)(x-3)}{24}$

473. $y = \dfrac{1}{1 + x^2}$ **474.** $y = \dfrac{x}{1 + x^2}$

475. $y = \dfrac{1}{1 - x^2}$ **476.** $y = \dfrac{1}{x} + \dfrac{1}{x - 1} + \dfrac{1}{x - 2}$

477. $y = x + \dfrac{2x}{x^2 - 1}$ **478.** $y = \dfrac{x^2 - 1}{x}$

479. $y = \dfrac{x^3}{x^2 - 1}$ **480.** $y = \dfrac{x^2}{x^4 - 4}$

481. $y = \dfrac{9 + x^3}{1 - x^2}$ **482.** $y = \dfrac{x(2x + 1)}{x + 1}$

483. $y = \dfrac{x}{(1 - x)^2}$ **484.** $y = \dfrac{x^3 - 2x}{x^2 - 3}$

485. $y = \dfrac{x^3}{(x + 2)^2}$ **486.** $y = \dfrac{5x^2 - 18x + 45}{x^2 - 9}$

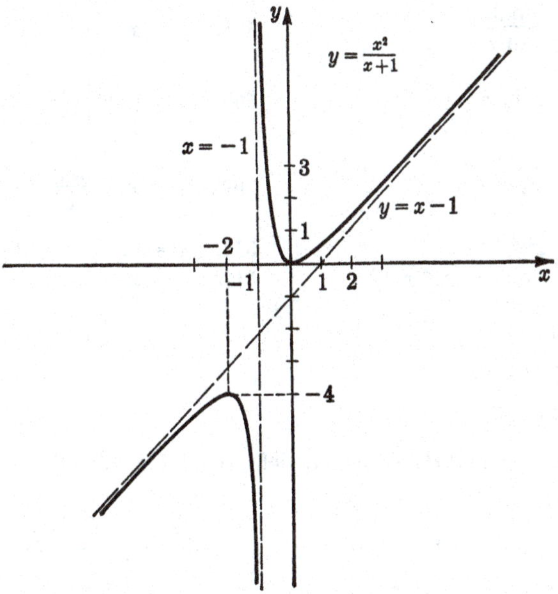

FIGURE 27

Irrational Functions

487. $y = \pm\sqrt{x-2}$

488. $y = \pm\frac{1}{2}\sqrt{25-x^2}$

489. $y = \pm\frac{1}{2}\sqrt{x^2-1}$

490. $y = \pm\frac{1}{2}x\sqrt{x}$

491. $y = \pm x\sqrt{\dfrac{x}{4-x}}$

492. $y = \pm\sqrt{\dfrac{x-1}{x+1}}$

493. $2y = \sqrt{x^2+x+1} + \sqrt{x^2-x+1}$

494. $y = \pm x^2\sqrt{x+1}$

495. $y = (x+1)^3\sqrt[3]{x^2}$

496. $y = \dfrac{x\sqrt{1-x}}{1+x}$

497. $y = \dfrac{x+3}{\sqrt{x+1}}$

498. $y = \sqrt{\dfrac{x-1}{x(x+1)}}$

499. $y = 2x - 3\sqrt[3]{x^2}$

Transcendental Functions

500. $y = e^{1/x}$

501. $y = \dfrac{\sin x}{x}$

502. $y = \sin x^2$

503. $y = \sin x + \frac{1}{2}\sin 2x + \frac{1}{3}\sin 3x$

504. $y = \frac{\tan 3x}{\tan x}$

505. $y = x \ln x$

506. $y = x^2 \ln x$

507. $y = \frac{1}{2} \ln \frac{1+x}{1-x}$

508. $y = x^2 e^{-x^2}$

509. $y = x + \frac{\ln x}{x}$

510. $y = x^2 e^{-x}$

511. $y = (1 + x^2)e^{-x^2}$

512. $y = \sin x + \cos^2 x$

513. $y = x - \ln x$

514. $y = xe^{-x^2/2}$

Implicit Functions

515. $x^2y + xy^2 = 2$

Solution. To find the domain of definition let us solve the equation with respect to y:

$$xy^2 + x^2y - 2 = 0; \qquad y = \frac{-x^2 \pm \sqrt{x^4 + 8x}}{2x}.$$

We need $x \neq 0$ and $x^4 + 8x \geqslant 0$, or $x(x + 2)(x^2 - 2x + 4) \geqslant 0$. The third factor is always positive, and we have to solve $x(x + 2) \geqslant 0$. We obtain

$$x \leqslant -2 \quad \text{or} \quad x > 0 \quad (\text{because } x \neq 0).$$

If in the given equation we interchange x and y nothing will change, i.e., together with any point (a,b) also (b,a) belongs to the curve, and this shows that the straight line $y = x$ is a symmetry axis.

To find asymptotes let us substitute $y = ax + b$ in the equation of the curve:

$$x^2(ax + b) + x(ax + b)^2 = 2,$$
$$ax^3 + bx^2 + a^2x^3 + 2abx^2 + b^2x = 2.$$

We require

$$a + a^2 = 0; \qquad a_1 = 0, \qquad a_2 = -1.$$
$$b + 2ab = 0; \qquad b = 0,$$

($a = -\frac{1}{2}$ does not agree with the values just obtained.)

We have two asymptotes: $y = 0$ and $y = -x$. From the symmetry it follows that together with $y = 0$ also $x = 0$ is an asymptote. *Note:* This asymptote cannot be obtained by the above method, because $x = 0$ is a straight line for which the slope is undefined. But we can find this asymptote by evaluating the limit:

$$\lim_{x\to 0+} \frac{-x^2 + \sqrt{x^4 + 8x}}{2x} = \lim_{x\to 0+} \frac{x^4 - x^4 - 8x}{2x(-x^2 - \sqrt{x^4 + 8x})} = \infty.$$

To investigate extrema let us differentiate:

$$x^2y + xy^2 = 2, \qquad 2xy + x^2y' + y^2 + 2xyy' = 0,$$

$$y' = -\frac{2xy + y^2}{x^2 + 2xy} = -\frac{y(2x + y)}{x(x + 2y)}.$$

$y' = 0$ gives $y = 0$; but this value of y leads to the contradiction $0 = 2$. Take $y = -2x$ and substitute this in the given equation:

$$-2x^3 + 4x^3 = 2, \qquad x^3 = 1; \qquad x = 1, \quad y = -2.$$

y' exists at this point. We find the second derivative:

$$2y + 2xy' + 2xy' + x^2y'' + 2yy' + 2yy' + 2x(y')^2 + 2xyy'' = 0.$$

We substitute $x = 1$, $y = -2$, $y' = 0$:

$$-4 + y'' - 4y'' = 0, \qquad y'' = -\tfrac{4}{3} < 0, \qquad \text{Max } (1, -2).$$

The points $x = 0$ and $x = -2y$ where y' is not defined give no extrema. (At $x = 0$, y is not defined, and $x = -2y$ leads to the point $(-2, 1)$, i.e., to a point symmetric (with respect to $y = x$) to the maximum. Hence the tangent is vertical there.)

Let us find the point of intersection of the curve with its symmetry axis:

$$\begin{cases} x^2y + xy^2 = 2, \\ x = y, \end{cases} \qquad 2x^3 = 2, \qquad x = 1, \qquad y = 1.$$

The graph is given in Figure 28.

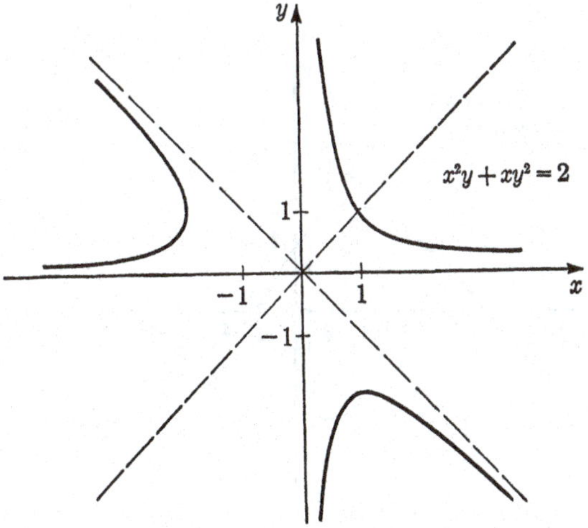

FIGURE 28

516. $y^2 - x^4 + x^6 = 0.$ **517.** $y^3 - x^3 + y - 2x = 0$

518. $y^2 = x^3 - 2x^2 + x$ **519.** $x^3 = (y - x)^2$

Curves Given in Parametric Form

520. $x = t^3 + 3t + 1,\ y = t^3 - 3t + 1$

Solution. x and y are defined for any t. The points of intersection with the axes can be found by solving cubic equations, but this will not be done here.

$$\lim_{t\to\infty} x = \infty \quad \text{and} \quad \lim_{t\to\infty} y = \infty.$$

$$\lim_{t\to\infty} y/x = 1; \quad \lim_{t\to\infty} (y - x) = \lim_{t\to\infty} (-6t) = -\infty.$$

There are, consequently, no asymptotes. ($t \to -\infty$ gives a similar result.) To find the extrema let us differentiate:

$$\frac{dy}{dx} = \frac{3t^2 - 3}{3t^2 + 3} = \frac{t^2 - 1}{t^2 + 1};$$

$$\frac{d^2y}{dx^2} = \frac{1}{3t^2 + 3} \cdot \frac{2t(t^2 + 1) - (t^2 - 1)2t}{(t^2 + 1)^2} = \frac{4t}{3(t^2 + 1)^3}.$$

$$y' = 0, \quad t = \pm 1, \qquad y''(t = 1) > 0; \qquad \text{Min } (5, -1).$$
$$y''(t = -1) < 0; \qquad \text{Max } (-3, 3).$$

$y'' = 0$, $t = 0$ ($x = 1$, $y = 1$); this is an inflection point.

A few more points: $t = 2$, $x = 15$, $y = 3$.
$t = -2$, $x = -13$, $y = -1$.

See Figure 29.

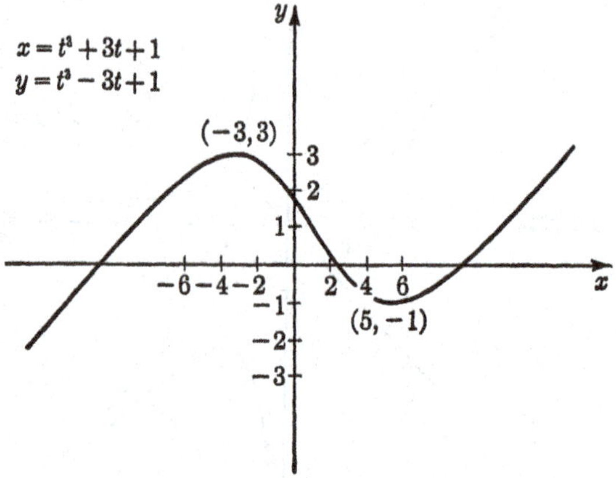

FIGURE 29

521. $x = \dfrac{3t}{1 + t^3}, \; y = \dfrac{3t^2}{1 + t^3}$

We shall take up the topic of parametric equations again in Section 6.8.

6.7 GRAPHS IN POLAR COORDINATES

Points in the plane are often located by means of polar coordinates. For this purpose consider an arbitrary (but fixed) ray emanating from a point O, called the pole. This particular ray will be referred to as the polar axis. We make correspond to any point M in the plane two numbers: r which is equal to the length of OM (called radius vector) and θ, the argument, which is equal to the angle from the polar axis to the radius vector OM (measured in the counterclockwise direction (see Fig. 30).

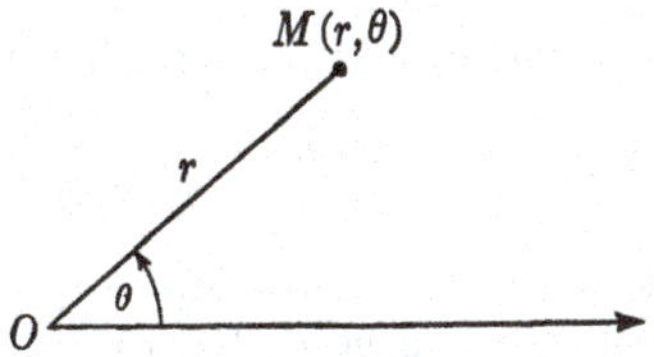

FIGURE 30

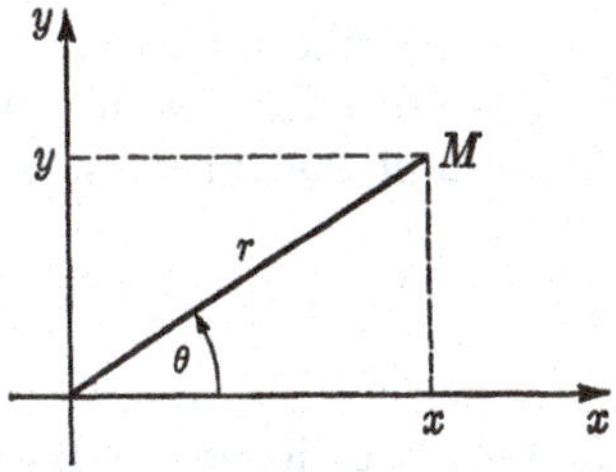

FIGURE 31

The only exception is the pole O, for which $r = 0$ but θ is arbitrary. By the above definition $0 \leqslant r < \infty$ and $0 \leqslant \theta < 2\pi$.

Sometimes also negative values for r are considered and θ may be negative or greater than 2π. In this case, for example, $(-r,\theta) \equiv (r, \theta - \pi)$ and so on. In general we shall restrict ourselves to nonnegative r and $0 \leqslant \theta < 2\pi$.

Together with the polar coordinates let rectangular coordinates also be defined in the plane, with origin at the pole and with the positive half of the x axis on the polar axis (Fig. 31). We have then the following connections between the various coordinates of the same point M in the plane:

$$x = r\cos\theta, \qquad y = r\sin\theta;$$

$$r = \sqrt{x^2 + y^2}, \qquad \cos\theta = \frac{x}{\sqrt{x^2 + y^2}}, \qquad \sin\theta = \frac{y}{\sqrt{x^2 + y^2}},$$

$$0 \leqslant \theta < 2\pi.$$

(Sometimes we compute θ from the equation $\tan\theta = y/x$, which together with the signs of x and y defines θ exactly.)

Equations of the form $r = f(\theta)$ or $F(r,\theta) = 0$ in general define curves in the plane, in the same manner as similar equations in cartesian coordinates. In this section we shall deal with curves given in polar coordinates. The methods are slightly different and will be illustrated in the examples that follow.

Sketch the graphs of the curves given by the following equations.

522. $r = a \sin 3\theta$

Solution. $\sin 3\theta$ is periodic with period $2\pi/3$. Thus, we may restrict ourselves to investigating $0 \leqslant \theta < 2\pi/3$. For $\pi/3 < \theta < 2\pi/3$, $\sin 3\theta$ is negative, and by our convention about r, the curve does not exist for these values of θ.

From the equality

$$\sin 3\left(\frac{\pi}{6} - \theta\right) = \sin 3\left(\frac{\pi}{6} + \theta\right)$$

it follows that the radius vector $\theta = \pi/6$ is an axis of symmetry of the curve. We therefore restrict ourselves to $0 \leqslant \theta < \pi/6$.

The form of the curve can now be found by computing a few points:

$$\theta = 0,\ r = 0; \qquad \theta = \frac{\pi}{18},\ r = \frac{a}{2}; \qquad \theta = \frac{\pi}{9},\ r = \frac{a\sqrt{3}}{2}; \qquad \theta = \frac{\pi}{6},\ r = a.$$

The last point is also clearly a maximum of r, because $\sin 3\theta \leqslant 1$. We can find also the direction of the tangent to the curve at the origin. To do this let us use rectangular coordinates:

$$y = r \sin \theta = a \sin 3\theta \sin \theta,$$
$$x = r \cos \theta = a \sin 3\theta \cos \theta.$$

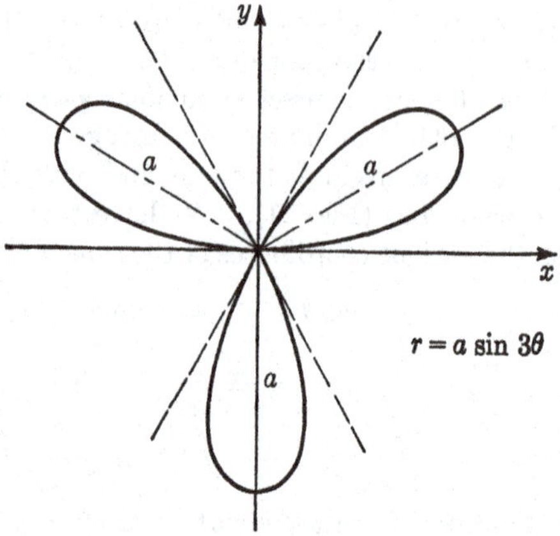

FIGURE 32

These equations can be regarded as parametric equations of the curve with θ as parameter.

$$\frac{dy}{dx} = \frac{dy/d\theta}{dx/d\theta} = \frac{3a \cos 3\theta \sin \theta + a \sin 3\theta \cos \theta}{3a \cos 3\theta \cos \theta - a \sin 3\theta \sin \theta};$$

$$\frac{dy}{dx}(\theta = 0) = 0,$$

i.e., the tangent to the curve at the origin is horizontal. It is clear by symmetry that also the radius vector $\theta = \pi/3$ is a tangent to the curve at the origin. The form of the graph for $0 \leqslant \theta \leqslant \pi/3$ is clear. For $\pi/3 < \theta < 2\pi/3$ it is not defined (r is negative). For $2\pi/3 \leqslant \theta < 4\pi/3$ and $4\pi/3 \leqslant \theta < 2\pi$ it repeats itself periodically. See Figure 32.

523. $(x^2 + y^2)^3 = 27x^2y^2$

Solution. This equation is given in cartesian coordinates but its graph is much easier to find using polar coordinates. Before passing to them, let us remark that the curve is symmetric with respect to both x and y axes and also with respect to the lines $y = x$ and $y = -x$ (and to the origin, of course). It is enough to find its form for $0 \leqslant \theta \leqslant \pi/4$ and then continue by symmetry. Now we shall transform the given equation by setting $x = r \cos \theta$, $y = r \sin \theta$:

$$r^6 = 27r^4 \cos^2 \theta \sin^2 \theta = \tfrac{27}{4} r^4 \sin^2 2\theta.$$

We divide by r^4, but it must be taken into account that $r = 0$ is a point of the graph. We obtain $r^2 = \frac{27}{4} \sin^2 2\theta$. (It can be seen that $r = 0$ will still appear.) We are interested in $0 \leqslant \theta \leqslant \pi/4$ and $r > 0$, so that extracting a square root is permitted: $r = (3\sqrt{3}/2) \sin 2\theta$. Now several points of the curve will be obtained:

$$\theta = 0,\ r = 0; \qquad \theta = \frac{\pi}{12},\ r = \frac{3\sqrt{3}}{4}; \qquad \theta = \frac{\pi}{8},\ r = \frac{3\sqrt{6}}{4};$$

$$\theta = \frac{\pi}{6},\ r = \frac{9}{4}; \qquad \theta = \frac{\pi}{4},\ r_{\max} = \frac{3\sqrt{3}}{2}.$$

Next,

$$y = r \sin \theta = \frac{3\sqrt{3}}{2} \sin \theta \sin 2\theta;$$

$$\frac{dy}{d\theta} = \frac{3\sqrt{3}}{2} \cos \theta \sin 2\theta + 3\sqrt{3} \sin \theta \cos 2\theta; \qquad \frac{dy}{d\theta}(\theta = 0) = 0,$$

i.e., the tangent to the curve at the origin is horizontal.

$$x = r\cos\theta = \frac{3\sqrt{3}}{2}\sin 2\theta\cos\theta;$$

$$\frac{dx}{d\theta} = \frac{3\sqrt{3}}{2}(2\cos 2\theta\cos\theta - \sin 2\theta\sin\theta);$$

$$\frac{dx}{d\theta} = 0; \qquad 2(\cos^2\theta - \sin^2\theta)\cos\theta - 2\sin^2\theta\cos\theta = 0;$$

$$\cos\theta = 0,\ \theta = \pi/2;$$

$$\cos^2\theta - 2\sin^2\theta = 0,\ \tan^2\theta = \tfrac{1}{2}; \qquad \tan\theta = \sqrt{\tfrac{1}{2}},\ \theta = \arctan\sqrt{\tfrac{1}{2}}.$$

The last value of θ is the only value in the interval $0 \leqslant \theta \leqslant \pi/4$ for which $dx/d\theta = 0$. For this θ, x obtains its maximum value, $x_{max} = 2$, i.e., at this point ($\theta = \arctan\sqrt{\frac{1}{2}}$, $r \approx 2.3$), the tangent to the curve is vertical.

The form of the curve in the above interval is now clear. By symmetry we obtain the whole curve (see Fig. 33).

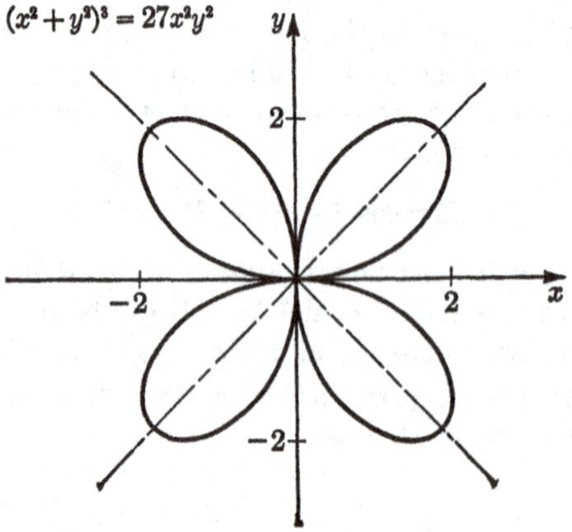

FIGURE 33

Remark. The graph of $r = (3\sqrt{3}/2)\sin 2\theta$ consists only of the parts in the first and third quadrants, because for $\pi/2 < \theta < \pi$ and $3\pi/2 < \theta < 2\pi$, $\sin 2\theta < 0$.

524. $r = a(1 + \cos\theta)$

525. $x^4 + y^4 = 2xy$

526. $(x^2 + y^2 - 4x)^2 = 16(x^2 + y^2)$ (cf. Prob. 524)

527. $x^4 + y^4 = x^2 + y^2$ (cf. Prob. 409)

528. $r = \sqrt{1 - t^2}$, $\theta = \arcsin t + \sqrt{1 - t^2}$

(a curve in polar coordinates given by parametric equations).

6.8 PARAMETRIC EQUATIONS

In this section we shall solve a number of problems using parametric representations of functions and curves. This representation has many important applications. Even a function $y = f(x)$ given explicitly can be regarded as represented parametrically by

$$x = t, \qquad y = f(t).$$

Differentiation and graphic representation of functions given parametrically have been met already on several occasions.

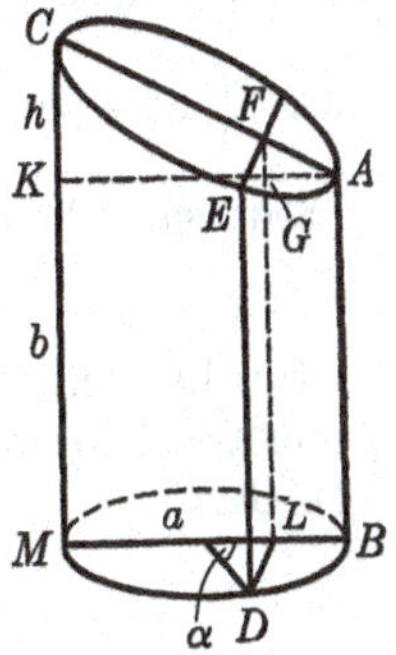

FIGURE 34

529. A truncated cylinder, covered with fresh paint, rolls on a plane. What is the equation of the curve which forms the boundary of the area painted by the cylinder?

Solution. We take α as a parameter (see Fig. 34). The value of α corresponding to the particular position when AB touches the plane is taken as zero. In addition let us suppose that the base of the cylinder will roll along the x axis, starting with B at the origin. We have to determine the coordinates of a point E on the oblique base of the cylinder when E is on the plane.

Let a be the radius of the cylinder. Then

$$x_E = a\alpha.$$

Now $y_E = DE = LF = MK + GF = b + \frac{h}{2a} \cdot AG = b + \frac{h}{2a} BL$

$$= b + \frac{h}{2a}(a - a\cos\alpha) = b + h\sin^2\frac{\alpha}{2}.$$

We can also obtain an explicit form of the equation of the curve by eliminating α from the two equations:

$$y = b + h\sin^2\frac{x}{2a}.$$

Determine the curves represented by the equations in the following exercises.

530. $x = t^2 - t + 1,\ y = t^2 + t + 1$

Solution. We shall eliminate the parameter:

$$y - x = 2t, \qquad t = \tfrac{1}{2}(y - x);$$
$$y = \tfrac{1}{4}(y - x)^2 + \tfrac{1}{2}(y - x) + 1$$

or

$$(y - x)^2 - 2(y + x) + 4 = 0.$$

By the methods of analytic geometry it can be seen that the curve is a parabola.

531. $x = t^2 - 2t + 3,\ y = t^2 - 2t + 1$

532. $x = a \sin^2 t,\ y = b \cos^2 t,\ a > 0,\ b > 0$

533. Find a parametric representation of the ellipse $\dfrac{x^2}{a^2} + \dfrac{y^2}{b^2} = 1$.

Solution. We may put

$$x = a \cos t, \qquad y = b \sin t.$$

It can be seen that these equations determine the whole ellipse. The geometrical meaning of t is clear from Figure 35. (It is *not* the angle $\sphericalangle MOx$.)

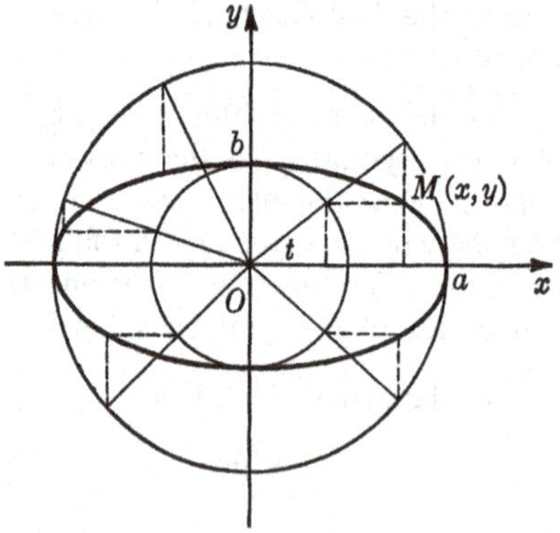

FIGURE 35

The same figure can also serve to demonstrate how to construct points of the curve by ruler and compass.

We remark that for a hyperbola $\dfrac{x^2}{a^2} - \dfrac{y^2}{b^2} = 1$ a parametrization can be obtained with $x = a \cosh t,\ y = b \sinh t$.

534. Represent parametrically the equation of the circle $x^2 + y^2 = a^2$ when the parameter is the slope t of the chord that goes through the point $(-a,0)$.

535. Find a parametric representation of $x^3 + y^3 = 3axy$.

536. Show that three different points of the curve

$$x = \frac{3at}{1+t^3}, \qquad y = \frac{3at^2}{1+t^3}$$

are on a straight line if and only if the corresponding values of the parameter, t_1, t_2 and t_3, satisfy $t_1t_2t_3 = -1$.

537. Give a parametric representation of the lemniscate

$$(x^2 + y^2)^2 = a^2(x^2 - y^2)$$

using rational functions.

538. Prove that the graph of $x^y = y^x$ consists of the straight line $y = x$ $(x > 0)$ and the curve

$$x = \left(1 + \frac{1}{t}\right)^t, \qquad y = \left(1 + \frac{1}{t}\right)^{t+1}.$$

6.9 TANGENT AND NORMAL

Let C be a curve in the plane and M a point on it. We define a straight line l to be a tangent to C at M if lines through M and P, where P is a neighboring point on C, tend to l when P tends to M along C.

If the curve is given by $y = f(x)$ and the derivative at $M(x_0,y_0)$ exists, then the tangent at M also exists and its equation is

$$y - y_0 = f'(x_0)(x - x_0).$$

It should be remarked that even if $f'(x_0)$ does not exist, a tangent may nevertheless exist at x_0.

The angle of intersection of two curves is defined as the angle between the tangents to the curves at the point of intersection.

A straight line perpendicular to the tangent at M is called the normal to the curve at M. If $f'(x_0) \neq 0$, the equation of the normal is

$$y - y_0 = -\frac{1}{f'(x_0)}(x - x_0).$$

The segment MA (see Fig. 36) is often called the length of the tangent, and is denoted by t.

The segment AN is called the subtangent and is denoted by sbt.

MB is called the length of the normal and is denoted by n.

NB is called the subnormal and is denoted by sbn.

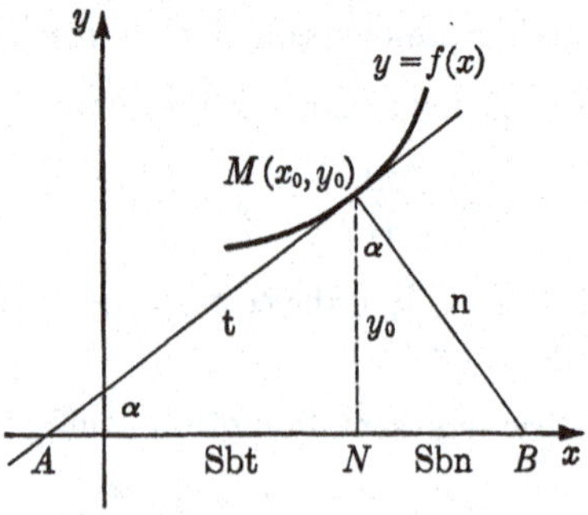

FIGURE 36

We find easily from the above figure:

$$\text{sbt} = \left|\frac{y_0}{\tan\alpha}\right| = \left|\frac{y_0}{f'(x_0)}\right|;$$

$$\text{t} = \sqrt{y_0^2 + (\text{sbt})^2} = \left|\frac{y_0}{f'(x_0)}\right|\sqrt{1 + [f'(x_0)]^2};$$

$$\text{sbn} = |y_0 f'(x_0)|;$$

$$\text{n} = \sqrt{y_0^2 + (\text{sbn})^2} = |y_0|\sqrt{1 + [f'(x_0)]^2}.$$

All these numbers can also be computed when the curve is given in implicit or parametric form.

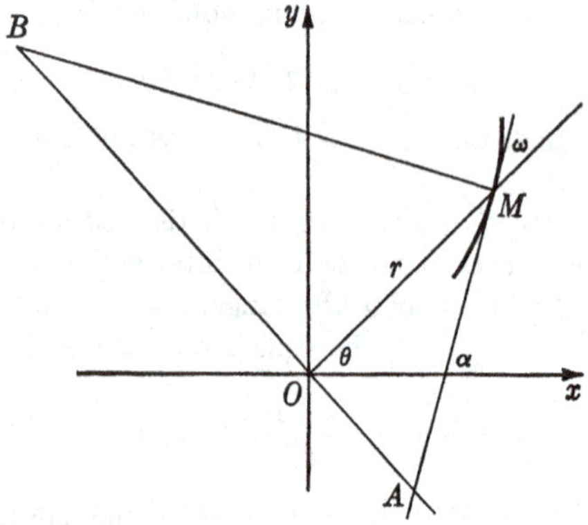

FIGURE 37

For curves given in polar coordinates, the above notions are often defined with respect to the radius vector of the point (cf. Fig. 37). We shall first find an expression for $\tan \omega$ where ω is the angle between the tangent at M and the radius vector of the point M. We have (for the situation as given in the figure)

$$\omega = \alpha - \theta, \qquad \tan \omega = \frac{\tan \alpha - \tan \theta}{1 + \tan \alpha \tan \theta}.$$

Now $\tan \alpha = dy/dx$ (at M). But $y = r \sin \theta$, $x = r \cos \theta$ and because r is a function of θ we can consider x and y as expressed parametrically in terms of θ. Then

$$\frac{dy}{dx} = \frac{dy/d\theta}{dx/d\theta} = \frac{r' \sin \theta + r \cos \theta}{r' \cos \theta - r \sin \theta},$$

where $r' = dr/d\theta$. We have

$$\tan \omega = \frac{\dfrac{r' \sin \theta + r \cos \theta}{r' \cos \theta - r \sin \theta} - \dfrac{\sin \theta}{\cos \theta}}{1 + \dfrac{r' \sin \theta + r \cos \theta}{r' \cos \theta - r \sin \theta} \cdot \dfrac{\sin \theta}{\cos \theta}}$$

$$= \frac{r' \sin \theta \cos \theta + r \cos^2 \theta - r' \sin \theta \cos \theta + r \sin^2 \theta}{r' \cos^2 \theta - r \sin \theta \cos \theta + r' \sin^2 \theta + r \sin \theta \cos \theta} = \frac{r}{r'}.$$

We now define:

Polar subtangent: $$OA \equiv \mathrm{sbt_p} = r|\tan \omega| = \frac{r^2}{|r'|};$$

Polar subnormal: $$OB \equiv \mathrm{sbn_p} = \frac{r}{|\tan \omega|} = |r'|;$$

$$MA \equiv \mathrm{t_p} = \frac{r}{|r'|} \sqrt{r^2 + (r')^2};$$

and $$MB \equiv \mathrm{n_p} = \sqrt{r^2 + (r')^2}.$$

All the above formulas, which were derived with the aid of Figure 37, hold for any position of the curve in the plane.

539. Discuss the properties of the subtangent and subnormal of the parabola $y^2 = 2px$. Find the equation of the tangent to the parabola at $M(x_0, y_0)$.

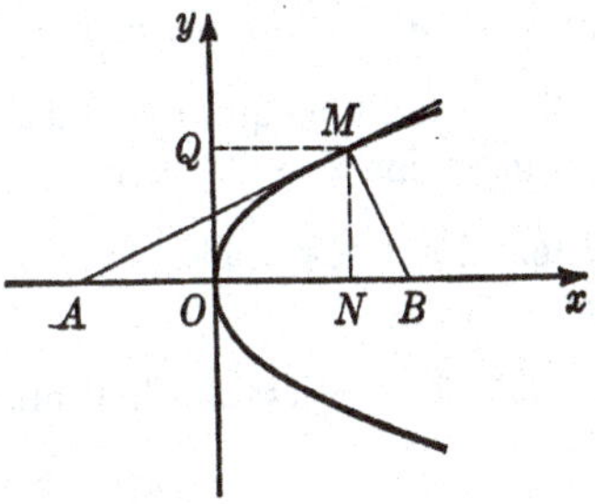

FIGURE 38

Solution. (See Fig. 38.) First we find the derivative:

$$2yy' = 2p, \qquad y' = p/y.$$

$\mathrm{sbn} = yy' = p$, i.e., the subnormal in the case of a parabola is constant and equal to the parameter p of the parabola.

$$\text{sbt} = \frac{y}{y'} = \frac{y}{p/y} = \frac{y^2}{p} = \frac{2px}{p} = 2x.$$

($x > 0$ for all points on the parabola except the vertex, for which sbt is clearly 0.)

We see that the tangent at (x_0,y_0) intersects the x axis in the point $(-x_0,0)$, i.e., the y axis bisects the segment AM. The equation of the tangent is given by

$$y - y_0 = \frac{p}{y_0}(x - x_0), \qquad yy_0 - y_0^2 = px - px_0,$$

$$yy_0 - 2px_0 = px - px_0, \qquad yy_0 = p(x + x_0).$$

The same holds also for (0,0) where the tangent is $x = 0$.

540. Find the equation of the tangent to the ellipse $\frac{x^2}{a^2} + \frac{y^2}{b^2} = 1$ at a point (x_0,y_0) on the curve.

541. Find the equation of the tangent and the length of the normal to the cycloid

$$x = a(t - \sin t), \qquad y = a(1 - \cos t)$$

at $t = t_0$.

542. Find the polar subnormal and the angle between tangent and radius vector of the spiral of Archimedes, $r = a\theta$.

543. The same question for the logarithmic spiral $r = ae^{b\theta}$.

544. Find the equations of the tangent and the normal to the curve

$$y = \frac{8a^3}{4a^2 + x^2},$$

at the point whose abscissa is $2a$.

545. Find the equation of a normal to the curve $y = x \ln x$ parallel to the straight line $2x - 2y + 3 = 0$.

546. Find the angle of intersection of the two curves $x^2 + y^2 = 8$ and $y^2 = 2x$.

Solution. First we shall find the points of intersection:

$$x^2 + 2x - 8 = 0; \qquad x_1 = 2,\ x_2 = -4.$$

The second solution is to be discarded because for the parabola $x \geqslant 0$ only. The points of intersection are $(2,\pm 2)$. By symmetry we can restrict ourselves to the angle of intersection at one point only, e.g., (2,2). We shall compute the slopes of the tangents to the curves at this point.

The parabola: $2yy' = 2;\qquad y_1'(2,2) = \frac{1}{2}.$
The circle: $2x + 2yy' = 0;\qquad y_2'(2,2) = -1.$
The (acute) angle between the curves is given by the formula

$$\tan\varphi = \left|\frac{y_1' - y_2'}{1 + y_1'y_2'}\right|;$$

$$\tan\varphi = \left|\frac{\frac{1}{2} + 1}{1 - \frac{1}{2}}\right| = 3; \qquad \varphi = \arctan 3.$$

547. The same question for $y = \sin x$ and $y = \cos x$ $(0 \leqslant x \leqslant \pi)$.

548. Prove that the segment of a tangent to $x^{2/3} + y^{2/3} = a^{2/3}$ between the axes of coordinates is of constant length a.

549. Prove the following construction of the tangent to the curve

$$y = \frac{a}{2}\left(e^{x/a} + e^{-x/a}\right) = a\cosh\frac{x}{a}.$$

On the ordinate MN of the point M (Fig. 39) we construct a semicircle and mark off the segment $NP = a$; the straight line MP is the tangent.

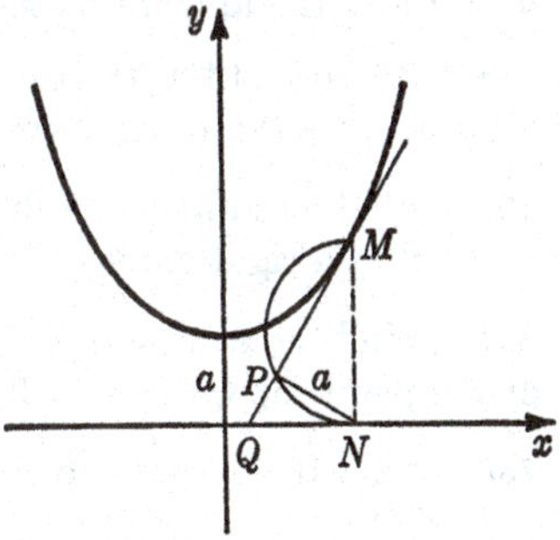

FIGURE 39

550. Find the equation of the tangent to the curve

$$x = t^3 + 1, \qquad y = t^2 + t + 1$$

at the point (1,1).

Solution. Let us find the corresponding value of t.

$$x = t^3 + 1 = 1, \qquad t^3 = 0, \qquad t = 0; \qquad y(t = 0) = 1.$$

$$\frac{dy}{dt} = 2t + 1, \qquad \frac{dy}{dt}(t = 0) = 1;$$

$$\frac{dx}{dt} = 3t^2, \qquad \frac{dx}{dt}(t = 0) = 0.$$

The derivative $\dfrac{dy}{dx} = \dfrac{dy/dt}{dx/dt}$ is not defined at $t = 0$. From the fact that $dx/dt = 0$ and $dy/dt \neq 0$, we conclude that the tangent is vertical and its equation is $x = 1$.

551. Find the angle of intersection of the curves

$$y = x^2 \quad \text{and} \quad \begin{cases} x = \frac{5}{3}\cos t \\ y = \frac{5}{4}\sin t. \end{cases}$$

552. The same question for the curves

$$\begin{cases} x = a\cos\varphi \\ y = a\sin\varphi \end{cases} \quad \text{and} \quad \begin{cases} x = \dfrac{at^2}{1+t^2} \\ y = \dfrac{at\sqrt{3}}{1+t^2}. \end{cases}$$

553. Show by computation that any normal to the curve

$$x = a(\cos t + t\sin t), \qquad y = a(\sin t - t\cos t)$$

(called involute of the circle) is a tangent to the circle $x^2 + y^2 = a^2$.

554. Given the curve $r = at^3$, $\theta = bt^2$; find the angle between the radius vector and the tangent to the curve at (r,θ).

555. Prove that the two cardioids $r_1 = a(1 + \cos\theta)$ and $r_2 = a(1 - \cos\theta)$ intersect at a right angle (we also say that the two curves are orthogonal).

556. Find a straight line through the origin that intersects the hyperbola $xy = a^2$ at a right angle.

557. Find the equation of the normal of the curve $x^3 + y^2 + 2x - 6 = 0$ at the point for which $y = 3$.

558. Find the tangent to the curve $x^3 - y^3 = 3x^2$ parallel to $y = x$.

559. Prove that the two curves $b^2x^2 + a^2y^2 = n$ and $y^{b^2} = mx^{a^2}$ are orthogonal.

560. Find the locus of the points from which it is possible to draw two orthogonal normals to a parabola.

561. Prove that the tangents to the curve $(3 + x^2)y = 1 + 3x^2$ at the points for which $y = 1$ intersect in the origin.

562. Find the angle of intersection of the parabolas

$$r = \frac{1}{\cos^2(\theta/2)}, \qquad r = \frac{1}{\sin^2(\theta/2)}.$$

563. Find the angle of intersection of $y = \tan x$ and $y = 2\cos x$ in the interval $(0,\pi/2)$.

564. Given the curve $y = 1/x^2$. Find on its right-hand branch (i.e., for $x > 0$) such a point that the tangent to the curve at this point will intersect the left-hand branch at a right angle.

565. Through $(2,3)$ draw two tangents to the ellipse $2x^2 + y^2 = 2$ and show that the straight line $y = \frac{3}{2}x$ goes through the middle of the segment joining the points of tangency.

566. Given the curve $x^2(4 - y) - y^3 = 0$. Find the equations of the tangent and normal to the curve at (2,2). Compute the length of the above tangent and normal. What is the angle between the given curve and the circle $x^2 + y^2 = 8$?

567. A circle of radius a rolls upon the x axis and a certain point on it describes the cycloid

$$x = a(t - \sin t) = \varphi(t), \qquad y = a(1 - \cos t) = \psi(t).$$

Show that the normal to the cycloid at the point $(\varphi(t),\psi(t))$ goes through the point where the circle touches the x axis.

6.10 THE ORDER OF CONTACT

Two curves determined by $y = f_1(x)$ and $y = f_2(x)$ are said to be in contact of order n at the point (x_0,y_0) if $f_1(x_0) = f_2(x_0)$, $f_1'(x_0) = f_2'(x_0)$, . . . , $f_1^{(n)}(x_0) = f_2^{(n)}(x_0)$, but $f_1^{(n+1)}(x_0) \neq f_2^{(n+1)}(x_0)$. If $f_1(x)$ and $f_2(x)$ are in contact of order n at (x_0,y_0), then

$$\lim_{\Delta x \to 0} \frac{f_1(x_0 + \Delta x) - f_2(x_0 + \Delta x)}{(\Delta x)^n} = 0,$$

i.e., the difference between the ordinates near the point (x_0,y_0) tends to zero faster than $(\Delta x)^n$. This statement can be verified by considering Taylor's expansion.

In this section we shall compute orders of contact and deal with some applications of this notion.

568. Show that the order of contact at $x = 1$ of $y_1 = \ln x$ and $y_2 = \dfrac{4x - 4}{x + 3}$ is equal to 1.

Solution:

$$y_1(1) = 0, \qquad y_2(1) = 0;$$

$$y_1' = \frac{1}{x}, \; y_1'(1) = 1; \qquad y_2' = 4\frac{x + 3 - x + 1}{(x + 3)^2} = \frac{16}{(x + 3)^2}, \; y_2'(1) = 1;$$

$$y_1'' = -\frac{1}{x^2}, \; y_1''(1) = -1; \quad y_2'' = -\frac{32}{(x + 3)^3}, \; y_2''(1) = -\frac{1}{2}.$$

The order of contact at $x = 1$ is indeed 1, since $y_1(1) = y_2(1)$, $y_1'(1) = y_2'(1)$, but $y_1''(1) \neq y_2''(1)$.

569. Find a cubic curve of the form $y_1 = x^3 + ax + b$ that will have contact of order 1 with $y_2 = e^x$ at the point $x = 0$.

Solution:

$$y_1(0) = b, \qquad y_2(0) = 1.$$

We require $b = 1$. Now

$$y_1' = 3x^2 + a,\ y_1'(0) = a; \qquad y_2' = e^x,\ y_2'(0) = 1.$$

To obtain order of contact 1 we require $a = 1$. Further,

$$y_1'' = 6x,\ y_1''(0) = 0; \qquad y_2'' = e^x,\ y_2''(0) = 1;$$

$y_1''(0) \neq y_2''(0)$, i.e., the order of contact is 1 when the equation is $y = x^3 + x + 1$.

570. Find a function $y_1 = A \sin(ax + b)$, $a > 0$, that has order of contact 2 with the parabola $y_2 = \sqrt{x}$ at $x = 1$.

571. Prove that the order of contact of the circle

$$x^2 + y^2 - 18x - 18y + 90 = 0$$

and the parabola $\sqrt{x} + \sqrt{y} = 2\sqrt{3}$ at $x = 3$ equals 3.

572. Prove that the cycloid $x = a(t - \sin t)$, $y = a(1 - \cos t)$ and the ellipse $3(x - \pi a)^2 + 4(y + a)^2 = 36a^2$ are in contact of order 4 at least, at the point $x = \pi a$ $(a > 0)$.

573. Find a parabola of degree 4 ($y = f(x)$, where $f(x)$ is a quartic polynomial) that has order of contact 4 with the curve $y_1 = a \cosh(x/a)$ at $x = 0$.

6.11 OSCULATING CIRCLE; RADIUS OF CURVATURE

DEFINITION. A circle is said to be an osculating circle or a circle of curvature of a given curve at a point P if its order of contact with the curve at P is at least 2.

There is at most one osculating circle to a given curve at a given point P. Its radius R and center are called the radius and center of curvature of the curve at P. $1/R$ is called the curvature at P.

Let the curve be given by an explicit equation $y = f(x)$ and let $P(x,y)$ be a point on it. Assume that the osculating circle is given by the equation

$$(x - a)^2 + (y - b)^2 = R^2.$$

We differentiate twice (and divide by 2):

$$x - a + (y - b)y' = 0,$$

$$1 + (y')^2 + (y - b)y'' = 0.$$

The last three equations may be considered as three equations involving the unknowns a, b, R and the known numbers x, y, y', y'', (x,y) being the coordinates of P, and y' and y'' the values of the first and second derivatives of y computed at P.

Let us compute a, b, and R from the above equations. Should an osculating circle exist at P, then from the third equation it follows that $y'' \neq 0$ (since $1 + (y')^2 > 0$), and so

$$b = y + \frac{1 + (y')^2}{y''}.$$

After substitution in the second equation we obtain

$$a = x - y'\,\frac{1 + (y')^2}{y''}.$$

Substitution of the two results into the first equation gives

$$R^2 = \left[\frac{1 + (y')^2}{y''}\,y'\right]^2 + \left[\frac{1 + (y')^2}{y''}\right]^2 = \frac{[1 + (y')^2]^3}{(y'')^2};$$

$$R = \left|\frac{[1 + (y')^2]^{3/2}}{y''}\right|.$$

For a curve given by parametric equations $x = f(t)$, $y = g(t)$, we obtain by substitution

$$\frac{dy}{dx} = \frac{g'(t)}{f'(t)}, \qquad \frac{d^2y}{dx^2} = \frac{g''(t)f'(t) - g'(t)f''(t)}{[f'(t)]^3},$$

$$R = \left|\frac{\{[f'(t)]^2 + [g'(t)]^2\}^{3/2}}{g''(t)f'(t) - g'(t)f''(t)}\right|.$$

For a curve given in polar coordinates, $r = r(\theta)$, we obtain

$$x = r\cos\theta,\ y = r\sin\theta; \qquad \frac{dy}{dx} = \frac{\dfrac{dy}{d\theta}}{\dfrac{dx}{d\theta}} = \frac{r'\sin\theta + r\cos\theta}{r'\cos\theta - r\sin\theta},$$

$$\frac{d^2y}{dx^2} = \frac{\dfrac{d}{d\theta}\left(\dfrac{dy}{dx}\right)}{\dfrac{dx}{d\theta}}$$

$$= \frac{1}{(r'\cos\theta - r\sin\theta)^3}\,[(r''\sin\theta + 2r'\cos\theta - r\sin\theta)(r'\cos\theta - r\sin\theta) - (r'\sin\theta + r\cos\theta)(r''\cos\theta - 2r'\sin\theta - r\cos\theta)]$$

$$= \frac{r^2 + 2(r')^2 - rr''}{(r'\cos\theta - r\sin\theta)^3}.$$

(Here $r' = dr/d\theta$, $r'' = d^2r/d\theta^2$.) Now

$$R = \left| \frac{\left[1 + \left(\dfrac{r' \sin\theta + r\cos\theta}{r'\cos\theta - r\sin\theta}\right)^2\right]^{3/2}}{\dfrac{r^2 + 2(r')^2 - rr''}{(r'\cos\theta - r\sin\theta)^3}} \right| = \left| \frac{[r^2 + (r')^2]^{3/2}}{r^2 + 2(r')^2 - rr''} \right|.$$

574. Find the osculating circle of the parabola $y = x^2$ at the point (1,1).

Solution. We shall solve this problem without using the above formulas.

$$x = 1, \qquad y = 1; \qquad y'(1) = 2, \qquad y''(2) = 2.$$

Let the equation of the circle be

$$x^2 + y^2 + dx + ey + f = 0;$$

then

$$2x + 2yy' + d + ey' = 0,$$
$$2 + 2(y')^2 + 2yy'' + ey'' = 0.$$

After substitution we obtain

$$d + e + f = -2, \qquad d + 2e = -6, \qquad 2e = -14;$$
$$e = -7, \qquad d = 8, \qquad f = -3.$$

The circle therefore is $x^2 + y^2 + 8x - 7y - 3 = 0$,

or

$$(x + 4)^2 + (y - \tfrac{7}{2})^2 = \tfrac{125}{4}.$$

575. The same question for $y = e^x$ at (0,1).

576. The same question for the cissoid $(x^2 + y^2)x - 2ey^2 = 0$ at the point (e,e).

In the following problems find the radius of curvature of the given curves at the given points.

577. $xy = 4$ at (2,2).

578. $y = \ln x$ at (1,0).

579. $\dfrac{x^2}{a^2} - \dfrac{y^2}{b^2} = 1$ at (x,y).

580. $x^{2/3} + y^{2/3} = a^{2/3}$ at (x,y).

581. $x = 3t^2$, $y = 3t - t^3$ for $t = 1$.

582. $x = 2a\cos t - a\cos 2t$, $y = 2a\sin t - a\sin 2t$ for any t.

583. $r = a^\theta$ at $(r = 1, \theta = 0)$.

584. $r = a\theta^k$ at (r,θ), $k > 0$.

585. Prove that the radius of curvature of the lemmiscate $r^2 = a^2 \cos 2\theta$ is proportional to the reciprocal of the radius vector of the corresponding point on the curve.

586. Find the point where the radius of curvature of the curve $y = \ln x$ attains an extremal value (such a point is called a vertex of the curve).

587. Prove that the curvature of the curve $y = f(x)$ at $P(x,y)$ equals $|y'' \cos^3 \alpha|$, where α is the angle between the tangent to the curve at P and the x axis.

588. The function $y = f(x)$ is defined as follows:

$$f(x) = x^3, \qquad -\infty < x \leqslant 1;$$
$$f(x) = ax^2 + bx + c, \qquad 1 < x < \infty.$$

Determine a, b, and c such that the function and its curvature will be continuous for any x.

589. Given a circular arc AM of radius 5 and center at (0,5), and a segment BC joining $B(1,3)$ and $C(11,66)$; join the point M (see Fig. 40) with B by a parabola of fifth degree, so that the curve $AMBC$ will have continuous curvature everywhere.

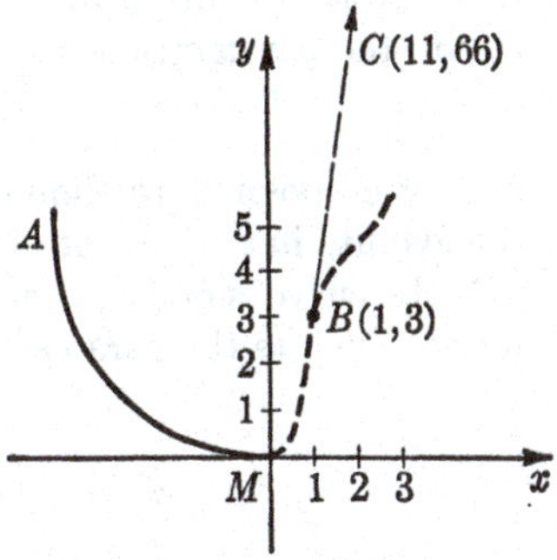

FIGURE 40

590. Find the point at which the radius of curvature of $y = 1 - 2\sqrt{x} + x$ attains an extremal value.

591. Prove that the radius of curvature of the cycloid $x = a(t - \sin t)$, $y = a(1 - \cos t)$ at every point equals twice the length of the normal to the curve at this point.

592. Find the center of curvature of the parabola $y^2 = ax$ at (0,0).

593. Find the equation of the osculating circle of $y = \sin x$ at $(\pi/2,1)$.

594. APB is a straight line tangent to the ellipse $\dfrac{x^2}{a^2} + \dfrac{y^2}{b^2} = 1$ at P. A and B are the points of intersection of the tangent with the axes. P is the middle of the segment AB. Find the radius of curvature of the ellipse at P.

595. Find the radius of curvature of the curve $y = x[\sqrt[3]{(1 + x)^4} - 1]$ at $x = 0$.

6.12 EVOLUTE AND INVOLUTE

The locus E of the centers of curvature of a given curve C is called the evolute of the curve C, and C is called the involute of E. The following theorem holds: A tangent to the evolute is normal to the involute.

To find the equation of the evolute we use the expressions for the coordinates of the center of curvature of the curve, obtained in the previous section·

$$a = x - y'\frac{1 + (y')^2}{y''}, \qquad b = y + \frac{1 + (y')^2}{y''}.$$

y, y', and y'' are functions of x and we can consider these expressions as parametric equations of the evolute (x is the parameter). It is convenient to substitute x and y for a and b respectively and t for x, and we then obtain the parametric equations of the evolute in the usual form:

$$x = x(t), \qquad y = y(t).$$

It is often possible to eliminate the parameter and obtain the equation of the evolute in an implicit or even in an explicit form.

If the curve is given in the parametric form $y = f(t)$, $x = g(t)$, we obtain for the evolute the parametric equations

$$\begin{aligned} a(=x) &= g(t) - \frac{f'(t)}{g'(t)} \cdot \frac{1 + \left[\dfrac{f'(t)}{g'(t)}\right]^2}{\dfrac{f''(t)g'(t) - f'(t)g''(t)}{[g'(t)]^3}} \\ &= g - f'\frac{(g')^2 + (f')^2}{f''g' - f'g''}; \\ b(=y) &= f + g'\frac{(g')^2 + (f')^2}{f''g' - f'g''}. \end{aligned}$$

When the curve is given in polar coordinates, $r = r(\theta)$, we have

$$\frac{1 + (y')^2}{y''} = \frac{1 + \left(\dfrac{r'\sin\theta + r\cos\theta}{r'\cos\theta - r\sin\theta}\right)^2}{\dfrac{r^2 + 2(r')^2 - rr''}{(r'\cos\theta - r\sin\theta)^3}} = \frac{r'\cos\theta - r\sin\theta}{r^2 + 2(r')^2 - rr''}[r^2 + (r')^2].$$

The cartesian parametric equations of the evolute (with θ as parameter) are in this case:

$$a(=x) = r\cos\theta - \frac{r'\sin\theta + r\cos\theta}{r^2 + 2(r')^2 - rr''}[r^2 + (r')^2],$$

$$b(=y) = r\sin\theta + \frac{r'\cos\theta - r\sin\theta}{r^2 + 2(r')^2 - rr''}[r^2 + (r')^2].$$

To find the equation of the involute requires integral calculus, and will be postponed.

596. Find the evolute of the curve $y = x^n$, $n \geqslant 2$.

Solution:

$$y' = nx^{n-1}, \qquad y'' = n(n-1)x^{n-2}.$$

We shall write down the parametric equations of the evolute. Let us change a and b to x and y, and x to t. We obtain

$$x = t - nt^{n-1}\frac{1 + n^2t^{2n-2}}{n(n-1)t^{n-2}} = t\left(1 - \frac{1 + n^2t^{2n-2}}{n-1}\right),$$

$$y = t^n + \frac{1 + n^2t^{2n-2}}{n(n-1)t^{n-2}}.$$

597. The same question for the astroid $x^{2/3} + y^{2/3} = c^{2/3}$.

598. Find the evolute of the curve $x = 3t$, $y = t^2 - 6$.

599. Find the evolute of the tractrix

$$x = -a\left(\ln\tan\frac{t}{2} + \cos t\right), \qquad y = a\sin t.$$

600. Prove that the curve

$$x = a(\cos t + t\sin t), \qquad y = a(\sin t - t\cos t)$$

is an involute of the circle $x^2 + y^2 = a^2$.

Solution. We shall show that the circle is an evolute of the given curve and this will be a proof of the statement of the problem.

$$x' = a(-\sin t + t\cos t + \sin t) = at\cos t, \qquad x'' = a\cos t - at\sin t,$$
$$y' = a(\cos t - \cos t + t\sin t) = at\sin t, \qquad y'' = a\sin t + at\cos t;$$
$$(x')^2 + (y')^2 = a^2t^2.$$
$$x'y'' - x''y' = a^2(t\sin t\cos t + t^2\cos^2 t - t\sin t\cos t + t^2\sin^2 t) = a^2t^2.$$

The parametric equations of the evolute are

$$x = a(\cos t + t\sin t) - at\sin t\,\frac{a^2t^2}{a^2t^2} = a\cos t,$$

$$y = a(\sin t - t\cos t) + at\cos t\,\frac{a^2t^2}{a^2t^2} = a\sin t,$$

and from this, $x^2 + y^2 = a^2$.

601. Show that the evolute of a logarithmic spiral $r = a^\theta$ is also such a spiral.

602. Find the evolute of the cycloid $x = t - \sin t$, $y = 1 - \cos t$.

6.13 SOLUTION OF EQUATIONS BY NEWTON'S APPROXIMATION METHOD

In this section we shall find approximate solutions of equations of the form $f(x) = 0$ by Newton's method. In Figure 41 we draw four curves $y = f(x)$

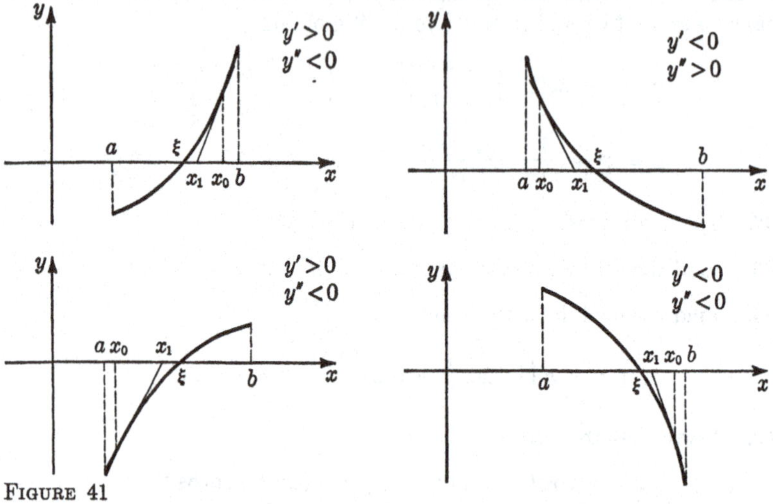

FIGURE 41

intersecting the x axis at points denoted by ξ. At points denoted by x_0 we draw tangents to the curves. The equations of the tangents are $y - f(x_0) = f'(x_0)(x - x_0)$. The tangents intersect the x axis at

$$x_1 = x_0 - \frac{f(x_0)}{f'(x_0)}.$$

In all cases illustrated in the figure, x_1 is nearer to ξ than x_0. We can repeat this procedure, i.e., draw a tangent at x_1. It will intersect the x axis at

$$x_2 = x_1 - \frac{f(x_1)}{f'(x_1)},$$

and this point will be nearer to ξ than x_1, and so on. In this way we can approximate ξ as closely as we like, the only difficulty being the cumbersome computations.

Now we shall introduce the conditions that ensure the success of the method and show how to begin and how to conclude the computations. The conditions become clear on inspection of the figure, assuming a function $y = f(x)$ which is twice differentiable.

(1) First we find an interval of isolation of a root. By this we mean an interval (a,b) fulfilling the following conditions:

(i) $f(a)\cdot f(b) < 0$, i.e., the values of the function have different signs at the ends of the interval.

(ii) $f'(x) \neq 0$, $a < x < b$, i.e., the curve is monotonic in the whole interval, and thus intersects the x axis once between a and b.

(iii) $f''(x) \neq 0$, $a < x < b$, i.e., the curve does not change the direction of its concavity in the interval.

The determination of an interval of isolation can be made by a trial and error process, or by graphical methods which give a rough initial approximation.

It may happen that $f'(\xi) = 0$ or $f''(\xi) = 0$, i.e., the first or second derivative vanishes exactly at the point of intersection ξ of the curve with the x axis. In this case of course it is not possible to find an interval of isolation for ξ fulfilling the above conditions, but it is not needed here, since ξ is more easily located as a root of the highest common factor of $f(x)$ and $f'(x)$ or $f''(x)$.

(2) We begin the computations from an arbitrary point x_0 of (a,b) with the one condition that $f(x_0)f''(x_0) > 0$, i.e., the sign of $f(x_0)$ must be equal to the sign of $f''(x)$ in the interval of isolation. We can arrange the computation in the tabular form shown.

x	$f(x)$	$f'(x)$	$-\dfrac{f(x)}{f'(x)}$
x_0	$f(x_0)$	$f'(x_0)$	$-\dfrac{f(x_0)}{f'(x_0)}$
x_1	$f(x_1)$	$f'(x_1)$	$-\dfrac{f(x_1)}{f'(x_1)}$
x_2	...	...	...

The entry in the last column in any line of the table is added to the entry in the first column in the same line to obtain the entry in the first column in the next line, that is,

$$x_{i+1} = x_i + \left(-\frac{f(x_i)}{f'(x_i)}\right), \qquad i = 0, 1, 2, \ldots.$$

In the computation, when the value of $-f(x_0)/f'(x_0)$ is rounded off, this must be done in the direction which decreases the absolute value of this number in order to avoid passing to the other side of ξ.

(3) The computation will be finished when the required accuracy is

obtained. If we require an accuracy to 10^{-n}, then such an x_i must be found that

$$|x_i - \xi| < \tfrac{1}{2}\cdot 10^{-n}.$$

Suppose we have obtained in the last column a correction with an absolute value of less than $\tfrac{1}{2}\cdot 10^{-n}$, and we have found the corresponding x_i. To check if it is satisfactory, we compute $f(x_i + \tfrac{1}{2}\cdot 10^{-n})$ or $f(x_i - \tfrac{1}{2}\cdot 10^{-n})$, where the sign must be chosen as in the previous correction. If this number has a sign different from $f(x_i)$ (or $f(x_0)$), we have ended the computation.

This will seldom fail to be the case. However, if it happens, we begin the computations once again from a new x_0' nearer to ξ than the last x_i found, and fulfilling of course $f(x_0')f''(x_0') > 0$.

We shall apply the above method to some examples. It should be noted that an equation may have several real solutions, and that if nothing is said to the contrary, we are to find all of them.

Solve the following equations by Newton's approximation method. The accuracy required is given in every case.

603. $x^3 + x + 1 = 0,\ 10^{-2}$

Solution. First we shall find an interval of isolation. We put

$$f(x) = x^3 + x + 1.$$

Now $$f(-1) = -1 - 1 + 1 = -1, \qquad f(0) = 1.$$

$f'(x) = 3x^2 + 1 > 0$ for any x, that is, the function increases everywhere, and there is only one real solution which is clearly in the interval $(-1,0)$.

$f''(x) = 6x < 0$ for $-1 < x < 0$. We see that $(-1,0)$ can serve as an interval of isolation.

Now let us arrange the table. We begin with $x_0 = -1$ because $f(-1) = -1$ is negative as is $f''(x)$ in our interval.

x	$f(x)$	$f'(x)$	$-\dfrac{f(x)}{f'(x)}$
-1	-1	4	0.25
-0.75	-0.172	2.69	0.064
-0.686	-0.009	2.41	0.0037
-0.682	0.001		

By rounding off -0.6823 to -0.682, we pass ξ. The solution with the required accuracy is clearly -0.68.

604. $2x - \cos x = 0,\ 10^{-3}$

Solution. We draw the graphs of $y_1 = \cos x$ and $y_2 = 2x$ (Fig. 42). The required solution is clearly between 0 and 1. Let us now consider the function

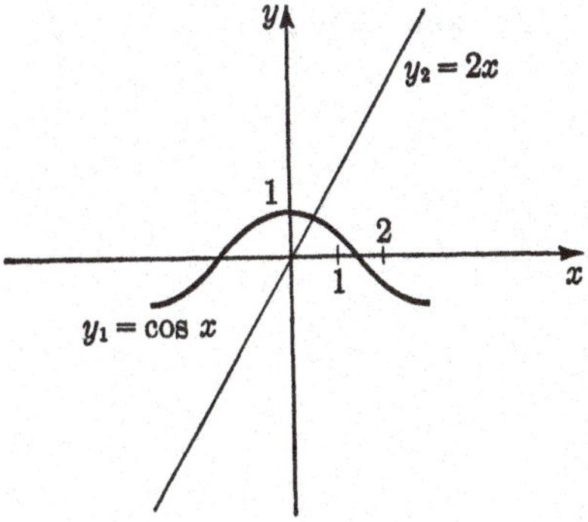

FIGURE 42

$$f(x) = 2x - \cos x, \quad f(0) = -1, \quad f(\tfrac{1}{2}) = 1 - \cos \tfrac{1}{2} > 0,$$
$$f'(x) = 2 + \sin x > 0 \quad \text{in the interval } (0, \tfrac{1}{2}),$$
$$f''(x) = \cos x > 0 \quad \text{in this interval.}$$

We begin with $x_0 = 0.5$ (of course, trigonometric tables may be used to find the sines and cosines), and compute the values as given in the table.

x	$f(x)$	$f'(x)$	$-\dfrac{f(x)}{f'(x)}$
0.5	0.1224	2.479	−0.049
0.451	0.002	2.436	−0.0008
0.4502			

Now $f(0.450) = -0.00045$, i.e., the solution with accuracy to 0.001 is 0.450.

605. $x^3 - 1.8x^2 - 10x + 17 = 0$. Find the solution near 1.6 with accuracy to 10^{-2}.

606. $x^4 + 96x = 80$. Find the positive solution with accuracy to 10^{-2}.

607. $x - \sin x = 2,\ 10^{-4}$

608. A sphere of radius 2 is made of material with specific gravity $\frac{2}{3}$. How deep will the sphere sink in water? Compute with accuracy to 10^{-2}.

609. $4x^3 - 11x^2 + 10x - 4 = 0$. Find a solution near 1.5 with accuracy to 10^{-3}.

610. $10^x = 20x$, 10^{-4}.

611. Find with accuracy to 10^{-2} the value of x where the function $y = \frac{1}{3}x^3 - 3 \ln x - 5x$ attains a minimum.

VII

THE DIFFERENTIAL

7.1 DEFINITION OF THE DIFFERENTIAL

THEOREM. If $y = f(x)$ has a derivative at x_0, the increment of the function at this point can be represented in the form

$$\Delta y = \Delta y(x_0,\Delta x) = f'(x_0)\Delta x + \epsilon\Delta x$$

where ϵ depends on Δx and tends to 0 when $\Delta x \to 0$ (i.e., $\lim_{\Delta x \to 0} \epsilon = 0$). A graphical illustration of this equality is given below (Fig. 43).

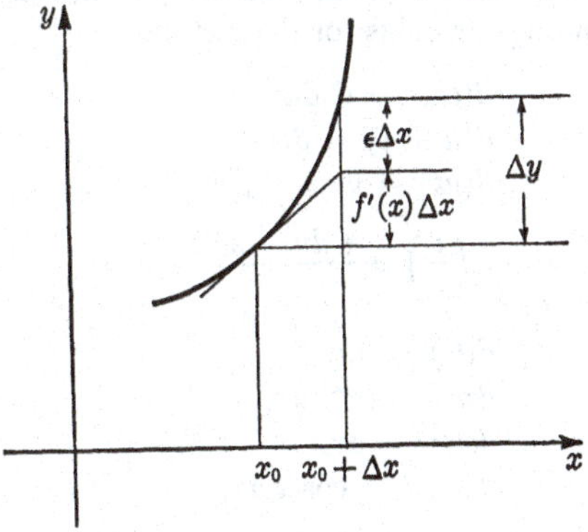

FIGURE 43

On the other hand, if the increment Δy can be represented in the form

$$\Delta y = \Delta y(x_0,\Delta x) = a\Delta x + \epsilon\Delta x$$

where a is a constant and ϵ tends to zero together with Δx, we see, after dividing by Δx and letting Δx tend to zero, that the function has a derivative at x_0 and $a = f'(x_0)$.

The differential of $y = f(x)$ is a function dy of two variables x and dx defined for *any* number dx and any x in the domain of the derived function f' by

$$dy(x,dx) = f'(x)dx.$$

It is convenient (and customary) to allow some ambiguity and also denote the number $f'(x)dx$ by dy.

The discussion above shows that if $dx = \Delta x$ then dy is a good approximation to Δy in the sense that

$$\lim_{\Delta x \to 0} \frac{\Delta y - dy}{\Delta x} = 0.$$

It must be understood that both $\Delta y = f(x + \Delta x) - f(x)$ and $dy = f'(x)dx$ are functions of two variables, x and Δx for Δy, x and dx for dy. If $dx = \Delta x$ and x is in the domain of f' then usually $\Delta y \neq dy$ even though $\Delta y/\Delta x$ and $dy/\Delta x$ have the same limit $f'(x)$ as Δx tends to 0.

The concept of the differential has many practical applications, which will be dealt with in the next sections.

We remark in passing that we have now obtained a new interpretation of the derivative, namely, that it is the quotient of dy by dx: $f'(x) = dy/dx$. The expression dy/dx is thus not only a symbol for the derivative but it represents a fraction, the numerator and denominator of which can be treated separately.

To compute differentials we use the following formulas, obtained directly from the corresponding formulas for derivatives:

$$d(Cu) = C\,du,$$
$$d(u \pm v) = du \pm dv,$$
$$d(uv) = u\,dv + v\,du,$$
$$d\left(\frac{u}{v}\right) = \frac{v\,du - u\,dv}{v^2},$$
$$d(C) = 0,$$
$$dx^\alpha = \alpha x^{\alpha-1}\,dx,$$
$$de^x = e^x\,dx,$$
$$d \sin x = \cos x\,dx,$$
$$d \arcsin x = \frac{1}{\sqrt{1 - x^2}}\,dx,$$

and so on.

612. Given the function $y = x^3 + 2x$; find Δy and dy at $x = 2$ and $\Delta x = 1$, $\Delta x = 0.1$, $\Delta x = 0.01$.

Solution:

$$\Delta y = f(x + \Delta x) - f(x), \qquad dy = f'(x)\,dx = (3x^2 + 2)\,dx.$$

For $\Delta x = 1$;

$$\begin{aligned}\Delta y &= f(3) - f(2)\\ &= 3^3 + 2\cdot 3 - (2^3 + 2\cdot 2) = 27 + 6 - 12 = 21,\\ dy &= (3\cdot 2^2 + 2)\,dx = 14\cdot 1 = 14\ (dx = \Delta x).\end{aligned}$$

For $\Delta x = 0.1$,

$$\begin{aligned}\Delta y &= f(2.1) - f(2)\\ &= 2.1^3 + 2\cdot 2.2 - 12 = 9.261 + 4.2 - 12 = 1.461,\\ dy &= 14\cdot 0.1 = 1.4.\end{aligned}$$

For $\Delta x = 0.01$,

$$\begin{aligned}\Delta y &= f(2.01) - f(2)\\ &= 2.01^3 + 2\cdot 2.01 - 12\\ &= 8.120601 + 4.02 - 12 = 0.140601,\\ dy &= 14\cdot 0.01 = 0.14.\end{aligned}$$

It can easily be seen that $\dfrac{\Delta y - dy}{\Delta x}$ tends to 0 about as fast as Δx.

613. Given the function $y = \sin x$; find Δy and dy at $x = 1.1$ and $\Delta x = -0.1$, $\Delta x = -0.01$.

Find the differentials of the following functions.

614. $y = 5/\sqrt{x}$.

Solution:

$$dy = f'(x)\,dx = -\frac{5}{2}\cdot\frac{1}{x\sqrt{x}}\,dx.$$

615. $y = (5 - 2x + x^5)^4$

616. $y = \sqrt{\arcsin x} + (\arctan x)^2$

617. $y = 5^{-1/x^2} + \dfrac{2}{x^2} - 5x^2$

618. Given $y = \dfrac{1 + \ln x}{x - x\ln x}$, prove that $2x^2\,dy = (x^2y^2 + 1)\,dx$.

619. Show that the function $f(x) = 1 + \sqrt[3]{(x-1)^2}$ is not differentiable at $x = 1$.

Solution. The function $f(x)$ has no derivative at $x = 1$ and by the above theorems it is not differentiable there. Let us show this directly:

$$\Delta y = f(1 + \Delta x) - f(1) = 1 + \sqrt[3]{(\Delta x)^2} - 1 = \sqrt[3]{(\Delta x)^2}.$$

We would like to see whether a can be found such that

$$\lim_{\Delta x \to 0} \frac{\Delta y - a\Delta x}{\Delta x} = 0.$$

But

$$\frac{\Delta y - a\Delta x}{\Delta x} = \frac{1}{(\Delta x)^{1/3}} - a,$$

and

$$\lim_{\Delta x \to 0} \left[\frac{1}{(\Delta x)^{1/3}} - a\right] = \begin{cases} \infty \\ \text{or} \\ -\infty \end{cases}$$

for any constant a; consequently the function $f(x)$ is not differentiable at $x = 1$.

620. Show that $y = |x|$ is not differentiable at $x = 0$.

7.2 THE INVARIANCE OF THE FORM OF THE DIFFERENTIAL

Let y be a composite function of x, e.g., $y = f(u)$, $u = g(x)$. By definition $dy = (dy/dx)\,dx$, but

$$\frac{dy}{dx} = \frac{dy}{du}\frac{du}{dx}, \quad \text{i.e.,} \quad dy = \frac{dy}{du}\frac{du}{dx}\,dx.$$

Now $(du/dx)\,dx = du$ by definition of the differential, thus finally

$$dy = \frac{dy}{du}\,du.$$

The form of the differential remains invariant also when the argument itself is a function of another independent variable.

At the same time it must be understood that for an independent variable x, dx is a free quantity, that is, it can take any value, and in case u depends on x, du is itself a function of dx.

621. Given $y = \sqrt[3]{x^2 + 5x}$, $x = t^3 + 2t + 1$; find dy.

Solution:

$$dy = \frac{dy}{dx}\frac{dx}{dt}\,dt = \frac{1}{3}\frac{2x + 5}{(x^2 + 5x)^{2/3}}(3t^2 + 2)\,dt$$

$$= \frac{(2t^3 + 4t + 7)(3t^2 + 2)}{3\sqrt[3]{[(t^3 + 2t + 1)^2 + 5(t^3 + 2t + 1)]^2}}\,dt.$$

622. $S = \cos^2 z$, $z = \frac{1}{4}(t^2 - 1)$; find dS.

623. $y = e^u$, $u = \frac{1}{2}\ln v$, $v = 2x^2 - 3x + 1$; find dy.

7.3 THE DIFFERENTIAL AS THE PRINCIPAL PART OF THE INCREMENT OF THE FUNCTION; APPLICATIONS TO APPROXIMATE CALCULATIONS

We have $\Delta y = dy + \epsilon\Delta x$, where ϵ tends to zero together with Δx. It follows that for sufficiently small Δx we can write $\Delta y \approx dy$. At the same time it must be emphasized that there is no estimate of the error obtained by writing dy instead of Δy.

A comparison with Taylor's formula shows that dy is the first term of the expansion (after $f(a)$, of course). The advantage of Taylor's formula is that it contains an expression which can be used to estimate the error. Nevertheless, in practical computations the use of the differential is very convenient, especially for finding absolute and relative errors of quantities computed from approximate data.

The absolute error of an approximate value x_1 of an exact number x is $|x_1 - x|$. The absolute error can in general be only estimated, and depends on the conditions of the measurement.

The absolute error $|\Delta y|$ of a computed quantity y, the computation of which is based on the result of measurement of x, can be approximated by $|dy|$.

The relative error equals the quotient $|\Delta x/x|$, i.e., the ratio of the absolute error and the quantity itself. In the same way $|\Delta y/y| \approx |dy/y|$ is the relative error of the computed quantity y. The relative error gives, in general, a better idea of the exactness of the measurements than the absolute error.

624. Find the increment and differential of the function $y = x^2 - x$ at $x = 10$ if $\Delta x = 0.1$. Find the absolute and relative errors when the exact increment is replaced by the differential.

Solution:

$$\begin{aligned} \Delta y &= (10.1)^2 - 10.1 - 10^2 + 10 = 102.01 - 10.1 - 100 + 10 \\ &= 1.91. \\ dy &= (2x - 1)\,dx = (20 - 1)\cdot 0.1 \\ &= 1.9. \end{aligned}$$

The absolute error is $|\Delta y - dy| = 0.01$.

The relative error is $\dfrac{|\Delta y - dy|}{|\Delta y|} = \dfrac{0.01}{1.91} = 0.0052.$

625. Compute arctan 1.02 approximately.

Solution:

$$\arctan 1.02 = \arctan 1 + \Delta y = \frac{\pi}{4} + \Delta y.$$

$$y = \arctan x; \quad \Delta y \approx dy = \frac{1}{1 + x^2}\, dx.$$

In our case $\quad dx = \Delta x = 0.02, \quad x = 1;$

$$dy = \frac{1}{1 + x^2}\, dx = \frac{1}{1 + 1} \cdot 0.02 = 0.01.$$

Now we obtain

$$\arctan 1.02 \approx \frac{\pi}{4} + 0.01 = 0.785 + 0.01 = 0.795.$$

626. Compute approximately $\sin 60°3'$.

627. Compute approximately $e^{0.1}$.

628. Find approximately formulas for $\sqrt{1 + x}$, e^x, $\ln(1 + x)$, $\sin x$ (for small $|x|$).

Solution. We shall use the expression

$$f(u_0 + \Delta u) = f(u_0) + \Delta y \approx f(u_0) + dy = f(u_0) + f'(u_0)\, du,$$
$$y = f(u).$$

In the case of $\sqrt{1 + x}$ we have $f(u) = \sqrt{u}$. $u_0 = 1$; $\Delta u = x$. Consequently

$$\sqrt{1 + x} \approx \sqrt{1} + \frac{1}{2\sqrt{1}}\, x = 1 + \frac{x}{2}.$$

In the same manner,

$$e^x \approx e^0 + e^0 \cdot x = 1 + x;$$

$$\ln(1 + x) \approx \ln 1 + \frac{1}{1}\, x = x;$$

$$\sin x \approx \sin 0 + \cos 0 \cdot x = x.$$

629. The measurement of the radius r of a circle involved an absolute error α. What is the absolute and relative error in the calculation of the area of the circle?

Solution. $S = \pi r^2$;

$$|\Delta S| \approx |dS| = |2\pi r\, dr| = 2\pi r\alpha.$$

This is the absolute error. The relative error will be found by dividing by S:

$$\left|\frac{dS}{S}\right| = \frac{2\pi r\alpha}{\pi r^2} = 2\,\frac{\alpha}{r}.$$

The relative error in the determination of S is twice the relative error in the measurement of the radius.

630. Find the relative error in the computation of the volume of a sphere the diameter of which was measured with a relative error α.

631. In a Wheatstone bridge (see Fig. 44) the unknown resistance R_x is computed by the formula $R_x = Rx/(a - x)$. Find the point on the ruler

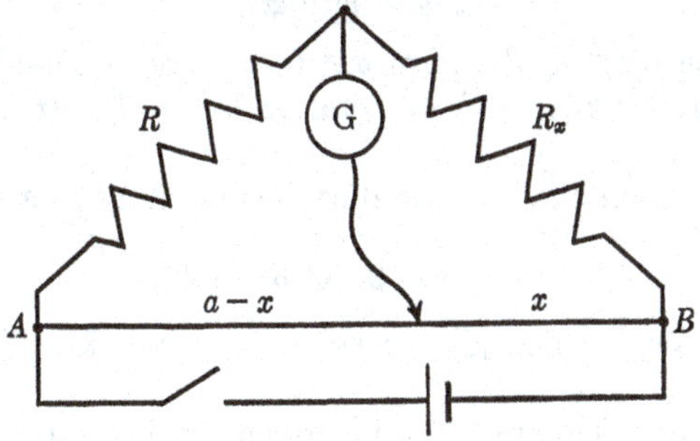

FIGURE 44

AB at which the relative error of R_x is minimal, assuming that the absolute error in reading x is the same everywhere.

632. Prove that the tables of $\log \tan x$ determine x more exactly than the tables of $\log \sin x$ with the same numbers of decimal places.

633. What is the maximal relative error that can be permitted in measuring the edge of a cube if its volume is to be evaluated with a relative error not exceeding 2%?

634. To find the length of a circular arc AB, the line interval $AB = L$ and the central angle $\measuredangle AOB = \alpha = 20°$ are measured. Assuming that L is measured exactly and that α is measured with an absolute error not exceeding 1°, find the relative error in the computation of the length of the arc AB.

7.4 HIGHER ORDER DIFFERENTIALS

The second differential d^2y at $x = x_0$ is defined as the differential of the (first) differential of the function at x_0:

$$d^2y = d(dy).$$

In an analogous way we define the third, fourth, etc., differentials,

$$d^3y = d(d^2y), \qquad d^4y = d(d^3y), \ldots$$

and in general

$$d^{n+1}y = d(d^n y).$$

When computing higher differentials it must be realized that dx *is independent of* x *and its derivative with respect to* x *is* 0. We obtain

$$d^2y = d(dy) = d(y'dx) = (y'dx)'dx = (y''dx)dx = y''(dx)^2,$$
$$d^3y = d(d^2y) = d[y''(dx)^2] = [y''(dx)^2]'dx = y'''(dx)^3,$$

and, in general, $$d^n y = y^{(n)}(dx)^n.$$

We often write $(dx)^n$ in the form dx^n (not to be confused with $d(x^n) = nx^{n-1}\,dx$). The already familiar notation $y^{(n)} = d^n y/dx^n$ now obtains a new meaning.

For the nth differential of a product, Leibniz's formula holds:

$$d^n(uv) = \sum_{i=0}^{n} {}_nC_i\, d^{n-i}u\, d^i v.$$

Here $d^0u = u$, $d^0v = v$, and ${}_nC_i = n(n-1)\cdots(n-i+1)/1\cdot 2\cdots i$. See Section 4.8.

For the higher differentials the invariance of form, true in the case of the first differential, no longer holds. Suppose $y = f(x)$, $x = g(t)$. We have seen

$$dy = f'(x)\,dx, \qquad dx = g'(t)\,dt.$$

Now $$d^2y = d(f'(x)\,dx) = df'(x)\,dx + f'(x)\,d(dx).$$

The first term on the right is, as above, $f''(x)\,dx^2$. But the second term now is not zero, because dx is not independent and $d(dx) = d^2x$. In the same way

$$d^3y = f'''(x)\,dx^3 + 3f''(x)\,dx\,d^2x + f'(x)\,d^3x,$$

and so on.

Assuming that x is independent, then $d^2x = 0$, $d^3x = 0, \ldots,$ and the above formula reduces back to

$$d^2y = y''dx^2, \qquad d^3y = y'''dx^3, \ldots .$$

635. Given $y = x^2$, find dy and d^2y, assuming (a) x an independent variable, (b) $x = \sin t$.

Solution:

(a) When x is the independent variable we have

$$dy = 2x\,dx, \qquad d^2y = 2\,dx^2.$$

(b) When $x = \sin t$ we can substitute $y = \sin^2 t$,

$$dy = 2\sin t\cos t\,dt = \sin 2t\,dt, \qquad d^2y = 2\cos 2t\,dt^2.$$

Now we shall obtain these expressions without substitution:

$$dx = \cos t\,dt, \qquad d^2x = -\sin t\,dt^2;$$
$$dy = 2x\,dx = 2\sin t\cos t\,dt = \sin 2t\,dt.$$

This result is equal to dy computed directly, and this illustrates the invariance of the form of the first differential. If we perform the same calculation for d^2y we obtain

$$2\,dx^2 = 2(\cos t\,dt)^2 = 2\cos^2 t\,dt^2,$$

and this expression is not equal to d^2y computed directly. The difference results from the fact that dx depends on t and it too must be differentiated:

$$d^2y = d(2x\,dx) = 2(dx)^2 + 2x\,d^2x.$$

After substitution we obtain

$$\begin{aligned} d^2y &= 2(\cos t\,dt)^2 + 2\sin t(-\sin t\,dt^2) = 2(\cos^2 t - \sin^2 t)\,dt^2 \\ &= 2\cos 2t\,dt^2, \end{aligned}$$

the correct expression.

In the following exercises find the indicated differentials.

636. $y = \sqrt[3]{x^2}$, $d^2y = ?$

637. $y = (x+1)^3(x-1)^2$, $d^4y = ?$

638. $y = \sin^2 x$, $d^3y = ?$

639. $r^2\cos^3\theta - a^2\sin^3\theta = 0$, $d^2r = ?$

640. $y = \ln\dfrac{1-x^2}{1+x^2}$, $x = \tan t$, $d^2y = ?$

Express d^2y in terms of (a) x and its differentials; (b) t and dt.

641. $y = \sin z$, $z = a^x$, $x = t^3$.

Express d^2y in terms of (a) z and its differentials, (b) x and its differentials, (c) t and dt.

VIII

THE INDEFINITE INTEGRAL

8.1 DEFINITION AND BASIC PROPERTIES

DEFINITION. A function $F(x)$ is said to be the indefinite integral (or the antiderivative) of the function $f(x)$ if $dF(x) = f(x)\,dx$, or in other words $F'(x) = f(x)$.

The same relation is also denoted by

$$F(x) = \int f(x)\,dx.$$

(The symbol $\int$ is read *integral.*)

From the above definition it follows that integration and differentiation are inverse operations, and consequently

$$d\left[\int f(x)\,dx\right] = f(x)\,dx,$$

$$\int df(x) = f(x) + C.$$

The arbitrary constant C in the right-hand side is necessary, because

$$d[f(x) + C] = df(x).$$

The same fact can also be formulated as follows:

If $F(x) = \int f(x)\,dx$, then for any C, $F(x) + C$ is also an antiderivative of $f(x)$.

The following theorems concerning integration can be proved easily by differentiating both sides of the equations.

(1) $$\int [f_1(x) + f_2(x) + \ldots + f_n(x)]\,dx = \int f_1(x)\,dx + \int f_2(x)\,dx + \ldots + \int f_n(x)\,dx.$$

(2) $\int Cf(x)\,dx = C\int f(x)\,dx.$

(3) Let $\int f(x)\,dx = F(x) + C$ and let $u(x)$ be any differentiable function. Then

$$\int f[u(x)]\,du(x) = F[u(x)] + C.$$

(4) If $\int f(x)\,dx = F(x) + C$, then

$$\int f(ax + b)\,dx = \frac{1}{a}F(ax + b) + C.$$

By inverting the formulas of differentiation (see Sect. 4.2), we obtain the following integration formulas:

(1) $\int 1 \cdot dx \left(= \int dx\right) = x + C$

(2) $\int x^\alpha\,dx = \dfrac{x^{\alpha+1}}{\alpha + 1} + C, \alpha \neq -1$

(3) $\int \dfrac{1}{x}\,dx = \ln|x| + C$ distinct C's allowed for $x > 0$ and $x < 0$.

(4) $\int e^x\,dx = e^x + C$

(5) $\int a^x\,dx = \dfrac{a^x}{\ln a} + C$

(6) $\int \cos x\,dx = \sin x + C$

(7) $\int \sin x\,dx = -\cos x + C$

(8) $\int \dfrac{1}{\cos^2 x}\,dx = \tan x + C$

(9) $\int \dfrac{1}{\sin^2 x}\,dx = -\cot x + C$

(8)–(9): distinct C's allowed for the distinct branches of $\tan x$ ($\cot x$).

(10) $\int \dfrac{1}{1 + x^2}\,dx = \arctan x + C$

(11) $\int \dfrac{1}{\sqrt{1 - x^2}}\,dx = \arcsin x + C$

(12) $\int \sinh x\,dx = \cosh x + C$

(13) $\int \cosh x\,dx = \sinh x + C$

(14) $\int \frac{1}{\cosh^2 x}\,dx = \tanh x + C$

(15) $\int \frac{1}{\sinh^2 x}\,dx = -\coth x + C$ distinct C's allowed as in (3).

Remark. The formula $\int \frac{1}{x}\,dx = \ln|x| + C$ can be proved as follows:

for $x > 0$: $d(\ln|x|) = d(\ln x) = \frac{1}{x}\,dx;$

for $x < 0$: $d(\ln|x|) = d\ln(-x) = \frac{1}{-x}(-1)\,dx = \frac{1}{x}\,dx.$

The above formulas and theorems can be used for a direct computation of a variety of integrals.

It should be remarked that in many cases preparatory steps have to be made in order to reduce the integral to one of the standard forms given above (also called immediate integrals). This is often an awkward task. Moreover, there exist elementary functions which have no integrals among the elementary functions. For example, there is no elementary function the derivative of which is e^{x^2} or $(\sin x)/x$, i.e., neither

$$\int e^{x^2}\,dx \quad \text{nor} \quad \int \frac{\sin x}{x}\,dx$$

can be represented by elementary functions.

The correctness of a given integration may be proved by differentiation. For example,

$$\int \sin^3 x \cos x\,dx = \frac{\sin^4 x}{4} + C,$$

because

$$\left(\frac{\sin^4 x}{4} + C\right)' = \sin^3 x \cos x.$$

It is advisable that at the beginning the student check every integration by differentiation.

We shall start with immediate integrals or integrals which can easily be brought to such a form. The following sections will deal with two basic methods of integration: substitution and integration by parts. After this we shall perform the integrations of various sorts of elementary functions.

8.2 IMMEDIATE INTEGRALS

Find the following integrals:

642. $I = \int (4x^3 + 5x - 7)\,dx.$

Solution:

$$I = \int 4x^3\,dx + \int 5x\,dx - \int 7\,dx$$

$$= 4\int x^3\,dx + 5\int x\,dx - 7\int dx = 4\cdot\frac{x^4}{4} + 5\cdot\frac{x^2}{2} - 7x + C$$

$$= x^4 + \tfrac{5}{2}x^2 - 7x + C.$$

643. $I = \int (x^3 + 2)^2\,dx$

644. $I = \int \frac{x^3 + 2x^2 - x + 3}{x}\,dx$

645. $I = \int \frac{\sqrt[3]{x^2} - \sqrt[4]{x}}{\sqrt{x}}\,dx$

Solution:

$$I = \int (x^{2/3-1/2} - x^{1/4-1/2})\,dx = \int (x^{1/6} - x^{-1/4})\,dx$$

$$= \tfrac{6}{7}x^{7/6} - \tfrac{4}{3}x^{3/4} + C = \tfrac{6}{7}x\sqrt[6]{x} - \tfrac{4}{3}\sqrt[4]{x^3} + C.$$

646. $I = \int \frac{dx}{\sqrt[3]{(b + ax)^2}}$

Solution:

$$I = \int (b + ax)^{-2/3}\,dx = \frac{3}{a}\sqrt[3]{b + ax} + C.$$

Here we made use of theorem (4).

647. $I = \int \left[\frac{1}{x - a} + \frac{1}{(x - a)^k}\right] dx$

Solution:

$$I = \ln|x - a| + \frac{1}{-k + 1}\cdot\frac{1}{(x - a)^{k-1}} + C$$

$$= \ln|x - a| - \frac{1}{(k - 1)(x - a)^{k-1}} + C.$$

648. $I = \int \sin 7x\,dx$

Solution. $I = -\tfrac{1}{7}\cos 7x + C.$ (We used theorem 4.)

649. $I = \int e^{-5x}\,dx$

650. $I = \int \frac{1 + 2x^2}{x^2(1 + x^2)}\,dx$

Solution:

$$I = \int \frac{1 + x^2}{x^2(1 + x^2)}\,dx + \int \frac{x^2}{x^2(1 + x^2)}\,dx = \int \frac{dx}{x^2} + \int \frac{dx}{1 + x^2}$$

$$= -\frac{1}{x} + \arctan x + C.$$

651. $I = \int \frac{e^{5x} + e^{4x} - 1}{e^{2x}}\,dx$

652. $I = \int \frac{3 \cdot 2^x - 2 \cdot 3^x}{2^x}\,dx$

653. $I = \int \cosh ax$

654. $I = \int \frac{dx}{\cosh^2 ax}$

In evaluating the following integrals we have to bring them to a standard form by various means.

655. $I = \int \tan^2 x\,dx$

Solution:

$$I = \int (\sec^2 x - 1)\,dx = \int \left(\frac{1}{\cos^2 x} - 1\right)dx = \tan x - x + C.$$

656. $I = \int \frac{dx}{a^2 + x^2}$

Solution:

$$I = \frac{1}{a^2}\int \frac{dx}{1 + (x/a)^2} = \frac{1}{a^2} \cdot a \cdot \arctan \frac{x}{a} + C = \frac{1}{a} \arctan \frac{x}{a} + C.$$

We used theorem (4). The formula

$$\int \frac{dx}{a^2 + x^2} = \frac{1}{a} \arctan \frac{x}{a} + C$$

occurs quite often and should be remembered.

657. $I = \int \frac{dx}{\sqrt{a^2 - x^2}}$

Solution:

$$I = \frac{1}{a}\int \frac{dx}{\sqrt{1 - (x/a)^2}} = \frac{a}{a} \arcsin \frac{x}{a} + C = \arcsin \frac{x}{a} + C.$$

The formula

$$\int \frac{dx}{\sqrt{a^2 - x^2}} = \arcsin \frac{x}{a} + C,$$

like the preceding one, is important.

658. $I = \int \frac{5x^2 + 2x - 3}{x + 2}\, dx$

Solution:

$$I = \int \left(5x - 8 + \frac{13}{x + 2}\right) dx = \frac{5}{2}x^2 - 8x + 13 \ln |x + 2| + C.$$

659. $I = \int \frac{dx}{\sin^2 (4x + 3)}$

660. $I = \int (\arcsin x + \arccos x)\, dx$

Solution:

$$I = \int \frac{\pi}{2}\, dx = \frac{\pi}{2} x + C \quad \text{(cf. Prob. 84).}$$

661. $I = \int \frac{x^4\, dx}{x - 1}$

In the following integrals we shall use theorem (3).

662. $I = \int e^x \sin e^x\, dx$

Solution:

$$I = \int \sin e^x\, d(e^x) = -\cos e^x + C.$$

663. $I = \int \tan x\, dx$

664. $I = \int \frac{x + 1}{9x^2 + 1}\, dx$

665. $I = \int \frac{e^x}{e^x + 1}\, dx$

666. $I = \int x\sqrt{1 - x^2}\, dx$

667. $I = \int x^2 \sqrt[5]{x^3 + 2}\, dx$

668. $I = \int \frac{6x - 5}{2\sqrt{3x^2 - 5x + 6}}\, dx$

669. $I = \int \sin^7 x \cos x\, dx$

670. $I = \int \frac{(\arctan x)^2}{1 + x^2}\, dx$

671. $I = \int e^{-x^3} x^2\, dx$

672. $I = \int \frac{(1 + x)^2}{1 + x^2}\, dx$

673. $I = \int \frac{dx}{(\arcsin x)^3 \sqrt{1 - x^2}}$

674. $I = \int \frac{x\, dx}{\sqrt{a^2 - x^4}}$

675. $I = \int \cos^3 x \sin 2x \, dx$

676. $I = \int \frac{dx}{(e^x - e^{-x})^2}$

677. $I = \int \frac{dx}{1 + \sin x}$

678. $I = \int \frac{1 + \sin x}{1 - \sin x} \, dx$

679. $I = \int (\cosh^2 ax + \sinh^2 ax) \, dx$

680. $I = \int \cos^2 ax \, dx$

681. $I = \int \frac{dx}{x \ln x}$

8.3 THE METHOD OF SUBSTITUTION

This method of integration is based on the following formula, the correctness of which can be immediately proved by differentiation:

$$\int f(x) \, dx = \int f[g(t)]g'(t) \, dt.$$

Here we substituted $x = g(t)$. It often happens that the integral

$$\int f[g(t)]g'(t) \, dt,$$

is simpler than the original.

In many cases we substitute t for a function of x, i.e., we assume $\varphi(x) = t$, and substitute this in the integral. In these cases we have to use $\varphi'(x) \, dx = dt$. For example, in the integral $\int f(ax + b) \, dx$ we assume $ax + b = t$, $a \, dx = dt$, i.e.,

$$\int f(ax + b) \, dx = \frac{1}{a} \int f(t) \, dt,$$

and this is theorem (4) of Section 8.1. When the integration is accomplished, the substitution leading back from t to x must be performed.

The choice of an appropriate substitution depends on the case in question.

We shall now give some examples of integration by substitution. We begin with a number of exercises of the former section and solve them by substitution.

682. $I = \int \sin^7 x \cos x \, dx$

Solution. We substitute $\sin x = t$, then $\cos x \, dx = dt$ and $I = \int t^7 \, dt = \frac{t^8}{8} + C$. We return to x:

$$I = \frac{\sin^8 x}{8} + C.$$

683. $I = \int \frac{dx}{\sqrt{a^2 - x^2}}$

Solution:

$x = a \sin t, \qquad dx = a \cos t \, dt.$

$$I = \int \frac{a \cos t \, dt}{\sqrt{a^2 - a^2 \sin^2 t}} = \int \frac{a \cos t \, dt}{a \cos t} = \int dt = t + C = \arcsin \frac{x}{a} + C.$$

684. $I = \int \frac{(6x - 5)\, dx}{2\sqrt{3x^2 - 5x + 6}}$

Solution:

$$3x^2 - 5x + 6 = t^2, \qquad (6x - 5)\, dx = 2t \, dt.$$

$$I = \int \frac{2t \, dt}{2t} = \int dt = t + C = \sqrt{3x^2 - 5x + 6} + C.$$

685. $I = \int \frac{dx}{1 + \sqrt{x + 1}}$

Solution:

$$x + 1 = t^2, \qquad dx = 2t \, dt.$$

$$I = \int \frac{2t \, dt}{1 + t} = 2 \int \frac{1 + t}{1 + t} dt - 2 \int \frac{dt}{1 + t} = 2(t - \ln |1 + t|) + C$$

$$= 2[\sqrt{x + 1} - \ln (1 + \sqrt{x + 1})] + C.$$

686. $I = \int \frac{e^{2x}\, dx}{\sqrt[4]{e^x + 1}}$

Solution:

$$e^x + 1 = t^4, \qquad e^x \, dx = 4t^3 \, dt.$$

$$I = \int \frac{4t^3 \, dt \cdot (t^4 - 1)}{t} = 4 \int (t^6 - t^2)\, dt = 4\left(\frac{t^7}{7} - \frac{t^3}{3}\right) + C$$

$$= 4t^3 \left(\frac{t^4}{7} - \frac{1}{3}\right) + C = 4\sqrt[4]{(e^x + 1)^3}\left(\frac{e^x + 1}{7} - \frac{1}{3}\right) + C$$

$$= \frac{4}{21}(3e^x - 4)\sqrt[4]{(e^x + 1)^3} + C.$$

687. $I = \int \frac{\sqrt{x}\, dx}{\sqrt{x} - \sqrt[3]{x}}$

Solution. We substitute $x = z^6$, $dx = 6z^5 \, dz$.

$$I = \int \frac{z^3 \cdot 6z^5 \, dz}{z^3 - z^2} = 6 \int \frac{z^6 \, dz}{z - 1}$$

$$= 6 \int \left(z^5 + z^4 + z^3 + z^2 + z + 1 + \frac{1}{z - 1} \right) dz$$

$$= 6 \left(\frac{z^6}{6} + \frac{z^5}{5} + \frac{z^4}{4} + \frac{z^3}{3} + \frac{z^2}{2} + z + \ln |z - 1| \right) + C$$

$$= x + \frac{6}{5} \sqrt[6]{x^5} + \frac{3}{2} \sqrt[3]{x^2} + 2\sqrt{x} + 3\sqrt[3]{x} + 6\sqrt[6]{x} + 6 \ln |\sqrt[6]{x} - 1| + C.$$

688. $I = \int \frac{dx}{x\sqrt{x^2 - 1}}$

Solution:

$$x^2 - 1 = t^2, \qquad 2x\,dx = 2t\,dt, \qquad x\,dx = t\,dt.$$

$$I = \int \frac{x\,dx}{x^2\sqrt{x^2 - 1}} = \int \frac{t\,dt}{(t^2 + 1)t} = \arctan t + C = \arctan \sqrt{x^2 - 1} + C.$$

The substitution $x = 1/t$ is also convenient here.

In integrals of the form $\int f(\ln x) \frac{dx}{x}$ we substitute

$$\ln x = t, \qquad \frac{dx}{x} = dt.$$

Two examples are given below.

689. $I = \int \frac{\ln^2 x}{x} dx$

Solution. By the above substitution,

$$I = \int t^2\,dt = \frac{t^3}{3} + C = \frac{(\ln x)^3}{3} + C.$$

690. $I = \int \frac{\cos (\ln x)}{x} dx$

Solution:

$$\ln x = t, \qquad \frac{1}{x} dx = dt.$$

$$I = \int \cos t\,dt = \sin t + C = \sin (\ln x) + C.$$

Integrals of the form

$$\int f(\sin x) \cos x\,dx, \qquad \int f(\cos x) \sin x\,dx, \qquad \int f(\tan x) \frac{dx}{\cos^2 x}$$

are to be solved by the respective substitutions

$$t = \sin x, \qquad t = \cos x, \qquad t = \tan x.$$

691. $I = \int (1 + \sin x)^5 \cos x \, dx$

Solution:

$$1 + \sin x = t, \qquad \cos x \, dx = dt.$$

$$I = \int t^5 \, dt = \frac{t^6}{6} + C = \frac{(1 + \sin x)^6}{6} + C.$$

692. $I = \int \frac{dx}{a^2 \sin^2 x + b^2 \cos^2 x}$

Solution:

$$I = \int \frac{dx}{\cos^2 x(a^2\tan^2 x + b^2)}.$$

We substitute $\tan x = t, \dfrac{dx}{\cos^2 x} = dt$:

$$I = \int \frac{dt}{a^2t^2 + b^2} = \frac{1}{a^2} \int \frac{dt}{t^2 + (b^2/a^2)} = \frac{1}{a^2} \cdot \frac{a}{b} \arctan \frac{at}{b} + C$$

$$= \frac{1}{ab} \arctan \left(\frac{a}{b} \tan x\right) + C.$$

693. $I = \int \ln (\cos x) \tan x \, dx$

694. $I = \int \frac{dx}{\sin x}$

695. $I = \int \frac{dx}{\cos 2x}$

696. $I = \int \sinh \frac{1}{x^2} \cdot \frac{dx}{x^3}$

697. $I = \int \frac{\sinh x}{e^x + e^{-x}} \, dx$

When there appear in the integrand (i.e., the expression under the integral sign) the expressions $a^2 - x^2$, $a^2 + x^2$ or $x^2 - a^2$, we often use trigonometric or hyperbolic substitutions, viz.,

for $a^2 - x^2$: $x = a \sin t$;
for $a^2 + x^2$: $x = a \tan t$ or $x = a \sinh t$;
for $x^2 - a^2$: $x = a \sec t$ or $x = a \cosh t$.

A few examples follow.

698. $I = \int \frac{dx}{(x^2 + a^2)^2}$

699. $I = \int \frac{x^2 \, dx}{\sqrt{a^2 - x^2}}$

700. $I = \int \frac{\sqrt{x^2 - a^2}}{x} \, dx$

701. $I = \int \frac{dx}{\sqrt{x^2 + a^2}}$

702. $I = \int \sqrt{x^2 - a^2} \, dx$

In the following exercises various substitutions will be used.

703. $I = \int \frac{x+1}{x\sqrt{x-2}}\,dx$ **704.** $I = \int \frac{\cos\sqrt{x}}{\sqrt{x}}\,dx$

705. $I = \int \frac{\sin^2 x \cos x}{1 + \sin^2 x}\,dx$ **706.** $I = \int \frac{e^x - 1}{e^x + 1}\,dx$

707. $I = \int \frac{\ln \tan x}{\sin x \cos x}\,dx$ **708.** $I = \int \frac{x^5\,dx}{\sqrt{a^3 - x^3}}$

709. $I = \int \frac{\sqrt{a^2 - x^2}}{x^2}\,dx$ **710.** $I = \int \frac{dx}{\sqrt{(x^2 - a^2)^3}}$

711. $I = \int \sqrt{e^x - 1}\,dx$ **712.** $I = \int \sqrt{\frac{x-1}{x+1}} \cdot \frac{1}{x^2}\,dx$

713. $I = \int \frac{x+1}{x(1 + xe^x)}\,dx$ **714.** $I = \int \frac{\ln(x+1) - \ln x}{x(x+1)}\,dx$

8.4 INTEGRATION BY PARTS

Given two functions $u = u(x)$ and $v = v(x)$; from the known formula of differentiation

$$d(uv) = u\,dv + v\,du, \quad \text{i.e.,} \quad u\,dv = d(uv) - v\,du,$$

by integration we have the identity

$$\int u\,dv = uv - \int v\,du.$$

This is the formula of integration by parts. Before any further discussion of this formula, we shall illustrate it by an example:

$$I = \int xe^x\,dx.$$

We take $\quad u = x, \quad dv = e^x\,dx;$

then $\quad du = dx, \quad v = e^x + C_1,$

and by the formula,

$$I = xe^x + C_1x - \int (e^x + C_1)\,dx$$

$$= xe^x + C_1x - e^x - C_1x + C = xe^x - e^x + C.$$

We see that the constant C_1 added to v does not appear in the final result. It is easy to observe that this is also true in general. Consequently, the integral for v will be taken without a constant.

If in the above example we take

$$u = e^x, \qquad dv = x\,dx,$$

$$du = e^x\,dx, \qquad v = x^2/2,$$

we obtain
$$I = \frac{x^2}{2} e^x - \int \frac{x^2}{2} e^x\,dx.$$

This integral is more complicated than the original one and we cannot integrate directly.

The example shows that in any integration by parts, care must be taken that (*a*) the intermediate integration for determining v can be performed; (*b*) the second integral $\int v\,du$ be not more difficult than the given one. One cannot always see immediately how to achieve this, and several trials may be needed for the proper choice of u and dv.

Integration by parts is very efficient in the following cases:

$$\int x^n e^{ax}\,dx, \qquad \int x^n \sin bx\,dx, \qquad \int x^n \cos bx\,dx,$$

$$\int x^n e^{ax} \sin bx\,dx, \qquad \int x^n e^{ax} \cos bx\,dx, \qquad \int x^\alpha \ln^n x\,dx.$$

In the first five cases we shall take $u = x^n$. In the last case it is necessary to put $u = \ln^n x$. We shall also show another method of solving some of the above integrals.

It should be observed that in other cases also, integration by parts can be very useful. Some of these will appear among the exercises of this section.

The formula of integration by parts can be generalized by several repetitions. We obtain

$$\text{(A)} \quad \int uv^{(n+1)}\,dx = uv^{(n)} - u'v^{(n-1)} + u''v^{(n-2)} - \ldots + (-1)^n u^{(n)} v$$

$$+ (-1)^{n+1} \int u^{(n+1)} v\,dx.$$

This formula will be referred to in this section as formula (A).

715. $I = \int x \sin x\,dx.$

Solution:

$$u = x, \qquad dv = \sin x\,dx,$$

$$du = dx, \qquad v = -\cos x.$$

$$I = -x \cos x + \int \cos x\,dx = \sin x - x \cos x + C.$$

716. $I = \int x^2 \sin x \, dx$

Solution:

$$u = x^2, \qquad dv = \sin x \, dx,$$
$$du = 2x \, dx, \qquad v = -\cos x.$$
$$I = -x^2 \cos x + 2 \int x \cos x \, dx.$$

We shall use integration by parts once more:

$$u = x, \qquad dv = \cos x \, dx,$$
$$du = dx, \qquad v = \sin x.$$
$$I = -x^2 \cos x + 2x \sin x - 2 \int \sin x \, dx$$
$$= -x^2 \cos x + 2x \sin x + 2 \cos x + C.$$

This is an appropriate example for the use of the general formula (A):

$$u = x^2, \qquad v''' \, dx = \sin x \, dx.$$

now $\qquad u' = 2x, \qquad u'' = 2, \qquad u''' = 0,$

and $\qquad v'' = \int v''' \, dx = \int \sin x \, dx = -\cos x,$

$$v' = \int v'' \, dx = -\sin x,$$
$$v = \int v' \, dx = \cos x.$$

And now by the formula,

$$I = uv'' - u'v' + u''v - \int u''' \, dv$$
$$= -x^2 \cos x + 2x \sin x + 2 \cos x + C.$$

717. $I = \int x \sin x \cos x \, dx$ **718.** $I = \int x^2 \cos^2 x \, dx$

719. $I = \int x^3 e^x \, dx$ **720.** $I = \int x^2 a^x \, dx$

721. $I = \int e^x \cos x \, dx$

Solution:

$$u = e^x, \ du = e^x \, dx; \qquad dv = \cos x \, dx, \ v = \sin x.$$
$$I = e^x \sin x - \int e^x \sin x \, dx.$$

The second integral is not simpler than the first, but if we integrate by parts once more, taking

$$u = e^x, \qquad du = e^x\,dx, \qquad dv = \sin x\,dx, \qquad v = -\cos x,$$

then $I = e^x \sin x + e^x \cos x - \int e^x \cos x\,dx = e^x \sin x + e^x \cos x - I.$

Therefore $$I = \tfrac{1}{2}e^x(\sin x + \cos x) + C.$$

We can also begin with $u = \cos x$, etc. But if for the first time we set $u = e^x$, and for the second $u = \sin x$, we obtain only the useless identity $0 = 0$.

When the exponent is multiplied by a constant a, we obtain similarly

$$\int e^{ax} \cos bx\,dx = \frac{1}{a} e^{ax} \cos bx + \frac{b}{a} \int e^{ax} \sin bx\,dx,$$

$$\int e^{ax} \sin bx\,dx = \frac{1}{a} e^{ax} \sin bx - \frac{b}{a} \int e^{ax} \cos bx\,dx;$$

thus

$$\int e^{ax} \cos bx\,dx = e^{ax}\frac{b \sin bx + a \cos bx}{a^2 + b^2} + C,$$

$$\int e^{ax} \sin bx\,dx = e^{ax}\frac{a \sin bx - b \cos bx}{a^2 + b^2} + C.$$

At the end of this section we shall demonstrate another procedure for solving these integrals.

722. $I = \int \ln x\,dx$

Solution:

$$u = \ln x,\ du = \frac{1}{x}\,dx; \qquad dv = dx,\ v = x.$$

$$I = x \ln x - \int dx = x \ln x - x + C.$$

723. $I = \int \sqrt{x} \ln x\,dx$

724. $I = \int \ln^2 x\,dx$

725. $I = \int x^\alpha \ln^n x\,dx$

Solution. For $\alpha = -1$ we obtain directly

$$\int \frac{\ln^n x}{x}\,dx = \frac{\ln^{n+1} x}{n + 1} + C.$$

Assume now $\alpha \neq -1$, and set

$$u = \ln^n x, \qquad dv = x^\alpha\,dx,$$

$$du = n \ln^{n-1} x \cdot \frac{1}{x}\,dx, \qquad v = \frac{x^{\alpha+1}}{\alpha+1}.$$

Then

$$I = \frac{x^{\alpha+1}}{\alpha+1} \ln^n x - \frac{n}{\alpha+1}\int x^\alpha \ln^{n-1} x\,dx.$$

As the integral on the right-hand side is of the same form as the original one, we can substitute for it the last expression for I with $n - 1$ replacing n. The process may be repeated with $n - 2$, etc., and it will terminate because the power of $\ln x$ decreases and after n steps becomes zero.

This is an instance of the so-called reduction formulas which will be discussed in more detail in further examples.

We now give an application of the formula just obtained.

$$\begin{aligned}
\int \sqrt{x} \ln^3 x\,dx &= \tfrac{2}{3}x^{3/2}\ln^3 x - 3\cdot\tfrac{2}{3}\int x^{1/2}\ln^2 x\,dx \\
&= \tfrac{2}{3}x^{3/2}\ln^3 x - 2\left(\tfrac{2}{3}x^{3/2}\ln^2 x - 2\cdot\tfrac{2}{3}\int x^{1/2}\ln x\,dx\right) \\
&= \tfrac{2}{3}x^{3/2}\ln^3 x - 2\left[\tfrac{2}{3}x^{3/2}\ln^2 x - \tfrac{4}{3}\left(\tfrac{2}{3}x^{3/2}\ln x - \tfrac{2}{3}\int x^{1/2}\,dx\right)\right] \\
&= \tfrac{2}{3}x^{3/2}\ln^3 x - \tfrac{4}{3}x^{3/2}\ln^2 x + \tfrac{16}{9}x^{3/2}\ln x - \tfrac{32}{27}x^{3/2} + C.
\end{aligned}$$

We proceed to solve some more problems requiring integration by parts.

726. $I = \int \arcsin x\,dx$

727. $I = \int \arctan x\,dx$

728. $I = \int x \cosh x\,dx$

729. $I = \int x \arctan x\,dx$

730. $I = \int (x^2 + x)\ln(x + 1)\,dx$

731. $I = \int \frac{\arcsin x}{\sqrt{x+1}}\,dx$

732. $I = \int x \tan^2 x\,dx$

733. $I = \int \frac{\ln \ln x}{x}\,dx$

734. $I = \int e^{-x^2}x^5\,dx$

735. $I = \int \cos \ln x\,dx$

736. $I = \int \frac{x \arcsin x}{\sqrt{1 - x^2}}\,dx$

737. $I = \int \frac{3x^2 - 1}{2x\sqrt{x}} \arctan x\,dx$

738. $I = \int \frac{\arcsin x}{\sqrt{(1 - x^2)^3}}\,dx$

739. $I = \int \frac{x^2 \arctan x}{1 + x^2}\,dx$

740. $I = \int \frac{e^x(1 + \sin x)}{1 + \cos x}\,dx$

741. $I_n = \int \frac{dx}{(x^2 + a^2)^n}$, $n = 1, 2, 3, \ldots$

Solution:

$$u = \frac{1}{(x^2 + a^2)^n}, \qquad du = -\frac{2nx}{(x^2 + a^2)^{n+1}}; \qquad dv = dx, \quad v = x.$$

$$I_n = \frac{x}{(x^2 + a^2)^n} + 2n \int \frac{x^2}{(x^2 + a^2)^{n+1}}\, dx$$

$$= \frac{x}{(x^2 + a^2)^n} + 2n\left[\int \frac{dx}{(x^2 + a^2)^n} - a^2 \int \frac{dx}{(x^2 + a^2)^{n+1}}\right]$$

$$= \frac{x}{(x^2 + a^2)^n} + 2nI_n - 2a^2nI_{n+1}.$$

We obtain from this equation

$$I_{n+1} = \frac{1}{2na^2} \cdot \frac{x}{(x^2 + a^2)^n} + \frac{2n - 1}{2na^2} I_n.$$

This is a reduction formula which reduces the computation of I_{n+1} to that of I_n. After a finite number of steps we arrive at

$$I_1 = \int \frac{dx}{a^2 + x^2} = \frac{1}{a} \arctan \frac{x}{a} + C.$$

Two examples follow:

$$I_2 = \int \frac{dx}{(x^2 + a^2)^2} = \frac{1}{2a^2} \cdot \frac{x}{x^2 + a^2} + \frac{1}{2a^3} \arctan \frac{x}{a} + C.$$

$$I_3 = \int \frac{dx}{(x^2 + a^2)^3}$$

$$= \frac{1}{4a^2} \frac{x}{(x^2 + a^2)^2} + \frac{3}{4a^2}\left(\frac{1}{2a^2} \cdot \frac{x}{x^2 + a^2} + \frac{1}{2a^3} \arctan \frac{x}{a}\right) + C.$$

Let us now exhibit another method of integration of

$$I = \int e^{ax}[P_n(x) \sin bx + Q_m(x) \cos bx]\, dx.$$

Here $P_n(x)$ and $Q_m(x)$ are polynomials of degree n and m respectively. Let us assume the result

$$I = e^{ax}[R_k(x) \sin bx + S_k(x) \cos bx] + C.$$

Here $R_k(x)$ and $S_k(x)$ are polynomials in x of degree $k = \max(m,n)$ and with undetermined coefficients.

We differentiate the last expression. It can be shown that by equating the coefficients of similar terms on both sides we obtain a system of linear equations which enables us in any case to determine the unknown coefficients uniquely.

Let us demonstrate the use of this method, finding

$$I = \int x^2e^x\,dx.$$

We set $$I = (Ax^2 + Bx + C)e^x + C_1.$$

Now $$I' = x^2e^x = e^x(Ax^2 + Bx + C + 2Ax + B)$$
$$= Ax^2e^x + (B + 2A)xe^x + (B + C)e^x,$$

whence

$$A = 1; \quad B + 2A = 0, B = -2; \quad B + C = 0, C = 2;$$

and $$I = (x^2 - 2x + 2)e^x + C_1.$$

Let us take another example,

$$I = \int e^x \sin x\,dx.$$

We set $$I = Ae^x \sin x + Be^x \cos x + C.$$

Now $$e^x \sin x = Ae^x \sin x + Ae^x \cos x + Be^x \cos x - Be^x \sin x.$$

$$A - B = 1; \quad A + B = 0; \quad A = \tfrac{1}{2}, B = -\tfrac{1}{2};$$

thus $$I = \tfrac{1}{2}e^x(\sin x - \cos x) + C.$$

This method is faster than the one used in Problem 721. It is also more convenient than integration by parts, in the following case:

742. $I = \int x^2 \cos x\, e^x\,dx$

Solution. We set

$$I = e^x[(Ax^2 + Bx + C) \sin x + (Dx^2 + Ex + F) \cos x] + C_1.$$

$$I' = x^2 \cos x\, e^x = e^x[(Ax^2 + Bx + C) \sin x + (Dx^2 + Ex + F) \cos x$$
$$+ (Ax^2 + Bx + C) \cos x - (Dx^2 + Ex + F) \sin x + (2Ax + B) \sin x$$
$$+ (2Dx + E) \cos x].$$

We equate the coefficients of similar terms:

(1)	$e^xx^2 \sin x$:	$A - D = 0$
(2)	$e^xx^2 \cos x$:	$D + A = 1$
(3)	$e^xx \sin x$:	$B - E + 2A = 0$
(4)	$e^xx \cos x$:	$E + B + 2D = 0$
(5)	$e^x \sin x$:	$C - F + B = 0$
(6)	$e^x \cos x$:	$F + C + E = 0$

From (1) and (2) we obtain $\quad A = \tfrac{1}{2}, \quad D = \tfrac{1}{2}.$

(3) and (4) give $\quad 2B = -2; \quad B = -1, E = 0.$

Now $\quad 2F - B = 0; \quad F = -\tfrac{1}{2}, C = \tfrac{1}{2}.$

Finally, $\quad I = \tfrac{1}{2}e^x[(x^2 - 2x + 1) \sin x + (x^2 - 1) \cos x] + C_1$

743. $I = \int (x^2 + 3x + 5) \cos 2x \, dx$

We remark, without entering into details, that integrals involving sines and cosines can often be computed more easily by the use of complex numbers through Euler's formulas

$$\cos x = \frac{e^{ix} + e^{-ix}}{2}, \qquad \sin x = \frac{e^{ix} - e^{-ix}}{2i}.$$

8.5 INTEGRALS OF RATIONAL FUNCTIONS

In this and in the following sections we shall deal with further important cases of integration of elementary functions. We begin with rational integrands, i.e., integrals of the form

$$\int \frac{P(x)}{Q(x)} \, dx,$$

where $P(x)$ and $Q(x)$ are polynomials in x. First we quote some facts from algebra:

(1) If the degree of the numerator is not less than the degree of the denominator, then division of $P(x)$ by $Q(x)$ gives

$$\frac{P(x)}{Q(x)} = S(x) + \frac{R(x)}{Q(x)},$$

where $S(x)$ is a polynomial and $R(x)$ is a polynomial of degree less than the degree of $Q(x)$. $S(x)$, $R(x)$, or both may be constants, i.e., polynomials of degree zero.

(2) In algebra it is shown that any polynomial $Q(x)$ with real coefficients can be represented as a product of binomials of the first degree and quadratic trinomials with real coefficients. (Clearly, repetition of factors in this decomposition is possible.) The trinomials may be assumed to have no real roots, since otherwise they could be decomposed into products of first degree binomials.

Hence we can write, as the first step,

$$Q(x) = (x - a_1)^{k_1} \cdots (x - a_i)^{k_i}(x^2 + b_1 x + c_1)^{m_1} \cdots (x^2 + b_j x + c_j)^{m_j}$$

(We assume that the term of $Q(x)$ of highest degree has coefficient 1.) Here

$$k_1 + \ldots + k_i + 2(m_1 + \ldots + m_j) = n,$$

and $\quad b_t^2 - 4c_t < 0, \ldots, \quad \text{for} \quad t = 1, 2, \ldots, j.$

We remark that there are no general methods for the actual decomposition; special techniques are used in each case.

The next step is to decompose $R(x)/Q(x)$ into partial fractions. ($R(x)$ is of lower degree than $Q(x)$.) By this is meant that we write $R(x)/Q(x)$ as a sum of fractions constructed as follows. For every factor of the form $(x-a)^k$ in the decomposition of $Q(X)$ we put the fractions

$$\frac{A_1}{x-a}+\frac{A_2}{(x-a)^2}+\frac{A_3}{(x-a)^3}+\ldots+\frac{A_k}{(x-a)^k},$$

where $A_1, A_2, \ldots, A_k$ are constants to be determined later. Further, for every factor in $Q(x)$ of the form $(x^2+bx+c)^m$, $(b^2-4c<0)$ we put the fractions

$$\frac{B_1x+C_1}{x^2+bx+c}+\frac{B_2x+C_2}{(x^2+bx+c)^2}+\ldots+\frac{B_mx+C_m}{(x^2+bx+c)^m}.$$

For example, if

$$Q(x)=(x-1)^4(x-2)(x^2+x+1)^3(x^2+1)^2(x^2-x+1),$$

we write

$$\frac{R(x)}{Q(x)}=\frac{A_1'}{x-1}+\frac{A_2'}{(x-1)^2}+\frac{A_3'}{(x-1)^3}+\frac{A_4'}{(x-1)^4}+\frac{A''}{x-2}$$

$$+\frac{B_1'x+C_1'}{x^2+x+1}+\frac{B_2'x+C_2'}{(x^2+x+1)^2}$$

$$+\frac{B_3'x+C_3'}{(x^2+x+1)^3}+\frac{B_1''x+C_1''}{x^2+1}+\frac{B_2''x+C_2''}{(x^2+1)^2}+\frac{B'''x+C'''}{x^2-x+1}.$$

All A's, B's, and C's are undetermined coefficients which are to be found. For this purpose we multiply both sides of the above equality by $Q(x)$, which is the common denominator of all fractions. By equating the coefficients of similar terms on both sides we obtain a system of linear equations with the undetermined coefficients as unknowns. By algebraic methods it can be proved that this system always has a unique solution, i.e., there exists one and only one decomposition of the above form.

In practice we often use slightly different procedures for the decomposition into partial fractions. This will be shown in the examples of this section.

In conclusion we see that the integral of any rational function can be represented as a sum of an integral of a polynomial and integrals of the following four types:

(1) $\displaystyle\int \frac{A}{x-a}\,dx = A\ln|x-a|+C;$

(2) $\displaystyle\int \frac{A\,dx}{(x-a)^k} = -\frac{A}{k-1}\cdot\frac{1}{(x-a)^{k-1}}+C;$

(3) $$\int \frac{Bx + C}{x^2 + bx + c}\,dx$$

$$= \frac{B}{2}\int \frac{2x + b}{x^2 + bx + c}\,dx + \left(C - \frac{Bb}{2}\right)\int \frac{dx}{x^2 + bx + c}$$

$$\left(\text{Here } c - \frac{b^2}{4} > 0.\right)$$

$$= \frac{B}{2}\ln\,(x^2 + bx + c) + \left(C - \frac{Bb}{2}\right)\int \frac{dx}{\left(x + \frac{b}{2}\right)^2 + c - \frac{b^2}{4}}$$

$$= \frac{B}{2}\ln\,(x^2 + bx + c) + \frac{C - \frac{Bb}{2}}{\sqrt{c - \frac{b^2}{4}}}\arctan \frac{x + \frac{b}{2}}{\sqrt{c - \frac{b^2}{4}}} + C$$

$$= \frac{B}{2}\ln\,(x^2 + bx + c) + \frac{2C - Bb}{\sqrt{4c - b^2}}\arctan \frac{2x + b}{\sqrt{4c - b^2}} + C;$$

(4) $$\int \frac{Bx + C}{(x^2 + bx + c)^m}\,dx$$

$$= \frac{B}{2}\int \frac{2x + b}{(x^2 + bx + c)^m}\,dx + \left(C - \frac{Bb}{2}\right)\int \frac{dx}{(x^2 + bx + c)^m}$$

$$\left(\text{Here } c - \frac{b^2}{4} > 0\right)$$

$$= -\frac{B}{2(m - 1)} \cdot \frac{1}{(x^2 + bx + c)^{m-1}}$$

$$+ \left(C - \frac{Bb}{2}\right)\int \frac{dx}{\left[\left(x + \frac{b}{2}\right)^2 + c - \frac{b^2}{4}\right]^m}.$$

The last integral is exactly of the form introduced in Problem 741 and can be solved by the reduction formula obtained there.

All four integrals have been computed, and as every rational function can be decomposed into a sum of such terms (partial fractions), the integral $\int P(x)/Q(x)\,dx$ can always be computed. Moreover, any such integral can be expressed in terms of rational functions, logarithms, and arctangents. As a result, every integral which can be transformed to an integral of rational functions can be computed by the methods of this section. We shall deal with such problems in the next sections.

Before beginning with examples, let us remark that in practice the computation of more involved rational integrals can be very cumbersome.

744. $I = \int \frac{x^5 + x^4 - 8}{x^3 - 4x}\, dx$

Solution. Divide the numerator by the denominator:

$$\begin{array}{l|l} x^5 + x^4 - 8 & x^3 - 4x \\ \underline{x^5 - 4x^3} & x^2 + x + 4 \\ \quad x^4 + 4x^3 - 8 & \\ \quad \underline{x^4 - 4x^2} & \\ \qquad 4x^3 + 4x^2 - 8 & \\ \qquad \underline{4x^3 - 16x} & \\ \qquad\quad 4x^2 + 16x - 8 & \end{array}$$

Thus $$\frac{x^5 + x^4 - 8}{x^3 - 4x} = x^2 + x + 4 + \frac{4x^2 + 16x - 8}{x^3 - 4x},$$

The resulting fraction will now be decomposed; let

$$\frac{4x^2 + 16x - 8}{x(x+2)(x-2)} = \frac{A}{x} + \frac{B}{x+2} + \frac{C}{x-2},$$

$$4x^2 + 16x - 8 = A(x^2 - 4) + B(x^2 - 2x) + C(x^2 + 2x).$$

We equate the coefficients of similar terms:

$$\begin{array}{lll} x^2: \quad 4 = A + B + C, & A = 2, & \\ x^1: \quad 16 = -2B + 2C, & B + C = 2, & B = -3, \qquad C = 5. \\ x^0: \quad -8 = -4A; & -B + C = 8, & \end{array}$$

Now $$I = \int \left(x^2 + x + 4 + \frac{2}{x} - \frac{3}{x+2} + \frac{5}{x-2}\right) dx$$

$$= \frac{x^3}{3} + \frac{x^2}{2} + 4x + \ln\left|\frac{x^2(x-2)^5}{(x+2)^3}\right| + C.$$

The coefficients can be found more conveniently as follows:

$$4x^2 + 16x - 8 = A(x-2)(x+2) + Bx(x-2) + Cx(x+2).$$

This is an identity and must hold for all x. We substitute

$$\begin{array}{llll} x = 0; & \text{then} & -8 = -4A, & A = 2. \\ x = 2; & \text{then} & 40 = 8C, & C = 5. \\ x = -2; & \text{then} & -24 = 8B, & B = -3. \end{array}$$

745. $I = \int \frac{x - 5}{x^3 - 3x^2 + 4}\, dx$ **746.** $I = \int \frac{4x + 3}{(x-2)^3}\, dx$

747. $I = \int \frac{x^3 - 2x^2 + 4}{x^3(x-2)^2}\, dx$ **748.** $I = \int \frac{3\, dx}{x^2 + 7x + 16}$

Solution. The given integrand is of the standard form, since the denominator has no real roots.

$$I = 3\int \frac{dx}{(x+\frac{7}{2})^2 + 16 - \frac{49}{4}} = 3\int \frac{dx}{(x+\frac{7}{2})^2 + \frac{15}{4}}.$$

This is an integral of the form $\int \frac{du}{u^2+a^2}$ and its solution was given in Problem 656,

$$\frac{1}{a}\arctan\frac{u}{a} + C.$$

In our case,

$$I = \frac{3}{\sqrt{\frac{15}{4}}}\arctan\frac{x+\frac{7}{2}}{\sqrt{\frac{15}{4}}} + C = \frac{6}{\sqrt{15}}\arctan\frac{2x+7}{\sqrt{15}} + C.$$

749. $I = \int \frac{5x-2}{3x^2+8x+10}\,dx$

Solution:

$$I = \frac{5}{6}\int \frac{6x+8}{3x^2+8x+10}\,dx - \frac{26}{3}\int \frac{dx}{3[(x+\frac{4}{3})^2+\frac{14}{9}]}$$

$$= \frac{5}{6}\ln(3x^2+8x+10) - \frac{26}{3\sqrt{14}}\arctan\frac{3x+4}{\sqrt{14}} + C.$$

Remark. In solving the following integral, one might write

$$\int \frac{3x+5}{x^2+4x+5}\,dx = \int \frac{2x+4}{x^2+4x+5}\,dx + \int \frac{x+1}{x^2+4x+5}\,dx = \ldots.$$

This, however, is inconvenient, since we gain nothing by the manipulation. The second integral is in no respect simpler than the original one. An efficient method should remove x from the numerator in a single operation. This can always be done as shown in the general case. In the present case,

$$\int \frac{3x+5}{x^2+4x+5}\,dx = \frac{3}{2}\int \frac{2x+4}{x^2+4x+5}\,dx - \int \frac{dx}{(x+2)^2+1}$$

$$= \tfrac{3}{2}\ln(x^2+4x+5) - \arctan(x+2) + C.$$

750. $I = \int \frac{12\,dx}{x^4+x^3-x-1}$

751. $I = \int \frac{dx}{x^4+1}$

752. $I = \int \frac{x^2-1}{x^4+3x^3+5x^2+3x+1}\,dx$

753. $I = \int \frac{x^2+1}{x^4+x^2+1}\,dx$

754. $I = \int \frac{dx}{(x+1)^2(x-2)^3}$

755. $I = \int \frac{3x^4 + x^3 + 4x^2 + 1}{x^5 + 2x^3 + x}\, dx$

756. $I = \int \frac{x^2 + 3x - 2}{(x-1)(x^2+x+1)^2}\, dx$

Solution. Partial fraction decomposition is cumbersome here. Also, if done we should still have to perform quite involved integrations. Somewhat easier is another method, which we first explain in the general case. It can be proved that for polynomials $P(x)$ and $Q(x)$ the following relationship holds:

$$\int \frac{P(x)}{Q(x)}\, dx = \frac{P_1(x)}{Q_1(x)} + \int \frac{P_2(x)}{Q_2(x)}\, dx,$$

where if $\quad Q(x) = (x-a)^k \cdots (x^2 + px + q)^m \cdots,$

then $\quad Q_1(x) = (x-a)^{k-1} \cdots (x^2 + px + q)^{m-1},$

and $\quad Q_2(x) = (x-a) \cdots (x^2 + px + q).$

$P_1(x)$ and $P_2(x)$ are polynomials the degree of which is lower by 1 than the degree of the corresponding denominator. We remark that in the above expression $p^2 - 4q < 0$.

We obtain in this way an integrand the denominator of which splits into factors of multiplicity one (i.e., each factor appears only once). The decomposition into partial fractions and the integration are now simpler.

As an illustration we solve the present problem. According to the above general formula we assume

$$\int \frac{x^2 + 3x - 2}{(x-1)(x^2+x+1)^2}\, dx = \frac{Ax + B}{x^2 + x + 1} + \int \left(\frac{C}{x-1} + \frac{Dx + F}{x^2 + x + 1} \right) dx.$$

Now we differentiate both sides,

$$\frac{x^2 + 3x - 2}{(x-1)(x^2+x+1)^2} = \frac{A(x^2 + x + 1) - (Ax + B)(2x + 1)}{(x^2 + x + 1)^2} + \frac{C}{x-1} + \frac{Dx + F}{x^2 + x + 1}.$$

We multiply by the common denominator and equate the coefficients of the same powers of x:

$$\begin{aligned} x^4&: \quad 0 = C + D, \\ x^3&: \quad 0 = -A + 2C + F, \\ x^2&: \quad 1 = A - 2B + 3C, \end{aligned}$$

$$x^1: \quad 3 = A + B + 2C - D,$$
$$x^0: \quad -2 = -A + B + C - F.$$

Hence $A = \frac{5}{3}, \quad B = \frac{2}{3}, \quad C = \frac{2}{9}, \quad D = -\frac{2}{9}, \quad F = \frac{11}{9}.$

Now
$$I = \frac{5x+2}{3(x^2+x+1)} + \frac{2}{9}\int \frac{dx}{x-1} - \frac{1}{9}\int \frac{2x-11}{x^2+x+1}\,dx$$
$$= \frac{5x+2}{3(x^2+x+1)} + \frac{2}{9}\ln|x-1| - \frac{1}{9}\int \frac{2x+1}{x^2+x+1}\,dx$$
$$+ \frac{12}{9}\int \frac{dx}{(x+\frac{1}{2})^2 + \frac{3}{4}}$$
$$= \frac{5x+2}{3(x^2+x+1)} + \frac{2}{9}\ln|x-1| - \frac{1}{9}\ln(x^2+x+1)$$
$$+ \frac{8}{3\sqrt{3}}\arctan\frac{2x+1}{\sqrt{3}} + C.$$

757. $I = \int \frac{5x^2+6x+9}{(x-3)^2(x+1)^2}\,dx$

8.6 IRRATIONAL INTEGRALS

The Integral $\int R\left(x, \sqrt[m]{\frac{ax+b}{cx+d}}, \sqrt[n]{\frac{ax+b}{cx+d}}, \ldots\right) dx$

If R is a rational function of its arguments, this form can be integrated using the substitution
$$\frac{ax+b}{cx+d} = u^p,$$
where p is the least common multiple of $m, n, \ldots$. This substitution leads to rationalization, as will be illustrated in the following examples. (Compare also Problems 685, 687, 703, 712.)

758. $I = \int \frac{x\,dx}{\sqrt{x+1} - \sqrt[3]{x+1}}$

Solution. We substitute $x + 1 = t^6$, $dx = 6t^5\,dt$, $x = t^6 - 1$:
$$I = \int \frac{(t^6-1)\cdot 6t^5\,dt}{t^3-t^2} = 6\int \frac{(t^6-1)t^3\,dt}{t-1}$$
$$= 6(t^8+t^7+t^6+t^5+t^4+t^3)\,dt = 6\left(\frac{t^9}{9}+\frac{t^8}{8}+\frac{t^7}{7}+\frac{t^6}{6}+\frac{t^5}{5}+\frac{t^4}{4}\right) + C.$$

Here $t = \sqrt[6]{x+1}$.

759. $\displaystyle\int \frac{dx}{\sqrt[4]{(x-1)^3(x+2)^5}}$

Binomial Integrals $\displaystyle\int x^m(a+bx^n)^p\,dx$

Here a, b are arbitrary and m, n, p are rational numbers. Integration in elementary functions can be performed in this case only when at least one of the three numbers

$$p, \qquad \frac{m+1}{n} \quad \text{or} \quad \frac{m+1}{n}+p$$

is an integer.

(1) When p is an integer we substitute $x = t^r$ where r is the common denominator of the fractions m and n.

(2) If $(m+1)/n$ is an integer we substitute $a + bx^n = t^s$ where s is the denominator of p.

(3) If $p + (m+1)/n$ is an integer we substitute $ax^{-n} + b = t^s$ where s is the denominator of p.

Remark. For p a positive integer we simply use Newton's binomial formula.

Let us illustrate each of the above cases by an example.

760. $\displaystyle I = \int \frac{dx}{\sqrt[3]{x^2}(\sqrt[3]{x}+1)^3}$

Solution:

$$I = \int x^{-2/3}(x^{1/3}+1)^{-3}\,dx.$$

Here $p = -3$, $m = -\frac{2}{3}$, $n = \frac{1}{3}$. This is the first case. We substitute

$$x = t^3, \qquad dx = 3t^2\,dt;$$

$$I = \int t^{-2}(t+1)^{-3}\,3t^2\,dt = \int \frac{3\,dt}{(t+1)^3} = \frac{-3}{2(t+1)^2} + C$$

$$= -\frac{3}{2}\frac{1}{(\sqrt[3]{x}+1)^2} + C.$$

761. $\displaystyle I = \int \frac{dx}{\sqrt{x}(\sqrt[3]{x}+1)^2}$

762. $\displaystyle I = \int \frac{x^5\,dx}{\sqrt{1-x^2}}$

Solution. Here $m = 5$, $n = 2$, $p = -\frac{1}{2}$, $\dfrac{m+1}{m} = \dfrac{5+1}{2} = 3$. This is the second case. We substitute

$$1 - x^2 = t^2,$$
$$-2x\,dx = 2t\,dt,$$

$$I = -\int \frac{t\,dt\cdot(1-t^2)^2}{t} = -\int (1 - 2t^2 + t^4)\,dt = -\left(t - \frac{2}{3}t^3 + \frac{t^5}{5}\right) + C.$$

Here $t = \sqrt{1-x^2}$.

763. $I = \displaystyle\int \frac{dx}{\sqrt[3]{1+x^3}}$

Solution. Here $m = 0$, $n = 3$, $p = -\frac{1}{3}$, $\frac{m+1}{n} + p = \frac{1}{3} - \frac{1}{3} = 0.$
This is the third case. We substitute

$$x^{-3} + 1 = t^3, \qquad x = (t^3 - 1)^{-1/3},$$
$$dx = -\tfrac{1}{3}(t^3 - 1)^{-4/3}\,3t^2\,dt = -(t^3-1)^{-4/3}t^2\,dt;$$

$$I = \int \frac{-(t^3-1)^{-4/3}t^2\,dt}{[1 + (t^3-1)^{-1}]^{1/3}} = -\int \frac{(t^3-1)^{-4/3}t^2\,dt}{(t^3)^{1/3}/(t^3-1)^{1/3}} = -\int \frac{t}{t^3-1}\,dt.$$

We continue by decomposing the integrand into partial fractions:

$$\frac{t}{t^3-1} = \frac{A}{t-1} + \frac{Bt+C}{t^2+t+1}; \qquad t = A(t^2+t+1) + (Bt+C)(t-1).$$

$t = 1$, $A = \frac{1}{3}$; $\quad t = 0$, $0 = A - C$, $C = \frac{1}{3}$; $\quad A + B = 0$, $B = -\frac{1}{3}$.

$$\begin{aligned}
I &= -\frac{1}{3}\int\left(\frac{1}{t-1} - \frac{t-1}{t^2+t+1}\right)dt \\
&= -\frac{1}{3}\left[\ln|t-1| - \frac{1}{2}\int\frac{2t+1}{t^2+t+1}\,dt + \frac{3}{2}\int\frac{dt}{(t+\frac{1}{2})^2+\frac{3}{4}}\right] \\
&= -\frac{1}{3}\left[\ln|t-1| - \frac{1}{2}\ln(t^2+t+1) + \sqrt{3}\arctan\frac{2t+1}{\sqrt{3}}\right] + C \\
&= \frac{1}{6}\ln\frac{t^2+t+1}{(t-1)^2} - \frac{\sqrt{3}}{3}\arctan\frac{2t+1}{\sqrt{3}} + C.
\end{aligned}$$

Here $t = \sqrt[3]{1+x^{-3}}$.

The Form $\displaystyle\int R(x, \sqrt{ax^2+bx+c})\,dx$

R is a rational function of its arguments. These integrals can always be brought to a rational form by one of the following substitutions due to Euler:

(1) If $a > 0$, we substitute

$$\sqrt{ax^2+bx+c} = t - \sqrt{a}x.$$

Then $\qquad ax^2 + bx + c = t^2 - 2\sqrt{a}tx + ax^2,$

$$x = \frac{t^2 - c}{2\sqrt{a}t + b}, \qquad dx = 2\frac{\sqrt{a}t^2 + bt + c\sqrt{a}}{(2\sqrt{a}t + b)^2}\,dt,$$

$$\sqrt{ax^2 + bx + c} = t - \sqrt{a}x = \frac{\sqrt{a}t^2 + bt + c\sqrt{a}}{2\sqrt{a}t + b}.$$

All these expressions are rational functions of t.

(2) If $c > 0$, we can substitute

$$\sqrt{ax^2 + bx + c} = xt + \sqrt{c},$$

and obtain

$$x = \frac{2\sqrt{c}t - b}{a - t^2}, \qquad dx = \frac{2\sqrt{c}t^2 - bt + \sqrt{c}a}{(a - t^2)^2}\,dt,$$

$$\sqrt{ax^2 + bx + c} = \frac{\sqrt{c}t^2 - bt + a\sqrt{c}}{a - t^2},$$

(3) If the trinomial $ax^2 + bx + c$ has distinct real solutions x_1 and x_2, we substitute

$$ax^2 + bx + c = a(x - x_1)(x - x_2) = t^2(x - x_1)^2,$$

$$a(x - x^2) = t^2(x - x_1), \qquad x = \frac{-ax_2 + x_1t^2}{t^2 - a};$$

$$\sqrt{ax^2 + bx + c} = \frac{a(x_1 - x_2)t}{t^2 - a},$$

$$dx = \frac{2a(x_2 - x_1)}{(t^2 - a)^2}\,tdt.$$

In all expressions only rational functions of t appear; that is, Euler's substitutions always lead to rationalization of the above integral. However, it should be noted that in general Euler's substitutions lead to cumbersome computations, and if possible other methods should be used.

The following exercises deal with such methods.

764. $I = \int \frac{dx}{\sqrt{ax^2 + bx + c}}$

Solution. If $a > 0$ we have

$$I = \frac{1}{\sqrt{a}}\int \frac{dx}{\sqrt{x^2 + \frac{b}{a}x + \frac{c}{a}}} = \frac{1}{\sqrt{a}}\int \frac{dx}{\sqrt{\left(x + \frac{b}{2a}\right)^2 + \frac{c}{a} - \left(\frac{b}{2a}\right)^2}}$$

$$= \frac{1}{\sqrt{a}}\ln\left|x + \frac{b}{2a} + \sqrt{x^2 + \frac{b}{a}x + \frac{c}{a}}\right| + C.$$

We used here the solution of Problem 701.

For $a < 0$ we have

$$I = \frac{1}{\sqrt{-a}} \int \frac{dx}{\sqrt{-x^2 - \frac{b}{a}x - \frac{c}{a}}} = \frac{1}{\sqrt{-a}} \int \frac{dx}{\sqrt{\left(\frac{b}{2a}\right)^2 - \frac{c}{a} - \left(x + \frac{b}{2a}\right)^2}}$$

$$= \frac{1}{\sqrt{-a}} \arcsin \frac{x + \frac{b}{2a}}{\sqrt{\left(\frac{b}{2a}\right)^2 - \frac{c}{a}}} + C = \frac{1}{\sqrt{-a}} \arcsin \frac{2ax + b}{\sqrt{b^2 - 4ac}} + C.$$

We used the integral of Problem 657.

Remark. $b^2 - 4ac > 0$ in this case, since if $a < 0$ and $b^2 - 4ac < 0$, then $ax^2 + bx + c < 0$ for all x and $\sqrt{ax^2 + bx + c}$ has no meaning.

765. $I = \int \frac{dx}{\sqrt{2x^2 + x - 3}}$ **766.** $I = \int \frac{dx}{\sqrt{1 + 2x - 3x^2}}$

767. $I = \int \frac{3x + 5}{\sqrt{x^2 + x + 1}} dx$

Solution:

$$I = \frac{3}{2} \int \frac{2x + 1}{\sqrt{x^2 + x + 1}} dx + \frac{7}{2} \int \frac{dx}{\sqrt{x^2 + x + 1}}$$

$$= \frac{3}{2} \cdot 2\sqrt{x^2 + x + 1} + \frac{7}{2} \int \frac{dx}{\sqrt{(x + \frac{1}{2})^2 + \frac{3}{4}}}$$

$$= 3\sqrt{x^2 + x + 1} + \frac{7}{2} \ln \left| x + \frac{1}{2} + \sqrt{x^2 + x + 1} \right| + C.$$

768. $I = \int \frac{x\,dx}{\sqrt{1 + x - x^2}}$

The Form $I = \int \frac{P(x)\,dx}{\sqrt{ax^2 + bx + c}}$

$P(x)$ is a polynomial of degree n, and the integral can be solved by putting

$$I = P_1(x)\sqrt{ax^2 + bx + c} + p \int \frac{dx}{\sqrt{ax^2 + bx + c}},$$

where $P_1(x)$ is a polynomial of degree $n - 1$ with undetermined coefficients. These coefficients and the number p are determined after differentiating the above equation and equating the coefficients of similar terms on both sides.

769. $I = \int \frac{(x^2 + 1)\,dx}{\sqrt{-x^2 + 3x - 2}}$

Solution. We write

$$I = (Ax + B)\sqrt{-x^2 + 3x - 2} + p\int \frac{dx}{\sqrt{-x^2 + 3x - 2}}.$$

We differentiate,

$$\frac{x^2 + 1}{\sqrt{-x^2 + 3x - 2}} = A\sqrt{-x^2 + 3x - 2} + \frac{(Ax + B)(-2x + 3)}{2\sqrt{-x^2 + 3x - 2}}$$

$$+ p\frac{1}{\sqrt{-x^2 + 3x - 2}},$$

then multiply by the common denominator:

$$2x^2 + 2 = 2A(-x^2 + 3x - 2) + (Ax + B)(-2x + 3) + 2p.$$

Now $-2A - 2A = 2, \qquad A = -\dfrac{1}{2},$

$$0 = 6A + 3A - 2B, \qquad B = \frac{9A}{2} = -\frac{9}{4},$$

$$2 = -4A + 3B + 2p, \quad p = 1 + 2A - \frac{3}{2}B = 1 - 1 + \frac{27}{8} = \frac{27}{8}.$$

We proceed to the computation of the integral.

$$\int \frac{dx}{\sqrt{-x^2 + 3x - 2}} = \int \frac{dx}{\sqrt{\frac{1}{4} - (x - \frac{3}{2})^2}} = \arcsin(2x - 3) + C.$$

Finally,

$$I = (-\tfrac{1}{2}x - \tfrac{9}{4})\sqrt{-x^2 + 3x - 2} + \tfrac{27}{8}\arcsin(2x - 3) + C.$$

770. $I = \displaystyle\int \frac{x^4}{\sqrt{1 - x^2}}\,dx$

The Form $I = \displaystyle\int \frac{P(x)\,dx}{(x - h)^n\sqrt{ax^2 + bx + c}}$

$P(x)$ is a polynomial, and the integral can be brought to the form just discussed by the following procedure. First divide $P(x)$ by $(x - h)^n$ and obtain

$$\frac{P(x)}{(x - h)^n} = Q(x) + \frac{R(x)}{(x - h)^n},$$

where $Q(x)$ and $R(x)$ are polynomials and the degree of the latter is less than n. Now in

$$I = \int \frac{Q(x)\,dx}{\sqrt{ax^2 + bx + c}} + \int \frac{R(x)\,dx}{(x - h)^n\sqrt{ax^2 + bx + c}}$$

the first integral is of the form discussed in the preceding section, and the second integral can be brought to that form by the substitution $x - h = 1/t$.

771. $I = \int \frac{dx}{x\sqrt{x^2 + x + 1}}$

Solution. We substitute $x = \frac{1}{t}$, $dx = -\frac{1}{t^2}\,dt$.

$$I = \int \frac{-\frac{1}{t^2}\,dt}{\frac{1}{t}\sqrt{\left(\frac{1}{t}\right)^2 + \frac{1}{t} + 1}} = -\int \frac{dt}{\sqrt{1 + t + t^2}}$$

$$= -\ln\left|t + \frac{1}{2} + \sqrt{t^2 + t + 1}\right| + C$$

$$= -\ln\left|\frac{1}{x} + \frac{1}{2} + \sqrt{\frac{1}{x^2} + \frac{1}{x} + 1}\right| + C.$$

772. $I = \int \frac{dx}{(x-1)^2\sqrt{x^2 - 1}}$

773. $I = \int \frac{x^3 + x - 1}{(x-1)\sqrt{x^2 + 2x - 1}}\,dx$

The Form $I = \int \sqrt{ax^2 + bx + c}\,dx$

These integrals can be brought to one of the three forms that appear in the following three exercises.

774. $I = \int \sqrt{a^2 - x^2}\,dx$

Solution. We substitute $x = a \sin t$, $dx = a \cos t\,dt$.

$$I = \int a \cos t \cdot a \cos t\,dt = \frac{a^2}{2}\int (1 + \cos 2t)\,dt$$

$$= \tfrac{1}{2} a^2 \left(t + \tfrac{1}{2}\sin 2t\right) + c = \tfrac{1}{2} a^2 \left(\arcsin \frac{x}{a} + \frac{x}{a}\sqrt{1 - \frac{x^2}{a^2}}\right) + C$$

$$= \tfrac{1}{2}\left(a^2 \arcsin \frac{x}{a} + x\sqrt{a^2 - x^2}\right) + C.$$

For $a = 1$ we have

$$\int \sqrt{1 - x^2}\,dx = \tfrac{1}{2}\left(\arcsin x + x\sqrt{1 - x^2}\right) + C.$$

775. $I = \int \sqrt{x^2 - a^2}\,dx$

Solution. By aid of the substitution $x = a \cosh t$ we obtain

$$I = \tfrac{1}{2}(x\sqrt{x^2 - a^2} - a^2 \ln |x + \sqrt{x^2 - a^2}|) + C.$$

For $a = 1$,

$$\int \sqrt{x^2 - 1}\, dx = \tfrac{1}{2}(x\sqrt{x^2 - 1} - \ln |x + \sqrt{x^2 - 1}|) + C.$$

776. $I = \int \sqrt{x^2 + a^2}\, dx$

Solution. $x = a \sinh t$, $dx = a \cosh t\, dt$.

$$I = a^2 \int \cosh^2 t\, dt = \frac{a^2}{2} \int (\cosh 2t + 1)\, dt = \frac{a^2}{2}\left(\frac{\sinh 2t}{2} + t\right) + C$$

$$= \frac{a^2}{2}\left(\frac{x\sqrt{x^2 + a^2}}{a^2} + \operatorname{arg\,sinh} \frac{x}{a}\right) + C$$

$$= \tfrac{1}{2}\,[x\sqrt{x^2 + a^2} + a^2 \ln (x + \sqrt{x^2 + a^2})] + C.$$

For $a = 1$,

$$\int \sqrt{x^2 + 1}\, dx = \tfrac{1}{2}[x\sqrt{x^2 + 1} + \ln (x + \sqrt{x^2 + 1})] + C.$$

Another way of handling the integral $I = \int \sqrt{ax^2 + bx + c}\, dx$ is to transform it to the form

$$I = \int \frac{ax^2 + bx + c}{\sqrt{ax^2 + bx + c}}\, dx,$$

which has already been discussed. The same transformation is useful in the case

$$I = \int P(x)\sqrt{ax^2 + bx + c}\, dx = \int \frac{P(x)(ax^2 + bx + c)}{\sqrt{ax^2 + bx + c}}\, dx,$$

where $P(x)$ is a polynomial. Integrals of the last form have also been discussed.

An additional procedure for $I = \int R(x, \sqrt{ax^2 + bx + c})$, R a rational function, is to bring it (after completing a square under the radical sign) by a linear substitution to one of the following three forms:

$$\int R(y, \sqrt{1 - y^2})\, dy, \qquad \int R(y, \sqrt{y^2 - 1})\, dy, \qquad \int R(y, \sqrt{1 + y^2})\, dy.$$

The respective substitutions

$$y = \sin t, \qquad y = 1/\cos t, \quad \text{and} \quad y = \tan t$$

transform the integrands to rational functions of trigonometric functions. By the additional substitution $\tan (t/2) = z$ (see the next section), we obtain a rationalization of these integrals. We can, of course, directly substitute

$$y = \sin t = \frac{2z}{1+z^2}, \qquad y = \frac{1}{\cos t} = \frac{1+z^2}{1-z^2} \quad \text{or} \quad y = \tan t = \frac{2z}{1-z^2}$$

respectively, but in many cases it is more convenient to compute the integrals in the trigonometric form.

777. $I = \int \sqrt{x^2 - 2x - 1}\, dx$

Solution:

$$I = \int \sqrt{(x-1)^2 - 2}\, dx$$
$$= \tfrac{1}{2}(x-1)\sqrt{x^2 - 2x - 1} - \ln|x - 1 + \sqrt{x^2 - 2x + 1}| + C.$$

We used the solution of Problem 775.

778. $I = \int \sqrt{\frac{a-x}{x-b}}\, dx$

779. $I = \int \frac{\sqrt{x^2+2x}}{x}\, dx$

780. $I = \int \frac{dx}{\sqrt{1-x^2} - 1}$

781. $I = \int \frac{dx}{1 + \sqrt{x^2 + 2x + 2}}$

8.7 TRIGONOMETRIC INTEGRALS

Every integral of the form $\int R(\sin x, \cos x)\, dx$, where R is a rational function of its arguments, can be transformed into a rational integral by the substitution $\tan(x/2) = t$. Indeed, in this case,

$$\sin x = \frac{2\tan(x/2)}{1+\tan^2(x/2)} = \frac{2t}{1+t^2}, \qquad \cos x = \frac{1-\tan^2(x/2)}{1+\tan^2(x/2)} = \frac{1-t^2}{1+t^2},$$

and

$$x = 2\arctan t, \qquad dx = \frac{2\,dt}{1+t^2}.$$

All expressions are rational in t.

However, this substitution actually is often very inconvenient as it leads to cumbersome, though standard, computations. In many cases other methods are preferable. There follow some special cases where such other methods can be used.

We first treat integrals of the form

$$\int \sin^m x \cos^n x\, dx.$$

If at least one of the numbers m or n is a positive odd integer, we proceed as follows. Suppose $n = 2p + 1$; then

$$\cos^n x = \cos^{2p} x \cos x = (1 - \sin^2 x)^p \cos x.$$

Using the substitution $\sin x = t$ we obtain, after removing the parentheses, an integral of a polynomial in t.

If m and n are both nonnegative even integers, we use the formulas

$$\sin x \cos x = \tfrac{1}{2}\sin 2x, \quad \sin^2 x = \tfrac{1}{2}(1 - \cos 2x), \quad \cos^2 x = \tfrac{1}{2}(1 + \cos 2x).$$

By repeated use of these formulas we lower the power of sin and cos as much as needed (while the argument becomes multiplied by a larger integer factor). The resulting integral is easily evaluated.

If m or n is a negative odd integer we substitute $\sin x = t$ or $\cos x = t$. (We substitute for the function whose power is even. If both powers are odd it makes no difference which substitution is used.)

If both m and n are even and one of them is negative, we can use the substitution $\tan x = t$. Then

$$\sin^2 x = \frac{t^2}{1+t^2}, \qquad \cos^2 x = \frac{1}{1+t^2}, \qquad dx = \frac{dt}{1+t^2}.$$

This substitution is useful in every case when the integrand does not change sign if the signs of $\sin x$ and $\cos x$ are changed simultaneously.

Reduction formulas for various trigonometric integrals are often useful. Let us derive one of them,

$$I_m = \int \cos^m x\,dx, \qquad m > 0.$$

Put $\quad u = \cos^{m-1} x, \qquad dv = \cos x\,dx;$

then $\quad du = -(m-1)\cos^{m-2} x \sin x\,dx, \qquad v = \sin x.$

$$\begin{aligned} I_m &= \sin x \cos^{m-1} x + (m-1)\int \cos^{m-2} x \sin^2 x\,dx \\ &= \sin x \cos^{m-1} x + (m-1)\int \cos^{m-2} x\,dx - (m-1)\int \cos^m x\,dx \\ &= \sin x \cos^{m-1} x + (m-1)I_{m-2} - (m-1)I_m. \end{aligned}$$

The reduction formula is

$$I_m = \frac{1}{m}\sin x \cos^{m-1} x + \frac{m-1}{m} I_{m-2}.$$

There follow four additional reduction formulas:
For $m > 0$ and $n > 0$:

$$\begin{aligned} &\int \sin^n ax \cos^m ax\,dx \\ &\qquad = -\frac{\sin^{n-1} ax \cos^{m+1} ax}{a(n+m)} + \frac{n-1}{n+m}\int \sin^{n-2} ax \cos^m ax\,dx \\ &\qquad = \frac{\sin^{n+1} ax \cos^{m-1} ax}{a(n+m)} + \frac{n-1}{n+m}\int \sin^n ax \cos^{m-2} ax\,dx. \end{aligned}$$

For $m > 0$, $n > 1$ (or for $m > 1$, $n > 0$):

$$\int \frac{dx}{\sin^n ax \cos^m ax}$$

$$= -\frac{1}{a(n-1)} \cdot \frac{1}{\sin^{n-1} ax \cos^{m-1} ax} + \frac{n+m-2}{n-1} \int \frac{dx}{\sin^{n-2} ax \cos^m ax}$$

$$= \frac{1}{a(m-1)} \cdot \frac{1}{\sin^{n-1} ax \cos^{m-1} ax} + \frac{n+m-2}{m-1} \int \frac{dx}{\sin^n ax \cos^{m-2} ax}.$$

For $m > 0$, $n > 0$, $m \neq 1$:

$$\int \frac{\sin^n ax}{\cos^m ax}\,dx = \frac{\sin^{n+1} ax}{a(m-1)\cos^{m-1} ax} - \frac{n-m+2}{m-1} \int \frac{\sin^n ax}{\cos^{m-2} ax}\,dx.$$

For $m > 0$, $n > 0$, $n \neq 1$:

$$\int \frac{\cos^m ax}{\sin^n ax}\,dx = \frac{-\cos^{m+1} ax}{a(n-1)\sin^{n-1} ax} + \frac{n-m-2}{n-1} \int \frac{\cos^m ax}{\sin^{n-2} ax}\,dx.$$

If the integrand is a product of trigonometric functions of various arguments we use the following formulas:

$$\cos a \cos b = \tfrac{1}{2}[\cos(a-b) + \cos(a+b)],$$
$$\sin a \sin b = \tfrac{1}{2}[\cos(a-b) - \cos(a+b)],$$
$$\sin a \cos b = \tfrac{1}{2}[\sin(a-b) + \sin(a+b)].$$

We shall often use the following integrals which were found previously:

$$\int \frac{dx}{\sin x} = \ln|\tan x| + C, \qquad \int \frac{dx}{\cos x} = \ln\left|\tan\left(\frac{x}{2} + \frac{\pi}{2}\right)\right| + C,$$

$$\int \sin x \cos x\,dx = \frac{\sin^2 x}{2} + C, \qquad \int \frac{dx}{\sin x \cos x} = \ln|\tan x| + C.$$

782. $I = \int \sin^6 x \cos^5 x\,dx$

783. $I = \int \frac{\sin^3 x}{\cos^2 x}\,dx$

784. $I = \int \frac{dx}{\cos^3 x}$

785. $I = \int \frac{dx}{\sin^3 x \cos x}$

786. $I = \int \sin^4 x \cos^2 x\,dx$

Solution. By trigonometric formulas,

$$\sin^4 x \cos^2 x = (\sin x \cos x)^2 \sin^2 x = \frac{1}{4}\sin^2 2x \cdot \frac{1 - \cos 2x}{2}$$

$$= \frac{1}{8} \cdot \frac{1}{2}(1 - \cos 4x) - \frac{1}{8} \cdot \sin^2 2x \cos 2x.$$

$$I = \tfrac{1}{16}\int (1 - \cos 4x)\,dx - \tfrac{1}{8}\int \sin^2 2x \cos 2x\,dx$$
$$= \tfrac{1}{16}x - \tfrac{1}{64}\sin 4x - \tfrac{1}{48}\sin^3 2x + C.$$

787. $I = \int \sin^6 x\,dx$

788. $I = \int \sin^4 x \cos^4 x\,dx$

789. $I = \int \dfrac{dx}{\sin^4 x}$

Solution. We substitute $\tan x = t$, $\sin^2 x = \dfrac{t}{1 + t^2}$, $dx = \dfrac{dt}{1 + t^2}$.

$$I = \int \frac{dt/(1 + t^2)}{t^4/(1 + t^2)^2} = \int \frac{1 + t^2}{t^4}\,dt = -\frac{1}{3t^3} - \frac{1}{t} + C$$
$$= -\frac{1}{3\tan^3 x} - \frac{1}{\tan x} + C.$$

790. $I = \int \dfrac{dx}{\sin^4 x \cos^4 x}$

791. $I = \int \cos x \cos 3x\,dx$

Solution:

$$I = \tfrac{1}{2}\int (\cos 2x + \cos 4x)\,dx = \tfrac{1}{4}\sin 2x + \tfrac{1}{8}\sin 4x + C.$$

We remark that this could also have been solved by integrating by parts.

792. $I = \int \sin x \sin 2x \sin 3x\,dx$

793. $I = \int \sin^2 x \cos^2 3x\,dx$

794. $I = \int \dfrac{dx}{3 + 5\cos x}$

Solution. In this case we shall substitute $\tan(x/2) = t$.

$$I = \int \frac{\dfrac{2\,dt}{1 + t^2}}{3 + 5\dfrac{1 - t^2}{1 + t^2}} = \int \frac{2\,dt}{8 - 2t^2} = \int \frac{dt}{4 - t^2} = \frac{1}{4}\int \left(\frac{1}{2 - t} + \frac{1}{2 + t}\right) dt$$
$$= \frac{1}{4}\ln\left|\frac{2 + t}{2 - t}\right| + C = \frac{1}{4}\ln\left|\frac{2 + \tan(x/2)}{2 - \tan(x/2)}\right| + C.$$

The result can be further reduced to the form

$$\tfrac{1}{4}\ln\left|\frac{5 + 3\cos x + 4\sin x}{3 + 5\cos x}\right| + C.$$

795. $I = \int \dfrac{dx}{1 + 3\cos^2 x}$

796. $I = \int \dfrac{dx}{1 + 2\tan x}$

797. $I = \int \frac{\sin x - \cos x}{\sin x + \cos x}\, dx$

798. $I = \int \frac{dx}{a \sin x + b \cos x}$

799. $I = \int \frac{\sqrt{\tan x}\, dx}{\sin x \cos x}$

800. $I = \int \frac{\cos 2x}{\sin^4 x\, dx}$

801. $I = \int \sqrt{\frac{\sin^3 x}{\cos^7 x}}\, dx$

8.8 INTEGRALS OF EXPONENTIAL AND HYPERBOLIC FUNCTIONS

We shall not give any special procedure for these integrals. Each case has to be handled according to its special nature.

802. $I = \int \frac{dx}{\sqrt{e^{2x} + e^x + 1}}$

Solution. We shall use the substitution $e^x = t$, $e^x\, dx = dt$. Then

$$I = \int \frac{dt}{t\sqrt{t^2 + t + 1}} = -\ln\left(\frac{1}{t} + \frac{1}{2} + \sqrt{\frac{1}{t^2} + \frac{1}{t} + 1}\right) + C$$

$$= -\ln\left(e^{-x} + \tfrac{1}{2} + \sqrt{e^{-2x} + e^{-x} + 1}\right) + C$$

$$= x - \ln\left[1 + \tfrac{1}{2}e^x + \sqrt{e^{2x} + e^x + 1}\right] + C.$$

We used the solution of Problem 771.

803. $I = \int \frac{xe^x}{(x + 1)^2}\, dx$

Solution. Here we shall use integration by parts.

$$u = xe^x, \qquad du = e^x(1 + x)\, dx,$$

$$dv = \frac{dx}{(x + 1)^2}, \qquad v = -\frac{1}{x + 1}.$$

$$I = -\frac{xe^x}{x + 1} + \int \frac{e^x(1 + x)}{1 + x}\, dx = -\frac{xe^x}{x + 1} + e^x + C = \frac{e^x}{x + 1} + C.$$

804. $I = \int \frac{x + \sinh x}{\cosh x - \sinh x}\, dx$

Solution. Here it is convenient to use the explicit expressions for the hyperbolic functions:

$$I = \int \frac{x + \frac{1}{2}(e^x - e^{-x})}{e^{-x}}\,dx = \int xe^x\,dx + \frac{1}{2}\int (e^{2x} - 1)\,dx$$

$$= xe^x - e^x + \frac{1}{4}e^{2x} - \frac{1}{2}x + C.$$

805. $I = \int \frac{\tanh x}{\sqrt{1 - \tanh x}}\,dx$

806. $I = \int (\cosh x + \sinh x)\sqrt{\cosh x - \sinh x}\,dx$

807. $I = \int \frac{dx}{(1 - \cosh x)^2}$

Solution. Here it is more convenient to retain the hyperbolic notation. We use the hyperbolic identities

$$\cosh x - 1 = 2\sinh^2(x/2) \text{ and } \cosh^2 u - \sinh^2 u = 1$$

$$I = \int \frac{dx}{4\sinh^4(x/2)} = \frac{1}{4}\int \frac{\cosh^2(x/2) - \sinh^2(x/2)}{\sinh^4(x/2)}\,dx$$

$$= \frac{1}{4}\int \frac{\coth^2(x/2)}{\sinh^2(x/2)}\,dx - \frac{1}{4}\int \frac{dx}{\sinh^2(x/2)} = -\frac{1}{6}\coth^3\frac{x}{2} + \frac{1}{2}\coth\frac{x}{2} + C.$$

808. $I = \int \frac{dx}{1 - \sinh^4 x}$

809. $I = \int \frac{\sinh^3 x\,dx}{\sqrt{\cosh x}}$

810. $I = \int \frac{x\,dx}{\cosh^2 x}$

811. $I = \int \cosh x \cosh 2x \cosh 3x\,dx$

8.9 MISCELLANEOUS INTEGRALS

In this section we treat an assortment of various integrals, not arranged according to types. Some of them will be solved in detail, for others only the answers are given. This section can be used as a recapitulation of the whole chapter.

812. $I = \int \frac{\ln(x+1)}{\sqrt{x+1}}\,dx$

813. $I = \int \frac{dx}{1 + \sin x + \cos x}$

814. $I = \int \frac{\arctan x}{(1+x)^3}\,dx$

815. $I = \int \frac{x + \sin x}{1 + \cos x}\,dx$

816. $I = \int \frac{x^2 - 1}{x^2 + 1} \cdot \frac{dx}{\sqrt{1 + x^4}}$

817. $I = \int e^{\sin x}\,\frac{x\cos^3 x - \sin x}{\cos^2 x}\,dx$

818. $I = \int \frac{2x^3 - 4}{(x^2 + 1)(x + 1)^2} \, dx$ **819.** $I = \int \frac{(x + 3)^2}{(x^2 + 4x + 5)^2} \, dx$

820. $I = \int \frac{2x^2 - 3x - 3}{(x - 1)(x^2 - 2x + 5)} \, dx$

821. $I = \int \frac{2x^3 + 1}{x^5 + x^3} \, dx$ **822.** $I = \int \frac{x^5}{1 + x^4} \, dx$

823. $I = \int (1 - x)\sqrt{2x - x^2} \, dx$ **824.** $I = \int \frac{x + 3}{x\sqrt{2x + 4}} \, dx$

825. $I = \int \frac{3x + 11}{\sqrt{3x^2 - 4x + 6}} \, dx$ **826.** $I = \int \frac{\sqrt{a^2 - x^2}}{x^4} \, dx$

827. $I = \int \sqrt{\frac{a + x}{a - x}} \frac{dx}{x}$ **828.** $I = \int \frac{x \, dx}{\sqrt{4x + 2 - x^2}}$

829. $I = \int \frac{dx}{\sin^3 x}$

830. $I = \int \frac{dx}{a \sin^2 x + 2b \sin x \cos x + c \cos^2 x}$

Investigate the three cases $b^2 - ac \gtreqless 0$.

831. $I = \int \frac{\sin^4 x}{\cos^2 x} \, dx$ **832.** $I = \int \frac{e^x + 1}{e^x - 1} \, dx$

833. $I = \int \ln \sqrt{x^2 + 1} \, dx$ **834.** $I = \int x \ln (x^2 - 1) \, dx$

835. $I = \int \frac{x \arctan x}{(1 + x^2)^2} \, dx$ **836.** $I = \int \frac{x^5}{\sqrt{1 - x^2}} \, dx$

837. $I = \int \frac{\sin x \cos^3 x}{1 + \cos^2 x} \, dx$ **838.** $I = \int x \cos^2 x \, dx$

839. $I = \int x^2(\sin x + \cos x) \, dx$ **840.** $I = \int \frac{x^4 \, dx}{(1 - x^2)^{3/2}}$

841. $I = \int \arctan \sqrt{x} \, dx$ **842.** $I = \int \frac{2\sqrt{x + 1} + 1}{(x + 1)^2 - \sqrt{x + 1}} \, dx$

843. $I = \int \frac{1 + \tan x}{\sin 2x} \, dx$

IX

THE DEFINITE INTEGRAL

9.1 DEFINITION

Given a function $y = f(x)$ defined and bounded in the interval $[a,b]$; divide this interval by the $n + 1$ points

$$a = x_0 < x_1 < x_2 < \ldots < x_i < \ldots < x_n = b$$

into n subintervals. Put

$$\Delta x_i = x_i - x_{i-1}, \quad i = 1, 2, \ldots, n.$$

In each subinterval take an arbitrary point x_i^*, $x_{i-1} \leqslant x_i^* \leqslant x_i$. Now consider the sum

$$S = \sum_{i=1}^{n} f(x_i^*) \, \Delta x_i.$$

Sums of such a type depend, of course, on the particular subdivision of $[a,b]$ and on the choice of the points x_i^* in each subinterval.

If when the subdivision is successively refined so that the maximal Δx_i tends to zero, it results that the corresponding sums tend to a limit which is independent of the form of the initial partition of $[a,b]$, of the particular choice of the x_i^*, and of the manner of successive refinements, then this limit is denoted by

$$\int_a^b f(x) \, dx,$$

which is called the (definite) integral of $f(x)$ from a to b. The number a is called the lower limit and b the upper limit of the integral.

Although the notion of the definite integral just defined may seem at first sight somewhat artificial, very many practical problems, such as the computation of areas and volumes and many quantities occurring in physics, lead in a natural way to this notion.

Various applications of the definite integral will be discussed in the next chapter. The present chapter is devoted to the study of the properties of the definite integral and of the means by which it may be evaluated.

844. Find the area bounded by the x axis, the straight lines $x = 2$ and $x = 4$, and the parabola $y = x^2/4$.

Solution. The area in question is shown in Figure 45. We divide the segment [2,4] on the x axis into n parts by the points

$$2 = x_0 < x_1 < x_2 \ldots < x_n = 4.$$

The parts can be chosen equal, i.e., $\Delta x_i = x_i - x_{i-1} = (4 - 2)/n = 2/n$, but this is not necessary. We erect ordinates at the points x_i and divide

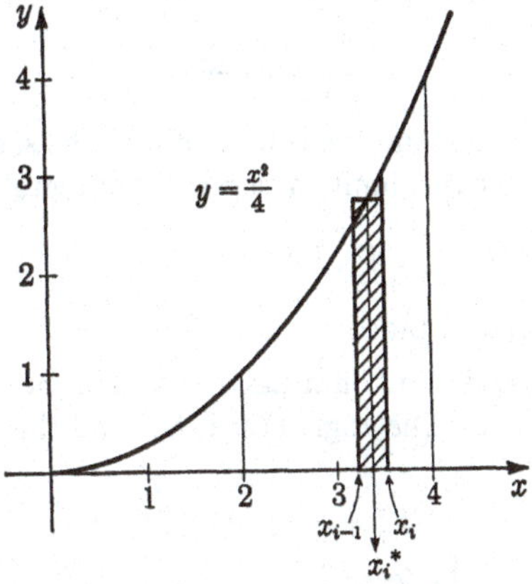

FIGURE 45

the area into n thin curvilinear trapezoids, one of which is represented in the figure by the crosshatched area.

Next we choose an arbitrary point in each small interval and denote it by x_i^* $i = 1, 2, \ldots, n$ $(x_{i-1} \leqslant x_i^* \leqslant x_i)$. This point may be chosen in the middle of the interval, but this is not essential. The product $f(x_i^*)\,\Delta x_i$ represents the area of a rectangle having the same base as the thin trapezoid, and altitude $f(x_i^*)$.

It can be shown that the sum of the areas of these rectangles,

$$\sum_{i=1}^{n} f(x_i^*)\,\Delta x_i,$$

will approach a well-defined limit if the condition that $\max \Delta x_i \to 0$ is fulfilled. This limit, which is denoted by

$$\int_2^4 \frac{x^2}{4}\,dx,$$

seems intuitively to be the number which measures the area. It is in fact *defined* to be the area. Its evaluation will be postponed.

845. A point moves along a straight line with the velocity $v = 3t + 1$ m/sec. Find the distance traversed by the moving point in the first ten seconds of its motion.

Solution. We divide the time interval [0,10] into n subintervals. We have good reason to believe that the velocity changes very little within each small subinterval and can be considered nearly constant in it. We choose a point t_i^* satisfying $t_{i-1} \leqslant t_i^* \leqslant t_i$, and compute $v(t_i^*)$. This velocity, multiplied by $\Delta t_i = t_i - t_{i-1}$, will yield an approximate value for the distance passed over by the point in the time interval Δt_i. Now we form the sum

$$\sum_{i=1}^{n} v(t_i^*)\,\Delta t_i.$$

It can be proved that this sum tends to a limit in the sense of the definition of this section and that this limit, which is the integral

$$\int_0^{10} (3t + 1)\,dt,$$

represents the required distance.

846. A directed force **F** acts on a particle which moves along the x axis from $x = -2$ to $x = 5$. The angle $\mathbf{F}Ox$ is 60° and the absolute value of **F** is a function of x,

$$|\mathbf{F}| = x^2 + 2x + 1.$$

Find the work done by the force.

9.2 BASIC PROPERTIES OF THE DEFINITE INTEGRAL

A function, for which the limit as described in Section 9.1 exists, is said to be integrable in the interval $[a,b]$. It is clear that a necessary condition for integrability is boundedness of the functions. The following two theorems give sufficient conditions:

(A) Every function continuous in the interval $[a,b]$ is integrable in this interval.

(B) A function bounded in $[a,b]$, and having in this interval only a finite number of points of discontinuity, is integrable in $[a,b]$.

We remark that the class of integrable functions is not exhausted by categories (A) and (B).

We now list several basic properties of the definite integral. All functions appearing below are assumed to be integrable in appropriate intervals.

(1) $\int_a^b [f(x) \pm g(x)]\,dx = \int_a^b f(x)\,dx \pm \int_a^b g(x)\,dx.$

(2) $\int_a^b Cf(x)\,dx = C\int_a^b f(x)\,dx$, C a constant.

(3) $\int_b^a f(x)\,dx = -\int_a^b f(x)\,dx.$

(4) $\int_a^a f(x)\,dx = 0.$

(5) $\int_a^c f(x)\,dx + \int_c^b f(x)\,dx = \int_a^b f(x)\,dx.$

(6) If in $[a,b]$ $f(x) \geqslant g(x)$, then

$$\int_a^b f(x)\,dx \geqslant \int_a^b g(x)\,dx.$$

(7) $\left|\int_a^b f(x)\,dx\right| \leqslant \int_a^b |f(x)|\,dx.$

It should be remarked that if $f(x)$ is integrable in $[a,b]$, then also $|f(x)|$ is integrable in this interval.

(8) Suppose that in $[a,b]$ $m \leqslant f(x) \leqslant M$, where m and M are two constants. Then

$$m(b-a) \leqslant \int_a^b f(x)\,dx \leqslant M(b-a).$$

Thus we can estimate the value of a definite integral.

(9) If $f(x)$ is continuous in $[a,b]$, then there exists in this interval at least one point c such that

$$\frac{\int_a^b f(x)\,dx}{b-a} = f(c).$$

The left-hand member of this relation is called the average value of $f(x)$ in $[a,b]$. This theorem is the mean value theorem of the integral calculus.

(10) Suppose the upper limit of the integral, denoted by u, is permitted to vary. Then the definite integral becomes a function of u:

$$I(u) = \int_a^u f(x)\,dx.$$

(u varies within an interval where $f(x)$ is integrable.) For values of u where $f(x)$ is continuous we can differentiate $I(u)$. We obtain

$$\frac{dI(u)}{du} = \left[\int_a^u f(x)\,dx\right]' = f(u),$$

the derivative of the integral with respect to its upper limit.

Using also property (3) we obtain

$$\left[\int_u^b f(x)\,dx\right]' = -f(u).$$

(11) Let $F(x)$ be an arbitrary indefinite integral (primitive function) of $f(x)$, i.e.,

$$\int f(x)\,dx = F(x) + C.$$

Then

$$\int_a^b f(x)\,dx = F(b) - F(a).$$

This formula, which is the Newton-Leibniz formula, is one of the most important in mathematics.

Remark. It follows clearly from the definition of the definite integral that

$$\int_a^b f(x)\,dx = \int_a^b f(u)\,du = \int_a^b f(t)\,dt, \text{ etc.}$$

In the remainder of this chapter we shall present various problems illustrating the above notions.

9.3 EVALUATION OF THE DEFINITE INTEGRAL FROM ITS DEFINITION

847. Evaluate by definition $\int_0^1 x^3\,dx$.

Solution. We divide the interval [0,1] into n equal parts by the points

$$0, \frac{1}{n}, \frac{2}{n}, \ldots, \frac{i-1}{n}, \frac{i}{n}, \ldots, \frac{n-1}{n}, 1.$$

Every subinterval is

$$\Delta x_i = \frac{i}{n} - \frac{i-1}{n} = \frac{1}{n}.$$

We choose in every subinterval the point $x_i^* = i/n$ (i.e., the right-hand end of this interval). The value of the function at x_i^* is $(x_i^*)^3 = i^3/n^3$. Now we form the sum

$$S_n = \left(\frac{1}{n}\right)^3 \cdot \frac{1}{n} + \left(\frac{2}{n}\right)^3 \cdot \frac{1}{n} + \ldots + \left(\frac{i}{n}\right)^3 \cdot \frac{1}{n} + \ldots + \left(\frac{n}{n}\right)^3 \cdot \frac{1}{n}$$

$$= \sum_{i=1}^{n} \left(\frac{i}{n}\right)^3 \cdot \frac{1}{n} = \frac{1}{n^4} \sum_{i=1}^{n} i^3 = \frac{1}{n^4} \cdot \frac{n^2(n+1)^2}{4} = \frac{(n+1)^2}{4n^2}.$$

We used here the known formula for the sum of the cubes of the first n positive integers. The next and last step is to find the limit,

$$\lim_{n\to\infty} S_n = \lim_{n\to\infty} \frac{(n+1)^2}{4n^2} = \frac{1}{4}.$$

(In case of division into equal parts $n \to \infty$ implies $\max \Delta x_i \to 0$.)

We may disregard the fact that we used a very special way of subdividing the interval [0,1] and choosing the points in the partial intervals. The function x^3 is continuous and, by theorem (A) of Section 9.2, the integral $\int_0^1 x^3\,dx$ exists; hence, if one of the sequences of sums tends to the limit 1/4 so will all other possible sequences. We can write

$$\int_0^1 x^3\,dx = \tfrac{1}{4}.$$

The same integral can be immediately calculated by the Newton-Leibniz formula:

$$\int_0^1 x^3\,dx = \frac{x^4}{4}\bigg|_0^1 = \frac{(1)^4}{4} - \frac{(0)^4}{4} = \frac{1}{4}.$$

This simple example illustrates the power of the formula.

Remark. Note that we write

$$\int_a^b f(x)\,dx = F(x)\Big|_a^b = F(b) - F(a).$$

848. Find $\int_0^1 e^x\,dx$, using the definition of the definite integral.

849. From the definition find $\int_1^2 \frac{dx}{x}$.

Solution. Here we shall divide the segment [1,2] into parts so that the points of division form a geometrical progression. We set $a_0 = 1$, $a_n = 2$. The number of terms is $n + 1$; the last term is $a_n = a_0 q^n$, i.e., $2 = 1 \cdot q^n$. Consequently $q = 2^{1/n}$ and $a_i = q^i = 2^{i/n}$. We choose $x_i^* = a_{i-1} = q^{i-1}$ and form the sum

$$S = \sum_{i=1}^{n} f(x_i^*)\,\Delta x_i = \sum_{i=1}^{n} \frac{1}{q^{i-1}}(a_i - a_{i-1}) = \sum_{i=1}^{n} \frac{1}{q^{i-1}}\,a_{i-1}(q-1)$$

$$= \sum_{i=1}^{n} \frac{1}{q^{i-1}}\,q^{i-1}(q-1) = \sum_{i=1}^{n} (q-1) = (q-1)n = (2^{1/n}-1)n.$$

Now

$$\lim_{x\to\infty} (2^{1/x}-1)x = \lim_{x\to\infty} \frac{2^{1/x}-1}{1/x} = \lim_{x\to\infty} \frac{2^{1/x}(-1/x^2)\ln 2}{-1/x^2} = \ln 2.$$

It follows that

$$\int_1^2 \frac{dx}{x} = \lim_{n\to\infty} (2^{1/n}-1)n = \ln 2.$$

By the Newton-Leibniz formula

$$\int_1^2 \frac{dx}{x} = \ln x \bigg|_1^2 = \ln 2 - \ln 1 = \ln 2.$$

850. Compute as above $\int_0^\pi \sin x\, dx$.

Solution. Let us divide the segment $[0,\pi]$ into n equal parts. We choose x_i^* at the right end of every subinterval. We have $\Delta x_i = \pi/n$ and $f(x_i^*) = \sin(\pi i/n)$.

$$S = \sum_{i=1}^{n} \sin\frac{\pi i}{n} \cdot \frac{\pi}{n} = \frac{\pi}{n}\sum_{i=1}^{n} \sin\frac{\pi i}{n} = \frac{\pi}{n} \cdot \frac{\cos(\pi/2n)}{\sin(\pi/2n)},$$

$$\lim_{x\to\infty} \frac{\frac{\pi}{x}\cos\frac{\pi}{2x}}{\sin\frac{\pi}{2x}} = \lim_{x\to\infty} \frac{\frac{\pi}{2x}\cdot 2}{\sin\frac{\pi}{2x}} = 2,$$

i.e.,

$$\int_0^\pi \sin x\, dx = 2.$$

By the Newton-Leibniz formula,

$$\int_0^\pi \sin x\, dx = -\cos x \bigg|_0^\pi = -\cos\pi - (-\cos 0) = 1 - (-1) = 2.$$

Now we shall prove the formula for the sum used in the problem just solved.

$$C = \sin\alpha + \sin 2\alpha + \ldots + \sin n\alpha;$$

$$2C\sin\frac{\alpha}{2}$$

$$= 2\sin\alpha\sin\frac{\alpha}{2} + 2\sin 2\alpha\sin\frac{\alpha}{2} + \ldots + 2\sin n\alpha\sin\frac{\alpha}{2}$$

$$= \cos\frac{\alpha}{2} - \cos\frac{3\alpha}{2} + \cos\frac{3\alpha}{2} - \cos\frac{5\alpha}{2} + \ldots + \cos\frac{2n-1}{2}\alpha - \cos\frac{2n+1}{2}\alpha$$

$$= \cos\frac{\alpha}{2} - \cos\left(n + \frac{1}{2}\right)\alpha = 2\sin\frac{n+1}{2}\alpha\sin\frac{n}{2}\alpha;$$

i.e.,

$$C = \frac{\sin\frac{n+1}{2}\alpha\sin\frac{n}{2}\alpha}{\sin\frac{\alpha}{2}}.$$

In our case $\alpha = \pi/n$ and we obtain

$$C\left(\alpha = \frac{\pi}{n}\right) = \frac{\sin\left(\frac{n+1}{2}\cdot\frac{\pi}{n}\right)\sin\left(\frac{n}{2}\cdot\frac{\pi}{n}\right)}{\sin\frac{\pi}{2n}} = \frac{\sin\left(\frac{\pi}{2}+\frac{\pi}{2n}\right)}{\sin\frac{\pi}{2n}} = \frac{\cos\frac{\pi}{2n}}{\sin\frac{\pi}{2n}}.$$

9.4 ESTIMATION OF DEFINITE INTEGRALS

851. Prove the inequality $\frac{2}{\sqrt[4]{e}} \leqslant \int_0^2 e^{x^2-x}\,dx \leqslant 2e^2$.

Solution. Let us find the greatest and smallest values of the integrand in [0,2].

$$f(x) = e^{x^2-x}, \qquad f'(x) = e^{x^2-x}(2x-1);$$

$2x - 1 = 0$, $x = \frac{1}{2}$; $f'(x)$ changes sign from $-$ to $+$ at $x = \frac{1}{2}$; i.e., $(\frac{1}{2}, e^{1/4-1/2} = e^{-1/4})$ is a minimum point. The greatest value of the function will thus be at one of the ends of the interval. We compute

$$f(0) = 1, \qquad f(2) = e^{4-2} = e^2.$$

Then clearly

$$M = e^2 \quad \text{and} \quad m = e^{-1/4},$$

with M denoting the greatest value and m the smallest value of the function in the interval of integration.

By property (8) of Section 9.2 we have

$$2\cdot e^{-1/4} \leqslant \int_0^2 e^{x^2-x}\,dx \leqslant 2\cdot e^2.$$

852. Estimate the integral $\int_0^2 \frac{x^2+5}{x^2+2}\,dx$.

853. Without evaluation compare the following pairs of integrals:

(a) $\int_0^1 x^2\,dx$ and $\int_0^1 x^3\,dx$.

(b) $\int_1^2 x^2\,dx$ and $\int_1^2 x^3\,dx$.

The inequality

$$\int_a^b f_1(x)f_2(x)\,dx \leqslant \sqrt{\int_a^b [f_1(x)]^2\,dx}\,\sqrt{\int_a^b [f_2(x)]^2\,dx}$$

is known as the Cauchy-Schwarz inequality for integral calculus. To prove it we consider the inequality

$$0 \leqslant \int_a^b [f_1(x) - tf_2(x)]^2\,dx$$
$$= \int_a^b [f_1(x)]^2 - 2t\int_a^b f_1(x)f_2(x)\,dx + t^2\int_a^b [f_2(x)]^2\,dx,$$

which holds for any t. Consequently the discriminant of this trinomial in t cannot be positive, and this is the Cauchy-Schwarz inequality.

854. Use the Cauchy-Schwarz inequality to estimate the integral

$$\int_0^{\pi/2} \sqrt{x \sin x}\,dx.$$

Solution. Put $f_1(x) = \sqrt{x}$, $f_2(x) = \sqrt{\sin x}$. Then

$$\int_0^{\pi/2} \sqrt{x \sin x}\,dx \leqslant \sqrt{\int_0^{\pi/2} x\,dx} \cdot \sqrt{\int_0^{\pi/2} \sin x\,dx} = \sqrt{\frac{\pi^2}{8}} \cdot 1 = \frac{\pi}{2\sqrt{2}} \approx 1.11.$$

The integrals on the right hand side were evaluated by the Newton-Leibniz formula.

Remark. The maximal value of $x \sin x$ in $[0,\pi/2]$ is easily seen to occur at $x = \pi/2$, i.e.,

$$\int_0^{\pi/2} \sqrt{x \sin x}\,dx \leqslant \frac{\pi}{2} \cdot \sqrt{\frac{\pi}{2}} \cdot 1 = \frac{\pi\sqrt{\pi}}{2\sqrt{2}} \approx 1.97.$$

The estimation by the Cauchy-Schwarz inequality is much better.

855. Estimate the integral $\int_0^1 \sqrt{1 + x^3}\,dx$.

9.5 THE MEAN VALUE THEOREM OF INTEGRAL CALCULUS

856. Find the average value $\overline{f(x)}$ of $f(x) = 2x^2 + 3x + 3$ in the interval [1.4]. (See property (9), Section 9.2.)

Solution:

$$\overline{f(x)} = \frac{1}{4-1}\int_1^4 f(x)\,dx = \frac{1}{3}\int_1^4 (2x^2 + 3x + 3)\,dx$$
$$= \tfrac{1}{3}\left[\tfrac{2}{3}x^3 + \tfrac{3}{2}x^2 + 3x\Big|_1^4\right] = \tfrac{1}{3}\left(\tfrac{128}{3} + 24 + 12 - \tfrac{2}{3} - \tfrac{3}{2} - 3\right) = 24.5.$$

The integrand is continuous, so that there exists at least one point c in the interval [1,4] such that $f(c) = 24.5$. Let us find such a point.

$$2x^2 + 3x + 3 = 24.5, \qquad 2x^2 + 3x - 21.5 = 0;$$
$$x_{1,2} = \frac{-3 \pm \sqrt{9 + 172}}{4} = \frac{-3 \pm \sqrt{181}}{4}.$$

The solution $\frac{-3+\sqrt{181}}{4}$ is inside the above interval.

857. Find the mean value $\bar{y}$ of $y = \sqrt{a^2 - x^2}$ in the interval $[-a,a]$.

9.6 INTEGRALS WITH VARIABLE LIMITS

858. Evaluate the integrals

$$\text{(a)} \int_a^x t^5\,dt \qquad \text{(b)} \int_a^x x^5\,dx \qquad \text{(c)} \int_x^a \frac{du}{\sqrt{u}}$$

Solution:

$$\text{(a)} \quad \int_a^x t^5\,dt = \frac{t^6}{6}\bigg|_a^x = \frac{x^6}{6} - \frac{a^6}{6}.$$

$$\text{(b)} \quad \int_a^x x^5\,dx = \frac{x^6}{6}\bigg|_a^x = \frac{x^6}{6} - \frac{a^6}{6}.$$

$$\text{(c)} \quad \int_x^a \frac{du}{\sqrt{u}} = 2\sqrt{u}\bigg|_x^a = 2\sqrt{a} - 2\sqrt{x}.$$

859. A curvilinear trapezoid is bounded by the curve $y = x^3$, the x axis, and the ordinate at a variable point x. Compute the increment ΔA of the area of this trapezoid, and the differential dA. Find also the absolute error α and the relative error $\beta = \alpha/\Delta A$ occurring when the increment is replaced by the differential, with $x = 4$, and $\Delta x = 1$ and 0.1.

860. Find the derivative of $y = \int_0^x \frac{1-t+t^2}{1+t+t^2}\,dt$ at $x = 1$.

Solution:

$$\frac{dy}{dx} = \frac{1-x+x^2}{1+x+x^2}, \qquad \frac{dy}{dx}(x=1) = \frac{1}{3}.$$

861. Find the derivatives of $y = \int_x^5 \sqrt{1+t^2}\,dt$ at $x = 0$ and $x = \frac{3}{4}$.

862. Find y' if $y = \int_1^{2x} \frac{\sin t}{t^2}\,dt$.

Solution:

$$\frac{dy}{dx} = \frac{d}{dx}\int_1^{2x}\frac{\sin t}{t^2}\,dt = \frac{d}{d(2x)}\int_1^{2x}\frac{\sin t}{t^2}\,dt \cdot \frac{d(2x)}{dx} = \frac{\sin 2x}{(2x)^2}\cdot 2 = \frac{\sin 2x}{2x^2}.$$

863. Find the derivatives of

(a) $\int_{x^3}^2 \ln x\,dx$

(b) $\int_x^{2x} \ln^2 t\,dt.$

Solution of (b):

$$\frac{d}{dx}\int_x^{2x} \ln^2 t\,dt = \frac{d}{dx}\left[\int_x^{a} \ln^2 t\,dt + \int_a^{2x} \ln^2 t\,dt\right] = -\ln^2 x + 2\ln^2 2x.$$

864. Find dy/dx of the implicit function given by

$$\int_0^y e^t\,dt + \int_0^x \cos t\,dt = 0.$$

Solution. We shall differentiate the equation with respect to x,

$$y'e^y + \cos x = 0,\quad y' = -\frac{\cos x}{e^y}.$$

865. Find dy/dx if

$$y = \int_{t^2}^1 t^2 \ln t\,dt,\qquad x = \int_1^{t^2} t\ln t\,dt.$$

9.7 EVALUATION OF DEFINITE INTEGRALS

Evaluate the following definite integrals.

866. $I = \int_{-2}^{-1} \frac{dx}{(11 + 5x)^3}$

867. $I = \int_0^1 (e^x - 1)^4 e^x\,dx$

868. $I = \int_0^\pi \cos^2 \frac{t}{2}\,dt$

869. $I = \int_0^{a/2} \frac{a\,dx}{(x - a)(x - 2a)}$

870. Find the average value of $y = \cos^3 x$ in the interval $[0,\pi]$.

Evaluation of a definite integral by parts can be performed using the formula

$$\int_a^b u\,dv = uv\Big|_a^b - \int_a^b v\,du.$$

One may also use the generalized formula

$$\int_a^b uv^{(n+1)}\,dx = \left[uv^{(n)} - u'v^{(n-1)} + \dots + (-1)^n u^{(n)}v\right]\Big|_a^b + (-1)^{n+1}\int_a^b u^{(n+1)}v\,dx.$$

Evaluate the following integrals using integration by parts.

871. $I = \int_0^1 xe^{-x}\,dx$

Solution:

$$u = x,\quad dv = e^{-x}\,dx;\qquad du = dx,\quad v = -e^{-x}.$$

$$I = -xe^{-x}\Big|_0^1 + \int_0^1 e^{-x}\,dx = -\frac{1}{e} - e^{-x}\Big|_0^1 = -\frac{2}{e} + 1.$$

872. $I = \int_0^\pi x^3 \sin x \, dx$

Solution. Here we shall use the generalized formula.

$$v''' = \sin x, \qquad v'' = -\cos x, \qquad v' = -\sin x, \qquad v = \cos x.$$

$$I = [-x^3 \cos x + 3x^2 \sin x + 6x \cos x]\Big|_0^\pi - 6 \int_0^\pi \cos x \, dx$$

$$= \pi^3 - 6\pi - 6 \sin x\Big|_0^\pi = \pi^3 - 6\pi.$$

873. $I = \int_0^{e-1} \ln (x + 1) \, dx$ **874.** $I = \int_0^{\pi/2} e^{2x} \cos x \, dx$

875. Find a reduction formula for the integral $I_n = \int_0^{\pi/2} \sin^n x \, dx$.

Solution:

$$u = \sin^{n-1} x, \qquad dv = \sin x \, dx;$$

$$du = (n-1) \sin^{n-2} x \cos x \, dx, \qquad v = -\cos x.$$

$$I_n = -\sin^{n-1} x \cos x\Big|_0^{\pi/2} + (n-1) \int_0^{\pi/2} \sin^{n-2} x \cos^2 x \, dx$$

$$= (n-1) \int_0^{\pi/2} (\sin^{n-2} x - \sin^n x) \, dx = (n-1)I_{n-2} - (n-1)I_n.$$

Consequently the reduction formula is $I_n = \dfrac{n-1}{n} I_{n-2}$. For example,

$$\int_0^{\pi/2} \sin^5 x \, dx = \frac{4}{5} \int_0^{\pi/2} \sin^3 x \, dx = \frac{4}{5} \cdot \frac{2}{3} \int_0^{\pi/2} \sin x \, dx$$

$$= \frac{4 \cdot 2}{5 \cdot 3} (-\cos x)\Big|_0^{\pi/2} = \frac{4 \cdot 2}{5 \cdot 3} = \frac{8}{15}.$$

In general we obtain

$$I_{2n} = \int_0^{\pi/2} \sin^{2n} x \, dx = \frac{(2n-1)(2n-3)\cdots 3 \cdot 1}{2n(2n-2)\cdots 4 \cdot 2} \cdot \frac{\pi}{2}$$

(because $I_0 = \int_0^{\pi/2} dx = \pi/2$), and

$$I_{2n+1} = \int_0^{\pi/2} \sin^{2n+1} x \, dx = \frac{2n(2n-2)\cdots 4 \cdot 2}{(2n+1)(2n-1)\cdots 5 \cdot 3}.$$

Remark. Using the two last expressions we can obtain the famous formula of Wallis. Let us find

$$\lim_{n \to \infty} \frac{I_{2n}}{I_{2n+1}}.$$

First we observe that in the interval $(0, \pi/2)$, $\sin^{2n-1} x > \sin^{2n} x > \sin^{2n+1} x$. Thus

$$I_{2n-1} > I_{2n} > I_{2n+1}.$$

This together with the reduction formula

$$I_{2n+1} = \frac{2n}{2n+1} I_{2n-1},$$

implies

$$1 < \frac{I_{2n}}{I_{2n+1}} < \frac{I_{2n-1}}{I_{2n+1}} = \frac{2n+1}{2n}.$$

But

$$\lim_{n\to\infty} \frac{2n+1}{2n} = 1, \qquad \text{i.e., } \lim_{n\to\infty} \frac{I_{2n}}{I_{2n+1}} = 1.$$

We substitute the above expressions for I_{2n} and I_{2n+1},

$$\lim_{n\to\infty} \frac{I_{2n}}{I_{2n+1}} = \lim_{n\to\infty} \frac{(2n+1)[(2n-1)\cdot(2n-3)\cdots 3\cdot 1]^2}{[2n\cdot(2n-2)\cdots 4\cdot 2]^2} \cdot \frac{\pi}{2} = 1.$$

From this we obtain the formula of Wallis

$$\frac{\pi}{2} = \lim_{n\to\infty} \frac{2\cdot 2\cdot 4\cdot 4\cdot 6\cdot 6\cdots 2n\cdot 2n}{1\cdot 1\cdot 3\cdot 3\cdot 5\cdot 5\cdots (2n-1)(2n-1)(2n+1)}.$$

9.8 CHANGING THE VARIABLE OF INTEGRATION

Suppose that $f(x)$ is a continuous function in $[a,b]$ and we have to evaluate

$$\int_a^b f(x)\,dx.$$

Suppose further that $x = g(t)$ is a continuous function with a continuous derivative $g'(t)$ in an interval $[t_1,t_2]$. If in addition

$$g(t_1) = a,\ g(t_2) = b, \quad \text{and} \quad t_1 \leqslant t \leqslant t_2 \Rightarrow a \leqslant g(t) \leqslant b,$$

then

$$\int_a^b f(x)\,dx = \int_{t_1}^{t_2} f[g(t)]g'(t)\,dt.$$

In other words, in order to change the variable of integration, we perform the substitution under the integral sign in exactly the same way as in the case of an indefinite integral, and transform the limits according to the equality given by the substitution formula.

We remark that care must be taken to ensure fulfillment of the above conditions. If this is not easy, one can perform the indefinite integration by substitution, return to the original variable, and then substitute the original limits.

Evaluate the following integrals by substitution.

876. $I = \int_{1/2}^{\sqrt{3}/2} \frac{dx}{x\sqrt{1-x^2}}$

Solution. We substitute $x = \sin t$, $dx = \cos t\, dt$. The new limits are obtained from

$$\sin t_1 = \frac{1}{2}, \quad t_1 = \frac{\pi}{6}; \qquad \sin t_2 = \frac{\sqrt{3}}{2}, \quad t_2 = \frac{\pi}{3} \cdot 1$$

There are infinitely many values of t satisfying these equations, but when t changes between $\pi/6$ and $\pi/3$, x changes between $1/2$ and $\sqrt{3}/2$; hence t_1 and t_2 satisfy the conditions given above. Now

$$I = \int_{\pi/6}^{\pi/3} \frac{\cos t\, dt}{\sin t \cos t} = \int_{\pi/6}^{\pi/3} \frac{dt}{\sin t} = \ln\left|\tan\frac{t}{2}\right|\Bigg|_{\pi/6}^{\pi/3}$$

$$= \ln\tan\frac{\pi}{6} - \ln\tan\frac{\pi}{12} = \ln\frac{\sqrt{3}}{3} - \ln(2-\sqrt{3})$$

$$= \ln\frac{\sqrt{3}(2+\sqrt{3})}{3} = \ln\frac{3+2\sqrt{3}}{3}.$$

The same integral can be solved using another substitution, $\sqrt{1-x^2} = u$. Then

$$1 - x^2 = u^2, \quad -2x\, dx = 2u\, dx, \quad \text{i.e.,} \ \frac{dx}{x} = \frac{-u\, du}{1-u^2}.$$

The new limits are

$$u_1 = \sqrt{1-\tfrac{1}{4}} = \sqrt{3}/2, \qquad u_2 = \sqrt{1-\tfrac{3}{4}} = \tfrac{1}{2}.$$

$$I = \int_{\sqrt{3}/2}^{1/2} \frac{-u\, du}{(1-u^2)u} = \int_{\sqrt{3}/2}^{1/2} \frac{du}{u^2-1} = \frac{1}{2}\int_{\sqrt{3}/2}^{1/2}\left(\frac{1}{u-1} - \frac{1}{u+1}\right) du$$

$$= \frac{1}{2}\ln\left|\frac{u-1}{u+1}\right|\Bigg|_{\sqrt{3}/2}^{1/2} = \frac{1}{2}\left(\ln\frac{1}{3} - \ln\frac{\sqrt{3}-2}{\sqrt{3}+2}\right) = \frac{1}{2}\ln\frac{\sqrt{3}+2}{3(\sqrt{3}-2)}$$

$$= \frac{1}{2}\ln\frac{7+4\sqrt{3}}{3}.$$

Both results agree, since

$$\left(\frac{3+2\sqrt{3}}{3}\right)^2 = \frac{7+4\sqrt{3}}{3}.$$

877. $I = \int_3^8 \frac{x\, dx}{\sqrt{1+x}}$

878. $I = \int_1^{e^3} \frac{dx}{x\sqrt{1+\ln x}}$

879. $I = \int_0^{\pi/2} \frac{dx}{2\cos x + 3}$

880. $I = \int_1^{\sqrt{3}} \frac{\sqrt{1+x^2}}{x^2}\, dx$

881. $I = \int_0^{\pi/2} \cos^n x \, dx$

Solution. We substitute $x = (\pi/2) - t$, $dx = -dt$. The new limits are $t_1 = \pi/2$ and $t_2 = 0$.

$$I = \int_{\pi/2}^{0} \cos^n\left(\frac{\pi}{2} - t\right)(-dt) = \int_0^{\pi/2} \sin^n t \, dt.$$

This is the integral of Problem 875.

882. $I = \int_0^{\pi} \sin^6 \frac{x}{2} \, dx$ **883.** $I = \int_0^{\pi/4} \cos^7 2x \, dx$

884. Prove that $\int_0^{\pi} xf(\sin x) \, dx = \pi \int_0^{\pi/2} f(\sin x) \, dx$.

Solution:

$$\int_0^{\pi} xf(\sin x) \, dx = \int_0^{\pi/2} xf(\sin x) \, dx + \int_{\pi/2}^{\pi} xf(\sin x) \, dx.$$

We substitute in the second integral:

$$x = \pi - t, \; dx = -dt; \qquad x = \pi/2, \; t = \pi/2; \qquad x = \pi, \; t = 0.$$

$$I = \int_0^{\pi/2} xf(\sin x) \, dx - \int_{\pi/2}^{0} (\pi - t)f[\sin(\pi - t)] \, dt$$

$$= \int_0^{\pi/2} xf(\sin x) \, dx + \pi \int_0^{\pi/2} f(\sin t) \, dt - \int_0^{\pi/2} tf(\sin t) \, dt$$

$$= \pi \int_0^{\pi/2} f(\sin t) \, dt = \pi \int_0^{\pi/2} f(\sin x) \, dx.$$

885. Prove that $\int_a^b f(x) \, dx = \int_a^b f(a + b - x) \, dx$.

886. Given that $I = \int_{-1}^{2} x^2 \, dx = \left.\frac{x^3}{3}\right|_{-1}^{2} = \frac{8}{3} + \frac{1}{3} = 3$, find the error in the following solution of the same integral:

$$x^2 = u, \quad x = \sqrt{u}, \quad dx = \frac{1}{2\sqrt{u}} \, du, \quad u_1 = 1, \quad u_2 = 4.$$

$$I = \int_1^4 \frac{u}{2\sqrt{u}} \, du = \frac{1}{2} \int_1^4 \sqrt{u} \, du = \frac{1}{2} \cdot \frac{2}{3} u^{3/2} \Big|_1^4 = \frac{1}{3}(8 - 1) = \frac{7}{3}.$$

887. (a) Prove that if $f(-x) = f(x)$ (i.e., $f(x)$ is even), then

$$\int_{-a}^{a} f(x) \, dx = 2 \int_0^a f(x) \, dx.$$

(b) Prove that if $f(-x) = -f(x)$ (i.e., $f(x)$ is odd), then

$$\int_{-a}^{a} f(x) \, dx = 0.$$

888. Prove the following:

(a) $\displaystyle\int_{-1}^{1} \frac{x^7 + 2x^5 + 8x}{\cos^4 x + 1}\, dx = 0.$

(b) $\displaystyle\int_{-\pi/12}^{\pi/12} (x^4 + x^2 - 3) \sin^7 2x\, dx = 0.$

(c) $\displaystyle\int_{-2}^{2} e^{\cosh x}\, dx = 2\int_{0}^{2} e^{\cosh x}\, dx.$

889. Given a periodic, integrable function $f(x)$ with a period T, show that

$$\int_a^{a+T} f(x)\, dx = \int_0^T f(x)\, dx.$$

Solution:

$$\int_a^{a+T} f(x)\, dx = \int_a^0 f(x)\, dx + \int_0^T f(x)\, dx + \int_T^{a+T} f(x)\, dx.$$

In the third integral we substitute

$$x = t + T, \quad dx = dt, \quad t_1 = T - T = 0, \quad t_2 = a + T - T = a.$$

Now

$$\begin{aligned}\int_a^{a+T} f(x)\, dx &= \int_a^0 f(x)\, dx + \int_0^T f(x)\, dx + \int_0^a f(t + T)\, dt \\ &= -\int_0^a f(x)\, dx + \int_0^T f(x)\, dx + \int_0^a f(t)\, dt = \int_0^T f(x)\, dx.\end{aligned}$$

$f(t + T) = f(t)$ by periodicity.

9.9 APPROXIMATE INTEGRATION

Situations often arise in which $I = \int_a^b f(x)\, dx$ has to be evaluated approximately. This is the case when $f(x)$ is given graphically or when the indefinite integral is either too involved or not expressible in elementary functions. Several approximating formulas will be mentioned here.

The Rectangle Formula. We divide the interval $[a,b]$ into n parts by the points

$$a = x_0, x_1, x_2, \ldots, x_n = b.$$

We find (either by computation or graphically) the ordinates at the points of division:

$$y_0 = f(x_0), y_1 = f(x_1), \ldots, y_n = f(x_n).$$

Either of the sums

$$\begin{aligned}S_1 &= (x_1 - x_0)y_0 + (x_2 - x_1)y_1 + \ldots + (x_n - x_{n-1})y_{n-1} \\ &= \sum_{i=1}^{n} (x_i - x_{i-1})y_{i-1}\end{aligned}$$

or

$$S_2 = (x_1 - x_0)y_1 + (x_2 - x_1)y_2 + \ldots + (x_n - {}_1)y_n$$
$$= \sum_{i=1}^{n} (x_i - x_{i-1})y_i$$

can serve as an approximate value of I.

In the majority of cases $[a,b]$ will be divided into equal parts, i.e., $x_i - x_{i-1} = (b - a)/n$. This number will be denoted by Δx. We obtain in this case

$$I \approx \Delta x \sum_{i=1}^{n} y_{i-1} \quad \text{or} \quad I \approx \Delta x \sum_{i=1}^{n} y_i.$$

We remark that these formulas can be interpreted as an approximation to the area of a curvilinear trapezoid represented by the integral. This area can be divided into n thin curvilinear trapezoids, each of which is

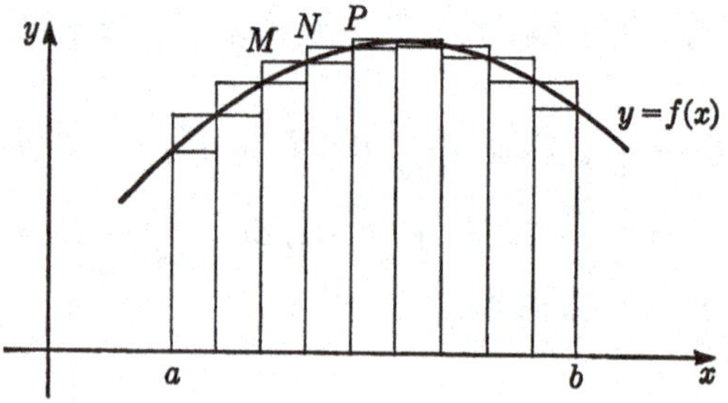

FIGURE 46

replaced by a rectangle with altitude equal to the left or right base of the particular thin trapezoid (see Fig. 46).

The Trapezoidal Formula. The arithmetic mean of the two above formulas gives the trapezoidal formula:

$$I \approx \sum_{i=1}^{n} (x_i - x_{i-1}) \frac{y_{i-1} + y_i}{2},$$

or in case of equal division

$$I \approx \Delta x \left(\frac{y_0 + y_n}{2} + y_1 + y_2 + \ldots + y_{n-1} \right).$$

The geometrical interpretation is clear: We replace every thin curvilinear trapezoid by a rectilinear trapezoid. This is a result of replacing the arc of the curve $y = f(x)$ joining the points $M(x_{i-1},y_{i-1})$ and $N(x_i,y_i)$ (see Fig. 46) by the segment joining these points.

Simpson's Formula. This formula can be obtained by dividing $[a,b]$ into an even number of parts $n = 2m$ and replacing the arc of $y = f(x)$ joining three consecutive points M, N, P (see Fig. 46) by an arc of a parabola passing through these points. (M is an even point of division, counting the first point as 0.) The area of such a "parabolic" trapezoid can be found exactly (cf. Fig. 47):

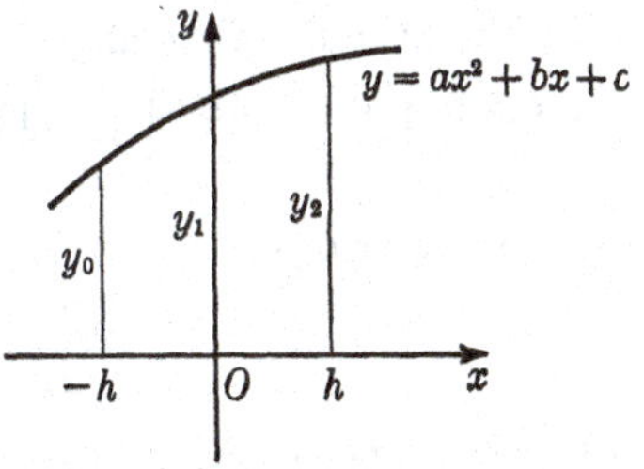

FIGURE 47

$$\int_{-h}^{h} (ax^2 + bx + c)\,dx = \frac{ax^3}{3} + \frac{bx^2}{2} + cx\Big|_{-h}^{h} = \frac{2}{3}ah^3 + 2ch$$

$$= \frac{h}{3}[(ah^2 - bh + c) + 4c + (ah^2 + bh + c)] = \frac{h}{3}(y_0 + 4y_1 + y_2).$$

Assuming that the division is into equal parts we obtain by summing up terms of the above form

$$I \approx \frac{\Delta x}{3}[y_0 + y_n + 4(y_1 + y_3 + \ldots + y_{n-1}) + 2(y_2 + y_4 + \ldots + y_{n-2})]$$

$$\left(\text{here } n = 2m \text{ and } \Delta x = \frac{b-a}{n}\right).$$

This is Simpson's formula for approximate integration.

Errors that occur when use is made of any of the above formulas can be estimated. We remark that in general Simpson's formula gives the most accurate results, and for any formula the accuracy of computation increases as Δx decreases.

890. Using the approximate formulas, evaluate the integral

$$I = \int_0^4 x^2\,dx = \frac{x^3}{3}\Big|_0^4 = \frac{64}{3} \approx 21.33,$$

and compare your result with this exact value.

Solution. We divide the interval of integration into ten equal parts, i.e., $\Delta x = 0.4$. We write the table as shown.

No.	0	1	2	3	4	5	6	7	8	9	10
x_i	0	0.4	0.8	1.2	1.6	2.0	2.4	2.8	3.2	3.6	4.0
y_i	0	0.16	0.64	1.44	2.56	4.0	5.76	7.84	10.24	12.96	16.0

By the formula of rectangles,

$$I \approx 0.4(0 + 0.16 + 0.64 + 1.44 + 2.56 + 4.0 + 5.76 + 7.84 + 10.24 + 12.96) = 0.4(45.6) = 18.24;$$

the relative error is

$$\alpha_1 = \frac{21.33 - 18.24}{21.33} \cdot 100\% = 14.5\%.$$

By the trapezoidal formula,

$$I \approx 0.4\left(\frac{0 + 16}{2} + 0.16 + 0.64 + 1.44 + 2.56 + 4.0 + 5.76 + 7.84 + 10.24 + 12.96\right) = 21.44;$$

the relative error is

$$\alpha_2 = \frac{21.44 - 21.33}{21.33} \cdot 100\% = 0.51\%.$$

The accuracy is very much better.

Simpson's formula will give in this case the exact value.

Remark. Simpson's rule gives the exact value when the integrand is a linear or quadratic or cubic function.

891. Using Simpson's formula, evaluate approximately the integral

$$\int_{0.2}^{1.4} \frac{\sin x}{x}\, dx.$$

Solution. We divide the interval [0.2,1.4] into $n = 6$ parts (an even number).

No.	0	1	2	3	4	5	6
x_i	0.2	0.4	0.6	0.8	1.0	1.2	1.4
y_i	0.9935	0.9735	0.9410	0.8968	0.8415	0.7767	0.7039

Now

$$I \approx \frac{0.2}{3}[0.9935 + 0.7039 + 4(0.9735 + 0.8968 + 0.7767) + 2(0.9410 + 0.8415)] = 1.0567.$$

Remark. The integral $\int (\sin x)/x\, dx$ cannot be expressed in terms of elementary functions. The function $\int_0^x (\sin t)/t\, dt$ is called the sine integral and is denoted by $\mathrm{Si}(x)$. Its values for various x can be found by approximate methods as above. These values are tabulated and the above integral, which can be expressed as Si(1.4) − Si(0.2), can be found in these tables.

892. Find approximately ln 1.6 by computing the integral $\int_1^{1.6} dx/x$ using Simpson's formula. (Divide the interval into six parts.)

9.10 IMPROPER INTEGRALS

Infinite Limits of Integration. The definite integral has been defined only for bounded functions and finite limits of integration. The definitions given below generalize the notion of the definite integral to cases where the integrand is not bounded or the limits are infinite. Such integrals are called improper. We begin with the definition of improper integration over an infinite interval.

We define

$$\int_a^{\infty} f(x)\,dx = \lim_{M\to\infty} \int_a^M f(x)\,dx.$$

$$\int_{-\infty}^{b} f(x)\,dx = \lim_{M\to\infty} \int_{-M}^{b} f(x)\,dx.$$

$$\int_{-\infty}^{\infty} f(x)\,dx = \lim_{\substack{M\to\infty \\ N\to\infty}} \int_{-N}^{M} f(x)\,dx.$$

The expressions on the left-hand sides of the formulas have meaning if and only if the limits on the right-hand sides exist. If so, we say that the corresponding improper integral exists or *converges*. If this is not the case, it makes no sense to speak about these expressions.

The above definitions also indicate a practical way to evaluate improper integrals. We now quote a theorem which makes it possible in many cases to establish the convergence of an improper integral:

If $f(x) \geqslant 0$ in $[a,\infty)$ and there exist numbers $m > 1$ and c such that for all x, $a \leqslant x < \infty$, $f(x) \leqslant c/x^m$, then $\int_a^{\infty} f(x)\,dx$ converges.

If $f(x) \geqslant 0$ in $[a,\infty)$ and there exist $m \leqslant 1$ and c such that $f(x) \geqslant c/x^m$ for any x, $a \leqslant x < \infty$, then $\int_a^{\infty} f(x)\,dx$ does not converge.

Evaluate the following improper integrals or show that they do not converge.

893. $I = \int_a^{\infty} \frac{dx}{x^m}$

Solution. According to the definition we begin with the definite integral

$$\int_a^M \frac{dx}{x^m} = -\frac{1}{m-1}\cdot\frac{1}{x^{m-1}}\bigg|_a^M = \frac{1}{m-1}\left(\frac{1}{a^{m-1}} - \frac{1}{M^{m-1}}\right).$$

Now we have to find

$$\lim_{M\to\infty} \frac{1}{m-1}\left(\frac{1}{a^{m-1}} - \frac{1}{M^{m-1}}\right).$$

For $m > 1$, $\lim_{M\to\infty} \frac{1}{M^{m-1}} = 0$, whence I exists and equals $\frac{1}{(m-1)a^{m-1}}$.

For $m < 1$, $\frac{1}{M^{m-1}}$ tends to infinity together with M; I has no meaning.

For $m = 1$, we obtain

$$\int_a^M \frac{dx}{x} = \ln x\bigg|_a^M = \ln\frac{M}{a}.$$

This expression tends to ∞ together with M, and consequently I does not converge. This example illustrates the above theorem.

894. (a) $I = \int_1^\infty \frac{dx}{x^4}$ (b) $I = \int_1^\infty \frac{dx}{\sqrt{x}}$

895. $I = \int_{-\infty}^\infty \frac{2x\,dx}{x^2+1}$

Solution:

$$\int_{-N}^M \frac{2x\,dx}{x^2+1} = \ln(x^2+1)\bigg|_{-N}^M = \ln\frac{M^2+1}{N^2+1}.$$

$$\lim_{\substack{M\to\infty \\ N\to-\infty}} \ln\frac{M^2+1}{N^2+1}$$

does not exist, because we know nothing about the manner in which M and N tend to ∞.

Remark. Improper integrals in the interval $(-\infty, +\infty)$ are defined by some writers as

$$\int_{-\infty}^\infty f(x)\,dx = \lim_{M\to\infty}\int_{-M}^M f(x)\,dx.$$

By this definition the above integral exists and is equal to zero. We shall use the former definition. As already remarked, according to it the integral does not exist.

896. $I = \int_1^\infty \frac{dx}{x^2(x+1)}$

897. $I = \int_0^\infty x^3e^{-x^2}\,dx$

898. $I = \int_0^\infty e^{-ax} \cos bx\, dx$ **899.** $I = \int_0^\infty \cos bx\, dx$

The following two improper integrals can be found by special methods:

$$\int_0^\infty e^{-x^2}\, dx = \frac{\sqrt{\pi}}{2}; \qquad \int_0^\infty \frac{\sin x}{x}\, dx = \frac{\pi}{2}.$$

We do not show the details here and shall only use the results in computing two other improper integrals:

900. $I = \int_0^\infty \frac{e^{-x}}{\sqrt{x}}\, dx$

Solution. We substitute $x = t^2$, $dx = 2t\, dt$.

$$I = \int_0^\infty \frac{e^{-t^2}}{t} \cdot 2t\, dt = 2\int_0^\infty e^{-t^2}\, dt = \sqrt{\pi}.$$

901. $I = \int_0^\infty \frac{\sin^2 x}{x^2}\, dx$

Solution. We use integration by parts:

$$u = \sin^2 x, \quad du = \sin 2x\, dx; \qquad dv = \frac{dx}{x^2}, \quad v = -\frac{1}{x}.$$

$$I = -\frac{\sin^2 x}{x}\Bigg|_0^\infty + \int_0^\infty \frac{\sin 2x}{x}\, dx.$$

The first term clearly vanishes. As for the second,

$$\int_0^\infty \frac{\sin 2x}{x}\, dx = \int_0^\infty \frac{\sin 2x}{2x}\, d(2x) = \frac{\pi}{2}.$$

Consequently $I = \frac{\pi}{2}$.

Improper Integrals with Unbounded Integrands. Let the function $y = f(x)$ have at c an infinite discontinuity, i.e.,

$$\lim_{x\to c} f(x) = +\infty \quad \text{or} \quad -\infty$$

(also when $\lim\limits_{x\to c+} f(x)$ or $\lim\limits_{x\to c-} f(x)$ is $+\infty$ or $-\infty$). Then the expression $I = \int_a^b f(x)\, dx$ is not yet defined if $a \leqslant c \leqslant b$. We define

$$\int_a^b f(x)\, dx = \lim_{\alpha\to 0}\int_a^{c-\alpha} f(x)\, dx + \lim_{\beta\to 0}\int_{c+\beta}^b f(x)\, dx,$$

and thus I has a clear meaning if and only if both limits at the right-hand side exist. If this is not the case, I does not converge and has no meaning.

If the function $f(x)$ tends to infinity at one of the limits of integration, say a, we have as above

$$\int_a^b f(x)\,dx = \lim_{\alpha\to 0}\int_{a+\alpha}^b f(x)\,dx.$$

The same procedure is used when this point is the upper limit, or when there are more than one (but a finite number) of points with infinite discontinuity of the integrand.

The following theorem makes it possible in many cases to establish the convergence of an improper integral of this kind:

Suppose $f(x) \geqslant 0$ in $[a,b)$ and $\lim_{x\to b} f(x) = \infty$; then (i) if there exist constant numbers $m < 1$ and c such that for any $a \leqslant x < b$, $f(x) \leqslant c/(b-x)^m$, then $\int_a^b f(x)\,dx$ converges; (ii) if there exist $m > 1$ and c such that for any $a \leqslant x < b$, $f(x) \geqslant c/(b-x)^m$, then $\int_a^b f(x)\,dx$ does not converge.

Evaluate the following improper integrals or establish their divergence (i.e., the fact that they do not converge).

902. $I = \displaystyle\int_a^b \frac{dx}{\sqrt[3]{x^2}}, \quad a < 0,\ b > 0$

Solution. $x = 0$ is a point of infinite discontinuity.

$$I = \lim_{\alpha\to 0}\int_a^{-\alpha}\frac{dx}{\sqrt[3]{x^2}} + \lim_{\beta\to 0}\int_\beta^b \frac{dx}{\sqrt[3]{x^2}} = \lim_{\alpha\to 0} 3\sqrt[3]{x}\Big|_a^{-\alpha} + \lim_{\beta\to 0} 3\sqrt[3]{x}\Big|_\beta^b$$

$$= -3\sqrt[3]{a} + 3\sqrt[3]{b} = 3(\sqrt[3]{b} - \sqrt[3]{a}).$$

In this example we could obtain the same result by performing the usual definite integration from a to b. But this procedure is clearly incorrect as can be seen in the following example:

$$\int_a^b \frac{dx}{x^2} = -\frac{1}{x}\Big|_a^b = \frac{1}{a} - \frac{1}{b}.$$

If $a < 0$ and $b > 0$ we obtain on the right-hand side a negative number, which is clearly nonsense because the integrand is always positive and $b > a$. The error is a result of the fact that this integral has no meaning at all because

$$\lim_{\alpha\to 0}\int_a^{\alpha}\frac{1}{x^2}\,dx$$

does not exist. (The second $\lim_{\beta\to 0}\int_\beta^b 1/x^2\,dx$ does not exist either, but this information is redundant after we know that the first one does not exist.)

We remark that the convergence of the first and divergence of the second integral of this example can be obtained from the preceding theorem.

903. $I = \int_0^a \frac{dx}{\sqrt{a^2 - x^2}}$

904. $I = \int_e^{10} \frac{dx}{x \ln x \ln \ln x}$

905. $I = \int_0^2 \frac{dx}{x^2 - 4x + 3}$

906. $I = \int_1^\infty \frac{dx}{x\sqrt{x - 1}}$

Solution. We have here a discontinuity at $x = 1$ and an infinite limit. We compute first

$$\int_{1+\alpha}^{M} \frac{dx}{x\sqrt{x-1}},$$

setting $\sqrt{x - 1} = t$, $dx = 2t\,dt$.

$$\int \frac{dx}{x\sqrt{x-1}} = \int \frac{2t\,dt}{(t^2+1)t} = 2 \arctan t + C = 2 \arctan \sqrt{x-1} + C.$$

$$\int_{1+\alpha}^{M} \frac{dx}{x\sqrt{x-1}} = 2(\arctan \sqrt{M-1} - \arctan \sqrt{\alpha}).$$

$$I = \lim_{\substack{M\to\infty \\ \alpha\to 0}} 2(\arctan \sqrt{M-1} - \arctan \sqrt{\alpha}) = 2 \cdot \frac{\pi}{2} = \pi.$$

9.11 MISCELLANEOUS PROBLEMS

Evaluate the following integrals:

907. $\int_2^3 \frac{dx}{3x^2 - 5x + 2}$

908. $\int_1^\infty \frac{1 - \ln x}{x^2}\,dx$

909. $\int_0^1 \frac{\ln x}{x}\,dx$

910. $\int_0^\infty xe^{-x^2}\,dx$

911. $\int_1^2 \frac{x\,dx}{\sqrt{x-1}}$

912. $\int_{-\infty}^{\infty} \frac{dx}{x^2 + 2x + 2}$

913. $\int_1^\infty \frac{\arctan x}{x^2}\,dx$

914. $\int_0^\pi \frac{1 - r^2}{1 - 2r\cos x + r^2}\,dx$

915. $\int_{\pi/2}^{0} e^{2x} \cos 3x\,dx$

916. $\int_0^\pi \frac{\cos t + a}{1 + 2a\cos t + a^2}\,dt$

917. Find the integrals

$$I_{m,n} = \int_0^{2\pi} \sin mx \sin nx\,dx,$$

$$J_{m,n} = \int_0^{2\pi} \sin mx \cos nx\,dx, \qquad K_{m,n} = \int_0^{2\pi} \cos mx \cos nx\,dx,$$

where m and n are positive integers or zero. Consider especially the cases where $m = n$ (also discuss $m = n = 0$).

918. Find $I = \int_0^1 \frac{dx}{(x^2+1)\sqrt{1-x^2}}$

919. Estimate the integral $I = \int_0^1 \sqrt{1+x^4}\,dx$ using the Cauchy-Schwarz inequality.

920. Find $\int_0^\infty \frac{\arctan(x-1)\,dx}{\sqrt[3]{(x-1)^4}}$.

921. Using the definite integral find

$$\lim_{n\to\infty}\left(\frac{n}{n^2+1^2}+\frac{n}{n^2+2^2}+\dots+\frac{n}{n^2+n^2}\right).$$

Solution:

$$\frac{n}{n^2+1^2}+\frac{n}{n^2+2^2}+\dots+\frac{n}{n^2+n^2}$$

$$=\frac{1}{n}\left[\frac{1}{1+(1/n)^2}+\frac{1}{1+(2/n)^2}+\dots+\frac{1}{1+(n/n)^2}\right].$$

The sum in brackets will be recognized as an appropriate sum for the direct evaluation of the integral

$$\int_0^1 \frac{1}{1+x^2}\,dx = \arctan x\Big|_0^1 = \frac{\pi}{4},$$

where the interval [0, 1] is divided into n subintervals. The desired limit thus equals $\pi/4$.

922. The same question for the limit

$$\lim_{n\to\infty}\left(\frac{1}{n+1}+\frac{1}{n+2}+\dots+\frac{1}{n+n}\right).$$

923. Find $\frac{d}{dx}\left(\int_{\sin x}^{\cos x} \sin t^3\,dt\right)$.

924. Find

$$\lim_{x\to\infty}\frac{\left(\int_0^x e^{x^2}\,dx\right)^2}{\left(\int_0^x e^{2x^2}\,dx\right)}.$$

Solution. We shall use l'Hôpital's rule:

$$\lim_{x\to+\infty}\frac{\left(\int_0^x e^{x^2}\,dx\right)^2}{\int_0^x e^{2x^2}\,dx} = \lim_{x\to+\infty}\frac{2\left(\int_0^x e^{x^2}\,dx\right)e^{x^2}}{e^{2x^2}}$$

$$= 2\lim_{x\to+\infty}\frac{\int_0^x e^{x^2}\,dx}{e^{x^2}} = 2\lim_{x\to\infty}\frac{e^{x^2}}{e^{x^2}\cdot 2x} = 0.$$

925. Find $I = \int_0^1 f(x)\,dx$ if $f(x) = \begin{matrix} x & \text{for } 0 < x \leqslant \frac{1}{2} \\ 2 & \text{for } \frac{1}{2} < x \leqslant 1. \end{matrix}$

Solution. Although $f(x)$ has a discontinuity at $x = \frac{1}{2}$, the integral exists and can be found by dividing the interval $[0,1]$ into two parts:

$$I = \int_0^{1/2} x\,dx + \int_{1/2}^1 2\,dx = \frac{x^2}{2}\bigg|_0^{1/2} + 2x\bigg|_{1/2}^1 = \frac{1}{8} + 2 - 1 = 1\frac{1}{8}.$$

926. Point out the difficulty in using the following substitutions, and how it can be avoided:

(a) $\displaystyle\int_{-2}^{2} \frac{dx}{1+x^2}, \quad x = \frac{1}{t}$

(b) $\displaystyle\int_{\pi/4}^{3\pi/4} \frac{dx}{a\sin^2 x + b\cos^2 x}, \quad \tan x = t.$

927. Compute $I_n = \int_0^1 (1 - x^2)^n\,dx.$

Solution. Take

$$u = (1 - x^2)^n,$$

$$du = -n(1 - x^2)^{n-1}\cdot 2x\,dx, \quad dv = dx,$$

$$v = x;$$

$$I_n = x(1 - x^2)^n\bigg|_0^1 + 2n\int_0^1 x^2(1 - x^2)^{n-1}\,dx$$

$$= 2n\int_0^1 -(1 - x^2)(1 - x^2)^{n-1}\,dx + 2n\int_0^1 (1 - x^2)^{n-1}\,dx$$

$$= -2nI_n + 2nI_{n-1}.$$

Consequently,

$$I_n = \frac{2n}{2n+1}I_{n-1} = \frac{2n(2n-2)}{(2n+1)(2n-1)}I_{n-2} = \ldots$$

$$= \frac{2^n n!}{(2n+1)(2n-1)\cdots 3\cdot 1}I_0 = \frac{2^n n!}{(2n+1)(2n-1)\cdots 3\cdot 1}.$$

By substitution $x = \cos t$ this integral can be transformed to the one of Problem 875.

928. The intensity of an alternating current is given by the formula $i = I_0 \sin(\omega t + \varphi)$, where t is the time and I_0, ω, and φ are constants. Find the average value of the square of the intensity of the current.

929. Prove the convergence of $\int_0^\infty e^{-x^2}\,dx.$

930. The gamma function $\Gamma(x)$ is defined by

$$\Gamma(x) = \int_0^\infty t^{x-1}e^{-t}\,dt.$$

Prove that $\Gamma(x+1) = x\Gamma(x)$, $x > 0$. Find $\Gamma(1)$ and prove $\Gamma(n+1) = n!$ for n a positive integer.

X

APPLICATIONS OF THE DEFINITE INTEGRAL

In this chapter we shall discuss various applications of the definite integral. The formulas which will be employed can often be obtained by a heuristic argument which will now be illustrated. Suppose the figure bounded by the x axis, the ordinates at $x = a$ and $x = b$, and the arc of a curve $y = f(x)$, from $x = a$ to $x = b$, revolves around the x axis (see Fig. 48). To find the volume of the resulting solid of revolution we consider a typical strip of width dx. The volume described by it can be approximated by a cylinder with radius of base $f(x)$ (x an arbitrary point in the interval dx) and altitude dx, i.e., this volume is approximately equal to $\pi[f(x)]^2\,dx$. It seems plausible that the whole volume v is the limit of the above sum when the width of the strips tends to zero. Hence

$$v = \pi \int_a^b [f(x)]^2\,dx.$$

It must be understood that this heuristic argument is lacking in rigor and may lead to errors if applied indiscriminately.

For example, suppose we wish to find the length of the arc of the curve in Figure 49 between the points A and B. We subdivide the arc and replace every small arc by its projection on the x axis, i.e., by dx. The error seems to be small (so is the arc itself), and by the same reasoning as above we obtain

$$L = \int_a^b dx = x\Big|_a^b = b - a.$$

This is clearly absurd: all curves with the same projection on the x axis would have the same length, and a straight line perpendicular to the x axis would have no length at all!

The error is a result of replacing dL by dx. In the previous case the volume of the cylinder was the principal part of the volume of the layer (corresponding to dx) of the body of revolution in the sense that the difference between these volumes tends to zero faster than dx (compare the notion of differential, Chap. VII). In the last case $(dL - dx)/dx$ does not tend to zero together with dx, and dx is not the principal part of dL. We have no right to replace dL by dx.

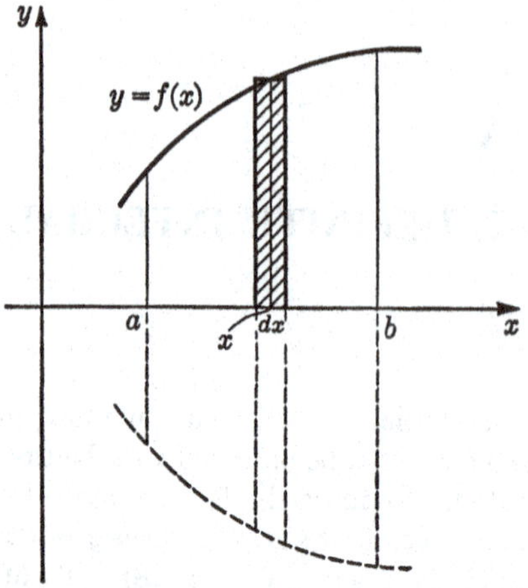

FIGURE 48

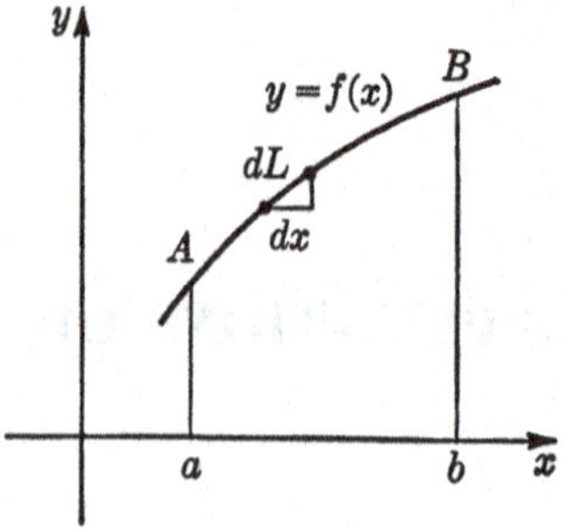

FIGURE 49

To conclude, we remark that all formulas of this chapter can of course be given rigorous proof. Nevertheless, the heuristic reasoning, when performed with sufficient care, is very useful in practical problems.

10.1 COMPUTATION OF PLANE AREAS

The shaded area in Figure 50 is given by

$$\int_a^b [f_1(x) - f_2(x)]\, dx.$$

The shaded area in Figure 51 is given by

$$\tfrac{1}{2} \int_{\theta_1}^{\theta_2} \{[g_1(\theta)]^2 - [g_2(\theta)]^2\}\, d\theta.$$

These are the two main cases of calculations of plane areas using cartesian and polar coordinates. Various special cases will be treated in the following examples.

931. Find the area bounded by the curve $y = \cos x$, the coordinate axes, and the straight line $x = \pi/6$.

Solution. See Figure 52.

$$A = \int_0^{\pi/6} (\cos x - 0)\, dx = \int_0^{\pi/6} \cos x\, dx = \sin x \Big|_0^{\pi/6} = \tfrac{1}{2}.$$

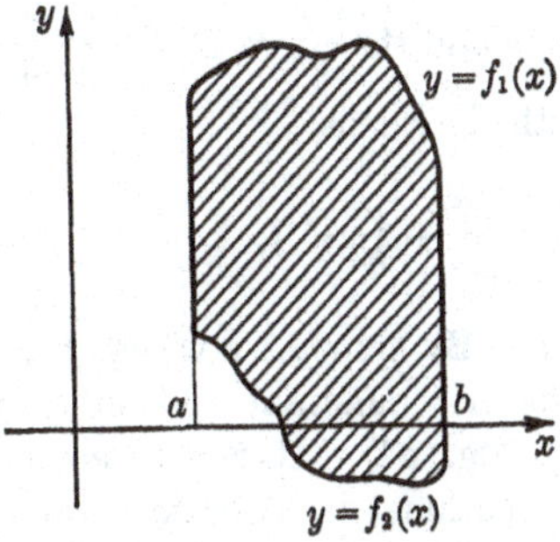

FIGURE 50

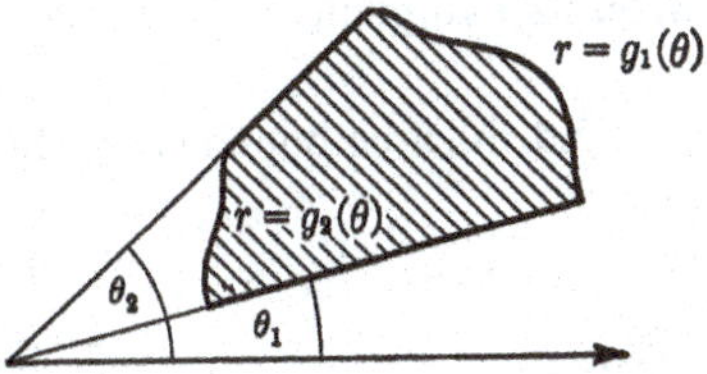

FIGURE 51

932. Find the area bounded by the curves $y^2 = 2x + 1$ and $x - y - 1 = 0$.

Solution. The points of intersection M and N (Fig. 53) can be found by solution of the system

$$\begin{cases} x = y + 1 \\ y^2 = 2x + 1; \end{cases}$$

$$y^2 = 2y + 2 + 1, \qquad y^2 - 2y - 3 = 0, \qquad y_1 = -1, \qquad y_2 = 3.$$

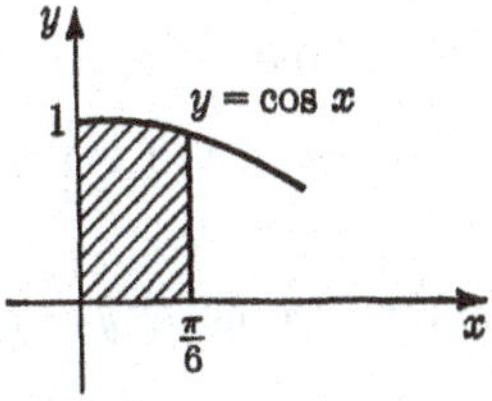

FIGURE 52

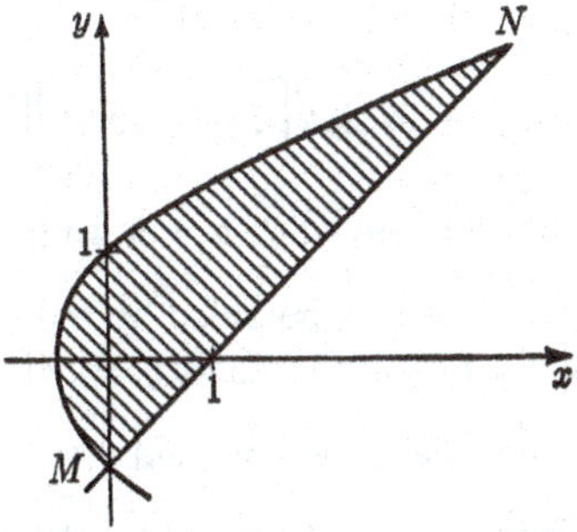

FIGURE 53

We find the area by integration along the y axis:

$$A = \int_{-1}^{3} \left(y + 1 - \frac{y^2 - 1}{2}\right) dy = \frac{y^2}{2} + \frac{3}{2}y - \frac{y^3}{6}\Big|_{-1}^{3}$$

$$= \frac{9}{2} + \frac{9}{2} - \frac{9}{2} - \frac{1}{2} + \frac{3}{2} - \frac{1}{6} = 5\frac{1}{3}.$$

933. Find the area bounded by the coordinate axes, the curve $y = \dfrac{\sin^4 x}{\cos^6 x}$, and the straight line $x = \pi/4$.

934. Find the area bounded by the x axis and the curve $y = \dfrac{x^2 - x - 2}{x^3 + 8}$ between its two points of intersection with the x axis.

935. Find the area bounded by the ellipse $\dfrac{x^2}{a^2} + \dfrac{y^2}{b^2} = 1$.

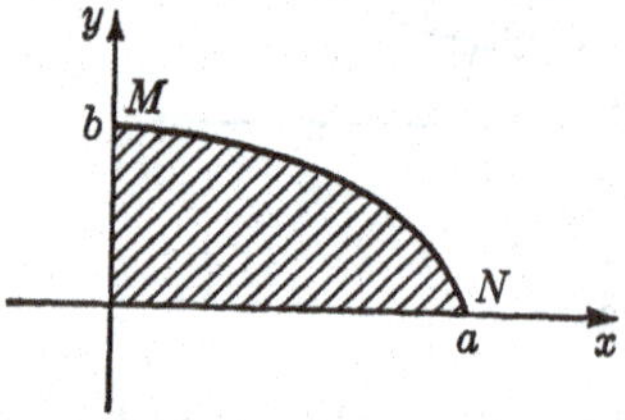

FIGURE 54

Solution. By symmetry we can find the area in the first quadrant and multiply it by 4 (see Fig. 54). It is convenient to use the parametric equations of the ellipse,

$$x = a \cos t, \qquad y = b \sin t.$$

The value of t corresponding to M is $t = \pi/2$ and that for N is $t = 0$. Consequently

$$A = 4 \int_0^a y\,dx = 4 \int_{\pi/2}^0 b \sin t\, d(a \cos t)$$

$$= -4ab \int_{\pi/2}^0 \sin^2 t\,dt = 4ab \int_0^{\pi/2} \frac{1 - \cos 2t}{2}\,dt$$

$$= 2ab \left[t - \frac{1}{2} \sin 2t \right] \Bigg|_0^{\pi/2} = \pi ab.$$

In the special case when $a = b$ we have the area of the circle, πa^2.

936. Find the area bounded by the x axis and by one arch ($0 \leqslant t \leqslant 2\pi$) of the cycloid $x = a(t - \sin t)$, $y = a(1 - \cos t)$.

937. Find the area bounded by the astroid $x = a \cos^3 t$, $y = a \sin^3 t$.

938. Find the area bounded by the closed curve $y^2 = x^2 - x^4$.

939. Find the area bounded by the closed curve $(y - \arcsin x)^2 = x - x^2$.

940. Find the area included inside the loop of the curve $x = t^2 - 1$, $y = t^3 - t$.

941. Find the area bounded by the curve $r = a \sin 2\theta$.

Solution. The curve consists of two equal loops. One of them is obtained when $0 \leqslant \theta \leqslant \pi/2$. Consequently,

$$A = 2 \cdot \frac{1}{2} \int_0^{\pi/2} r^2\, d\theta = \int_0^{\pi/2} a^2 \sin^2 2\theta\, d\theta = a^2 \int_0^{\pi/2} \frac{1 - \cos 2\theta}{2}\, d\theta = \frac{\pi a^2}{4}.$$

942. Find the area bounded by the curve $r = 2a(2 + \cos \theta)$.

943. Find the area between the outer and inner parts of the curve $r = a \sin^3 (\theta/3)$ $(0 \leqslant \theta \leqslant 3\pi)$.

944. Find the area bounded by the curve $r = \sqrt{1 - t^2}$, $\theta = \arcsin t + \sqrt{1 + t^2}$.

945. Find the area bounded by $x^4 + y^4 = x^2 + y^2$.

946. Find the area bounded by the loop of the curve $x^3 + y^3 = 3axy$.

947. Compute the area enclosed between the curve $y = \dfrac{1}{1 + x^2}$ and its asymptote.

Solution. The asymptote is $y = 0$ $(\lim_{x\to\infty} y = 0)$. The curve is symmetric with respect to the y axis and the required area is given by

$$A = 2 \int_0^{\infty} \frac{1}{1 + x^2}\, dx = 2 \lim_{M\to\infty} \arctan x \Big|_0^M = \pi.$$

948. Compute the area between the curve $xy^2 = 8 - 4x$ and its asymptote.

949. Compute the area between the curve $y = x^2 e^{-x^2}$ and its asymptote.

950. Find the area bounded by the curve $\theta = r \arctan r$ and the rays $\theta = 0, \theta = \pi/\sqrt{3}$.

Solution:

$$A = \frac{1}{2} \int_{\theta=0}^{\pi/\sqrt{3}} r^2\, d\theta; \qquad d\theta = \left(\arctan r + \frac{r}{1 + r^2}\right) dr.$$

For $\theta = 0$ we have $0 = r \arctan r$, i.e., $r = 0$.
For $\theta = \pi/\sqrt{3}$ we have $\pi/\sqrt{3} = r \arctan r$ and $r = \sqrt{3}$. Now

$$A = \frac{1}{2} \int_0^{\sqrt{3}} \left(r^2 \arctan r + \frac{r^3}{1 + r^2}\right) dr$$

$$= \frac{1}{2}\left[\frac{r^3}{3} \arctan r \Big|_0^{\sqrt{3}} - \frac{1}{3} \int_0^{\sqrt{3}} \frac{r^3\, dr}{1 + r^2} + \int_0^{\sqrt{3}} \frac{r^3\, dr}{1 + r^2}\right]$$

$$= \frac{1}{2}\left[\sqrt{3}\,\frac{\pi}{3} + \frac{2}{3}\left(\frac{r^2}{2} - \frac{1}{2} \ln (r^2 + 1)\right)\Big|_0^{\sqrt{3}}\right] = \frac{1}{2}\left(\frac{\pi}{\sqrt{3}} + 1 - \frac{2}{3} \ln 2\right).$$

951. Find the ratio of the areas of the two parts into which the circle $x^2 + y^2 = 8$ is divided by the parabola $y^2 = 2x$.

952. Find the area bounded by the lemniscate $(x^2 + y^2)^2 = 2a^2 xy$.

953. Find the area bounded by the curve $y = 1/(1 + x^2)$ and the parabola $y = \frac{1}{2}x^2$.

954. Find the area bounded by the curve $y = |\ln x|$ and the straight lines $y = 0$, $x = e^{-1}$, $x = e$.

955. Find the area bounded by the curves $r = 2a(1 + \cos\theta)$ and $r = a\cos\theta$.

956. Find the area inside one loop of the curve $x = a \sin 2t$, $y = a \sin t$, situated above the x axis.

10.2 COMPUTATION OF ARC LENGTH

Let $y = f(x)$ be a differentiable function with a continuous derivative. The length of an arc of the curve $y = f(x)$ between the points for $x = a$ and $x = b$ is given by the integral

$$L = \int_a^b \sqrt{1 + (y')^2}\, dx.$$

When b is replaced by a variable limit x we obtain

$$L(x) = \int_a^x \sqrt{1 + (y')^2}\, dx \quad \text{and} \quad dL = \sqrt{1 + (y')^2}\, dx = \sqrt{(dx)^2 + (dy)^2}.$$

The latter is the differential of the arc length in cartesian coordinates. (Heuristically it can be considered as a form of the Pythagorean theorem.)

The expression $\sqrt{1 + (y')^2}\, dx$ was computed in Section 6.11 for parametric representation and for polar coordinates. In the first case,

$$dL = \sqrt{(x_t')^2 + (y_t')^2}\, dt,$$
$$\text{i.e.,} \quad L = \int_{t_1}^{t_2} \sqrt{(x_t')^2 + (y_t')^2}\, dt.$$

Here x_t' stands for dx/dt. For polar coordinates,

$$dL = \sqrt{r^2 + (r')^2}\, d\theta,$$
$$L = \int_{\theta_1}^{\theta_2} \sqrt{r^2 + (r')^2}\, d\theta.$$

In all cases the integrand is positive, and to obtain a positive number for the arc length we must take care that the lower limit is a smaller number than the upper.

Find the arc lengths of the following curves between the given limits.

957. $y = \ln x$, $x_1 = \sqrt{3}$, $x_2 = \sqrt{8}$

Solution. $y' = 1/x$;

$$L = \int_{\sqrt{3}}^{\sqrt{8}} \sqrt{1 + \frac{1}{x^2}}\, dx = \int_{\sqrt{3}}^{\sqrt{8}} \frac{\sqrt{1 + x^2}}{x}\, dx.$$

We substitute $\sqrt{1 + x^2} = z$, i.e., $1 + x^2 = z^2$, $x\,dx = z\,dz$; then the limits are

$$z_1 = \sqrt{1 + 3} = 2, \qquad z_2 = \sqrt{1 + 8} = 3.$$

$$L = \int_2^3 \frac{z^2\,dz}{z^2 - 1} = \int_2^3 \left(dz + \frac{dz}{z^2 - 1}\right) = \left[z + \frac{1}{2}\ln\left|\frac{z - 1}{z + 1}\right|\right]\Bigg|_2^3$$

$$= 3 - 2 + \frac{1}{2}\left(\ln\frac{1}{2} - \ln\frac{1}{3}\right) = 1 + \frac{1}{2}\ln\frac{3}{2}.$$

958. $y = \frac{1}{2}[x\sqrt{x^2 - 1} - \ln(x + \sqrt{x^2 - 1})]$, $x_1 = 1$, $x_2 = a + 1$

959. $y = \ln(1 - x^2)$, $x_1 = 0$, $x_2 = \frac{1}{2}$

960. $y = \dfrac{x^5}{5} + \dfrac{1}{12x^3}$, $x_1 = \dfrac{1}{2}$, $x_2 = 1$

961. $y = x^2/2p$, $x_1 = 0$, $x_2 = x$

962. Find on the first arc of the cycloid $x = a(t - \sin t)$, $y = a(1 - \cos t)$ a point which divides this arc into two parts such that the ratio of their lengths equals 1/3.

Solution. First we shall find dL:

$$dx = a(1 - \cos t)\,dt, \quad dy = a\sin t\,dt,$$

$$dL = a\sqrt{1 - 2\cos t + \cos^2 t + \sin^2 t}\,dt = a\sqrt{2}\sqrt{1 - \cos t}\,dt$$

$$= a\sqrt{2}\cdot\sqrt{2}\sin\frac{t}{2}\,dt = 2a\sin\frac{t}{2}\,dt.$$

$$L(t) = 2a\int_0^t \sin\frac{t}{2}\,dt = 4a\left(-\cos\frac{t}{2}\right)\Bigg|_0^t = 4a\left(1 - \cos\frac{t}{2}\right).$$

The length of the whole arc is $4a\left(1 - \cos\dfrac{2\pi}{2}\right) = 8a$. We have to find a point t such that

$$4a\left(1 - \cos\frac{t}{2}\right) = \frac{1}{4}\cdot 8a = 2a,$$

i.e., $\cos\dfrac{t}{2} = \dfrac{1}{2}$. But $0 \leqslant \dfrac{t}{2} \leqslant \pi$ and $\dfrac{t}{2} = \dfrac{\pi}{3}$ or $t = \dfrac{2\pi}{3}$. The corresponding point is

$$x = a\left(\frac{2\pi}{3} - \frac{\sqrt{3}}{2}\right), \qquad y = \frac{3a}{2}.$$

963. Prove that the length of the arc of the curve given by the parametric equations

$$x = f''(t)\cos t + f'(t)\sin t, \qquad y = -f''(t)\sin t + f'(t)\cos t$$

between the points corresponding to $t = t_1$ and $t = t_2$ is equal to

$$f(t_2) + f''(t_2) - f(t_1) - f''(t_1).$$

964. Find the length of the curve $\left(\frac{x}{a}\right)^{2/3} + \left(\frac{y}{b}\right)^{2/3} = 1.$

965. Find the length of the arc of the tractrix

$$x = a\left(\cos t + \ln \tan \frac{t}{2}\right), \qquad y = a \sin t,$$

between the points $(0,a)$ and $(x > 0,y)$.

966. Find the length of the cardioid $r = a(1 + \cos\theta)$.

Solution:

$$r' = a\sin\theta;$$

$$\sqrt{r^2 + (r')^2} = a\sqrt{1 + 2\cos\theta + \cos^2\theta + \sin^2\theta}$$

$$= a\sqrt{2}\sqrt{1+\cos\theta} = 2a\cos\frac{\theta}{2};$$

$$L = 2\int_0^\pi 2a\cos\frac{\theta}{2}\,d\theta = 4a\cdot 2\sin\frac{\theta}{2}\Big|_0^\pi = 8a.$$

967. Find the length of the arc of the spiral of Archimedes $r = a\theta$ from $\theta = 0$ to $\theta = 2$.

968. Find the length of the arc of the curve $y = \int_{-\pi/2}^{x} \sqrt{\cos x}\,dx$, in $-\pi/2 \leqslant x \leqslant \pi/2$.

969. Find the length of the arc of $y = \sin x$ when $0 \leqslant x \leqslant \pi$.

Solution:

$$L = \int_0^\pi \sqrt{1 + \cos^2 x}\,dx.$$

This integral cannot be expressed in terms of elementary functions. By symmetry,

$$L = 2\int_0^{\pi/2} \sqrt{1+\cos^2 x}\,dx.$$

We shall compute L approximately using Simpson's formula. We divide $[0,\pi/2]$ into two equal intervals and find the values shown in the table.

No.	0	1	2
x	0	$\pi/4$	$\pi/2$
$y = \sqrt{1 + \cos^2 x}$	$\sqrt{2}$	$\sqrt{3/2}$	1

Thus

$$L = 2 \cdot \frac{\pi}{12}\left(\sqrt{2} + 4\sqrt{\frac{3}{2}} + 1\right) \approx 3.8.$$

970. Find the length of the ellipse $\frac{x^2}{a^2} + \frac{y^2}{b^2} = 1$. Show that the length of one "wave" of the curve $y = \sin x$ equals the length of an ellipse with semi-axes $\sqrt{2}$ and 1.

971. Find the length of the arc of the curve $x = \frac{1}{4}y^2 - \frac{1}{2}\ln y$ from $y = 1$ to $y = e$.

972. Find the length of the curve $\theta = \frac{1}{2}\left(r + \frac{1}{r}\right)$ from $r = 1$ to $r = 3$.

973. Find the length of the arc of the curve $y = \frac{x - 3}{3}\sqrt{x}$ from $x_1 = 0$ to $x_2 = 4$.

974. Find the length of the arc of the curve

$$x = a(\cos t + t \sin t), \qquad y = a(\sin t - t \cos t)$$

from $t_1 = 0$ to $t_2 = 2\pi$.

975. Find the length of $r = a \tanh(\theta/2)$ from $\theta = 0$ to $\theta = 2\pi$.

10.3 COMPUTATION OF VOLUMES

THEOREM. If for a given body there can be found an axis (taken as the x axis) such that when a plane perpendicular to this axis intersects the body the area of the figure obtained can be expressed as a function of $x, A(x)$, then the integral

$$\int_a^b A(x)\, dx$$

determines the volume of the part of this body between the planes $x = a$ and $x = b$.

It must be remarked that in general some additional conditions are needed for this theorem to hold. If, for instance, the volume is known to exist, as is the case in all practical problems, the above integral can be shown to determine it

The so-called solids of revolution are particular cases. If the area bounded by the curves $y = f_1(x)$ and $y = f_2(x)$ and the straight lines $x = a, x = b$ [$a < b$ and $f_1(x) < f_2(x)$ when $a < x < b$] rotates around the x axis, we have

$$A(x) = \pi\{[f_2(x)]^2 - [f_1(x)]^2\}$$

$$\text{and} \quad v = \pi \int_a^b \{[f_2(x)]^2 - [f_1(x)]^2\}\, dx.$$

976. Find the volume bounded by the ellipsoid

$$\frac{x^2}{a^2} + \frac{y^2}{b^2} + \frac{z^2}{c^2} = 1.$$

Solution. Let us intersect the ellipsoid by a plane $x = x_1, -a \leqslant x_1 \leqslant a$. We obtain

$$\frac{x_1^2}{a^2} + \frac{y^2}{b^2} + \frac{z^2}{c^2} = 1,$$

i.e.,

$$\frac{y^2}{b^2} + \frac{z^2}{c^2} = 1 - \frac{x_1^2}{a^2}$$

or

$$\frac{y^2}{\left(b\sqrt{1 - \dfrac{x_1^2}{a^2}}\right)^2} + \frac{z^2}{\left(c\sqrt{1 - \dfrac{x_1^2}{a^2}}\right)^2} = 1.$$

This is the projection of the curve of intersection of the ellipsoid and the plane onto the yz plane. This curve is an ellipse with semiaxes $b\sqrt{1 - (x_1^2/a^2)}$, $c\sqrt{1 - (x_1^2/a^2)}$. The area $A(x_1)$ of the figure obtained by the above intersection is consequently $\pi bc[1 - (x_1^2/a^2)]$ (cf. Prob. 935). Thus the volume of the ellipsoid is given by the integral

$$v = \int_{-a}^{a} \pi bc\left(1 - \frac{x^2}{a^2}\right) dx.$$

By symmetry,

$$v = 2\pi bc \int_0^a \left(1 - \frac{x^2}{a^2}\right) dx = 2\pi bc\left(x - \frac{x^3}{3a^2}\right)\Big|_0^a$$

$$= 2\pi bc \cdot \frac{2}{3} a = \frac{4}{3}\pi abc.$$

For $a = b = c$ (i.e., in the case of a sphere) we obtain $v = \frac{4}{3}\pi a^3$.

977. The axes of two equal cylinders of radius a intersect and are perpendicular. Find the volume of the part common to the two cylinders.

Solution. In Figure 55 is sketched one eighth of the required volume.

By symmetry, the shaded area is a square, one side of which equals $\sqrt{a^2 - x^2}$. Consequently $A(x) = a^2 - x^2$ and the volume is

$$v = 8 \int_0^a (a^2 - x^2)\, dx = 8 \left(a^2 x - \frac{x^3}{3} \right) \Bigg|_0^a = \frac{16}{3} a^3.$$

978. The area bounded by the parabolas $y = x^2$ and $y^2 = x$ rotates around Ox. Find the volume of the body so formed.

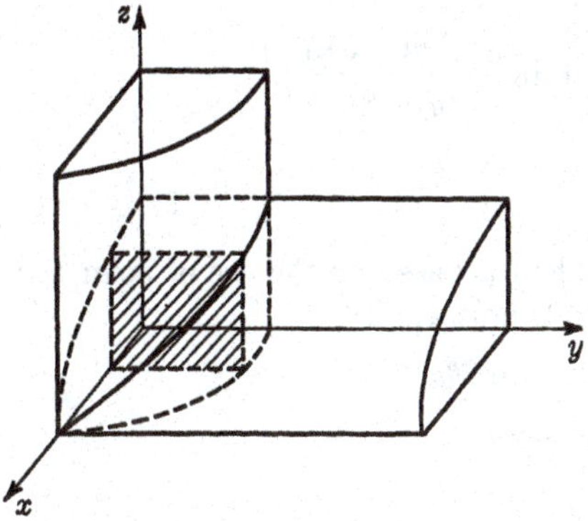

FIGURE 55

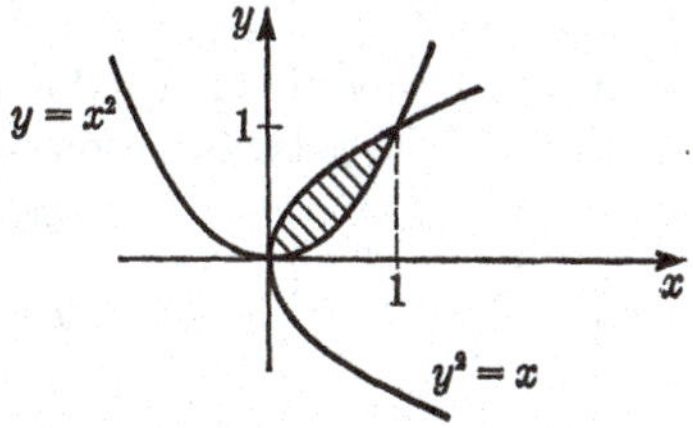

FIGURE 56

Solution. See Figure 56. The required volume is

$$v = \pi \int_0^1 (x - x^4)\, dx = \pi \left(\frac{x^2}{2} - \frac{x^5}{5} \right) \Bigg|_0^1 = \frac{3\pi}{10}.$$

Find the volumes of bodies of revolution obtained by revolving the following areas around the given axes.

979. The area bounded by

$$y = x^2 + \sin x, \qquad y = 0, \qquad x = 0, \qquad x = \pi/2, \quad \text{around } Ox.$$

980. The area bounded by the lemniscate

$$(x^2 + y^2)^2 = a^2(x^2 - y^2), \quad \text{around } Ox.$$

Solution. First we shall find y^2:

$$y^4 + y^2(2x^2 + a^2) + x^4 - a^2x^2 = 0,$$

$$y^2 = -x^2 - \frac{a^2}{2} + \sqrt{x^4 + a^2x^2 + \frac{a^4}{4} - x^4 + a^2x^2}$$

$$= -x^2 - \frac{a^2}{2} + \sqrt{2a^2x^2 + \frac{a^4}{4}}.$$

Now

$$v = 2\pi \int_0^a \left(-x^2 - \frac{a^2}{2} + a\sqrt{2}\sqrt{x^2 + \frac{a^2}{8}}\right) dx$$

$$= 2\pi \left\{-\frac{x^3}{3} - \frac{a^2}{2}x + a\sqrt{2}\frac{a^2}{16}\left[\frac{x\sqrt{x^2 + (a^2/8)}}{(a^2/8)} + \ln\left(x + \sqrt{x^2 + (a^2/8)}\right)\right]\right\}\Bigg|_0^a$$

$$= 2\pi\left[-\frac{a^3}{3} - \frac{a^3}{2} + \frac{a^3\sqrt{2}}{16}\left(3\sqrt{8} + \ln\frac{a + (3a/\sqrt{8})}{(a/\sqrt{8})}\right)\right]$$

$$= \pi a^3\left[\frac{\sqrt{2}}{8}\ln(2\sqrt{2} + 3) - \frac{1}{6}\right].$$

Another way of solving this problem is based on the use of polar coordinates. The polar equation of the lemniscate is

$$r^4 = a^2r^2 \cos 2\theta,$$

i.e.,

$$r^2 = a^2 \cos 2\theta.$$

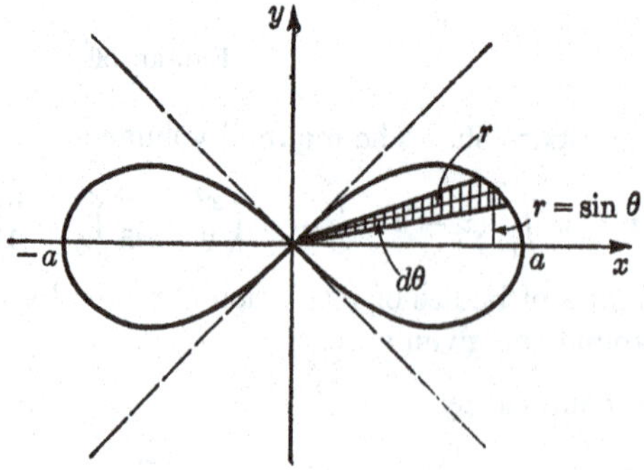

FIGURE 57

When the sector shaded in Figure 57 rotates about the x axis it describes a volume the principal part of which can be shown to be

$$dv = \tfrac{1}{3}r \cdot r\,d\theta \cdot 2\pi r \sin\theta = \tfrac{2}{3}\pi r^3 \sin\theta\,d\theta.$$

(This volume can be looked upon as composed of conical shells with altitude r and total base area $r\,d\theta \cdot 2\pi r \sin\theta$.) In our case

$$v = 2\int_0^{\pi/4} \frac{2\pi}{3} a^3(\cos 2\theta)^{3/2} \sin\theta\,d\theta = \frac{4\pi}{3}a^3 \int_0^{\pi/4} (2\cos^2\theta - 1)^{3/2} \sin\theta\,d\theta.$$

We put $\sqrt{2}\cos\theta = u$, $-\sqrt{2}\sin\theta\,d\theta = du$. The new limits are $u_1 = \sqrt{2}$, $u_2 = 1$.

$$v = -\frac{4\pi a^3}{3}\frac{1}{\sqrt{2}}\int_{\sqrt{2}}^{1} (u^2-1)^{3/2}\,du = \frac{2\sqrt{2}\pi a^3}{3}\int_{1}^{\sqrt{2}} (u^2-1)^{3/2}\,du$$

$$= \frac{2\sqrt{2}\pi a^3}{3}\left[\frac{u}{4}(u^2-1)^{3/2} - \frac{3}{8}u\sqrt{u^2-1} + \frac{3}{8}\ln|u+\sqrt{u^2-1}|\right]_1^{\sqrt{2}}$$

$$= \frac{2\sqrt{2}\pi a^3}{3}\left[\frac{\sqrt{2}}{4} - \frac{3\sqrt{2}}{8} + \frac{3}{8}\ln(\sqrt{2}+1)\right]$$

$$= \pi a^3\left[\frac{\sqrt{2}}{4}\ln(\sqrt{2}+1) - \frac{1}{6}\right].$$

The identity $(\sqrt{2}+1)^2 = 3 + 2\sqrt{2}$ proves that the results agree. In this example the first method was a bit simpler.

981. The area bounded by one arc of the cycloid $x = a(t - \sin t)$, $y = a(1 - \cos t)$ and the x axis, revolved around the x axis.

982. The area bounded by one arc of the cycloid $x = a(t - \sin t)$, $y = a(1 - \cos t)$ and the x axis, revolved around its axis of symmetry, i.e., the straight line $x = \pi a$.

Solution. We translate the coordinate system so that the new origin is situated at $x = \pi a$, $y = 0$. The equation of the cycloid in the new system of coordinates will be

$$x_1 = a(t - \sin t) - \pi a, \qquad y_1 = a(1 - \cos t).$$

The axis of rotation will be the y_1 axis.

$$v = \pi\int_0^{2a} x_1^2\,dy_1 = \pi\int_0^{\pi} a^2(t - \sin t - \pi)^2 a\sin t\,dt$$

$$= \pi a^3\int_0^{\pi}(t - \sin t - \pi)^2\sin t\,dt.$$

We substitute $t = u + \pi$, $dt = du$. Then

$$v = \pi a^3\int_{-\pi}^{0}(u + \sin u)^2(-\sin u)\,du$$

$$= \pi a^3\int_0^{-\pi}(u^2\sin u + 2u\sin^2 u + \sin^3 u)\,du$$

$$= \pi a^3\Big[-u^2\cos u + 2u\sin u + 2\cos u + \frac{u^2}{2} - \frac{u}{2}\sin 2u$$

$$\left.- \frac{1}{4}\cos 2u - \cos u + \frac{\cos^3 u}{3}\right]_0^{-\pi}$$

$$= \pi a^3\left(\pi^2 - 2 + \frac{\pi^2}{2} - \frac{1}{4} + 1 - \frac{1}{3} - 2 + \frac{1}{4} + 1 - \frac{1}{3}\right)$$

$$= \pi a^3\left(\frac{3\pi^2}{2} - \frac{8}{3}\right).$$

983. Find the volume of the torus formed by the circle $x^2 + (y - b)^2 = a^2$, $a < b$, rotating around the x axis.

984. The area bounded by the parabola $\sqrt{x} + \sqrt{y} = 2$ and the line $x + y = 4$ rotates around the line $y = x$. Find the volume of the body formed by this rotation.

985. Find the volume of the body obtained by rotation of the cardioid $r = a(1 + \cos\theta)$ around the polar axis.

986. The area between the curve $y = e^{-x^2}$ and its asymptote rotates around the y axis. Find the volume of the solid thus obtained.

Solution:

$$v = \pi \int_0^1 x^2\,dy = -\pi \int_0^1 \ln y\,dy$$

$$= -\pi \lim_{\alpha\to 0} (y \ln y - y)\Big|_\alpha^1 = -\pi \lim_{\alpha\to 0} (-1 - \alpha \ln \alpha + \alpha) = \pi,$$

because

$$\lim_{\alpha\to 0} \alpha \ln \alpha = \lim_{\alpha\to 0} \frac{\ln \alpha}{1/\alpha} = \lim_{\alpha\to 0} \frac{1/\alpha}{-1/\alpha^2} = 0.$$

987. The area described in Problem 986 rotates around the x axis. Find the volume of the solid of revolution.

988. Find the volume bounded by the cylinder $x^2 + y^2 = 1$ and the planes $y + z = 1$, $y = 0$, $z = 0$ $(y \geqslant 0)$.

Solution. In Figure 58 one half of the required volume is shown. We intersect it by planes perpendicular to the y axis and obtain rectangles with area $A(y) = xz = \sqrt{1 - y^2}(1 - y)$. Then

$$v = 2\int_0^1 \sqrt{1 - y^2}(1 - y)\,dy = 2\left[\frac{\pi}{4} + \frac{1}{3}(1 - y^2)^{3/2}\Big|_0^1\right] = \frac{\pi}{2} - \frac{2}{3}.$$

989. Consider the chords of a circle with radius R, parallel to a given direction. On each chord, in a plane perpendicular to the plane of the circle, a parabolic segment of constant altitude H is erected. Find the volume of the body so obtained.

990. The area bounded by the curve $y = x/(x^2 + 1)$, the x axis, and the straight lines $x = 1$, $x = 2$ rotates around the x axis. Find the volume of the resulting body.

991. Find the volume cut from the paraboloid $\dfrac{x^2}{a^2} + \dfrac{y^2}{b^2} = \dfrac{z}{h}$ by the plane $z = h$.

992. An elliptic cylinder is cut by a plane which cuts its base along the minor axis (see Fig. 59). Find the volume of the wedge-shaped body shown in the figure.

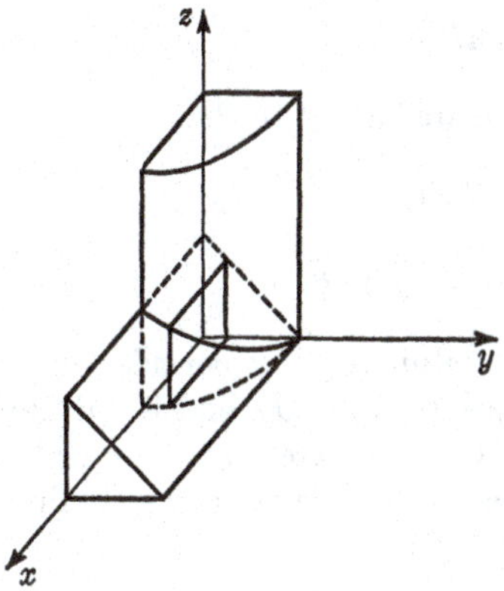

FIGURE 58

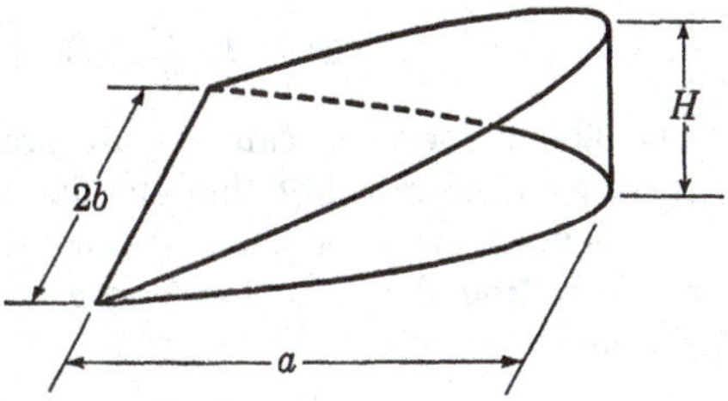

FIGURE 59

993. Find the volume of the body bounded by the three planes $x = 0$, $y = 0$, $z = 0$ and the surface formed by a straight line moving always parallel to the plane $z = 0$ and cutting the straight lines

$$\begin{cases} x = 1 \\ y = 0 \end{cases} \text{ and } \begin{cases} x = 0 \\ y + z = 1. \end{cases}$$

994. Find the volume of the body bounded by the surfaces

$$\frac{x^2}{a^2} + \frac{y^2}{b^2} - \frac{z^2}{c^2} = 1, \qquad z = 0, \qquad z = c.$$

995. Find the volume of an elliptic cone, the base of which is an ellipse with semiaxes a and b. The altitude of the cone is H.

996. Find the volume of the body formed by revolution of the curve $r = a \cos^2 \theta$ around the polar axis.

997. Find the volume of the body formed by revolution of the astroid

$$x^{2/3} + y^{2/3} = a^{2/3}$$

around the x axis.

10.4 AREA OF A SURFACE OF REVOLUTION

If the arc of the curve $y = f(x)$ from $x = a$ to $x = b$ rotates around the x axis, it describes a surface of revolution the area of which equals

$$P = 2\pi \int_a^b y\,dL.$$

The expression dL is the differential of the arc length, so that

$$P = 2\pi \int_a^b y\sqrt{1 + (y')^2}\,dx,$$

or

$$P = 2\pi \int_{t_1}^{t_2} y(t)\sqrt{(x_t')^2 + (y_t')^2}\,dt.$$

The above formulas can be interpreted intuitively by considering the surface as composed of thin circular stripes of width dL and length $2\pi y$.

When an arc of a curve given in polar coordinates by the equation $r = a(\theta)$ from $\theta = \theta_1$ to $\theta = \theta_2$ rotates around the polar axis, we obtain as above

$$P = 2\pi \int_{\theta_1}^{\theta_2} r \sin\theta \sqrt{r^2 + (r')^2}\,d\theta.$$

Find the areas of the surfaces of revolution obtained by rotating the following arcs around the given axes.

998. $y = \sin x$, $0 \leqslant x \leqslant \pi$, around Ox.

Solution:

$$P = 2\pi \int_0^\pi \sin x\sqrt{1 + \cos^2 x}\,dx.$$

We substitute $\cos x = t$, $-\sin x\,dx = dt$.

$$P = 2\pi \int_1^{-1} - dt\,\sqrt{1 + t^2} = 4\pi \int_0^1 \sqrt{t^2 + 1}\,dt$$

$$= 4\pi \cdot \tfrac{1}{2}[t\sqrt{t^2 + 1} + \ln(t + \sqrt{t^2 + 1})]\big|_0^1 = 2\pi \cdot [\sqrt{2} \pm \ln(1 + \sqrt{2})].$$

999. $y^2 = 4ax$, $0 \leqslant x \leqslant 3a$, around Ox.

1000. The arc of the circle $x^2 + y^2 = R^2$ from $x = a$ to $x = b$ around the x axis $(-R \leqslant a < b \leqslant R)$.

1001. The loop of the curve $9ay^2 = x(3a - x)^2$ around the x axis.

1002. The astroid $x = a\cos^3 t$, $y = a\sin^3 t$ around Ox.

1003. The cardioid $r = a(1 + \cos\theta)$ around the polar axis.

1004. $y = e^{-x}$, $x \geqslant 0$ around Ox.

1005. $x = e^t \sin t$, $y = e^t \cos t$ $(0 \leqslant t \leqslant \pi/2)$ around Ox.

1006. $y = x^3$ $(0 \leqslant x \leqslant 1)$ around Ox.

1007. $x = a(t - \sin t)$, $y = a(1 - \cos t)$, $0 \leqslant t \leqslant 2\pi$, around Ox.

1008. $y = x^2$, $0 \leqslant x \leqslant 1$, around Oy.

1009. $r^2 = a^2 \cos 2\theta$ around the polar axis.

10.5 MOMENT OF MASS; CENTROIDS

Given a plane figure bounded by the curves $y = f(x)$, $y = g(x)$ and the straight lines $x = a$, $x = b$ (see Fig. 60). We can consider this figure to

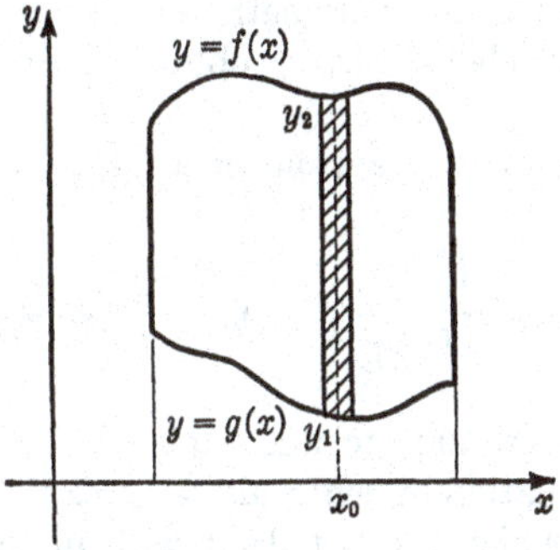

FIGURE 60

represent a plate with constant surface density 1, and therefore we may speak about the mass and centroid of this figure. This mass clearly is equal numerically to the area, and in what follows we shall simply speak about the centroid of the area. In the same way we shall deal with centroids of curves (representing thin material wires of constant linear density 1) or centroids of volumes (representing solids of constant density 1).

With reference to Figure 60, we note that the mass of the thin strip shaded there is approximately $(y_2 - y_1)\,dx$ and its centroid is at $(x_0, (y_1 + y_2)/2)$. (Here $y_1 = g(x_0)$, $y_2 = f(x_0)$.) The moment of mass of this strip with respect to the x axis is thus

$$(y_2 - y_1)\,dx\,\frac{y_1 + y_2}{2} = \frac{1}{2}(y_2^2 - y_1^2)\,dx.$$

The moment of mass of the whole figure with respect to Ox is consequently

$$M_x = \tfrac{1}{2}\int_a^b (y_2^2 - y_1^2)\,dx.$$

In the same way the moment of the mass with respect to Oy is

$$M_y = \int_a^b (y_2 - y_1)x\,dx.$$

Now the coordinates of the centroid of the whole figure will be

$$\bar{x} = \frac{M_y}{A} = \frac{\int_a^b (y_2 - y_1)x\,dx}{\int_a^b (y_2 - y_1)\,dx},$$

$$\bar{y} = \frac{M_x}{A} = \frac{1}{2}\frac{\int_a^b (y_2^2 - y_1^2)\,dx}{\int_a^b (y_2 - y_1)\,dx}.$$

Clearly, in special cases it may be more convenient to perform the integration along Oy or to use parametric equations or polar coordinates. The corresponding formulas are easily obtainable by an approach similar to the above.

For the coordinates of the centroid of a plane curve we obtain in an analogous way

$$\bar{x} = \frac{M_y}{L} = \frac{\int_a^b x\,dL}{\int_a^b dL}; \qquad \bar{y} = \frac{M_x}{L} = \frac{\int_a^b y\,dL}{\int_a^b dL}.$$

For some volumes we are able to find centroids by a similar procedure.

In general, the calculation of centroids requires the use of multiple and curvilinear intervals, which are not dealt with in this book.

1010. Find the centroid of the areas bounded by the semicircle $y = \sqrt{R^2 - x^2}$ and the x axis.

Solution. By symmetry, $\bar{x} = 0$. Now

$$M_x = \frac{1}{2}\int_{-R}^{R} y^2\,dx = 2 \cdot \frac{1}{2}\int_0^R (R^2 - x^2)\,dx$$

$$= R^3 - \frac{1}{3}R^3 = \frac{2}{3}R^3,$$

and

$$\bar{y} = \frac{M_x}{A} = \frac{\frac{2}{3}R^3}{\frac{\pi R^2}{2}} = \frac{4R}{3\pi}.$$

1011. Find the centroid of the arc of the semicircle $y = \sqrt{R^2 - x^2}$.

Solution. By symmetry, $\bar{x} = 0$.

$$M_x = 2\int_0^R y\,dL = 2\int_0^R \sqrt{y^2 + (yy')^2}\,dx.$$

$$2yy' = -2x, \quad yy' = -x;$$

$$M_x = 2\int_0^R \sqrt{R^2 - x^2 + x^2}\,dx = 2R^2.$$

$$\bar{y} = \frac{M_x}{L} = \frac{2R^2}{\pi R} = \frac{2R}{\pi}.$$

1012. Find the centroid of the area bounded by $y = \sin x$ $(0 \leqslant x \leqslant \pi)$ and the x axis.

1013. Find the centroid of the area bounded by the cycloid

$$x = a(t - \sin t),\ y = a(1 - \cos t),\quad 0 \leqslant t \leqslant 2\pi,$$

and the x axis.

1014. Find the centroid of the arc of the cycloid described in Problem 1013.

1015. Find the cartesian coordinates of the centroid of an area bounded by the rays $\theta = \theta_1$ and $\theta = \theta_2$ and the curve $r = r(\theta)$.

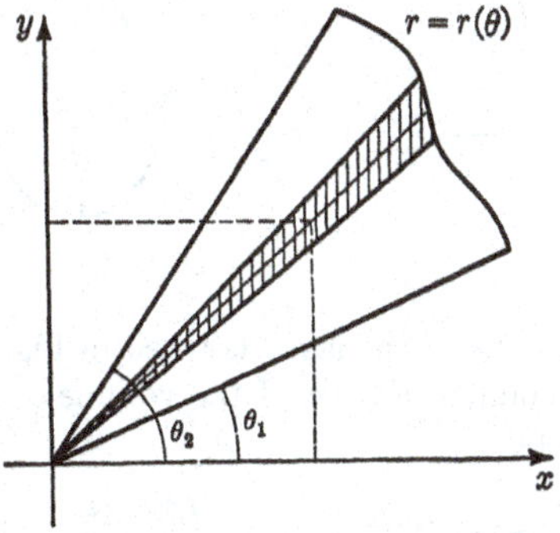

FIGURE 61

Solution. See Figure 61. The area of the shaded sector is $\frac{1}{2}r^2\,d\theta$, its centroid is on its altitude at the distance from the vertex equal to $\frac{2}{3}$ of this altitude, i.e., its cartesian coordinates are $(\frac{2}{3}r\cos\theta, \frac{2}{3}r\sin\theta)$. It follows that

$$\bar{x} = \frac{\int_{\theta_1}^{\theta_2} \frac{1}{2}r^2\,d\theta \cdot \frac{2}{3}r\cos\theta}{\int_{\theta_1}^{\theta_2} \frac{1}{2}r^2\,d\theta} = \frac{2}{3}\frac{\int_{\theta_1}^{\theta_2} r^3\cos\theta\,d\theta}{\int_{\theta_1}^{\theta_2} r^2\,d\theta},$$

and

$$\bar{y} = \frac{\int_{\theta_1}^{\theta_2} \frac{1}{2}r^2\,d\theta \cdot \frac{2}{3}r\sin\theta}{\int_{\theta_1}^{\theta_2} \frac{1}{2}r^2\,d\theta} = \frac{2}{3}\frac{\int_{\theta_1}^{\theta_2} r^3\sin\theta\,d\theta}{\int_{\theta_1}^{\theta_2} r^2\,d\theta}.$$

1016. Find the cartesian coordinates of the centroid of the area bounded by the cardioid $r = a(1 + \cos\theta)$.

1017. Find the cartesian coordinates of the centroid of an arc of the curve $r = r(\theta)$ from $\theta = \theta_1$ to $\theta = \theta_2$.

Solution. These coordinates are clearly

$$\bar{x} = \frac{\int_{\theta_1}^{\theta_2} r\cos\theta\sqrt{r^2 + (r')^2}\,d\theta}{\int_{\theta_1}^{\theta_2}\sqrt{r^2 + (r')^2}\,d\theta}, \qquad \bar{y} = \frac{\int_{\theta_1}^{\theta_2} r\sin\theta\sqrt{r^2 + (r')^2}\,d\theta}{\int_{\theta_1}^{\theta_2}\sqrt{r^2 + (r')^2}\,d\theta}.$$

1018. Find the cartesian coordinates of the centroid of the arc of the logarithmic spiral $r = ae^{\theta}$ from $\theta = \pi/2$ to $\theta = \pi$.

1019. Find the distance of the centroid of a circular cone from its base.

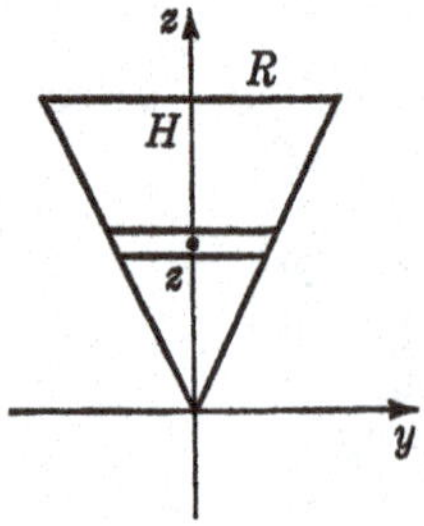

FIGURE 62

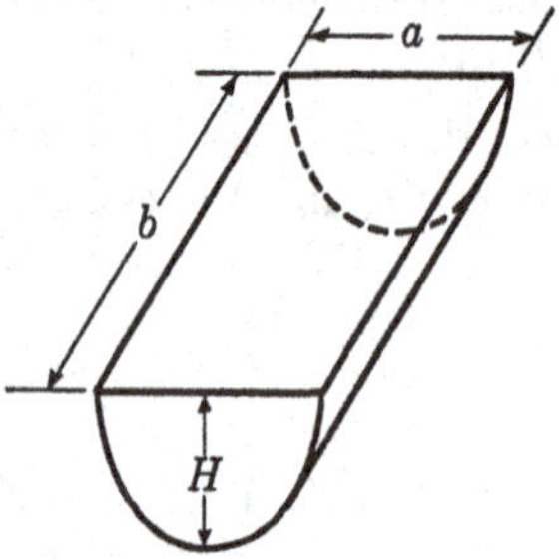

FIGURE 63

Solution. The volume of the layer marked in Figure 62 is $\pi(Rz/H)^2\,dz$. Consequently, the moment of mass of the volume of the cone with respect to the xy plane will be

$$\int_{z=0}^{H} \pi(Rz/H)^2\,dz\cdot z = \frac{\pi R^2}{H^2}\cdot\frac{H^4}{4} = \frac{\pi R^2 H^2}{4},$$

whence

$$\bar{z} = \frac{\pi R^2 H^2}{4} \div \frac{1}{3}\pi R^2 H = \frac{3}{4}H.$$

The required distance is $\frac{1}{4}H$.

1020. The curved surface bounding the solid in Figure 63 is part of a parabolic cylinder. Find the centroid of this solid.

1021. Find the centroid of the area bounded by the curves

$$y = (x+1)^2, \qquad y + x^2 = 1, \qquad y = 0.$$

1022. Find the centroid of the volume bounded by the four surfaces

$$x^2 + y^2 = 1, \qquad z + y = 1, \qquad z - y = 1, \qquad z = 0.$$

1023. Find the distance from the centroid of the solid bounded by a paraboloid of revolution to its plane base. The altitude of the solid is H.

1024. Find the centroid of the area bounded by the two circles

$$x^2 + y^2 - 2x = 0 \quad \text{and} \quad x^2 + y^2 - 4x = 0.$$

1025. Two spheres with radii r and $r + \alpha$ have their centers at the origin. Find the centroid of the solid bounded by the two upper hemispheres and the xy plane.

10.6 PAPPUS' THEOREMS

The following two theorems can be easily proved:

(1) If a surface of revolution is obtained by rotating an arc of a plane curve around an axis in its plane such that the arc lies only on one side of the axis, then the surface area is equal to the length of the arc multiplied by the length of the circumference described by its centroid.

(2) If a solid of revolution is obtained by rotating a plane area around an axis in its plane such that the area lies only on one side of the axis, then the volume of the solid equals the area of the figure multiplied by the length of the circumference described by its centroid.

Both theorems, known as Pappus' theorems, serve for determining the centroids of arcs and areas or, on the other hand, for calculating areas of surfaces of revolution or volumes of solids of revolution.

1026. Find the centroid of the arc of a semicircle $y = \sqrt{R^2 - x^2}$ and the centroid of the area bounded by it and by the x axis.

Solution. The semicircle rotating around the x axis describes a sphere, the area of which is $4\pi R^2$. By the first theorem

$$4\pi R^2 = 2\pi\bar{y}\cdot L = 2\pi\bar{y}\cdot\pi R;$$

hence

$$\bar{y} = \frac{4\pi R^2}{2\pi^2 R} = \frac{2R}{\pi}.$$

The second centroid is obtained using the second theorem:

$$v = \tfrac{4}{3}\pi R^3 = 2\pi\bar{y}\cdot A = 2\pi\bar{y}\cdot\tfrac{1}{2}\pi R^2;$$

$$\bar{y} = \frac{\frac{4}{3}\pi R^3}{\pi^2 R^2} = \frac{4R}{3\pi}.$$

Compare Problems 1010 and 1011.

1027. Compute the volume and surface area of the torus obtained by the revolution of a circle of radius a around an axis in the plane of the circle. The distance from the center of the circle to the axis is $d (d \geqslant a)$.

Solution. The centroid of both the rotating arc (i.e., the circle) and the rotating area (i.e., the area bounded by the circle) is the center of the circle, which is at distance d from the axis. By the above theorems,

$$P = 2\pi d \cdot L = 2\pi d \cdot 2\pi a = 4\pi^2 ad,$$
$$v = 2\pi d \cdot A = 2\pi d \cdot \pi a^2 = 2\pi^2 a^2 d.$$

Compare Problem 983.

1028. The figure bounded by the two cycloids

$$x = a(t - \sin t), \qquad y = a(1 - \cos t)$$

and $\quad x = a(t - \sin t), \quad y = -a(1 - \cos t) \quad (0 \leqslant t \leqslant 2\pi)$
rotates around the y axis. Find the surface area and volume of the body of revolution.

10.7 MOMENT OF INERTIA

The moment of inertia of a particle of mass m with respect to an axis placed at a distance d from the particle is md^2. Considering, as in Section 10.5, areas, arcs, and volumes as physical bodies with corresponding plane, line, and volume densities of 1, we can speak about the moments of inertia of an area, an arc, or a volume with respect to the given axis. Consider for example Figure 64; the moment of inertia of the rectangle shown there with respect to its base b (on the x axis) can be computed by summing up the moments of inertia of the thin horizontal strips, one of which is shaded in the figure. Every point of such a strip can be taken as being placed at an approximate distance y from the axis, i.e., the moment of inertia of the strip is approximately equal to its mass $b\,dy$ multiplied by y^2. Now the moment of inertia of the whole rectangle is

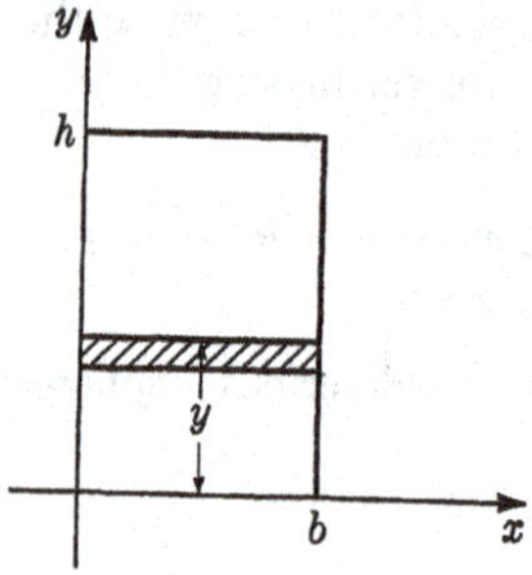

FIGURE 64

$$I_x = \int_0^h by^2\,dy = \frac{bh^3}{3}.$$

This result can be used to obtain the formula

$$I_x = \tfrac{1}{3}\int_a^b y^3\,dx,$$

the moment of inertia of a curvilinear trapezoid bounded by $y = f(x)$, $x = a$, $x = b$, and the x axis, with respect to this axis. The formula

$$I_y = \int_a^b yx^2\,dx$$

serves for computing the moment of inertia of the above trapezoid with respect to the y axis.

Corresponding formulas for moments of inertia of arcs and volumes can be obtained by similar arguments.

1029. Find the moment of inertia of the area bounded by a circle of radius R with respect to the diameter of this circle.

Solution. We shall find the moment of inertia with respect to the x axis. Using symmetry we obtain

$$I = \tfrac{1}{3}\cdot 4\int_0^R y^3\,dx.$$

We use the parametric equations of the circle, $x = R\cos t$, $y = R\sin t$:

$$I = \frac{1}{3}\cdot 4\int_{\pi/2}^{0} R^3\sin^3 t\,(-R\sin t)\,dt = \frac{4}{3}R^4\int_0^{\pi/2}\sin^4 t\,dt$$

$$= \frac{4}{3}R^4\cdot\frac{3\cdot 1}{4\cdot 2}\cdot\frac{\pi}{2} = \frac{\pi R^4}{4}.$$

Here we made use of the result of Problem 875.

1030. Find the moment of inertia of the area bounded by the ellipse $x^2 + 4y^2 = 4$ with respect to its major axis.

1031. Find the moment of inertia of the area bounded by a semicircle of radius R and its diameter with respect to an axis parallel to this diameter and passing through the centroid of the given area.

Solution. We shall use here a theorem of Steiner: Let there be given a system of particles of total mass M and two parallel axes l_1 and l_2, the last passing through the centroid of the system. If the distance between l_1 and l_2 is d, then $I_{l_1} = I_{l_2} + Md^2$. (I_{l_1}, I_{l_2} are the moments of inertia with respect to l_1 and l_2 correspondingly.) In our case we know $I_{\text{diam}} = \pi R^4/8$ (see Prob. 1029); the distance between the axes, by Problem 1010, is $4R/3\pi$, hence the required moment of inertia is

$$I = \frac{\pi R^4}{8} - \frac{\pi R^2}{2}\left(\frac{4R}{3\pi}\right)^2 = R^4\,\frac{9t^2 - 64}{72\pi}.$$

1032. Find the moment of inertia of the area of a square with side a with respect to its diagonal.

1033. Find the moment of inertia of a circular cylinder with respect to its axis of revolution.

Solution. See Figure 65. The volume of the shaded ring is approximately $2\pi xH\,dx$. The required moment of inertia is

$$I_z = \int_0^R 2\pi x H\, dx \cdot x^2 = \frac{2\pi R^4 H}{4} = \frac{\pi R^4 H}{2} = \pi H R^2 \cdot \frac{R^2}{2} = v\,\frac{R^2}{2}.$$

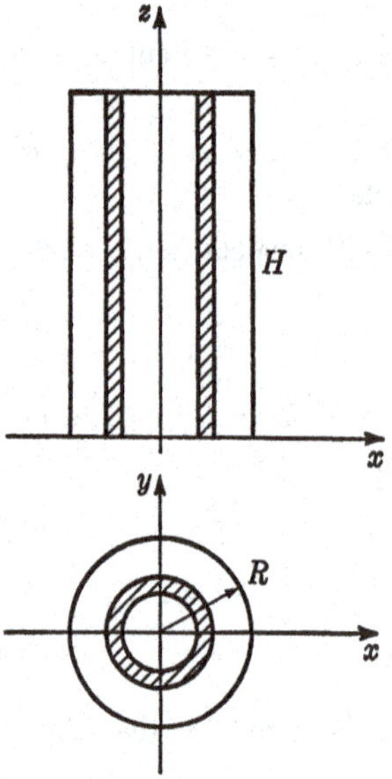

FIGURE 65

Here v denotes the volume (mass) of the cylinder.

1034. Find the moment of inertia of an ellipsoid of revolution with respect to its axis.

1035. Find the moment of inertia of the volume bounded by the surface $x^2 + y^2 = 4az$ and the plane $z = 2a$ with respect to the z axis.

10.8 PHYSICS PROBLEMS

In this section we shall solve some problems in physical variables, using integration.

1036. A rod of length a rotates n times per second around one of its ends. The mass per unit length is constant and equals γ. Find the force (caused by the rotation) acting on the fixed end.

Solution. See Figure 66. The mass of the interval dx of the rod is $\gamma\, dx$. The radial force acting on it is $\gamma\, dx \cdot x\omega^2$, where ω is the angular velocity of the rod, i.e., $\omega = 2\pi n$. The force in question is obtained by integration,

$$F = \int_0^a \gamma 4\pi^2 n^2 x\, dx = 2\pi^2 n^2 \gamma a^2.$$

1037. A wire of length l, cross section a, and modulus of elasticity E

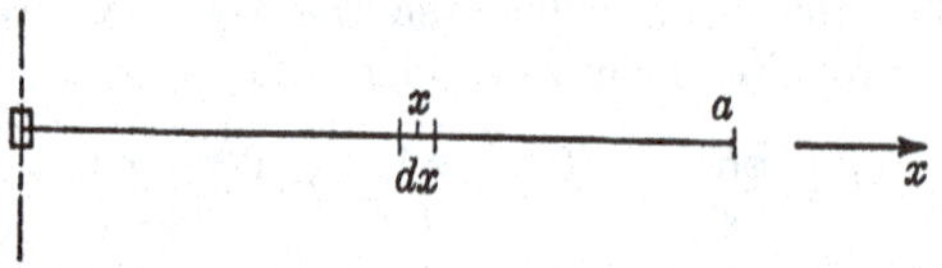

FIGURE 66

under the action of a force F elongates by $\Delta l = Fl/Ea$. Assuming the wire to hang vertically, find the elongation caused by its own weight. The specific weight of the wire is γ.

1038. Two positive electrical charges e_1 and e_2 repel each other with the force $e_1 e_2/r^2$, where r is the distance between them. Find the work needed to bring a charge $e_2 = +1$ from infinity to the distance a from the charge $+e_1$.

1039. Ey Torricelli's law, the velocity of water flowing out from a hole in a water container is $k\sqrt{2gh}$, where k is an empirical coefficient, g is the acceleration due to gravity, and h is the height of the water surface above the hole. Find the time needed to empty a water-filled cylinder of height H and cross section A after a hole of area a is made in its base.

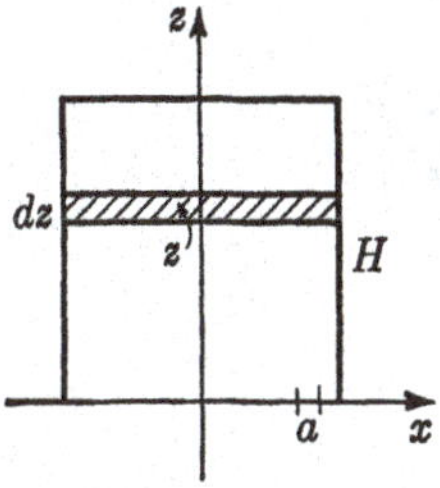

FIGURE 67

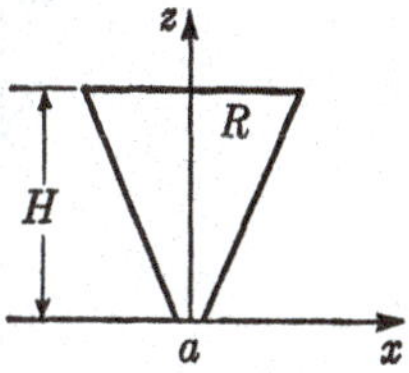

FIGURE 68

Solution. See Figure 67. Suppose the surface of the water is at height z. Then the amount of water flowing out from the cylinder in time dt is $av\,dt = ak\sqrt{2gz}\,dt$. The same amount can also be computed by $-A\,dz$ (negative because dz is negative and the amount of water positive). We have thus

$$-A\,dz = \sqrt{2gz}ka\,dt$$

or

$$dt = -\frac{A}{ka\sqrt{2g}} \cdot \frac{dz}{\sqrt{z}}.$$

Between $t = 0$ and $t = T$, z changes from H to 0 (T is the time of emptying the cylinder). Consequently

$$\int_0^T dt = \int_H^0 -\frac{A}{ka\sqrt{2g}} \cdot \frac{dz}{\sqrt{z}},$$

$$T = -\frac{A}{ka\sqrt{2g}} 2\sqrt{z}\Big|_H^0 = \frac{A}{ka}\sqrt{\frac{2H}{g}}.$$

1040. The above question for the conical funnel shown in Figure 68.

1041. Find the force exerted by water on a vertical triangular wall. The altitude of the triangle is h and its base (on the water surface) is a.

Solution. See Figure 69. The force acting on the shaded strip of the wall is equal to its area $(a/h)(h - y)\,dy$ multiplied by the pressure of the water at depth y, i.e., by γy (γ is the specific weight of the water). All such forces act in the same direction (perpendicular to the wall) and they add arithmetically:

$$F = \int_0^h \frac{a}{h}(h - y)\gamma y\, dy = \frac{a}{h}\gamma\left(\frac{h^3}{2} - \frac{h^3}{3}\right) = \frac{ah^2\gamma}{6}.$$

Remark. The area of the triangle is $\frac{1}{2}ah$, the depth of its centroid $\frac{1}{3}h$. We have obtained $F = \frac{1}{2}ah \cdot \frac{1}{3}h\gamma$.

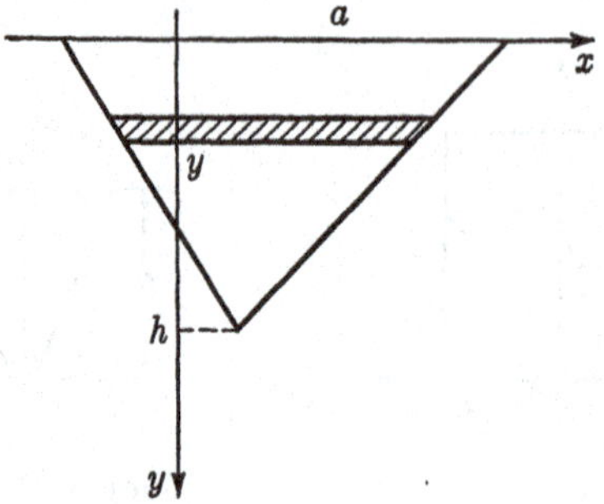

FIGURE 69

1042. Find the work needed to pour sand for an embankment in the form of a cone of height H and base radius R. The specific weight of the sand is γ.

1043. The current and voltage in an electric circuit are given by

$$I = I_0 \sin\left(\frac{2\pi t}{T} - \varphi_0\right) \quad \text{and} \quad V = V_0 \sin\frac{2\pi t}{T}.$$

I_0 and V_0 are the corresponding amplitudes, t is the time, T is the period, and φ_0 the phase angle. Find the work done by the current during one period and show that this work is maximal when $\varphi_0 = 0$. Find the average power.

Solution. The work is given by the integral

$$W = \int_0^T IV\, dt = \int_0^T I_0 V_0 \sin\frac{2\pi t}{T} \sin\left(\frac{2\pi t}{T} - \varphi_0\right) dt$$

$$= \frac{I_0 V_0}{2} \int_0^T \left[\cos\varphi_0 - \cos\left(\frac{4\pi t}{T} - \varphi_0\right)\right] dt = \frac{I_0 V_0 T}{2}\cos\varphi_0.$$

The work will be clearly maximal when $\cos\varphi_0 = 1$, i.e., at $\varphi_0 = 0$. The average power equals

$$N_{\text{av}} = \frac{W}{T} = \frac{I_0 V_0}{2}\cos\varphi_0.$$

The effective current of the circuit is defined by $I_{\text{eff}} = I_0/\sqrt{2}$. In the same way $V_{\text{eff}} = V_0/\sqrt{2}$. Then

$$N_{\text{av}} = I_{\text{eff}} V_{\text{eff}} \cos \varphi_0.$$

The instantaneous power is equal to IV.

1044. A cylinder of volume v is filled with steam under a pressure p. Find the work needed to compress the steam to the volume $\frac{1}{2}V$, if the temperature of the steam is to remain constant.

1045. Find the kinetic energy of a sphere with radius R rotating around its diameter with angular velocity ω. The density of the material of the sphere is γ.

1046. Find the gravitational force on a material point having mass m due to an infinite straight line with length density γ. The perpendicular distance from the point to the line is a. The constant of gravity is f.

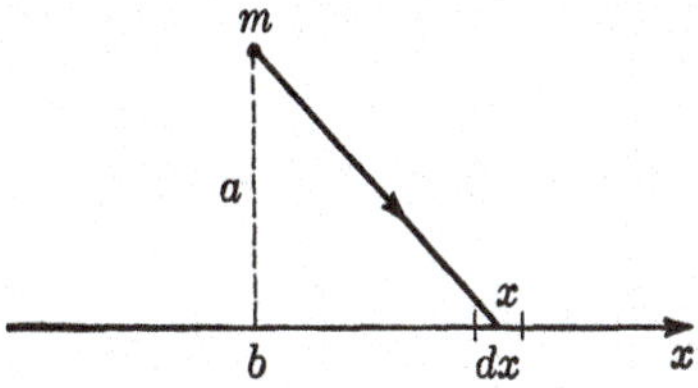

FIGURE 70

Solution. The force and its vertical component, indicated in Figure 70, are

$$\frac{f\gamma\, dx \cdot m}{a^2 + x^2}, \quad \frac{f\gamma\, dx \cdot m}{a^2 + x^2} \cdot \frac{a}{\sqrt{a^2 + x^2}}.$$

By symmetry, the sum of the horizontal components of these particular forces is 0. The sum of the vertical components is given by

$$F = 2\int_0^\infty \frac{fa\gamma m\, dx}{(x^2 + a^2)^{3/2}} = 2a\gamma mf \int_0^\infty \frac{dx}{(x^2 + a^2)^{3/2}}.$$

Put $x = a \tan t$, $dx = \dfrac{a}{\cos^2 t}\, dt$. Then

$$F = 2a\gamma mf \int_0^{\pi/2} \frac{a\, dt}{\cos^2 t\, a^3 \sec^3 t} = \frac{2m\gamma f}{a} \sin t \Big|_0^{\pi/2} = \frac{2m\gamma f}{a}.$$

1047. Find the work needed to pump the water out of a horizontal cylinder of length L and base radius R. The specific weight of the water is γ.

1048. The acceleration a of a rocket moving vertically is related to the time t by the equation $a(m - nt) = c$ (c, m, n constants and $t < m/n$). Find the velocity of the rocket at the moment t and its height at this moment, if the velocity and height at $t = 0$ are 0 and h_0 respectively.

XI

INFINITE SERIES

11.1 THE GENERAL NOTION OF A NUMBER SERIES

Given a finite sequence of numbers $u_1, u_2, \ldots, u_n$. The sum

$$u_1 + u_2 + \ldots + u_n = \sum_{i=1}^{n} u_i$$

is called a finite series. This notion can be generalized to the case when the sequence $u_1, u_2, \ldots, u_n, \ldots$ does not end. The corresponding expression

$$u_1 + u_2 + \ldots + u_n + \ldots = \sum_{n=1}^{\infty} u_n$$

is called an infinite series.

A series whose elements all are constant numbers is called a series of numbers. In later sections of this chapter series of functions will also be considered.

The series is given when the element u_n is uniquely determined by n. We shall often speak of the series $\sum_{n=1}^{\infty} u_n$ simply as the series u_n.

Denote by S_n the finite sum

$$S_n = u_1 + u_2 + u_3 + \ldots + u_n.$$

If the sequence

$$S_1 = u_1, \qquad S_2 = u_1 + u_2, \qquad S_3 = u_1 + u_2 + u_3, \qquad \ldots$$

has a limit, i.e., if there exists a number S such that

$$S = \lim_{n \to \infty} S_n,$$

we say that the corresponding series *converges* and its sum is S. If the above limit does not exist we say that the corresponding series *diverges*, and then it makes no sense to speak about the sum of such series.

In the case of convergent series we consider also the remainder

$$R_{n+1} = S - S_n = u_{n+1} + u_{n+2} + \dots .$$

Clearly,
$$\lim_{n\to\infty} R_{n+1} = 0.$$

The partial sum S_n can serve as an approximate value of S and the remainder R_{n+1} gives the corresponding error.

Determining the convergence or divergence of a given series is one of the main problems of the theory of series, and in what follows we shall devote much space to this question.

For any series we have the following necessary condition for convergence: For a series u_n to converge it is necessary that

$$\lim_{n\to\infty} u_n = 0.$$

We remark that when testing convergence we may disregard any finite number of terms occurring in the series. Deleting finitely many terms will in general alter S; however, the convergence or divergence will not be affected.

11.2 CONVERGENCE OF SERIES WITH POSITIVE TERMS

Here we shall deal with number series whose terms satisfy $u_n > 0$. We list a number of tests which are useful in determining the convergence or divergence of such series.

Test by Definition. If S_n can be obtained as a function of n, we try to find out whether $\lim_{n\to\infty} S_n$ exists. In this case, in addition to determining the convergence, we often find the sum S of the series as well.

Comparison Test. If every element of the series u_n is not larger than the corresponding element of a convergent series v_n, then the series u_n also converges. On the other hand, if every element of the series u_n is not smaller than the corresponding element of a divergent series v_n, then the series u_n diverges.

The two series most convenient for comparison are:

(a) The geometric series with the ratio q satisfying $0 < q < 1$. As is well known, the geometric series converges and has the sum $S = a/(1 - q)$ where a is its first term.

(b) The series

$$a + \frac{a}{2^p} + \frac{a}{3^p} + \frac{a}{4^p} + \ldots + \frac{a}{n^p} + \ldots$$

where a is a positive constant. For $p > 1$ this series converges, for $p \leqslant 1$ it diverges. For $p = 1$ and $a = 1$ it is the harmonic series, which is known to be divergent.

The geometric series comparison test yields two more very useful tests, as follows.

D'Alembert's Test (The Ratio Test). If a fixed number $q < 1$ can be found such that beginning with a certain n the ratio $r_n = u_{n+1}/u_n$ is no larger than q (i.e., $r_n \leqslant q$), then the series u_n converges. $\left(\text{The series } \sum_{n=1}^{\infty} \frac{1}{n} \text{ shows that the condition } r_n < 1 \text{ does not suffice for convergence.}\right)$ On the other hand, if beginning with a certain n, $r_n \geqslant 1$, then the series u_n diverges.

D'Alembert's test can in many cases be used in a more convenient form. The limit $r = \lim_{n\to\infty} u_{n+1}/u_n$ is computed (if it exists), and if $r < 1$ the series converges, or if $r > 1$ it diverges. In the case $r = 1$ no decision can be made on the basis of this test alone. Although this form is more convenient, there do exist series whose behavior can be determined with the first form of the test but not with the latter.

Cauchy's Test. In place of the expression r_n the expression $t_n = \sqrt[n]{u_n}$, and in place of $r = \lim_{n\to\infty} = u_{n+1}/u_n$ the limit $t = \lim_{n\to\infty} \sqrt[n]{u_n}$, are investigated. All results stated above with regard to r_n and r hold in exactly the same form for t_n and t.

Cauchy's Integral Test. This test uses the convergence of improper integrals to show the convergence of series. Suppose we have found a continuous monotonic decreasing function $y = f(x)$ such that, beginning with a certain n_0, $u_n = f(n)$; then the series u_n converges if the improper integral $\int_{n_0}^{\infty} f(x)\,dx$ exists, and it diverges if the integral does not exist.

There follow various examples of series tested for convergence. In the following five exercises find S_n and investigate the $\lim_{n\to\infty} S_n$.

1049. $\frac{1}{1\cdot 4} + \frac{1}{4\cdot 7} + \ldots + \frac{1}{(3n-2)(3n+1)} + \ldots$

Solution. We can decompose u_n into partial fractions,

$$u_n = \frac{1}{3}\left(\frac{1}{3n-2} - \frac{1}{3n+1}\right).$$

Now we find S_n,

$$S_n = \frac{1}{3}\left(1 - \frac{1}{4} + \frac{1}{4} - \frac{1}{7} + \frac{1}{7} - \frac{1}{10} + \ldots + \frac{1}{3n-2} - \frac{1}{3n+1}\right)$$
$$= \frac{1}{3}\left(1 - \frac{1}{3n+1}\right);$$

$$\lim_{n\to\infty} S_n = \lim_{n\to\infty} \frac{1}{3}\left(1 - \frac{1}{3n+1}\right) = \frac{1}{3}.$$

We have proved directly that the given series converges, and moreover we have found its sum, $S = \frac{1}{3}$.

1050. $\frac{1}{1\cdot 7} + \frac{1}{3\cdot 9} + \ldots + \frac{1}{(2n-1)(2n+5)} + \ldots$

1051. $\frac{1}{1\cdot 2\cdot 3} + \frac{1}{2\cdot 3\cdot 4} + \ldots + \frac{1}{n(n+1)(n+2)} + \ldots$

1052. $\frac{5}{6} + \frac{13}{36} + \ldots + \frac{3^n + 2^n}{6^n} + \ldots$

1053. $\frac{3}{4} + \frac{5}{36} + \ldots + \frac{2n+1}{n^2(n+1)^2} + \ldots$

Test the convergence of the following series by comparison.

1054. $\frac{1}{1\cdot 2} + \frac{1}{3\cdot 2^3} + \ldots + \frac{1}{(2n-1)\cdot 2^{2n-1}} + \ldots$

Solution. We compare this series with the convergent geometric series

$$\frac{1}{2} + \frac{1}{2^3} + \ldots + \frac{1}{2^{2n-1}} + \ldots.$$

We have
$$\frac{1}{(2n-1)2^{2n-1}} \leqslant \frac{1}{2^{2n-1}},$$

whence the given series converges.

1055. $\sin\frac{\pi}{2} + \sin\frac{\pi}{4} + \ldots + \sin\frac{\pi}{2^n} + \ldots$

1056. $\frac{1}{2\cdot 5} + \frac{1}{3\cdot 6} + \ldots + \frac{1}{(n+1)(n+4)} + \ldots$

1057. $1 + \frac{1+2}{1+2^2} + \ldots + \frac{1+n}{1+n^2} + \ldots$

Solution. We shall compare this series with the harmonic series:

$$\frac{1+n}{1+n^2} - \frac{1}{n} = \frac{n+n^2-1-n^2}{(1+n^2)n} = \frac{n-1}{(1+n^2)n} \geqslant 0.$$

Consequently $\frac{1+n}{1+n^2} \geqslant \frac{1}{n}$ and the given series diverges.

1058. $\frac{1}{2} + \frac{1}{5} + \ldots + \frac{1}{3n-1} + \ldots$

1059. $\frac{1}{\ln 2} + \frac{1}{\ln 3} + \ldots + \frac{1}{\ln n} + \ldots$

1060. $\sum_{n=1}^{\infty} \frac{1}{n^2 - 4n + 5}$

Solution. We shall use in this case a theorem which will prove convenient also in other cases. If a number $k \neq 0$ exists where $\lim_{n\to\infty} u_n/v_n = k \neq 0$ $(v_n \neq 0)$, then the series u_n and v_n converge or diverge together.

In our case let us choose for v_n the series $v_n = 1/n^2$. Then

$$\lim_{n\to\infty} \frac{u_n}{v_n} = \lim_{n\to\infty} \frac{n^2}{n^2 - 4n + 5} = 1.$$

Here $k = 1 \neq 0$, and v_n is known to converge. Therefore also u_n converges. We remark that it is also not difficult to show, by comparison, that u_n converges: we have

$$\frac{1}{n^2 - 4n + 5} < \frac{1}{n^{3/2}}$$

beginning with a certain n, $\sum_{n=1}^{\infty} \frac{1}{n^{3/2}}$ being known to converge.

1061. $\sum_{n=1}^{\infty} \frac{1}{\sqrt{n^2 + 2n}}$

Find the behavior of the following series using d'Alembert's test.

1062. $\sum_{n=1}^{\infty} \frac{1}{(2n+1)!}$

Solution. We shall use the second form of d'Alembert's test:

$$r = \lim_{n\to\infty} \frac{u_{n+1}}{u_n} = \lim_{n\to\infty} \frac{(2n+1)!}{(2n+3)!} = \lim_{n\to\infty} \frac{1}{(2n+2)(2n+3)} = 0.$$

The given series converges.

1063. $\tan\frac{\pi}{4} + 2\tan\frac{\pi}{8} + \ldots + n\tan\frac{\pi}{2^{n+1}} + \ldots$

1064. $\sum_{n=1}^{\infty} \frac{n^2}{3^n}$

1065. $\sum_{n=1}^{\infty} n^2 \sin \frac{\pi}{2^n}$

1066. $\sum_{n=1}^{\infty} \frac{(n+1)^5}{(1.01)^n}$

1067. $\sum_{n=1}^{\infty} \frac{2^n \cdot n!}{n^n}$

Solution:

$$\lim_{n\to\infty} \frac{u_{n+1}}{u_n} = \lim_{n\to\infty} \frac{2^{n+1}(n+1)!n^n}{(n+1)^{n+1}\cdot 2^n \cdot n!}$$

$$= \lim_{n\to\infty} \frac{2n^n}{(n+1)^n} = 2 \lim_{n\to\infty} \frac{1}{\left(1+\frac{1}{n}\right)^n} = \frac{2}{e} < 1.$$

The series converges.

1068. $\sum_{n=1}^{\infty} \frac{3^n n!}{n^n}$

1069. $\sum_{n=1}^{\infty} \frac{e^n n!}{n^n}$

Solution. This is a case intermediate between the two preceding series; we have

$$r = \lim_{n\to\infty} \frac{u_{n+1}}{u_n} = \frac{e}{e} = 1.$$

This result does not decide the question of convergence. Let us compute

$$r_n = \frac{u_{n+1}}{u_n} = \frac{e}{\left(1+\frac{1}{n}\right)^n}.$$

The sequence $\left(1+\frac{1}{n}\right)^n$, the limit of which is e, is monotone increasing, i.e., $\left(1+\frac{1}{n}\right)^n < e$ for any n, and $r_n > 1$. Consequently the series does not converge. This series demonstrates that the first form of d'Alembert's test is "stronger" than the second.

Test the convergence of the following series by Cauchy's test.

1070. $\frac{1}{\ln 2} + \frac{1}{\ln^2 3} + \ldots + \frac{1}{\ln^n (n+1)} + \ldots$

Solution. We compute

$$t = \lim_{n\to\infty} \sqrt[n]{u_n} = \lim_{n\to\infty} \frac{1}{\ln (n+1)} = 0.$$

The series converges.

1071. $\frac{1}{3} + \left(\frac{2}{5}\right)^2 + \ldots + \left(\frac{n}{2n+1}\right)^n + \ldots$

1072. $\sum_{n=1}^{\infty} \left(\arcsin \frac{1}{n}\right)^n$

1073. $\sum_{n=1}^{\infty} \frac{\left(\frac{n+1}{n}\right)^{n^2}}{3^n}$

1074. $\sum_{n=1}^{\infty} a^n \sin^2 n\alpha,\ 0 < a < 1$

Solution:

$$t = \lim a\sqrt[n]{\sin^2 n\alpha} \leqslant a < 1,$$

whence the series converges. It is interesting to remark that d'Alembert's test does not lead in this case to a definite answer, since

$$r_n = \frac{u_{n+1}}{u_n} = a\left[\frac{\sin (n+1)\alpha}{\sin n\alpha}\right]^2,$$

and this expression does not become smaller than 1 for all n greater than a certain n_0.

Investigate the convergence of the following series using Cauchy's integral test.

1075. $\frac{1}{2 \ln^2 2} + \frac{1}{3 \ln^2 3} + \ldots + \frac{1}{(n+1) \ln^2 (n+1)} + \ldots$

Solution. We shall use the function

$$f(x) = \frac{1}{(x+1) \ln^2 (x+1)}.$$

We try

$$I = \int_1^{\infty} f(x)\,dx = \int_1^{\infty} \frac{dx}{(x+1) \ln^2 (x+1)}$$
$$= \lim_{M\to\infty} \left(-\frac{1}{\ln (x+1)}\bigg|_1^M\right) = \frac{1}{\ln 2}.$$

The improper integral exists, hence the series converges.

1076. $\sum_1^{\infty} \frac{a}{n^p},\ a > 0$

Solution. $f(x) = a/x^p$;

$$\int_1^{\infty} f(x)\,dx = \lim_{M\to\infty} \int_1^M \frac{a}{x^p}\,dx = a \lim_{M\to\infty} \left(-\frac{1}{p-1} \cdot \frac{1}{x^{p-1}}\bigg|_1^M\right)$$
$$= a \lim_{M\to\infty} \frac{1}{p-1}\left(1 - \frac{1}{M^{p-1}}\right).$$

For $p > 1$ the integral exists and the series converges.
For $p < 1$ there is no limit and the series diverges.
For $p = 1$ we have to integrate separately:

$$\lim_{M\to\infty} \int_1^M \frac{a}{x}\,dx = a \lim_{M\to\infty} \ln M.$$

This limit does not exist and the series diverges.

Thus we have proved the assertion about this series made in the beginning of this section.

1077. $\sum_{n=2}^{\infty} \frac{1}{n \ln n}$

1078. $\sum_{n=1}^{\infty} \left(\frac{1+n}{1+n^2}\right)^2$

Investigate the convergence of the following series.

1079. $\frac{1}{\sqrt{8}} + \frac{1}{\sqrt{27}} + \ldots + \frac{1}{\sqrt{(n+1)^3}} + \ldots$

1080. $1 + \frac{2}{3} + \ldots + \frac{n}{2n-1} + \ldots$

Solution:

$$\lim_{n\to\infty} u_n = \lim_{n\to\infty} \frac{n}{2n-1} = \frac{1}{2}.$$

The necessary condition for convergence is not fulfilled and the series diverges.

1081. $1 + \frac{2!}{10^5} + \frac{3!}{10^{10}} + \ldots + \frac{n!}{10^{5(n-1)}} + \ldots$

1082. $\frac{1}{1001} + \frac{2}{2001} + \ldots + \frac{n}{1000n+1} + \ldots$

1083. $\sum_{n=1}^{\infty} (2n^2 + 3n + 5)(\frac{2}{3})^n$

1084. $\arctan 1 + \left(\arctan \frac{1}{2}\right)^2 + \ldots + \left(\arctan \frac{1}{n}\right)^n + \ldots$

1085. $\sin \frac{\pi}{2} + \sin \frac{\pi}{4} + \ldots + \sin \frac{\pi}{2n} + \ldots$

1086. Prove that $\lim_{n\to\infty} \frac{a^n}{n!} = 0$

Solution. We shall show that the series $\sum_{n=1}^{\infty} a^n/n!$ converges, whence its general term $u_n = a^n/n!$ must tend to zero. By d'Alembert's test,

$$\lim_{n\to\infty} \frac{a^{n+1}}{(n+1)!} \cdot \frac{n!}{a^n} = \lim_{n\to\infty} \frac{a}{n+1} = 0.$$

1087. Prove that $\lim_{n\to\infty} \frac{n^n}{(n!)^2} = 0.$

1088. Prove that $\lim_{n\to\infty} \frac{(n!)^n}{n^{n^2}} = 0.$

Investigate the convergence of the following series.

1089. $\sum_{n=1}^{\infty} \frac{1}{n\sqrt{n+2}}$

1090. $\sum_{n=1}^{\infty} a_n$ if $\lim_{n\to\infty} na_n = b \neq 0$

1091. $\sum_{n=2}^{\infty} \frac{1}{\sqrt[n]{\ln n}}$

1092. $\sum_{n=1}^{\infty} \frac{n^{\left(n+\frac{1}{n}\right)}}{\left(n+\frac{1}{n}\right)^n}$

1093. $\sum_{n=2}^{\infty} \frac{1}{n \ln^p n}$

1094. $\frac{4}{2} + \frac{4\cdot 7}{2\cdot 6} + \frac{4\cdot 7\cdot 10}{2\cdot 6\cdot 10} + \ldots$

1095. $\sum_{n=1}^{\infty} n^2 e^{-\sqrt{n}}$

1096. $\sum_{n=2}^{\infty} \frac{1}{n\sqrt[3]{n} - \sqrt{n}}$

11.3 CONVERGENCE OF SERIES WITH POSITIVE AND NEGATIVE TERMS

Given a series

$$(1) \qquad a_1 + a_2 + a_3 + \ldots + a_n + \ldots$$

with a_n an arbitrary number. We may assume that $a_n \neq 0$ for all n. If the series

$$(2) \qquad |a_1| + |a_2| + |a_3| + \ldots + |a_n| + \ldots,$$

composed of the absolute values of the numbers in (1), converges, then the series (1) also converges, and we say in this case that (1) converges *absolutely*. The converse is not true; the series (1) can converge and (2) not converge. We say in that case that (1) converges *conditionally* (but not absolutely). Absolute convergence can be tested by the methods of the preceding section, because in (2) we have positive terms only.

A general criterion due to Cauchy is valid for any series: The series converges if and only if for an arbitrary $\epsilon > 0$ there exists an n_0 such that for any pair $m,n > n_0$,

$$|S_{n+m} - S_n| = |a_{n+1} + a_{n+2} + \ldots + a_{n+m}| < \epsilon.$$

In a case often encountered the signs of the terms in (1) alternate, i.e., the series is of the form

$$b_1 - b_2 + b_3 - b_4 + \ldots + (-1)^{n-1}b_n + \ldots$$

or $$-b_1 + b_2 - b_3 + b_4 + \ldots + (-1)^n b_n + \ldots, \quad b_n > 0.$$

Here a special test of Leibniz can be used: If the conditions

(1) $$b_{n+1} < b_n, \quad n = 1, 2, \ldots,$$

(2) $$\lim_{n\to\infty} b_n = 0$$

are fulfilled, then the series converges.

It is very easy to see that the sum of this series has the sign of its first term and an absolute value less than it. The remainder R_{n+1} then satisfies the inequality $|R_{n+1}| < b_{n+1}$.

The following theorem of Dirichlet can also serve to establish the conditional convergence of a given series. The series

$$u_1v_1 + u_2v_2 + \ldots + u_nv_n + \ldots$$

converges if the sum $u_1 + u_2 + \ldots + u_n$ remains bounded when n increases, and the numbers $v_1, v_2, v_3, \ldots, v_n, \ldots$ decrease and tend to zero as $n \to \infty$.

Investigate the convergence of the following series.

1097. $1 - \frac{1}{2} + \frac{1}{3} - \frac{1}{4} + \ldots$

Solution. This series does not converge absolutely, because

$$1 + \tfrac{1}{2} + \tfrac{1}{3} + \tfrac{1}{4} + \ldots$$

is the harmonic series which is known to diverge. The conditional convergence can be proved by Leibniz's test since the series alternates:

$$\frac{1}{n+1} < \frac{1}{n} \quad \text{and} \quad \lim_{n\to\infty} \frac{1}{n} = 0.$$

Both conditions ($b_{n+1} < b_n$ and $\lim\limits_{n\to\infty} b_n = 0$) are fulfilled.

1098. $1 - \dfrac{1}{3^3} + \ldots + (-1)^{n+1}\dfrac{1}{(2n-1)^3} + \ldots$

1099. $1 - \dfrac{100}{2} + \dfrac{400}{2^2} - \dfrac{900}{2^3} + \ldots + (-1)^n \dfrac{100(n-1)^2}{2^{n-1}} + \ldots$

1100. $\dfrac{2}{4} - \dfrac{3}{8} + \dfrac{4}{12} - \ldots + (-1)^{n+1}\dfrac{n+1}{4n} + \ldots$

1101. $\dfrac{1}{\ln 2} - \dfrac{1}{\ln 3} + \ldots + (-1)^{n+1}\dfrac{1}{\ln(n+1)} + \ldots$

1102. $-1 + \frac{1}{\sqrt{2}} - \ldots + (-1)^n \frac{1}{\sqrt{n}} + \ldots$

1103. $\frac{1}{2} - \frac{8}{4} + \ldots + (-1)^{n+1} \frac{n^3}{2^n} + \ldots$

1104. $\sum_{n=1}^{\infty} \frac{(-1)^n}{n - \ln n}$

1105. $\frac{\sin \alpha}{1} + \frac{\sin 2\alpha}{2} + \frac{\sin 3\alpha}{3} + \ldots$

Solution. We proved in Problem 850 the formula

$$\sin \alpha + \sin 2\alpha + \ldots + \sin n\alpha = \frac{\sin \frac{n+1}{2} \alpha \sin \frac{n}{2} \alpha}{\sin \frac{\alpha}{2}}, \quad \alpha \neq 2\pi k.$$

For $n \to \infty$ this sum remains bounded, because the numerator is bounded and the denominator is a constant. Let us put

$$u_n = \sin n\alpha \quad \text{and} \quad v_n = \frac{1}{n}.$$

Then $u_1 + u_2 + \ldots + u_n$ is bounded when n increases and $v_1, v_2, \ldots, v_n, \ldots$ is a decreasing sequence tending to zero. By the above mentioned theorem of Dirichlet, the conditional convergence of the given series is proved. For $\alpha = 2\pi k$ every term of the series equals zero and it converges trivially.

1106. $\sum_{n=1}^{\infty} \frac{1}{n^2} \cos n\alpha$

1107. $\sum_{n=1}^{\infty} \frac{(-1)^n}{\sqrt[n]{n}}$

1108. $\sum_{n=2}^{\infty} \frac{\cos n\pi}{\ln n}$

1109. $\sum_{n=1}^{\infty} (-1)^n \frac{n-1}{n+1} \frac{1}{\sqrt[100]{n}}$

11.4 ARITHMETIC OPERATIONS ON SERIES

The following theorems hold:

(1) The series obtained by multiplying every term of a convergent (divergent) series by a constant number is also a convergent (divergent) series.

(2) If the series $\sum_{n=1}^{\infty} u_n$ and $\sum_{n=1}^{\infty} v_n$ converge, then so does $\sum_{n=1}^{\infty} (u_n \pm v_n)$.

(3) In a convergent series we may replace any number of consecutive terms by their sum without altering the fact of convergence or the sum of the series.

(4) In an absolutely convergent series the order of the terms may be altered arbitrarily without affecting either the convergence or the sum of the given series.

(5) In a conditionally convergent series, the order of the terms can always be changed so that the new series will converge to an arbitrarily given number, or will diverge.

This surprising result, which is Riemann's theorem, is due to the fact that in a series which converges conditionally the two partial series consisting of the positive and the negative terms tend to $+\infty$ and $-\infty$ respectively.

(6) The product $\left(\sum_{n=1}^{\infty} u_n\right)\left(\sum_{n=1}^{\infty} v_n\right)$ is defined as $\sum_{i,j=1}^{\infty} u_i v_j$, i.e., the sum of all possible products of elements of both series (one from each series). The following theorem holds:

The product of two absolutely convergent series is an absolutely convergent series, the sum of which equals the product of the sums of the two series.

The following exercises illustrate the above theorems.

1110. Prove that

$$1 - \frac{1}{2} + \frac{1}{3} - \frac{1}{4} + \frac{1}{5} - \frac{1}{6} + \ldots + \frac{1}{2n-1} - \frac{1}{2n} + \ldots$$
$$= 2\left(1 - \frac{1}{2} - \frac{1}{4} + \frac{1}{3} - \frac{1}{6} - \frac{1}{8} + \frac{1}{5} - \frac{1}{10} - \frac{1}{12} + \ldots + \frac{1}{2n-1} - \frac{1}{4n-2} - \frac{1}{4n} + \ldots\right).$$

Solution. By theorem (3) we can write the left-hand side in the form

$$\left(1 - \frac{1}{2}\right) + \left(\frac{1}{3} - \frac{1}{4}\right) + \left(\frac{1}{5} - \frac{1}{6}\right) + \ldots + \left(\frac{1}{2n-1} - \frac{1}{2n}\right) + \ldots$$
$$= \frac{1}{2} + \frac{1}{12} + \frac{1}{30} + \ldots + \frac{1}{(2n-1)\cdot 2n} + \ldots$$

and the right-hand side as

$$2\left[\left(1-\frac{1}{2}-\frac{1}{4}\right)+\left(\frac{1}{3}-\frac{1}{6}-\frac{1}{8}\right)+\left(\frac{1}{5}-\frac{1}{10}-\frac{1}{12}\right)+\dots\right.$$
$$\left.+\left(\frac{1}{2n-1}-\frac{1}{4n-2}-\frac{1}{4n}\right)+\dots\right]$$
$$=2\left(\frac{1}{4}+\frac{1}{24}+\frac{1}{60}+\dots+\frac{4n-2n-2n+1}{(2n-1)\cdot 4n}+\dots\right)$$
$$=\frac{1}{2}+\frac{1}{12}+\frac{1}{30}+\dots+\frac{1}{(2n-1)\cdot 2n}+\dots.$$

The sums are equal. This exercise illustrates Riemann's theorem.

1111. Prove that the series

$$1-\frac{1}{\sqrt{2}}+\frac{1}{\sqrt{3}}-\frac{1}{\sqrt{4}}+\frac{1}{\sqrt{5}}-\frac{1}{\sqrt{6}}+\dots$$

converges, and that the series

$$1+\frac{1}{\sqrt{3}}-\frac{1}{\sqrt{2}}+\frac{1}{\sqrt{5}}+\frac{1}{\sqrt{7}}-\frac{1}{\sqrt{4}}+\frac{1}{\sqrt{9}}+\frac{1}{\sqrt{11}}-\frac{1}{\sqrt{6}}+\dots,$$

consisting of the same terms written in another order, diverges.

Solution. The first series converges conditionally by Leibniz's theorem (cf. Prob. 1102). The second can be rewritten in the form

$$\left(1+\frac{1}{\sqrt{3}}-\frac{1}{\sqrt{2}}\right)+\left(\frac{1}{\sqrt{5}}+\frac{1}{\sqrt{7}}-\frac{1}{\sqrt{4}}\right)+\dots$$
$$+\left(\frac{1}{\sqrt{4n-3}}+\frac{1}{\sqrt{4n-1}}-\frac{1}{\sqrt{2n}}\right)+\dots.$$

Now

$$\frac{1}{\sqrt{4n-3}}+\frac{1}{\sqrt{4n-1}}-\frac{1}{\sqrt{2n}}>\frac{1}{\sqrt{4n}}+\frac{1}{\sqrt{4n}}-\frac{1}{\sqrt{2n}}=\frac{1}{\sqrt{n}}-\frac{1}{\sqrt{2n}}$$
$$=\frac{\sqrt{2}-1}{\sqrt{2n}},$$

but $\sum_{n=1}^{\infty}\frac{\sqrt{2}-1}{\sqrt{2n}}$ is clearly a divergent series. Hence our series diverges.

1112. Multiply $\left(\sum_{n=0}^{\infty} a_nx^n\right)\left(\sum_{m=0}^{\infty} b_mx^m\right)$.

Solution:

$$\left(\sum_{n=0}^{\infty} a_nx^n\right)\left(\sum_{m=0}^{\infty} b_mx^m\right)=a_0b_0+(a_0b_1+a_1b_0)x+(a_0b_2+a_1b_1+a_2b_0)x^2$$
$$+(a_0b_3+a_1b_2+a_2b_1+a_3b_0)x^3+\dots.$$

1113. Write $\left(\sum_{n=0}^{\infty} x^n\right)^2$ in the form $\sum_{m=0}^{\infty} a_m x^m$.

Solution. Using the solution of the former example we obtain

$$\left(\sum_{n=0}^{\infty} x^n\right)^2 = 1 + 2x + 3x^2 + 4x^3 + \ldots = \sum_{n=0}^{\infty} (n+1)x^n.$$

1114. Find the sum of the series

$$\sum_{n=1}^{\infty}\left(\frac{1}{5^n} + \frac{(-1)^n}{n^2}\right) + \sum_{n=1}^{\infty}\left(\frac{1}{5^{n+1}} + \frac{(-1)^{n+1}}{n^2}\right).$$

Solution. Both series clearly converge. Their sum is

$$\sum_{n=1}^{\infty}\left[\frac{1}{5^n} + \frac{1}{5^{n+1}} + \frac{(-1)^n}{n^2} + \frac{(-1)^{n+1}}{n^2}\right]$$

$$= \sum_{n=1}^{\infty} \frac{6}{5^{n+1}} = \frac{6}{5}\sum_{n=1}^{\infty}\frac{1}{5^n} = \frac{6}{5}\,\frac{\frac{1}{5}}{1-\frac{1}{5}} = \frac{3}{10}.$$

1115. Investigate the convergence of the differences

(a) $\sum_{n=1}^{\infty} \frac{1}{n} - \sum_{n=1}^{\infty} \frac{1}{n-1}$

(b) $\sum_{n=1}^{\infty} \frac{1}{n} - \sum_{n=1}^{\infty} \frac{1}{2n}$.

1116. Find the product $\left(\sum_{n=1}^{\infty} \frac{1}{3n\sqrt{n}}\right)\left(\sum_{n=1}^{\infty} \frac{1}{3^{n-1}}\right)$ and investigate its convergence.

11.5 SERIES OF FUNCTIONS

Uniform Convergence. A series of functions is a series each term of which is a function of the same independent variables. The series is defined only in the domain common to the domains of definition of all functions.

The general term $u_n(x)$ of the series

(1) $$u_1(x) + u_2(x) + \ldots + u_n(x) + \ldots$$

is a function of two variables, x and n. For $x = x_0$ in the domain of definition of the series (i.e., x_0 in the domain of definition of each $u_n(x)$), we obtain

$$u_1(x_0) + u_2(x_0) + \ldots + u_n(x_0) + \ldots$$

which is a number series. If this number series converges, we say that the series (1) converges at $x = x_0$, or that x_0 is a point of convergence of the series (1).

The set of points of convergence of a given series of functions is called the domain of convergence of this series. In its domain of convergence the series defines a function:

$$f(x) = S(x) = u_1(x) + u_2(x) + \ldots + u_n(x) + \ldots$$

In many cases we can find the domain of convergence of a given series of functions using the various tests from the previous sections.

For some series of functions, such as power series, special results are of importance; they will be listed in the next section.

A very important notion is that of *uniform convergence of a series.* We say: The series

$$u_1(x) + u_2(x) + \ldots + u_n(x) + \ldots$$

converges uniformly in a certain domain if for every $\epsilon > 0$, no matter how small, an n_0 can be found such that the remainder

$$R_n(x) = u_n(x) + u_{n+1}(x) + \ldots$$

satisfies $|R_n(x)| < \epsilon$ for every x in this domain and for every $n > n_0$.

We emphasize that the convergence of a series at x_0 requires that for every $\epsilon > 0$ there must exist an n_0 such that $|R_n(x_0)| < \epsilon$ if $n > n_0$. In general this n_0 is a function of ϵ and x_0.

The series converges uniformly in a certain domain if there can be found a "universal" n_0 depending only on ϵ and not on x (in this domain), and satisfying the above conditions. If the domain of convergence of a series consists of a finite number of points, such a universal n_0 can always be found: it is the greatest among all the n_0's corresponding to the various points of convergence. The situation is more involved if the domain of convergence consists of infinitely many points. In this case such a "greatest" n_0 may not exist, so that the convergence is not always uniform. Some examples will be given in the exercises of this section.

A test of uniform convergence convenient in many cases is due to Weierstrass: If in a certain domain the terms of a series of functions fulfill the inequality

$$|u_n(x)| \leqslant a_n \quad (n = 1, 2, \ldots)$$

where the a_n are terms of a convergent number series, then the series $\sum_{n=1}^{\infty} u_n(x)$ converges uniformly in this domain.

We remark that the a_n's must be nonnegative and the convergence of $\sum_{n=1}^{\infty} a_n$ is absolute.

The importance of the notion of uniform convergence will be made clear by the following theorems.

(1) A series of continuous functions uniformly convergent in some interval defines a continuous function in this interval.

(2) If $u_1(x) + u_2(x) + \ldots + u_n(x) + \ldots$ converges uniformly in $[a,b]$ and if every function in the series is integrable in the above interval, then

$$\int_a^b [u_1(x) + u_2(x) + \ldots + u_n(x) + \ldots]\, dx$$
$$= \int_a^b u_1(x)\, dx + \int_a^b u_2(x)\, dx + \ldots + \int_a^b u_n(x)\, dx + \ldots.$$

(3) If all functions in the convergent series

$$u_1(x) + u_2(x) + \ldots + u_n(x) + \ldots$$

are differentiable and if the series of derivatives

$$u_1'(x) + u_2'(x) + \ldots + u_n'(x) + \ldots$$

converges uniformly in a certain domain, then in that domain

$$[u_1(x) + u_2(x) + \ldots + u_n(x) + \ldots]'$$
$$= u_1'(x) + u_2'(x) + \ldots + u_n'(x) + \ldots.$$

The theorems of the last two sections can be restated in the following loose formulation: Absolutely convergent series permit arithmetic operations, whereas uniformly convergent series permit analytic operations.

Find the domain of convergence of the following series.

1117. $1 + x + \ldots + x^n + \ldots$

Solution. $S_n = \dfrac{1 - x^{n+1}}{1 - x}$. For $|x| < 1$,

$$\lim_{n\to\infty} S_n = \frac{1}{1 - x},$$

and the series converges. For $|x| \geqslant 1$ it clearly diverges.

1118. $\ln x + \ln^2 x + \ldots + \ln^n x + \ldots$

Solution. We require in this geometric series

$$|\ln x| < 1, \quad \text{i.e.,} \quad -1 < \ln x < 1$$

or $1/e < x < e$. This is the domain of convergence.

1119. $x + \dfrac{x^2}{2^2} + \ldots + \dfrac{x^n}{n^2} + \ldots$

1120. $x + \dfrac{x^2}{\sqrt{2}} + \ldots + \dfrac{x^n}{\sqrt{n}} + \ldots$

1121. $\sum_{n=1}^{\infty} \frac{1}{1+x^n}$, $x \neq -1$

Solution. For $|x| \leqslant 1$ $(x \neq -1)$, $\lim_{n\to\infty} \frac{1}{1+x^n}$ is not zero and the series diverges. For $x > 1$ we have

$$\frac{1}{1+x^n} < \frac{1}{x^n}.$$

$1/x^n$ is in this case a term of a convergent geometric series, so that the given series also converges. For $x < -1$, $1/(1+x^n)$ is positive for even and negative for odd n's, $\lim_{n\to\infty} 1/(1+x^n) = 0$ clearly.

Now we shall show that beginning with a certain n,

$$\left|\frac{1}{1+x^n}\right| > \left|\frac{1}{1+x^{n+1}}\right|.$$

If n is odd, this is evident. For an even $n = 2k$ we obtain

$$\frac{1}{1+x^{2k}} > \frac{1}{-x^{2k+1}-1} \Leftrightarrow -x^{2k+1} - 1 > 1 + x^{2k} \Leftrightarrow -x^{2k}(x+1) > 2,$$

and this is clearly true for $x < -1$ and k sufficiently large. By Leibniz's test the series converges for $x < -1$. The domain of convergence is thus $|x| > 1$.

1122. $\sum_{n=1}^{\infty} \frac{x^n}{1+x^{2n}}$

1123. $\sum_{n=1}^{\infty} \sin \frac{x}{2^n}$

1124. $\sum_{n=1}^{\infty} \frac{nx}{e^{nx}}$

1125. $\sum_{n=1}^{\infty} \frac{n \cos^{2n} x}{(1+\cos^2 x)(1+2\cos^2 x)\cdots(1+n\cos^2 x)}$

1126. $\sum_{n=1}^{\infty} \frac{2^n \sin^n x}{n^2}$

1127. $\sum_{n=1}^{\infty} \frac{1}{n!x^n}$

1128. $\sum_{n=1}^{\infty} \frac{(2n)!}{x^{2n}}$

1129. $\sum_{n=1}^{\infty} \frac{2n-3}{n^3x^n}$

In the following exercises find the domains of uniform convergence of the given series.

1130. $\frac{\sin x}{1!} + \frac{\sin 2x}{2!} + \ldots + \frac{\sin nx}{n!} + \ldots$

Solution. For any x we can write

$$\left|\frac{\sin nx}{n!}\right| \leqslant \frac{1}{n!}.$$

The series $\sum_{n=1}^{\infty} 1/n!$ converges, and according to the theorem of Weierstrass the given series converges uniformly for all x, $-\infty < x < \infty$.

1131. $\sum_{n=1}^{\infty} \frac{1}{n^2(1 + n^2x^2)}$ **1132.** $\sum_{n=1}^{\infty} \frac{\sin nx}{2^n}$

1133. $\sum_{n=1}^{\infty} \frac{e^{-n^2x^2}}{n^2}$

1134. $\sum_{n=1}^{\infty} \frac{1}{2^{n-1}\sqrt{1 + nx}}$. Assume $\epsilon > 0$ and find an n_0 such that for any x in the domain of uniform convergence $|R_{n+1}| < \epsilon$ if $n > n_0$. Find n_0 for $\epsilon = 0.001$.

Solution. For $x < 0$ and n sufficiently large, $1 + nx < 0$ and $\sqrt{1 + nx}$ is not defined there. For $x \geqslant 0$,

$$\frac{1}{2^{n-1}\sqrt{1 + nx}} \leqslant \frac{1}{2^{n-1}}.$$

$\sum_{n=1}^{\infty} 1/2^{n-1}$ is a convergent geometric series, and by Weierstrass's theorem the given series converges uniformly in the interval $0 \leqslant x < \infty$.

Now the remainder is

$$\begin{aligned} R_{n+1}(x) &= u_{n+1}(x) + u_{n+2}(x) + \dots \\ &= \frac{1}{2^n\sqrt{1 + (n + 1)x}} + \frac{1}{2^{n+1}\sqrt{1 + (n + 2)x}} + \dots . \end{aligned}$$

We want a universal n_0 such that for any $x \geqslant 0$, $|R_{n+1}(x)| < \epsilon$ if $n > n_0$. First of all $|R_{n+1}(x)| = R_{n+1}(x)$, and secondly $R_{n+1}(0) > R_{n+1}(x)$ for any x in the interval $0 \leqslant x < \infty$, because

$$\frac{1}{2^{n+i}\sqrt{1 + (n + i + 1)x}} \leqslant \frac{1}{2^{n+i}} \quad \text{for any } x \geqslant 0, \quad i = 0, 1, 2, \dots .$$

Consequently the n_0 found for $x = 0$ will suffice for the whole interval $0 \leqslant x < \infty$, and

$$R_{n+1}(0) = \frac{1}{2^n} + \frac{1}{2^{n+1}} + \dots = \frac{1/2^n}{1 - \frac{1}{2}} = \frac{1}{2^{n-1}}.$$

We require $1/2^{n-1} < \epsilon$, i.e., $2^{n-1} > 1/\epsilon$:

$$(n - 1) \ln 2 > -\ln \epsilon, \qquad n > 1 - \frac{\ln \epsilon}{\ln 2}.$$

For $\epsilon = 0.001$ we find

$$n > 1 - \frac{\ln 0.001}{\ln 2} = 1 - \frac{\log 0.001}{\log 2} = 1 + \frac{3}{0.3010},$$

whence $n_0 = 11$.

1135. Investigate the convergence of the series

$$x + (x^2 - x) + \ldots + (x^n - x^{n-1}) + \ldots$$

in $0 \leqslant x \leqslant 1$.

Solution. We have clearly $S_n(x) = x^n$,

$$\lim_{n\to\infty} S_n(x) = \lim_{n\to\infty} x^n.$$

This limit exists when $0 \leqslant x \leqslant 1$. To find where the series converges uniformly, let us compute the remainder:

$$R_{n+1}(x) = (x^{n+1} - x^n) + (x^{n+2} - x^{n+1}) + (x^{n+3} - x^{n+2}) + \ldots = -x^n,$$
$$|R_{n+1}(x)| = x^n.$$

For a given ϵ we need an n_0 such that $x^{n_0} < \epsilon$ for any x in the above interval. But $x^{n_0} < \epsilon$ is equivalent to $x < \epsilon^{1/n_0}$, and for an arbitrary n_0, no matter how large, we can find an x closer to 1 than ϵ^{1/n_0}, i.e., $1 > x > \epsilon^{1/n_0}$. Consequently there is no uniform convergence in the whole interval $0 \leqslant x \leqslant 1$.

We remark that in any subinterval $0 \leqslant x \leqslant q < 1$ the series converges uniformly. To prove this, it suffices to take n_0 satisfying $\epsilon^{1/n_0} > q$, i.e., $q^{n_0} < \epsilon$.

1136. The same question for

$$1 + x + \frac{x^2}{1\cdot 2} + \ldots + \frac{x^{n+1}}{(n+1)!} + \ldots, \quad 0 \leqslant x < \infty.$$

1137. Prove the convergence of the series

$$\sum_{n=1}^{\infty} 2^n \sin \frac{1}{3^n x}$$

in the interval $0 < x < \infty$. Is the convergence uniform in this interval?

1138. Prove that the function

$$f(x) = \sum_{n=1}^{\infty} \frac{\sin nx}{n^3}$$

is continuous and has a continuous derivative in the whole interval $-\infty < x < \infty$.

Solution. The series $\sum_{n=1}^{\infty} \frac{\sin nx}{n^3}$ converges uniformly in the whole interval $-\infty < x < \infty$, because $|(\sin nx)/n^3| \leqslant 1/n^3$ there and the Weierstrass theorem is applicable.

The functions $(\sin nx)/n^3$ are continuous and by theorem (1) of this section, $f(x)$ is continuous.

Let us consider the series

$$\sum_{n=1}^{\infty} \left(\frac{\sin nx}{n^3}\right)' = \sum_{n=1}^{\infty} \frac{\cos nx}{n^2}.$$

By the same argument as above we see that this series converges uniformly in the interval $-\infty < x < \infty$ and represents there a continuous function. But moreover, all conditions of theorem (3) are fulfilled and consequently $f'(x)$ exists and is equal to this series, i.e., it is continuous.

1139. Prove that the series of derivatives of the terms of the convergent series $\sum_{n=1}^{\infty} \frac{\sin n^3x}{n^2}$ does not converge.

Solution. The given series converges uniformly in the whole interval $-\infty < x < \infty$ ($|\sin n^3x/n^2| \leqslant 1/n^2$). The series

$$\sum_{n=1}^{\infty} \left(\frac{\sin n^3x}{n^2}\right)' = \sum_{n=1}^{\infty} n \cos n^3x$$

does not converge because $\lim_{n\to\infty} n \cos n^3x$ is not zero. The derivative (if any) of the function $f(x) = \sum_{n=1}^{\infty} (\sin n^3x)/n^2$ cannot be represented by the series of derivatives of the terms of its series.

1140. Represent the function

$$f(x) = x + \frac{x^5}{5} + \ldots + \frac{x^{4n-3}}{4n-3} + \ldots$$

using elementary functions.

Solution. This series converges for any $|x| < 1$, as can be easily seen by comparing it with a convergent geometric series. For $|x| \geqslant 1$ it does not converge. (Compare with the series $\frac{1}{4}(1 + \frac{1}{2} + \frac{1}{3} + \ldots + 1/n + \ldots)$.)

The series of derivatives

$$1 + x^4 + \ldots + x^{4n-4} + \ldots$$

converges uniformly in every interval $-q \leqslant x \leqslant q < 1$ (apply Weierstrass's theorem). Consequently $f'(x) = 1 + x^4 + \ldots + x^{4n-4} + \ldots$ in every such interval. Now $f'(x) = 1/(1 - x^4)$ and

$$f(x) = \int \frac{dx}{1 - x^4} + C = \frac{1}{2}\int \left(\frac{1}{1 + x^2} + \frac{1}{1 - x^2}\right) dx + C$$

$$= \frac{1}{2} \arctan x + \frac{1}{4} \ln \frac{1 + x}{1 - x} + C.$$

To find C we compute $f(0)$. From the series we find $f(0) = 0$ and by the above expression $f(0) = C$, i.e., $C = 0$ and

$$f(x) = \frac{1}{2}\arctan x + \frac{1}{4}\ln\frac{1+x}{1-x}.$$

We emphasize once more that this is true only for $-1 < x < 1$.

1141. Prove that

$$f(x) = e^{-x} + 2e^{-2x} + \ldots + ne^{-nx} + \ldots$$

is a continuous function in the interval $\epsilon \leqslant x < \infty$ ($\epsilon > 0$ arbitrary small). Compute $\int_{\ln 2}^{\ln 3} f(x)\,dx$.

Solution. If $x \geqslant \epsilon > 0$, $ne^{-nx} \leqslant ne^{-n\epsilon}$. But the series $\sum_{n=1}^{\infty} n/e^{n\epsilon}$ converges for any positive ϵ, as can be seen by applying d'Alembert's test:

$$\lim_{n\to\infty}\frac{(n+1)e^{n\epsilon}}{e^{(n+1)\epsilon}n} = \lim_{n\to\infty}\frac{n+1}{n}\cdot\frac{1}{e^{\epsilon}} = \frac{1}{e^{\epsilon}} < 1.$$

Consequently the given series converges uniformly in the interval $\epsilon \leqslant x < \infty$. Thus we can perform term by term integration:

$$\int_{\ln 2}^{\ln 3} f(x)\,dx = (-e^{-x} - e^{-2x} - \ldots - e^{-nx} + \ldots)\Big|_{\ln 2}^{\ln 3}$$

$$= \frac{-e^{-x}}{1-e^{-x}}\Big|_{\ln 2}^{\ln 3} = \frac{1}{e^{x}-1}\Big|_{\ln 3}^{\ln 2} = \frac{1}{2-1} - \frac{1}{3-1} = \frac{1}{2}.$$

1142. Find the sum of the series

$$q - \frac{q^4}{4} + \ldots + \frac{(-1)^{n+1}q^{3n-2}}{3n-2} + \ldots, \quad q = \frac{1}{2}.$$

Solution. We define

$$f(x) = x - \frac{x^4}{4} + \ldots + \frac{(-1)^{n+1}x^{3n-2}}{3n-2} + \ldots.$$

This series and the series of its derivatives converge uniformly in the interval $-\frac{3}{4} \leqslant x \leqslant \frac{3}{4}$, e.g. Thus in this interval

$$f'(x) = 1 - x^3 + \ldots + (-1)^{n+1}x^{3(n-1)} + \ldots$$

$$= \frac{1}{1+x^3}.$$

$$f(x) = \int\frac{dx}{1+x^3} + C = \frac{1}{6}\ln\frac{(1+x)^2}{x^2-x+1} + \frac{\sqrt{3}}{3}\arctan\frac{2x-1}{\sqrt{3}} + C.$$

For $x = 0$ we have

$$f(0) = \frac{\sqrt{3}}{3}\arctan\left(-\frac{1}{\sqrt{3}}\right) + C = \frac{\sqrt{3}}{3}\cdot\left(-\frac{\pi}{6}\right) + C = 0,$$

i.e., $C = \sqrt{3}\pi/18$. The sum of the given series is $f(\frac{1}{2})$, i.e.,

$$\frac{1}{6}\ln\frac{(\frac{3}{2})^2}{\frac{1}{4}-\frac{1}{2}+1}+\frac{\sqrt{3}}{3}\arctan 0+\frac{\sqrt{3}}{18}\pi=\frac{1}{6}\ln 3+\frac{\sqrt{3}}{18}\pi.$$

1143. Find the sums

(a) $1 + 2x + 3x^2 + \ldots + nx^{n-1} + \ldots, |x| < 1$

(b) $1 + 3x + \ldots + \dfrac{n(n+1)}{2}x^{n-1} + \ldots, |x| < 1.$

1144. Prove that the Bessel function of order zero, defined by

$$J_0(x) = 1 + \sum_{n=1}^{\infty} (-1)^n \frac{x^{2n}}{(n!)^2 2^{2n}},$$

is a solution of the differential equation

$$y'' + \frac{1}{x}y' + y = 0.$$

11.6 POWER SERIES; RADIUS OF CONVERGENCE

Power series are a very important particular case of series of functions. By a power series we mean the series

(1) $$a_0 + a_1x + a_2x^2 + a_3x^3 + \ldots + a_nx^n + \ldots,$$

or

(2) $$b_0 + b_1(x-a) + b_2(x-a)^2 + \ldots + b_n(x-a)^n + \ldots.$$

The form (2) can be converted to (1) by setting $x - a = t$; therefore we shall discuss the form (1) only.

The domain of definition of the power series can be found using the following theorem:

If the power series (1) converges at x_0, it converges also in the interval $-|x_0| < x < |x_0|$; if the power series (1) does not converge at $x = x_0$, it does not converge at any x for which $|x| > |x_0|$.

In other words, the domain of convergence of the power series (1) is always an interval on the x axis symmetric with respect to $x = 0$. The series (1) may converge at both ends of the interval, or at one end only, or at neither end. Particular cases of such intervals are the whole x axis, and the point $x = 0$ only.

The above can be restated as follows: All power series converge for $x = 0$. For those which converge elsewhere also, a positive R exists such that for $|x| < R$ the series converges and for $|x| > R$ it diverges. (For completeness we allow also $R = \infty$ and $R = 0$ in the above sense.) The situation for $x = R$ and $x = -R$ depends on the particular series. R is called the radius of convergence of the series.

To find R we can use the following theorem that follows directly from d'Alembert's test:

If the limit

$$\lim_{n\to\infty}\left|\frac{a_{n+1}}{a_n}\right| = r$$

exists, then $R = 1/r$.

(If $r = 0$ we write $R = \infty$ and if $r = \infty$ we write $R = 0$.)

Using Cauchy's test we obtain: If the limit

$$\lim_{n\to\infty}\sqrt[n]{|a_n|} = t$$

exists, then $R = 1/t$ (with $R = 0$ if $t = \infty$ and $R = \infty$ if $t = 0$).

The following theorem is easy to prove: A power series converges absolutely and uniformly in every closed subinterval of its domain of convergence.

Find the radii of convergence of the following power series.

1145. $10 + 100x^2 + \ldots + 10^n x^n + \ldots$

Solution:

$$r = \lim_{n\to\infty}\left|\frac{a_{n+1}}{a_n}\right| = \lim_{n\to\infty}\frac{10^{n+1}}{10^n} = 10.$$

The radius of convergence is

$$R = \frac{1}{r} = \frac{1}{10}.$$

At $x = \frac{1}{10}$ and $x = -\frac{1}{10}$ the series does not converge, because $u_n(\frac{1}{10})$ and $u_n(-\frac{1}{10})$ do not tend to zero when $n \to \infty$.

1146. $x - \dfrac{x^2}{2} + \ldots + (-1)\dfrac{x^n}{n} + \ldots$

1147. $1 + x + \ldots + n!x^n + \ldots$

1148. $1 + 2x^2 + \ldots + 2^{n-1}x^{2(n-1)} + \ldots$

Solution. By d'Alembert's test,

$$\lim_{n\to\infty}\left|\frac{u_{n+1}}{u_n}\right| = \lim_{n\to\infty}\left|\frac{2^n x^{2n}}{2^{n-1}x^{2(n-1)}}\right| = 2|x|^2.$$

We require $2|x|^2 < 1$, i.e., $|x| < 1/\sqrt{2}$. At $x = \pm 1/\sqrt{2}$ the series clearly does not converge, i.e., the domain of convergence is $-\sqrt{2}/2 < x < \sqrt{2}/2$.

This example shows that if the difference between the exponents of x in consecutive terms of a power series is k, then the radius of convergence equals

$$\left(\sqrt[k]{\lim_{n\to\infty}\left|\frac{a_{n+1}}{a_n}\right|}\right)^{-1},$$

assuming that this limit exists.

1149. $x - \frac{x^3}{3\cdot 3!} + \ldots + (-1)^{n+1}\frac{x^{2n-1}}{(2n-1)(2n-1)!} + \ldots$

1150. $x + \frac{(2x)^2}{2!} + \ldots + \frac{(nx)^n}{n!} + \ldots$

1151. $x + 4x^2 + \ldots + (nx)^n + \ldots$

1152. $\frac{\ln 2}{2}x^2 + \frac{\ln 3}{3}x^3 + \ldots + \frac{\ln(n+1)}{n+1}x^{n+1} + \ldots$

1153. $2x + \left(\frac{9}{4}x\right)^2 + \ldots + \left[\left(\frac{n+1}{n}\right)^n x\right]^n + \ldots$

1154. $\frac{x-2}{2\cdot 2} + \frac{(x-2)^2}{4\cdot 2^2} + \frac{(x-2)^3}{6\cdot 2^3} + \ldots + \frac{(x-2)^n}{2n\cdot 2^n} + \ldots$

Solution. We substitute $y = x - 2$ and obtain the series

$$\sum_{n=1}^{\infty}\frac{y^n}{2n\cdot 2^n}.$$

Then

$$r = \lim_{n\to\infty}\frac{2n\cdot 2^n}{2(n+1)2^{n+1}} = \frac{1}{2}; \qquad R = 2.$$

At $y = 2$ the series diverges and at $y = -2$ it converges. The domain of convergence is thus $-2 \leqslant y < 2$ or $-2 \leqslant x - 2 < 2$, i.e., $0 \leqslant x < 4$.

1155. $\sum_{n=1}^{\infty}\frac{(x+5)^n}{n^2}$

1156. $\sum_{n=1}^{\infty}\frac{x^n}{3^n + 2^n}$

1157. $\sum_{n=1}^{\infty}\frac{x^n}{2^{\sqrt{n}}}$

1158. $\sum_{n=1}^{\infty}\frac{5^n + (-3)^n}{n}(x+2)^n$

11.7 TAYLOR'S AND MACLAURIN'S SERIES; OPERATIONS ON POWER SERIES

A power series

(1) $$a_0 + a_1x + a_2x^2 + \ldots + a_nx^n + \ldots,$$

or $$b_0 + b_1(x-a) + b_2(x-a)^2 + \ldots + b_n(x-a)^n + \ldots$$

determines in its domain of convergence a function $f(x) = \sum_{n=0}^{\infty} a_n x^n$. From the last theorem of the preceding section it follows that $f(x)$ is continuous in every closed interval belonging to the domain of convergence of the series. Moreover, it can be shown that if x_0 is an end point of the domain of convergence and if the series converges there, then

$$\lim_{x \to x_0} f(x) = f(x_0)$$

($x \to x_0$ from the interior of the domain of convergence). Denote by R the radius of convergence of (1). The following formulas hold:

$$f'(x) = \sum_{n=1}^{\infty} n a_n x^{n-1}, \quad \text{and} \quad \int_0^x f(x)\,dx = \sum_{n=\infty}^{\infty} \frac{a_n x^{n+1}}{n+1}, \quad |x| < R;$$

i.e., the series can be differentiated and integrated term by term. Both series so obtained have exactly the same radius of convergence R and their sums are $f'(x)$ and $\int_0^x f(x)\,dx$ respectively.

Given

$$f(x) = \sum_{n=0}^{\infty} a_n x^n, \qquad g(x) = \sum_{n=0}^{\infty} b_n x^n,$$

let $R \neq 0$ be the smaller of the radii of convergence of these series. Then for any $-R < x < R$ we have

$$\sum_{n=0}^{\infty} (a_n x^n \pm b_n x^n) = f(x) \pm g(x).$$

$$\left(\sum_{n=0}^{\infty} a_n x^n\right)\left(\sum_{n=0}^{\infty} b_n x^n\right)$$
$$= a_0 b_0 + (a_1 b_0 + a_0 b_1)x + (a_2 b_0 + a_1 b_1 + a_0 b_2)x^2 + \cdots = f(x)g(x).$$

The following theorem is often useful: Let there be given

$$y = f(x) = \sum_{n=0}^{\infty} a_n x^n$$

with R_1 as radius of convergence. In addition, let the power series $g(y) = \sum_{n=0}^{\infty} b_n y^n$ be given, with the radius of convergence R_2. If $|f(0)| = |a_0| < R_2$, then there exists an interval around 0 on the x axis (within $(-R_1, R_1)$) where $|f(x)| < R_2$. In this interval the series

$$\sum_{n=0}^{\infty} \left(b_n \sum_{n=0}^{\infty} a_n x^n\right),$$

obtained by substitution of the first series into the second, represents the function $g[f(x)]$.

To represent a given function $f(x)$ by a power series we can use Taylor's expansion (see Section 5.2) in which n takes all positive integral values. (This is possible only if $f(x)$ has derivatives of all orders.) This procedure gives a power series. In the interval where $\lim_{n\to\infty} R_{n+1}(x) = 0$ ($R_{n+1}(x)$ is the remainder term of Taylor's formula), this power series converges and represents the function $f(x)$.

It should be realized that it is not sufficient to prove the convergence of the power series so obtained. Examples exist (see Prob. 1284) in which convergent power series serve as Taylor's series of a given function $f(x)$, (i.e. they are obtained from $f(x)$ by Taylor's procedure), but their sum $S(x)$ differs from $f(x)$. Only the fact that $\lim_{n\to\infty} R_{n+1}(x_0) = 0$ ensures that the sum $S(x_0)$ of Taylor's series of $f(x)$ equals $f(x_0)$ at x_0.

On the other hand, any power series is the Taylor's series of the function determined by it. In other words, a function can be represented by a power series only in one way, by its Taylor's series. It follows that there can be various functions with the same Taylor's series, but only one of them is represented by this series.

A function that can be represented in a certain interval by a power series is called an analytic function in this interval. The above can be now paraphrased: No two analytic functions have the same Taylor's series. A nonanalytic function can have a Taylor's series but this series does not represent it.

Power series representing a number of basic functions may be used to obtain many other such series by the operations described in the beginning of this section. Various examples will be given in what follows.

1159. Compute the Taylor's series of $f(x) = e^x$ around $x = 0$, and find the domain where this series represents this function.

Solution. In Problem 309 we found

$$e^x = 1 + \frac{x}{1!} + \frac{x^2}{2!} + \ldots + \frac{x^n}{n!} + R_{n+1},$$

$$R_{n+1} = \frac{e^{\theta x}}{(n+1)!} x^{n+1}.$$

For $x > 0$,

$$|R_{n+1}| \leqslant \frac{e^x}{(n+1)!} x^{n+1},$$

$$\lim_{n\to\infty} \frac{e^x x^{n+1}}{(n+1)!} = e^x \lim_{n\to\infty} \frac{x^{n+1}}{(n+1)!} = 0 \quad \text{for any } x.$$

(since $x^{n+1}/(n+1)!$ is a term of a series everywhere convergent; see Prob. 1136).

For $x < 0$,

$$|R_{n+1}| \leqslant e \frac{|x|^{n+1}}{(n+1)!} \to 0 \quad \text{as } n \to \infty,$$

as above. Consequently,

$$e^x = 1 + x + \frac{x^2}{2!} + \ldots + \frac{x^n}{n!} + \ldots$$

for any x.

1160. The same question for $f(x) = \sin x$.

Solution. See Problem 311.

$$|R_{2k+1}| \leqslant \frac{|x|^{2k+1}}{(2k+1)!} \to 0 \quad \text{as } k \to \infty, \text{ for any } x.$$

Consequently for $-\infty < x < \infty$,

$$\sin x = x - \frac{x^3}{3!} + \frac{x^5}{5!} - \frac{x^7}{7!} + \ldots + (-1)^{k-1} \frac{x^{2k-1}}{(2k-1)!} + \ldots.$$

In the same interval we obtain

$$\cos x = 1 - \frac{x^2}{2!} + \frac{x^4}{4!} - \frac{x^6}{6!} + \ldots + (-1)^k \frac{x^{2k}}{(2k)!} + \ldots.$$

Remark. In using these trigonometric formulas in computation, one must remember to take angle x in radians.

1161. The same question for $f(x) = \ln(1 + x)$.

Solution. In Problem 316 we found

$$\ln(1 + x) = x - \frac{x^2}{2} + \frac{x^3}{3} - \ldots + (-1)^{n-1} \frac{x^n}{n} + R_{n+1},$$

$$R_{n+1}(x) = \frac{x^{n+1}}{(n+1)!} f^{(n+1)}(\theta x) = \frac{x^{n+1}}{(n+1)!} \frac{(-1)^n n!}{(1+\theta x)^{n+1}}$$

$$= (-1)^n \frac{x^{n+1}}{(n+1)(1+\theta x)^{n+1}}.$$

The radius of convergence of the series is clearly 1. For $0 \leqslant x \leqslant 1$,

$$|R_{n+1}(x)| \leqslant \frac{1}{n+1}, \quad \text{i.e.,} \quad \lim_{n\to\infty} |R_{n+1}(x)| = 0.$$

For $x = -1$ the series does not converge. We now have to investigate $|R_{n+1}(x)|$ for $-1 < x < 0$. The above (Lagrange's) form of R_{n+1} cannot be used, because for $-1 < x < 0$ $1 + \theta x$ can be arbitrary small. We use Cauchy's form of the remainder:

$$R_{n+1}(x) = \frac{f^{(n+1)}(\theta x)}{n!}(1-\theta)^n x^{n+1} = \frac{(-1)^n n!}{(1+\theta x)^{n+1}} \cdot \frac{1}{n!}(1-\theta)^n x^{n+1}$$

$$= (-1)^n x^{n+1} \frac{(1-\theta)^n}{(1+\theta x)^{n+1}}.$$

Now $$|R_{n+1}(x)| = \frac{|x|^{n+1}}{|1+\theta x|}\left|\frac{1-\theta}{1+\theta x}\right|^n \leqslant \frac{|x|^{n+1}}{1-|x|}\left(\frac{1-\theta}{1+\theta x}\right)^n$$

(because $(1-\theta)/(1+\theta x) > 0$ for any x in the interval $-1 < x < 0$ and $|1+\theta x| \geqslant 1 - |\theta x| > 1 - |x|$). Since for $-1 < x < 0$, $1 + \theta x > 1 - \theta$, we have

$$\frac{1-\theta}{1+\theta x} < 1 \quad \text{and} \quad \lim_{n\to\infty} |R_{n+1}(x)| = 0.$$

Consequently,

$$\ln(1+x) = x - \frac{x^2}{2} + \frac{x^3}{3} - \frac{x^4}{4} + \ldots + (-1)^{n-1}\frac{x^n}{n} + \ldots$$

in the interval $-1 < x \leqslant 1$.

1162. The same question for $f(x) = (1+x)^m$.

Solution. For m a positive integer or zero we use Newton's binomial formula, and obtain a finite series for $f(x)$.

Now let us suppose m is different from 0, 1, 2, ... We found in Problem 317,

$$(1+x)^m = 1 + mx + \frac{m(m-1)}{2!}x^2 + \ldots$$

$$+ \frac{m(m-1)(m-2)\cdots(m-n+1)}{n!}x^n + R_{n+1}.$$

The remainder term in Lagrange's form is

$$R_{n+1} = \frac{m(m-1)\cdots(m-n)}{(n+1)!}(1+\theta x)^{m-n-1}x^{n+1}.$$

For $x \geqslant 0$ we have, for n large enough,

$$(1+\theta x)^{m-n-1} \leqslant 1$$

(because $m - n - 1 < 0$ for such n). Consequently, for n large enough,

$$|R_{n+1}| \leqslant \frac{|m(m-1)\cdots(m-n)|}{(n+1)!}x^{n+1}.$$

For $0 \leqslant x < 1$ the last expression tends to zero. This can be shown by considering it as a general term of a series $u_n(x)$. D'Alembert's test gives

$$\lim_{n\to\infty}\left|\frac{u_{n+1}(x)}{u_n(x)}\right| = \lim_{n\to\infty}\frac{|m-n|}{n+1}x = x < 1,$$

i.e., this series converges and $\lim\limits_{n\to\infty} u_n(x) = 0$.

Further investigation of $R_{n+1}(x)$ is a little cumbersome, and we shall give only the results:

$$(1 + x)^m = 1 + mx + \frac{m(m-1)}{2!}x^2 + \dots + \frac{m(m-1)\cdots(m-n+1)}{n!}x^n + \dots$$

in the interval $-1 < x < 1$. For special m's the series converges to the function $(1 + x)^m$ also at the points $x = -1$ and $x = 1$ (at $x = -1$ for $m > 0$ and at $x = 1$ for $m > -1$).

The above five expansions are basic. Many others can be obtained from them by using the theorems of this section.

1163. Find the Taylor's series of $y = \ln x$ around the point $x = 1$ (i.e., in powers of $x - 1$).

Solution. Let us set $x = 1 + z$. Then

$$y = \ln x = \ln(1 + z) = z - \frac{z^2}{2} + \frac{z^3}{3} - \dots + (-1)^{n-1}\frac{z^n}{n} + \dots$$

$$= (x - 1) - \frac{(x-1)^2}{2} + \frac{(x-1)^3}{3} - \dots + (-1)^{n-1}\frac{(x-1)^n}{n} + \dots.$$

We used here Problem 1161. The series converges and represents $\ln x$ for $-1 < x - 1 \leqslant 1$, i.e., $0 < x \leqslant 2$.

1164. Find the Taylor's series of $y = \sqrt{x^3}$ around $x = 1$.

1165. Find the Taylor's series of $y = 1/x$ around $x = 3$.

Solution. Compare Problem 1162.

$$y = \frac{1}{x} = \frac{1}{3 + x - 3} = \frac{1}{3}\cdot\frac{1}{1 + \dfrac{x-3}{3}} = \frac{1}{3}\left(1 + \frac{x-3}{3}\right)^{-1}$$

$$= \frac{1}{3}\left[1 - \frac{x-3}{3} + \frac{(x-3)^2}{9} - \frac{(x-3)^3}{27} + \dots + (-1)^n\frac{(x-3)^n}{3^n} + \dots\right].$$

This series represents $\frac{1}{x}$ when $-1 < \frac{x-3}{3} < 1$, i.e., in $0 < x < 6$.

1166. Find the Taylor's series of $y = \sin\frac{\pi x}{4}$ around $x = 2$.

Solution:

$$\sin\frac{\pi x}{4} = \sin\frac{\pi}{4}[(x-2) + 2] = \sin\left[\frac{\pi}{4}(x-2) + \frac{\pi}{2}\right]$$

$$= \cos\frac{\pi}{4}(x-2)$$

$$= 1 - \frac{1}{2!}\left(\frac{\pi}{4}\right)^2(x-2)^2 + \frac{1}{4!}\left(\frac{\pi}{4}\right)^4(x-2)^4 - \ldots$$

$$+ (-1)^n\frac{1}{(2n)!}\left(\frac{\pi}{4}\right)^{2n}(x-2)^{2n} + \ldots$$

Compare Problem 1160. The series converges and represents $\sin(\pi x/4)$ at every point.

1167. Find the first three terms of the Maclaurin series (i.e., the Taylor's series around $x = 0$) of $y = \cos x \cosh x$.

Solution. Let us begin with the series of $\cosh x$.

$$\cosh x = \frac{e^x + e^{-x}}{2} = \frac{1}{2}\left[\left(1 + \frac{x}{1} + \frac{x^2}{2!} + \frac{x^3}{3!} + \ldots + \frac{x^n}{n!} + \ldots\right)\right.$$

$$\left. + \left(1 - \frac{x}{1} + \frac{x^2}{2!} - \frac{x^3}{3!} + \ldots + (-1)^n\frac{x^n}{n!} + \ldots\right)\right]$$

$$= 1 + \frac{x^2}{2!} + \frac{x^4}{4!} + \ldots + \frac{x^{2n}}{(2n)!} + \ldots.$$

The series for e^{-x} was obtained by substituting $-x$ for x in the series of e^x. The series of e^x converges and represents e^x for any x; consequently the series obtained for $\cosh x$ represents this function everywhere.

Now we multiply the series for $\cos x$ and $\cosh x$. The resulting series will represent $\cos x \cosh x$ everywhere.

$$\cos x \cosh x = \left[1 - \frac{x^2}{2!} + \frac{x^4}{4!} - \ldots + (-1)^n\frac{x^{2n}}{(2n)!} + \ldots\right]$$

$$\times\left[1 + \frac{x^2}{2!} + \frac{x^4}{4!} + \ldots + \frac{x^{2n}}{(2n)!} + \ldots\right]$$

$$= 1 - \frac{4x^4}{4!} + \frac{4^2x^8}{8!} + \ldots$$

1168. Write down the three first terms of the Maclaurin series of $y = e^{\cos x}$.

Solution. We shall use the fact that there is only one power series representation of a given function. We assume

$$e^{\cos x} = a_0 + a_1x + a_2x^2 + a_3x^3 + a_4x^4 + \ldots$$

The a's are as yet undetermined. Now we substitute $x = 0$ and obtain $e^{\cos 0} = e = a_0$. We differentiate $e^{\cos x}$:

$$-\sin x\, e^{\cos x} = a_1 + 2a_2x + 3a_3x^2 + 4a_4x^3 + \ldots;$$

$$x = 0, \qquad a_1 = 0.$$

Differentiating again to find a_2,

$$\sin^2 x\, e^{\cos x} - \cos x\, e^{\cos x} = 2a_2 + 6a_3x + 12a_4x^2 + \ldots;$$
$$x = 0, \qquad -e = 2a_2, \qquad a_2 = -e/2.$$

Further,

$$(-\sin^3 x + \sin 2x + \sin x \cos x + \sin x)e^{\cos x} = 6a_3 + 24a_4x + \ldots;$$
$$x = 0, \qquad 0 = 6a_3, \qquad a_3 = 0;$$

and

$$\begin{pmatrix}\sin^4 x - \sin 2x \sin x - \sin^2 x \cos x - \sin^2 x \\ - 3 \sin^2 x \cos x + 3 \cos 2x + \cos x\end{pmatrix} e^{\cos x} = 24a_4 + \ldots;$$
$$x = 0, \qquad 4e = 24a_4, \qquad a_4 = e/6.$$

Now $$e^{\cos x} = e\left(1 - \frac{x^2}{2} + \frac{x^4}{6} - \ldots\right).$$

The method used in this solution is often very convenient.

1169. Find the four first terms of the Maclaurin series of $y = (1 + x)^x$.

Solution. Here we shall not use the above method because the differentiation is cumbersome. Instead, let us write

$$y = (1 + x)^x = e^{x \ln (1+x)},$$

and

$$x \ln (1 + x) = x\left(x - \frac{x^2}{2} + \frac{x^3}{3} + \frac{x^4}{4} - \ldots\right) = x^2\left(1 - \frac{x}{2} + \frac{x^2}{3} + \frac{x^3}{4} - \ldots\right).$$

Then

$$e^{x \ln (1+x)} = 1 + x \ln (1 + x) + \frac{[x \ln (1 + x)]^2}{2!} + \frac{[x \ln (1 + x)]^3}{3!} + \ldots$$
$$= 1 + x^2\left(1 - \frac{x}{2} + \frac{x^2}{3} - \ldots\right) + \frac{x^4}{2}\left(1 - \frac{x}{2} + \frac{x^2}{3} - \ldots\right)^2 + \ldots$$
$$= 1 + x^2 - \frac{x^3}{2} + \frac{x^4}{3} + \frac{x^4}{2} + \ldots$$
$$= 1 + x^2 - \frac{x^3}{2} + \frac{5x^4}{6} + \ldots.$$

We limited ourselves everywhere to terms not beyond x^4.

1170. Find the Maclaurin series of $y = (x - \tan x) \cos x$.

1171. Find the power series expansion of $y = \ln (x + 10)$ in powers of x.

1172. Expand the function $y = \sqrt{1 + x^2}$ around $x = 0$.

1173. Find the Maclaurin expansion of $y = \dfrac{x}{(1 - x)(1 - x^2)}$.

Solution. Let us decompose y into partial fractions.

$$y = \frac{x}{(1-x)^2(1+x)} = \frac{A}{(1-x)^2} + \frac{B}{1-x} + \frac{C}{1+x},$$

$$x = A(1+x) + B(1-x)(1+x) + C(1-x)^2;$$

$$x = 1,\ A = \tfrac{1}{2};\quad x = -1,\ C = -\tfrac{1}{4};\quad -B + C = 0,\ B = \tfrac{1}{4};$$

$$y = \frac{1}{2}\frac{1}{(1-x)^2} + \frac{1}{4(1-x)} - \frac{1}{4(1+x)}.$$

Now

$$\begin{aligned} y &= \tfrac{1}{2}(1-x)^{-2} + \tfrac{1}{4}(1-x)^{-1} - \tfrac{1}{4}(1+x)^{-1} \\ &= \tfrac{1}{2}(1 + 2x + 3x^2 + 4x^3 + \ldots + (n+1)x^n + \ldots) \\ &\quad + \tfrac{1}{4}(1 + x + x^2 + x^3 + \ldots + x^n + \ldots) \\ &\quad - \tfrac{1}{4}(1 - x + x^2 - x^3 + \ldots + (-1)^n x^n + \ldots) \\ &= \tfrac{1}{2}\sum_{n=0}^{\infty}(n+1)x^n + \tfrac{1}{2}\sum_{n=1}^{\infty} x^{2n-1}, \quad -1 < x < 1. \end{aligned}$$

1174. Find the Maclaurin expansion of $y = \arctan x$.

Solution. First we differentiate: $y' = \dfrac{1}{1+x^2}$. Now

$$y' = (1+x^2)^{-1} = 1 - x^2 + x^4 - x^6 + \ldots + (-1)^n x^{2n} + \ldots.$$

This series converges and represents y' in the interval $x^2 < 1$, i.e., $-1 < x < 1$. For any x in this interval we thus have

$$\arctan x = \int_0^x \frac{1}{1+x^2}\,dx = x - \frac{x^3}{3} + \frac{x^5}{5} - \frac{x^7}{7} + \ldots + \frac{(-1)^n x^{2n+1}}{2n+1} + \ldots.$$

By setting $x = 1$ we obtain the following series for $\pi/4$:

$$\frac{\pi}{4} = 1 - \frac{1}{3} + \frac{1}{5} - \ldots + \frac{(-1)^n}{2n+1} + \ldots.$$

The series obtained for $\arctan x$ converges also at the endpoints $x = \pm 1$ (as Leibniz's test proves). Hence, by the theorem mentioned in this section, the series represents $\arctan x$ also for $x = \pm 1$.

1175. The same question for $y = \arcsin x$.

1176. The same question for $y = \dfrac{1}{2}\ln\dfrac{1+x}{1-x}$.

1177. Find the first three terms of the Maclaurin series of $\tanh x$.

Solution. We could compute this series by dividing the series of $\sinh x$ by that of $\cosh x$, but prefer to compute the required terms by differentiation:

$y = \tanh x, \qquad y(0) = 0;$

$$y' = \frac{1}{\cosh^2 x}, \qquad y'(0) = 1;$$

$$y'' = -2\frac{\sinh x}{\cosh^3 x}, \qquad y''(0) = 0;$$

$$y''' = -2\left(\frac{1}{\cosh^2 x} - 3\frac{\sinh^2 x}{\cosh^4 x}\right), \qquad y'''(0) = -2;$$

$$y^{iv} = -2\left(-2\frac{\sinh x}{\cosh^3 x} - 6\frac{\sinh x}{\cosh^3 x} + 12\frac{\sinh^3 x}{\cosh^5 x}\right)$$

$$= -8\left(-2\frac{\sinh x}{\cosh^3 x} + 3\frac{\sinh^3 x}{\cosh^5 x}\right), \qquad y^{iv}(0) = 0;$$

$$y^{v} = -8\left(-\frac{2}{\cosh^2 x} + \text{terms containing } \sinh x \text{ in the numerator}\right),$$

$$y^{v}(0) = 16.$$

We obtain

$$\tanh x = x - \frac{2x^3}{3!} + \frac{16x^5}{5!} - \ldots = x - \frac{x^3}{3} + \frac{2x^5}{15} - \ldots.$$

1178. Find the expansion of $y = e^x$ around the point $x = -2$.

Solution. We find $y^{(k)} = e^x$; $y^{(k)}(-2) = e^{-2}$. Consequently

$$e^x = e^{-2}\left(1 + \frac{x+2}{1} + \frac{(x+2)^2}{2!} + \ldots + \frac{(x+2)^n}{n!}\right) + R_{n+1},$$

$$R_{n+1} = \frac{(x+2)^{n+1}}{(n+1)!}\, e^{-2+\theta(x+2)}.$$

$|R_{n+1}| \to 0$ as $n \to \infty$ for any x, and the series $e^{-2}\sum_{n=0}^{\infty}\frac{(x+2)^n}{n!}$ represents e^x everywhere.

1179. Find the Maclaurin expansion of $y = e^x \cos x$.

1180. Find the Taylor's expansion of $y = x^3 \ln x$ around $x = 1$.

1181. Find the Maclaurin expansion of $y = \frac{e^x}{1-x}$.

1182. Find the Maclaurin expansion of $y = \operatorname{argsinh} x$.

11.8 APPLICATIONS OF TAYLOR'S AND MACLAURIN'S EXPANSIONS

The expansion of a given function $f(x)$ in a power series has many applications to various problems occurring in analysis. We illustrate some of them in the following exercises.

1183. Given $f(x) = \dfrac{x}{1+x^2}$; find $f^{(5)}(0)$ using the Maclaurin expansion of this function.

Solution:

$$y = \frac{x}{1+x^2} = x(1 - x^2 + x^4 - x^6 + x^8 - \ldots)$$
$$= x - x^3 + x^5 - x^7 + x^9 - \ldots.$$

The representation of a function in a power series is unique, i.e., the coefficient $+1$ of the term with x^5 equals $f^{(5)}(0)/5!$ (the coefficient of x^5 in the Maclaurin expansion). Consequently

$$f^{(5)}(0) = 5!.$$

1184. Given $f(x) = x^6e^x$, find $f^{(11)}(0)$.

1185. Using Maclaurin's expansion find

$$\lim_{x\to 0} \frac{x + \ln(\sqrt{1+x^2} - x)}{x^3}.$$

Solution. We shall find the Maclaurin expansion of the numerator up to x^3; all other terms after dividing by x^3 still tend to zero. We will need to use the result of Problem 1161.

$$\ln(\sqrt{1+x^2} - x) = \ln\left(-x + 1 + \frac{1}{2}x^2 - \frac{1}{8}x^4 + \ldots\right)$$
$$= -x + \frac{x^2}{2} - \ldots - \frac{1}{2}\left(-x + \frac{x^2}{2} - \ldots\right)^2$$
$$+ \frac{1}{3}\left(-x + \frac{x^2}{2} - \ldots\right)^3 - \ldots$$
$$= -x + \frac{x^2}{2} - \frac{1}{2}x^2 + \frac{1}{2}x^3 - \frac{1}{3}x^3 + \ldots$$
$$= -x + \frac{1}{6}x^3 + \ldots.$$

$$\lim_{x\to 0} \frac{x + \ln(\sqrt{1+x^2} - x)}{x^3} = \lim_{x\to 0} \frac{x - x + \frac{1}{6}x^3}{x^3} = \frac{1}{6}.$$

1186. Using Maclaurin's expansion find $\lim\limits_{x\to 0}\left(\dfrac{1}{x^2} - \cot^2 x\right)$.

1187. Find approximately $\sqrt[3]{e}$, by computing three terms of the expansion of e^x around 0. Estimate the error.

Solution:

$$e^x = 1 + x + \frac{x^2}{2!} + R_3, \qquad R_3 = \frac{x^3}{3!}e^{\theta x}.$$

In our case $x = \frac{1}{3}$. Let us first estimate the error:

$$|R_3| \leqslant \frac{1}{3!} e^{1/3} \cdot \left(\frac{1}{3}\right)^3 \approx \frac{1}{6} \cdot 1.4 \cdot \frac{1}{27} = 0.0086.$$

The error is not bigger than 0.01. The terms of the expansion will be computed with the same accuracy. Now

$$\sqrt[3]{e} \approx 1 + \tfrac{1}{3} + \tfrac{1}{2} \cdot \tfrac{1}{9} \approx 1 + 0.333 + 0.056 \approx 1.39.$$

1188. Compute cos 10° with accuracy to 0.0001.

Solution:

$$\cos x = 1 - \frac{x^2}{2!} + \frac{x^4}{4!} - \ldots + (-1)^n \frac{x^{2n}}{2n!} + R_{2n+2}.$$

(We can assume that the $(2n + 1)$th term is computed, and as it is 0, the remainder is R_{2n+2}.)

$$R_{2n+2} = \frac{x^{2n+2}}{(2n+2)!} \cos\left[\theta x + (2n+2)\frac{\pi}{2}\right]$$

$$|R_{2n+2}| \leqslant \frac{x^{2n+2}}{(2n+2)!}.$$

We require $\frac{(0.17453)^{2n+2}}{(2n+2)!} < 0.00005$. Let us try $n = 2$. Then

$$\frac{(0.17453)^6}{6!} = \frac{2.8 \cdot 10^{-5}}{720} < 0.00005.$$

This is the required accuracy. (It can be shown that $n = 1$ is not enough.) Here 0.17453 is radian measure for 10°. Thus

$$\cos 10^0 \approx 1 - \frac{(0.17453)^2}{2!} + \frac{(0.17453)^4}{4!}$$

$$\approx 1 - 0.01522 + 0.00004 \approx 0.9848.$$

1189. Compute $\sqrt[3]{70}$ with accuracy to 0.001.

1190. Compute ln 3 with accuracy to 0.0001.

Solution. The direct use of the expansion of $\ln(1 + x)$ is impossible here, because the expansion converges only for $-1 < x \leqslant 1$, and $1 + 1 = 2 < 3$. We could first find ln 2 by the above expansion and then compute

$$\ln 3 = \ln(2 + 1) = \ln 2 + \ln(1 + \tfrac{1}{2});$$

in general, however, the use of the expansion of $\ln(1 + x)$ is not advisable, because the corresponding series converges very slowly.

Let us use the expansion of Problem 1176, because $(1 + x)/(1 - x)$ takes on every positive value when x varies in the interval $-1 < x < 1$.

This can be easily shown by drawing the graph of $y = (1 + x)/(1 - x)$. Thus this expansion can be used directly to compute ln a for any a (clearly, $a > 0$). Secondly, the series of Problem 1176 converges much faster than that of ln $(1 + x)$, and this is of vital importance in computational practice.

In our case we need

$$\frac{1+x}{1-x} = 3, \qquad 1 + x = 3 - 3x, \qquad x = \frac{1}{2}.$$

Then

$$\begin{aligned}\ln 3 &= 2(\tfrac{1}{2} + \tfrac{1}{3}\cdot\tfrac{1}{8} + \tfrac{1}{5}\cdot\tfrac{1}{32} + \tfrac{1}{7}\cdot\tfrac{1}{128} + \tfrac{1}{9}\cdot\tfrac{1}{512} + \tfrac{1}{11}\cdot\tfrac{1}{2048} + \dots)\\ &\approx 2(0.5 + 0.04167 + 0.00625 + 0.00112 + 0.00022 + 0.00004)\\ &= 2\cdot 0.54930 = 1.0986.\end{aligned}$$

To estimate the error we write

$$\begin{aligned}2\cdot\left[\frac{(\frac{1}{2})^{13}}{13} + \frac{(\frac{1}{2})^{15}}{15} + \frac{(\frac{1}{2})^{17}}{17} + \dots\right] &< \frac{2}{13}\left[\left(\frac{1}{2}\right)^{13} + \left(\frac{1}{2}\right)^{15} + \left(\frac{1}{2}\right)^{17} + \dots\right]\\ &= \frac{2}{13}\cdot\frac{(\frac{1}{2})^{13}}{1 - \frac{1}{4}} = \frac{1}{13}\cdot\frac{1}{3}\cdot\left(\frac{1}{2}\right)^{10}\\ &= \frac{1}{39\cdot 1024} < 0.00005.\end{aligned}$$

1191. Find the two first terms of the expansion of y given as a function of x by the equation

$$2\sin x + \sin y = x - y.$$

Solution. Let us assume

$$y = a_0 + a_1x + a_2x^2 + a_3x^3 + \dots.$$

For $x = 0$ we have $\sin y = -y$, and the unique solution of this equation is $y = 0$. Consequently $a_0 = 0$.

We substitute in the given equation, and using the expansion of $\sin x$ and $\sin y$ obtain

$$2\left(x - \frac{x^3}{3!} + \frac{x^5}{5!} - \dots\right) + (a_1x + a_2x^2 + a_3x^3 + \dots)$$

$$-\frac{1}{3}(a_1x + a_2x^2 + a_3x^3 + \dots)^3 + \dots = x - a_1x - a_2x^2 - a_3x^3 - \dots.$$

We equate the coefficients of equal powers of x on both sides:

x^1: $2 + a_1 = 1 - a_1, \quad 2a_1 = -1, \quad a_1 = -\frac{1}{2};$

x^2: $a_2 = -a_2, \quad a_2 = 0;$

x^3: $-\dfrac{2}{3!} + a_3 - \dfrac{1}{3!}a_1^3 = -a_3, \quad 2a_3 = \frac{1}{3} - \frac{1}{48} = \frac{15}{48} = \frac{5}{16}, \quad a_3 = \frac{5}{32}.$

The required two first terms are

$$y = -\frac{x}{2} + \frac{5}{32}x^3 + \ldots .$$

An alternative method can be used. We find the Maclaurin expansion of y by calculating the derivatives of y at $x = 0$.

$2 \sin x + \sin y = x - y; \quad x = 0, \quad y = 0.$

$2 \cos x + \cos y \cdot y' = 1 - y'; \quad 2 + y'(0) = 1 - y'(0), \quad y'(0) = -\frac{1}{2}.$

$-2 \sin x - \sin y \cdot (y')^2 + \cos y \cdot y'' = -y''; \quad y''(0) = -y''(0), \quad y''(0) = 0.$

$-2 \cos x - \cos y \cdot (y')^3 - 2 \sin y \cdot y' \cdot y'' - \sin y \cdot y' \cdot y'' + \cos y \cdot y''' = -y''';$

$$-2 - (-\tfrac{1}{2})^3 + y'''(0) = -y'''(0), \quad y'''(0) = \tfrac{15}{16}.$$

Now $\quad y = -\frac{1}{2}x + \frac{1}{3!} \cdot \frac{15}{16}x^3 + \ldots = -\frac{1}{2}x + \frac{5}{32}x^3 + \ldots .$

1192. Find the expansion of $\mathrm{Si}(x) = \int_0^x \frac{\sin x}{x}\,dx$ and determine its domain of convergence.

Solution. Compare Problem 891.

$$\mathrm{Si}(x) = \int_0^x \frac{1}{x}\left[x - \frac{x^3}{3!} + \frac{x^5}{5!} - \ldots + (-1)^{n+1}\frac{x^{2n-1}}{(2n-1)!} + \ldots\right]dx$$

$$= \int_0^x \left[1 - \frac{x^2}{3!} + \frac{x^4}{5!} - \ldots + \frac{(-1)^{n+1}x^{2n-2}}{(2n-1)!} + \ldots\right]dx$$

$$= x - \frac{x^3}{3 \cdot 3!} + \frac{x^5}{5 \cdot 5!} - \ldots + (-1)^{n+1}\frac{x^{2n-1}}{(2n-1)(2n-1)!} + \ldots .$$

From the theorems of the preceding section it follows that this series represents everywhere the integral in question.

1193. The same question for $\int_0^x e^{-x^2}\,dx$.

1194. Compute approximately the integral $\int_0^{1/2} \frac{dx}{\sqrt{1 + x^4}}$ (take two terms in the expansion).

1195. Compute the integral $\int_0^{1/2} \frac{dx}{1 + x^4}$ with accuracy to 0.001.

1196. Find the area bounded by the curve $y^2 = x^3 + 1$ and the straight lines $x = 0$, $x = \frac{1}{2}$. Compute with accuracy to 0.001.

1197. Given

$$f(x) = x^8 - 2x^7 + 5x^6 - x + 3;$$

find $f(2.02)$ using the three first terms of a Taylor's expansion.

1198. Find $\ln \frac{10}{9}$ with accuracy to 0.000001.

1199. Using two terms of a Maclaurin's expansion find tan 0.1.

1200. Prove that

$$\frac{\pi}{4} = 4 \arctan \frac{1}{5} - \arctan \frac{1}{239}$$

and use this to determine $\pi/4$ with accuracy to 0.001.

1201. Find the expansion of $f(x) = \int_0^x \sin t^2 \, dt$.

XII

VARIOUS PROBLEMS

In this chapter are collected a number of problems on some of the topics treated in this book. The problems are of various degrees of difficulty. The reader who has studied the text carefully will presumably be able to solve all the problems in this chapter. The problems may serve to check his understanding and skill.

The problems will be given without detailed solutions. Furthermore, they are not arranged in the order in which the relevant topics were presented, so that the reader will solve them without special clues as to the methods he is supposed to use.

1202. Given $y = \dfrac{\arcsin x}{\sqrt{1 - x^2}}$; prove that

$$(1 - x^2)y' = xy + 1,$$

and further, obtain the formula

$$(1 - x^2)y^{(n+1)} - (2n + 1)xy^{(n)} - n^2y^{(n-1)} = 0.$$

1203. Find the Maclaurin expansion of $f(x) = \dfrac{x + 2}{x^2 - 5x + 6}$ and indicate the domain of convergence of the resulting series.

1204. Compute $\displaystyle\int_0^{\pi/4} \frac{dx}{\sin^2 x + 4\cos^2 x}$.

1205. Draw the graph of $y = \dfrac{x^3}{(x + 2)^2}$. Investigate the domain of definition, the intercepts, the asymptotes, and extremum points.

1206. Compute $\displaystyle\int \frac{x\,dx}{(x - 1)^2(x^2 + 2)}$.

1207. Draw the graph of $x^4 + y^4 - 8xy^2 = 0$ using parametric representation. Find the symmetry, extrema of x and y, and the slope of the curve at the origin.

1208. Given an ellipse by means of the parametric equations $x = a \cos t$, $y = b \sin t$ and two points $P_1(t_1 = \pi/2)$ and $P_2(t_2 = \pi)$ on it; find a third point P_3 on the ellipse such that the area of the triangle $P_1P_2P_3$ will be (a) maximal (b) minimal. Find these maximal and minimal areas.

1209. A container formed by the revolution of the curve $y = \sqrt{2px}$ around the y axis is filled with water up to the height h. Find the increase in the height of the water if V cubic units of water are added.

1210. Draw the curve represented by

$$y = \sqrt{x^2 + 2x + 3} + \sqrt{x^2 - 2x + 3}.$$

Find the domain of definition, symmetry, asymptotes.

1211. Find $\lim\limits_{x \to 4} \left(\dfrac{2x^2 + x - 2}{x^2 + 3x + 6}\right)^{\frac{1}{x-4}}$.

1212. Given the equation

$$(x + a)^3 y''' + 3(x + a)^2 y'' + (x + a)y' + y = 0.$$

Substitute $\ln (x + a) = t$ and obtain the differential equation with t as argument.

1213. Compute $\displaystyle\int \frac{dx}{(x + 2)\sqrt{x^2 + 4x - 5}}$.

1214. Prove, using Lagrange's theorem, that

$$\frac{x}{1 + x} < \ln (1 + x) < x, \quad x > 0.$$

1215. Find the extrema of $y = e^{2x} - 8e^x + 6x$.

1216. Draw the graph of

$$y = \frac{x^2 - x - 6}{x^2 + x + 1}.$$

Find the intercepts, asymptotes, extrema.

1217. Compute the subtangent and subnormal of $y = x^3 + 2x^2 - 4x + 2$ at $x = 2$.

1218. The two curves $xy(3x - y) = 24$ and $xy(x + 2) = 15$ intersect at $x = 3$. Prove this, and find the angle of intersection.

1219. Compute $\int x \arcsin x \, dx$. Check the result by differentiation.

1220. Inscribe inside a sphere of radius R (a) a cylinder of maximal volume, and (b) a cylinder of maximal surface area (including the bases). Find in each case the radius and altitude of the cylinder.

1221. Given $x^3 + y^3 - 3axy = 0$, find y' and y''.

1222. Find the volume of a maximal cone inscribed in a sphere of radius R.

1223. Find $\int_{-1}^{1} \frac{1+x}{1-x}\,dx$.

1224. Compute $\lim_{x\to 0} \frac{(1+2x)^{1/2} - (1+3x)^{1/3}}{(1+4x)^{1/4} - (1+5x)^{1/5}}$.

1225. Find $\int \frac{4x\,dx}{1-x^4}$.

1226. Draw the graph of $y = x^2e^{-x}$. Find the domain of definition, intercepts, asymptotes, extrema, and inflection points.

1227. Compute sin 32° with accuracy to 0.001 using a Taylor's expansion around $x = \pi/6$.

1228. Compute $\int_0^1 \frac{x^3 - 7}{x^3 - 8}\,dx$.

1229. The volume and radius of a sphere were computed from its measured weight P and specific weight q. Find by the means of the differential the relative error in the computation of the radius, if an error of 1% occurs (a) in the measure of P, (b) in the measure of q.

1230. Find the length of the axes of an ellipse of minimal area which can be circumscribed around a rectangle of constant sides $2p$ and $2q$.

1231. Draw the graph of $y = x2^{-x}$. Find the domain of definition, intercepts, asymptotes, and extrema.

1232. Find $\int \sqrt{x - x^2}\,dx$.

1233. A vehicle leaves A and moves towards B with the velocity 80 miles/hr. Simultaneously a train leaves B and moves towards C with the velocity of 50 miles/hr. Given $AB = 200$ miles and $\measuredangle ABC = \pi/3$, at what time will the distance between the vehicle and the train be minimal?

1234. Given $p^2 = a^2 \cos^2 \varphi + b^2 \sin^2 \varphi$; prove that

$$p + \frac{d^2p}{d\varphi^2} = \frac{a^2b^2}{p^3}.$$

1235. Find the first four terms of the Maclaurin expansion of

$$y = \frac{\ln(1+x)}{1+\sin x}.$$

1236. Given $y = x\frac{x^2 + 3D}{3x^2 + D}$, $x > 0$, $D > 0$. Prove that

(a) $y > x$ and $y^2 < D$ if $x^2 < D$.

(b) $y < x$, $y > 0$, and $y^2 > D$ if $x^2 > D$.

1237. Draw the graph of $y = x^2\sqrt{x + 1}$. Find the domain of definition, symmetry, extrema, inflection points, and $\lim\limits_{x \to -1+} y'$.

1238. Find $\int x^2 \sin x \, e^x \, dx$.

1239. Find a rectangular parallelepiped with a prescribed sum P of all 12 edges, and prescribed surface area S, such that its volume will be (a) maximal, (b) minimal. Find in each case the length of the edges. What condition must P and S fulfill for the results to be real and positive?

1240. Given $y = \frac{x^4 + 1}{x^3 - x}$, find $y^{(k)}$.

1241. Given $y = 2x(x + 1) \ln (x + 1)$, find $y^{(n)}$.

1242. Prove that the first derivative of $y = x^3 \sin (1/x)$, $y(0) = 0$, is continuous at $x = 0$ but that there is no second derivative at $x = 0$.

1243. Draw the graph of $y = (\ln x)/x$. Find the domain of definition, symmetry, extrema, inflection points, and $\lim\limits_{x \to \infty} y$ and $\lim\limits_{x \to 0} y$.

1244. Find $\int \frac{dx}{a + b \cos x}$, $0 < a < b$.

1245. Given $H \ln (p + \sqrt{1 + p^2}) = ax$; express p as a function of x.

1246. Compute $\int_0^1 \frac{dx}{x^3 + x + 2}$.

1247. Draw the graph of $y = \frac{1}{8}x^2 \sin x$. Find the symmetry and extrema (find graphically the values of x of one minimum and one maximum with accuracy to 0.1).

1248. Prove that the function $y = (\arcsin x)^2$ is a solution of the differential equation $(1 - x^2)y'' - xy' = 2$.

1249. Find on the parabola $y^2 = 2px$ the point nearest to $(a,0)$. Find this minimal distance.

1250. Compute $\int \sin^2 2x \cos^4 2x \, dx$.

1251. The tangent to the curve $y = 0.1x^3$ at $x = 2$, intersects this curve in an additional point. Find this point.

1252. Draw the graph of

$$y = \frac{x^2 - 4x + 3}{x^2 - 6x + 8}$$

in the interval $-3 < x < 7$. Find intercepts, asymptotes, extrema. Draw in the same system of coordinates, and in the same interval, the graph of

$$y = \sqrt{\frac{x^2 - 4x + 3}{x^2 - 6x + 8}}.$$

1253. Given $y_1 = a \sin (x + \alpha)$, $y_2 = b \sin (x + \beta)$; compute

$$\int_0^{2\pi} y_1 y_2 \, dx.$$

1254. Compute the maximum of $y = \tan \alpha / \tan (\alpha + \beta)$ if β is constant and α variable. What is the maximum for $\beta = 10°$, and what is the corresponding α?

1255. The smaller base of a trapezoid is 2 and each of the nonparallel sides is 0.5. Find the angles of the trapezoid if its area is maximal. What is this maximal area?

1256. Draw the graph of $y = \dfrac{x^3 + 3x + 4}{x - 1}$ in the interval $-4 < x < 4$. Find extrema.

1257. Prove that $\ln (1 + e^x) = \ln 2 + \frac{1}{2}x + \frac{1}{8}x^2 - \frac{1}{192}x^4 + \cdots$.

1258. A right triangle with two sides a and b is drawn in such a manner that its hypotenuse is tangent to the curve $y = e^x$ and the side a is on the x axis. Find (a) the coordinate of the point of tangency; (b) the difference $\Delta y - dy$ for the right-hand end of the side a.

1259. Compute $\displaystyle\int_0^{\pi/4} x^2 \cos^2 x \, dx$.

1260. Draw the graph of $y = 5 \cosh x + 3 \sinh x$ in the interval $-3 < x < 3$. Find the domain of definition, extrema, inflection points.

1261. Given $e^{y/x} = \dfrac{y - x}{y + x} e^x$; prove that $y' = \dfrac{x(y^2 - x^2)}{y^2 - 3x^2} + \dfrac{y}{x}$.

1262. The displacement s is given as a function of the time t by the equation $s = e^{\sin t}$, $0 < t < 2$. At what time and in what place does the velocity $v = ds/dt$ attain its maximum value? Draw the graph of s in the said interval.

1263. Compute $\lim_{x \to 0} \dfrac{2 \sin x + \frac{1}{2} \ln \frac{1+x}{1-x} - 3x}{x^5}$.

1264. The curves $y = x^3$ and $y = 7x^2 - 36$ intersect in three points. Prove that the tangents to $y = x^3$ at these points meet in one point, and find its coordinates.

1265. Prove that the first derivatives of the two functions

$$y_1 = 2 \arcsin \sqrt{\frac{x - \beta}{\alpha - \beta}} \quad \text{and} \quad y_2 = 2 \arctan \sqrt{\frac{x - \beta}{\alpha - x}}$$

are equal. Explain why, and prove.

1266. Given $x = a \cos \varphi - \sqrt{1 - a^2 \sin^2 \varphi}$. Prove that if a is small, x can be computed from

$$x = \frac{a^4}{4} - 1 + a \cos \varphi - \frac{a^2}{4} \cos 2\varphi.$$

Estimate the error for $a = \frac{1}{5}$ if $0 \leqslant \varphi < 2\pi$.

1267. Compute $\int \cos x \sin 2x \cos 3x \, dx$.

1268. Find the area of the sector of the ellipse

$$\frac{x^2}{a^2} + \frac{y^2}{b^2} = 1$$

bounded by two rays emanating from its center and making angles α and β ($\beta > \alpha$) with the positive x axis.

1269. Prove without calculating each integral separately, that

$$\int_{-1/2}^{1/2} \cos x \log \frac{1+x}{1-x} \, dx = \int_{-1/2}^{1/2} \log \frac{1+x}{1-x} \, dx.$$

1270. Given the series $\sum_{n=0}^{\infty} e^{-|n-x|}$. Prove that (a) the series converges uniformly for $x \leqslant a$, where a is an arbitrary constant, and (b) the function defined by the series is continuous for any x.

1271. Investigate the convergence of the integral

$$\int_0^1 \frac{\ln x^{-1}}{x^p} \, dx$$

according to various values of the parameter p.

1272. Compute $\displaystyle\int \frac{2x^2 - 2x + 1}{(x-1)^2(x^2+1)} \, dx$.

1273. Find the radius of curvature of the curve given by the equations

$$x = \int_0^t \frac{\cos u}{\sqrt{u}}\, du, \qquad y = \int_0^t \frac{\sin u}{\sqrt{u}}\, du.$$

1274. Prove that the equation $x^4 - 4x^3 + 7x^2 - 3x + 2 = 0$ has no real solutions. (*Hint:* Draw the graphs of $y = \frac{1}{5}(x^2 - 4x + 7)$ and $y = (3x - 2)/5x^2$.)

1275. Given $(x^2 + y^2 - 5x)^2 = 9x^2 + 9y^2$, compute y' at (0,0).

1276. Among all cones of given volume V, which has the minimal surface area (including the base)?

1277. Compute, using the definition of the definite integral,

$$\lim_{n\to\infty} \frac{1}{n^2} \sqrt[n]{\frac{(3n)!}{n!}}.$$

(*Hint:* Use logarithms.)

1278. Compute

$$\lim_{x\to 0}\left(\sqrt{\frac{1}{x} + \sqrt{\frac{1}{x} + \sqrt{\frac{1}{x} + \sqrt{\frac{1}{x}}}}} - \sqrt{\frac{1}{x}}\right).$$

1279. Prove that if the sequence $s_1, s_2, \ldots s_n, \ldots$ tends to a limit $s = 0$ ($\lim\limits_{n\to\infty} s_n = 0$), then also the sequence

$$t_n = \frac{s_1 + s_2 + \ldots + s_n}{n}, \quad n = 1, 2, 3, \ldots$$

tends to zero (i.e., $\lim\limits_{n\to\infty} t_n = 0$).

1280. Find a function satisfying the following conditions:

(a) In any interval of its domain of definition a point exists where the value of the function is 0;

(b) In any interval of its domain of definition a point exists where the value of the function is different from zero;

(c) The function is defined for any real x and satisfies $-x^2 \leqslant f(x) \leqslant x^2$.

1281. Find the Taylor's expansion of $y = \sin 3x$ around the point $x = \pi/3$.

1282. Given a number series $\sum\limits_{n=1}^{\infty} a_n$ with $a_n > 0$ and $a_{n+1} < a_n$. Show that a necessary condition for convergence of the series is

$$\lim_{n\to\infty} na_n = 0.$$

Give an example showing that this condition is not sufficient.

1283. Prove that if a power series $\sum_{n=0}^{\infty} a_n x^n$ converges for $x = r$, it converges uniformly in any interval $-\rho \leqslant x \leqslant \rho$ $(0 < \rho < r)$.

1284. Show that in the Maclaurin expansion of the function $f(x) = e^{-1/x^2}$, $f(0) = 0$, all terms equal zero. Does the corresponding series represent the given function in all points of its domain of convergence?

SOLUTIONS, HINTS, ANSWERS

In this last part of the book will be found solutions to a large proportion of the problems stated but not solved in the first part. For others, answers are given, and in some cases hints as to how the solution could be started.

4. $u_1 = 2$, $u_2 = \frac{1}{3}$, $u_3 = \frac{8}{29}$, $u_4 = \frac{7}{26}$, $u_5 = \frac{22}{81}$.

We shall prove that the limit of the sequence is $\frac{1}{3}$.

$$\left|u_n - \frac{1}{3}\right| = \left|\frac{n^2 - n + 2}{3n^2 + 2n - 4} - \frac{1}{3}\right| = \left|\frac{-5n + 10}{3(3n^2 + 2n - 4)}\right|.$$

Already for $n = 3$ the expression is negative (the numerator is negative and the denominator positive); consequently for $n > 2$ we can write

$$\left|\frac{-5n + 10}{3(3n^2 + 2n - 4)}\right| = \frac{5n - 10}{3(3n^2 + 2n - 4)} < \frac{5n}{3(3n^2 - 4)} < \frac{5n}{3\cdot 2n^2} < \frac{5}{6n} < \frac{1}{n}$$

We use here the notion of the absolute value, $|-a^2| = a^2$. For $n > 2$ we have $n^2 - 4 > 0$, and this was used when we passed from $5n/3(3n^2 - 4)$ to $5n/3\cdot 2n^2$.

We have shown that $|u_n - \frac{1}{3}| < 1/n$, and by Problem 3 we can have for every $\epsilon > 0$, $|u_n - \frac{1}{3}| < \epsilon$ if $n > N = 1/\epsilon$. This proves that $\lim\limits_{n\to\infty} u_n = \frac{1}{3}$.

Remark. The last two examples show that evaluating the limit by its definition is in general cumbersome. We must also first "guess" the value of the limit. In what follows we shall learn faster and simpler methods for evaluation of limits.

5. (1)e 1; (2)e 1; (3)e the sequence has no limit.

(4)c $\frac{1}{2} + \frac{\sqrt{3}}{2}, -\frac{1}{2} + \frac{\sqrt{3}}{2}, -1, -\frac{1}{2} - \frac{\sqrt{3}}{2}, \frac{1}{2} - \frac{\sqrt{3}}{2}, 1.$

(5) $u_1 = 0$, $u_2 = 4$, $u_3 = 0$, $u_4 = 4$, $u_5 = 0$, $u_6 = 4$. The sequence is not monotone. 4 is the least upper and 0 the greatest lower bound. 0 and 4 are also points of accumulation and the lower and upper limits respectively. The sequence has no limit.

6.
$$u_n = \frac{n - \dfrac{(k-1)(k-2)}{2}}{k},$$

where k is the integer satisfying

$$\frac{1+\sqrt{8n+1}}{2} \leqslant k < \frac{3+\sqrt{8n+1}}{2}.$$

The sequence does not converge. (b) 1 and 0; (c) every real number in the closed interval [0, 1]; (d) 0, 1.

7. We shall prove the existence of the stated four conditions:

(1) $u_{n+1} - u_n$

$$= \frac{(n+1)^2 - 1}{3(n+1)^2 + 2} - \frac{n^2 - 1}{3n^2 + 2} = \frac{10n + 5}{(3n^2 + 2)[3(n+1)^2 + 2]} > 0;$$

i.e., $u_{n+1} > u_n$.

(2) $$v_{n+1} - v_n = \frac{n+1+3}{3(n+1) - 2} - \frac{n+3}{3n-2} = \frac{-11}{(3n+1)(3n-2)} < 0;$$

i.e., $v_{n+1} < v_n$.

(3) $$v_n - u_n = \frac{n+3}{3n-2} - \frac{n^2-1}{3n^2+2} = \frac{11n^2 + 5n + 4}{(3n-2)(3n^2+2)} > 0;$$

i.e., $v_n > u_n$.

(4) For large n we have

$$v_n - u_n = \frac{11n^2 + 5n + 4}{9n^3 - 6n^2 + 6n - 4} < \frac{12n^2}{8n^3} = \frac{\frac{3}{2}}{n}.$$

The inequality follows, because for $n > 10$, $n^2 > 5n + 4$ and $n^3 > 6n^2 - 6n + 4$. (We have increased the numerator and decreased the denominator; as a result the fraction increases.) But $\lim_{n\to\infty} \frac{3}{2}/n = 0$, consequently $\lim_{n\to\infty} (v_n - u_n) = 0$. All four conditions are fulfilled. It is now easy to check that the common limit of the two sequences is $\frac{1}{3}$.

9. Let us prove $a_1 > b_1$. In Problem 8 we saw that $(a+b)/2 > \sqrt{ab}$. Both sides of the inequality are positive and hence can be squared,

$$\frac{(a+b)^2}{4} > ab, \qquad \frac{a+b}{2} > \frac{2ab}{a+b} \qquad (\text{note } a + b > 0).$$

Now we prove that $b_1 > b$:

$$\frac{2ab}{a+b} - b = \frac{2ab - ab - b^2}{2} = \frac{b(a-b)}{2} > 0 \qquad (\text{as } a > b).$$

As in Problem 8, we have $a_n > a_{n+1} > b_{n+1} > b_n$. In a similar way we see that both sequences converge and have a common limit α. To find it we compute the product $a_1 b_1$,

$$a_1 b_1 = \frac{a+b}{2} \cdot \frac{2ab}{a+b} = ab.$$

In the same way,

$$a_2 b_2 = a_1 b_1 = ab,$$

and in general

$$a_n b_n = ab.$$

By the theorem about the limit of a product of two sequences we find

$$\lim_{n\to\infty} a_n b_n = \lim_{n\to\infty} a_n \cdot \lim_{n\to\infty} b_n = \alpha \cdot \alpha = \alpha^2.$$

On the other hand, $a_n b_n$ constantly equals ab, i.e., $\lim_{n\to\infty} a_n b_n = ab$. The arithmetic-harmonic mean of two positive numbers a and b is thus equal to their geometric mean: $\alpha = \sqrt{ab}$.

11. $$\left|u_n - \frac{1}{3}\right| = \left|\frac{n^2+n-1}{3n^2+1} - \frac{1}{3}\right| = \left|\frac{3n-4}{3(3n^2+1)}\right| = \frac{3n-4}{3(3n^2+1)} \quad (\text{for } n > 1).$$

We have to solve the quadratic inequality $\dfrac{3n-4}{3(3n^2+1)} < \epsilon$,

$$3n - 4 - 9\epsilon n^2 - 3\epsilon < 0,$$
$$9\epsilon n^2 - 3n + 3\epsilon + 4 > 0.$$

To solve this inequality we first find the solutions of the equation

$$9\epsilon n^2 - 3n + 3\epsilon + 4 = 0;$$

$$n_{1,2} = \frac{3 \pm \sqrt{9 - 4\cdot 9\epsilon(3\epsilon+4)}}{18\epsilon} = \frac{1}{6\epsilon}\left[1 \pm \sqrt{1 - 4\epsilon(3\epsilon+4)}\right].$$

We draw schematically (Fig. 71) the graph of the parabola

$$y = f(n) = 9\epsilon n^2 - 3n + 3\epsilon + 4.$$

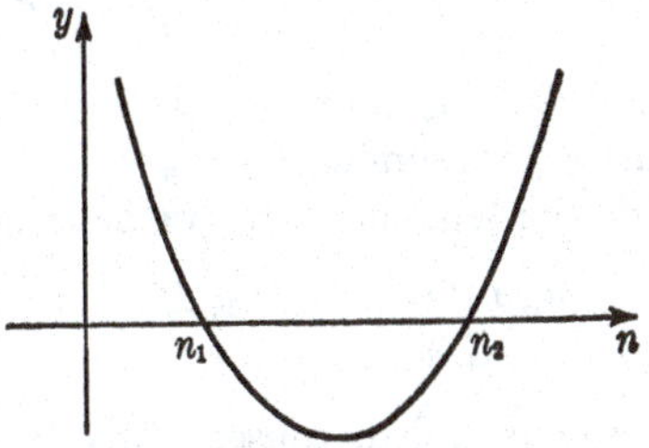

FIGURE 71

We see that $y > 0$ when $n < n_1$ or $n > n_2$. $n < n_1$ does not interest us because n is an increasing natural number. The solution of the problem is

$$n > n_2 = N = \frac{1}{6\epsilon}\left[1 + \sqrt{1 - 4\epsilon(3\epsilon + 4)}\right].$$

Neglecting the expression $4\epsilon(3\epsilon + 4)$ under the radical (i.e., putting zero in place of it) increases N, and we have a simpler expression, i.e., $N < N' = 1/3\epsilon$. However, the same result can be obtained more simply:

$$\frac{3n - 4}{3(3n^2 + 1)} < \frac{3n}{9n^2} = \frac{1}{3n} < \epsilon; \qquad n > \frac{1}{3\epsilon}.$$

If the only purpose of the computation is to prove that $\lim_{n\to\infty} u_n = \frac{1}{3}$, then the second method is preferable.

12. (a) $N = 1/\sqrt{\epsilon}$ (b) $N = (2/\epsilon)^{3/5}$

(c) $\left|\dfrac{\sin n + 2\cos^2 n}{\sqrt{n}}\right| < \dfrac{3}{\sqrt{n}}, \; n > \dfrac{9}{\epsilon^2}$

(d) $|\sqrt{n+1} - \sqrt{n}| = \dfrac{1}{\sqrt{n+1} + \sqrt{n}} < \dfrac{1}{2\sqrt{n}} < \epsilon, \; n > \dfrac{1}{4\epsilon^2}.$

14. Clearly $u_n > 0$. It follows that $u_{n+1} = 6/(1 + u_n) < 6$; i.e., the sequence is bounded. Assume that $u_n < 2$. Then

$$u_{n+1} = \frac{6}{1 + u_n} > \frac{6}{1 + 2} = 2; \quad \text{i.e.,} \quad u_{n+1} > 2.$$

But

$$u_{n+2} = \frac{6}{1 + u_{n+1}} < \frac{6}{1 + 2} = 2, \quad \text{i.e.,} \quad u_{n+2} < 2.$$

The sequence is not monotone; its elements oscillate around 2. We shall now prove that if $|u_n - 2| = \epsilon$, then $|u_{n+1} - 2| < \epsilon$; i.e., that the sequence elements approach 2 from both sides. Assume $u_n > 2$. Then

$$(u_n - 2) - (2 - u_{n+1}) = u_n - 4 + \frac{6}{1 + u_n} = \frac{u_n^2 - 3u_n + 2}{1 + u_n}$$
$$= \frac{(u_n - 2)(u_n - 1)}{1 + u_n} > 0.$$

Consequently $u_n - 2 > 2 - u_{n+1}$, i.e., u_{n+1} is closer to 2 than u_n. We can prove this in the same way assuming $1 < u_n < 2$.

Now we shall divide the sequence into two subsequences

$$u_1, u_3, u_5, \ldots, u_{2n-1}, \ldots;$$
$$u_2, u_4, u_6, \ldots, u_{2n}, \ldots.$$

Both are monotone and bounded, consequently both subsequences converge. Assume that the limit of the second is l. Let us write

$$\lim_{n\to\infty} u_{2n+2} = \lim_{n\to\infty} \frac{6}{1+u_{2n+1}} = \lim_{n\to\infty} \frac{6}{1+\dfrac{6}{1+u_{2n}}};$$

i.e., $$l = \frac{6}{1+\dfrac{6}{1+l}}; \qquad l^2+l-6=0; \qquad l_1 = 2,\ l_2 = -3.$$

Only the positive solution is suitable. The same calculation holds also for the first subsequence, so that 2 is the common limit of both subsequences and consequently also of the given sequence. (Prove this assertion!)

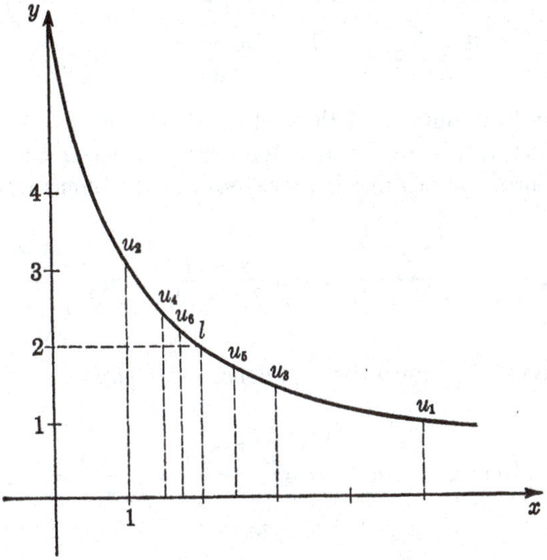

FIGURE 72

Let us draw the graph of $y = \dfrac{6}{1+x}$ for $x > 0$ (Fig. 72). The sequence elements can be found as follows: $u_1 = 1$ is the intersection point of the curve $y = 6/(1+x)$ and the straight line $y = 1$. Take now the ordinate of u_1 as an abscissa and obtain $y_2 = u_2 = 6/(1+u_1) = 3$. Now take the ordinate of u_2 as abscissa and obtain u_3 on the graph, and so on. This construction shows clearly the way u_n tends to 2.

15. (a) 2 (b) $\dfrac{1+\sqrt{1+4k}}{2}$.

18. $$\frac{n^2 - \dfrac{5}{n}}{1 - \dfrac{3}{n} + \dfrac{1}{n^2}} \to \infty \text{ when } n \to \infty.$$

19. The theorem on limit of a sum is of no use here, because neither of the two summands has a limit. However, after subtraction,

$$u_n = \frac{2n^4 + n^3 - 2n^4 + n^2}{(2n^2 - 1)(2n + 1)} = \frac{n^3 + n^2}{(2n^2 - 1)(2n + 1)},$$

and

$$\lim_{n\to\infty} u_n = \lim_{n\to\infty} \frac{1 + \frac{1}{n}}{\left(2 - \frac{1}{n^2}\right)\left(2 + \frac{1}{n}\right)} = \frac{1}{2\cdot 2} = \frac{1}{4}.$$

20. After division by n, the terms $1/n$, $2/n$, . . . tend to 0; but

$$\frac{n-1}{n} = 1 - \frac{1}{n}, \qquad \frac{n-2}{n} = 1 - \frac{2}{n}, \ldots$$

tend to 1 and the number of those elements increases without limit. We cannot decide in this way what is the limit. A better idea is to evaluate first the sum of the arithmetic progression in the numerator; we have

$$\lim_{n\to\infty} u_n = \lim_{n\to\infty} \frac{\frac{n(n+1)}{2}}{n(n+2)} = \lim_{n\to\infty} \frac{n+1}{2(n+2)} = \lim_{n\to\infty} \frac{1 + \frac{1}{n}}{2\left(1 + \frac{2}{n}\right)} = \frac{1}{2}.$$

21. We divide the numerator and denominator by n:

$$\lim_{n\to\infty} u_n = \lim_{n\to\infty} \frac{\sqrt{1 + \frac{1}{n^2}} + \sqrt{\frac{1}{n}}}{\sqrt[4]{\frac{1}{n} + \frac{1}{n^3}} - 1} = \frac{1+0}{0-1} = -1.$$

22. Here we divide by n^2:

$$\lim_{n\to\infty} u_n = \lim_{n\to\infty} \frac{\left(\sqrt{1 + \frac{1}{n^2}} + 1\right)^2}{\sqrt[3]{1 + \frac{1}{n^6}}} = \frac{(1+1)^2}{1} = 4.$$

24. $$\lim_{n\to\infty} u_n = \lim_{n\to\infty} \frac{n(n^2 + 1 - n^2)}{\sqrt{n^2+1} + n} = \lim_{n\to\infty} \frac{1}{\sqrt{1 + \frac{1}{n^2}} + 1} = \frac{1}{2}.$$

25. Here we use the formula $(a - b)(a^2 + ab + b^2) = a^3 - b^3$:

$$\lim_{n\to\infty} u_n = \lim_{n\to\infty} \frac{(n+1)^2 - (n-1)^2}{\sqrt[3]{(n+1)^4} + \sqrt[3]{(n+1)^2(n-1)^2} + \sqrt[3]{(n-1)^4}}$$

$$= \lim_{n\to\infty} \frac{4n}{\sqrt[3]{(n+1)^4} + \sqrt[3]{(n^2-1)^2} + \sqrt[3]{(n-1)^4}}$$

$$= \lim_{n\to\infty} \frac{4}{\sqrt[3]{(n^{1/4} + n^{-3/4})^4} + \sqrt[3]{(n^{1/2} - n^{-3/2})^2} + \sqrt[3]{(n^{1/4} - n^{-3/4})^4}}$$

$$= 0.$$

26. $$\lim_{n\to\infty} u_n = \lim_{n\to\infty} \frac{n^3(n^2 + \sqrt{n^4+1} - 2n^2)}{\sqrt{n^2 + \sqrt{n^4+1}} + n\sqrt{2}}$$

$$= \lim_{n\to\infty} \frac{n^3(n^4 + 1 - n^4)}{(\sqrt{n^2 + \sqrt{n^4+1}} + n\sqrt{2})(\sqrt{n^4+1} + n^2)}$$

$$= \lim_{n\to\infty} \frac{1}{\left(\sqrt{1 + \sqrt{1 + \frac{1}{n^4}}} + \sqrt{2}\right)\left(\sqrt{1 + \frac{1}{n^4}} + 1\right)}$$

$$= \frac{1}{2\sqrt{2}\cdot 2} = \frac{1}{4\sqrt{2}} = \frac{\sqrt{2}}{8}.$$

27. We use here the identity

$$1^2 + 2^2 + \ldots + n^2 = \frac{n(2n+1)(n+1)}{6}$$

which can be easily proved by mathematical induction. Then

$$\lim_{n\to\infty} u_n = \lim_{n\to\infty} \frac{n(2n+1)(n+1)}{6n^3} = \lim_{n\to\infty} \frac{\left(2 + \frac{1}{n}\right)\left(1 + \frac{1}{n}\right)}{6} = \frac{2}{6} = \frac{1}{3}.$$

28. 0. **29.** 2. **30.** 0. **31.** $\frac{1}{2}$.

32. a when $a > 1$, and $1/a$ when $0 < a < 1$.

34. $$f(a+b) = \frac{a+b+a}{a+b-a} = \frac{2a+b}{b};$$

$$f\left(\frac{1}{a}\right) + f\left(\frac{1}{b}\right) = \frac{\frac{1}{a} + a}{\frac{1}{a} - a} + \frac{\frac{1}{b} + a}{\frac{1}{b} - a} = \frac{1+a^2}{1-a^2} + \frac{1+ab}{1-ab}$$

$$= \frac{2(1-a^3b)}{(1-a^2)(1-ab)}.$$

35. $$\frac{f(x+h) - f(x)}{h} = \frac{1}{h}[(x+h)^3 - 4(x+h) - (x^3 - 4x)]$$

$$= 3x^2 - 4 + 3hx + h^2.$$

36. $\varphi(4) = \dfrac{|4-2|}{4+1} = \dfrac{2}{5}, \qquad \varphi(0) = \dfrac{|0-2|}{0+1} = 2,$

$$\varphi(-2) = \frac{|-2-2|}{-2+1} = \frac{4}{-1} = -4.$$

37. We have to solve the equation $x^2 + 6 = 5|x|$. First let us solve the equation $x^2 + 6 = 5x$. The solutions are $x_1 = 2$, $x_2 = 3$. Now, $(-x)^2 = x^2$ and $|-x| = |x|$, i.e., also $x_3 = -2$ and $x_4 = -4$ are solutions of the given equation.

38. $x + 3 > 0$, $x > -3$.

39. $5 - 2x \geqslant 0$, $2x \leqslant 5$, $x \leqslant \frac{5}{2}$.

40. $x^2 - 1 \neq 0$, $x \neq \pm 1$. The function is defined for every x except $x = \pm 1$.

41. Our problem reduces to that of solving the inequality $x^2 - 3x + 2 > 0$. To this end we shall sketch the graph of $y = x^2 - 3x + 2$. We are interested mainly in the points of intersection of the graph with the x axis:

$$y = 0,\ x^2 - 3x + 2 = 0; \qquad x_1 = 1,\ x_2 = 2.$$

The parabola $y = ax^2 + bx + c$ has a minimum if $a > 0$ and a maximum if $a < 0$. In our case it is a minimum, and we can sketch the graph as in Figure 73. The domain of x's for which $y > 0$ is $x < 1$ and $x > 2$. Hence this is the domain of definition of the given function.

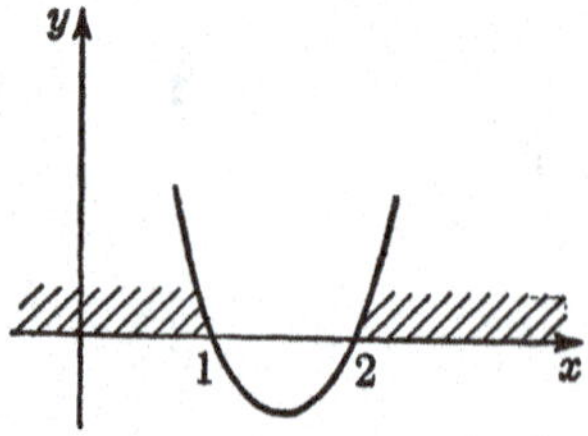

FIGURE 73

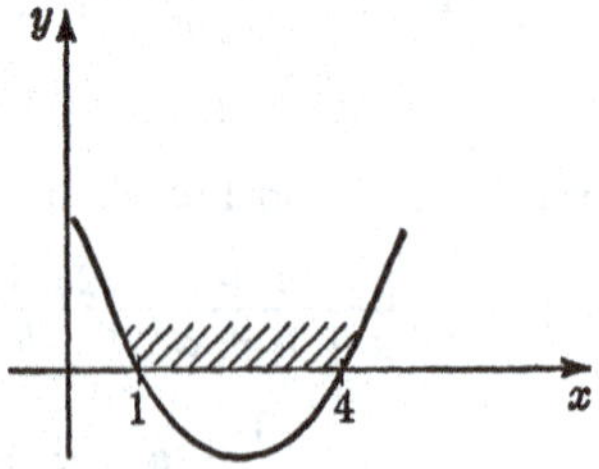

FIGURE 74

Remark. The domain of definition of $y = \sqrt{x^2 - 3x + 2}$ is $x \leqslant 1$, $x \geqslant 2$ because here we need $x^2 - 3x + 2 \geqslant 0$.

42. $\ln \dfrac{5x - x^2}{4} \geqslant 0 \Rightarrow \dfrac{5x - x^2}{4} \geqslant 1;$

$$5x - x^2 \geqslant 4,\ x^2 - 5x + 4 \leqslant 0,\ (x-1)(x-4) \leqslant 0;$$

$$y \leqslant 0; 1 \leqslant x \leqslant 4.$$ See Figure 74.

43. x must satisfy the following inequalities:

$$3 - x \geqslant 0, \ -1 \leqslant \frac{3-2x}{5} \leqslant 1; \qquad 3 - x \geqslant 0, \ x \leqslant 3;$$

$$-1 \leqslant \frac{3-2x}{5}, \ 2x \leqslant 8, \ x \leqslant 4; \qquad \frac{3-2x}{5} \leqslant 1, \ 2x \geqslant -2, \ x \geqslant -1.$$

The domain $-1 \leqslant x \leqslant 3$ satisfies all requirements.

44. $x \geqslant 0$; $x \neq 2$; $2x - 3 > 0$, $x > \frac{3}{2}$. The domain of definition is $x > \frac{3}{2}$, $x \neq 2$.

45. $\sin x \geqslant 0$, $2\pi n \leqslant x \leqslant 2\pi n + \pi$, $n = 0, \pm 1, \pm 2, \ldots$;
$16 - x^2 \geqslant 0$, $x^2 \leqslant 16$, $-4 \leqslant x \leqslant 4$.

Values of x satisfying both inequalities must lie in one of the intervals

$$-4 \leqslant x \leqslant -\pi, \qquad 0 \leqslant x \leqslant \pi.$$

46. The radical is defined for any x. We require

$$\frac{x-5}{x^2 - 10x + 24} > 0, \quad \text{i.e.,} \quad \frac{x-5}{(x-4)(x-6)} > 0.$$

We shall solve this inequality graphically (Fig. 75). Let us draw three number axes corresponding to the three linear expressions. On every axis

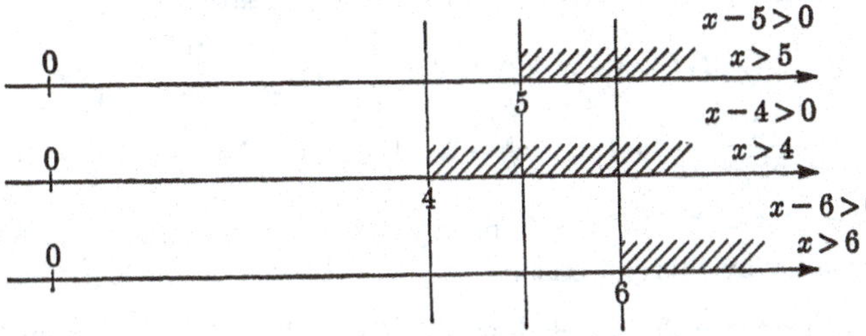

FIGURE 75

we indicate the interval in which the corresponding expression is positive. The whole expression is positive when the number of negative factors is even, i.e., in the intervals $4 < x < 5$ and $x > 6$.

47. x must satisfy the following requirements:

$$\frac{x-2}{x+2} \geqslant 0 \Rightarrow x < -2 \quad \text{or} \quad x \geqslant 2;$$

$$1 + x > 0 \Rightarrow x > -1;$$

$$1 - x \geqslant 0 \Rightarrow x \leqslant 1,$$

(by a generally accepted convention $\sqrt{a}$ is taken nonnegative, hence $\sqrt{1+x}$ is always positive). Let us draw the corresponding intervals on three axes (Fig. 76):

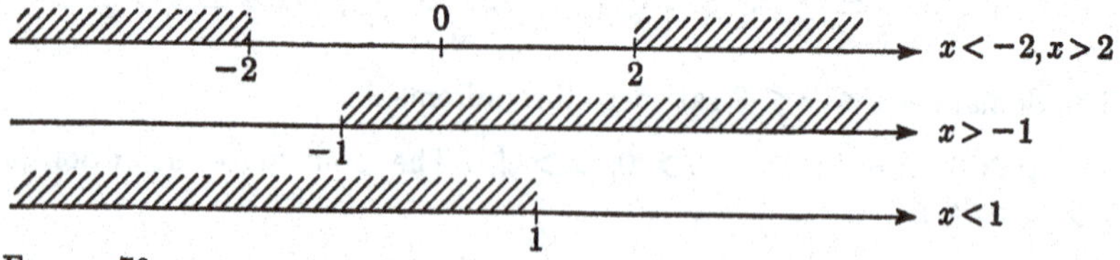

FIGURE 76

No x satisfies simultaneously the three requirements, and hence the function is not defined for any x.

48. We have to satisfy $\left|\dfrac{2}{2+\sin x}\right| \leqslant 1$. $2+\sin x \geqslant 0$ for any x, and thus

$$\left|\frac{2}{2+\sin x}\right| = \frac{2}{2+\sin x}.$$

$$\frac{2}{2+\sin x} \leqslant 1,\ 0 \leqslant \sin x; \qquad 2n\pi \leqslant x \leqslant (2n+1)\pi,\ n = 0, \pm 1, \ldots.$$

49. $x^2 - 5x + 16 = (x - \frac{5}{2})^2 + 16 - \frac{25}{4} = (x - \frac{5}{2})^2 + \frac{39}{4} > 0.$
This inequality holds for any x. Now we have to satisfy

$$1 - \log_{10}(x^2 - 5x + 16) > 0, \qquad \log_{10}\frac{10}{x^2 - 5x + 16} > 0,$$

$$\frac{10}{x^2 - 5x + 16} > 1, \qquad x^2 - 5x + 6 < 0, \quad \text{i.e.,} \quad 2 < x < 3.$$

Remark. $x^2 - 5x + 16 > 0$ for any x and it is permissible to multiply the inequality by this expression.

50. (a) $(-x)^4 - 2(-x)^2 + 1 = x^4 - 2x^2 + 1$; even.
(b) $(-x) - (-x)^2 = -x - x^2$; neither even nor odd.
(c) $\cos(-x) = \cos x$; even.
(d) $\sin(-x) = -\sin x$; odd.
(e) $2^{-x} = 1/2^x$; neither even nor odd.
(f) $-\dfrac{(-x)^3}{6} + \dfrac{(-x)^5}{120} = -\left(-\dfrac{x^3}{6} + \dfrac{x^5}{120}\right)$; odd.
(g) $2^{-(-x)^2} = 2^{-x^2}$; even.
(h) $\cosh(-x) = \dfrac{e^{-x} + e^x}{2} = \dfrac{e^x + e^{-x}}{2} = \cosh x$; even.

(k) $\sinh(-x) = \dfrac{e^{-x} - e^{x}}{2} = -\dfrac{e^{x} - e^{-x}}{2} = -\sinh x;$ odd.

(l) $\tanh(-x) = \dfrac{\sinh(-x)}{\cosh(-x)} = \dfrac{-\sinh x}{\cosh x} = -\tanh x;$ odd.

(m) $$\frac{-x \sin(-x)}{[(-x)^2 + 5] \tanh(-x)} = \frac{-x\cdot(-\sin x)}{(x^2 + 5)(-\tanh x)}$$

$$= -\frac{x \sin x}{(x^2 + 5)\tanh x} \qquad \text{odd.}$$

52. (a) $(ax^2 + c) + bx$ (b) $\cosh x + \sinh x$

(c) $A \cos kx \sin t + A \sin kx \cos t$

(d) $$\frac{\sqrt{1 - (x - a)^2} + \sqrt{1 - (x + a)^2}}{2} + \frac{\sqrt{1 - (x - a)^2} - \sqrt{1 - (x + a)^2}}{2}.$$

53. (a) We divide the x axis into parts according to the signs of the expressions which have to be taken in absolute value:

$x < -1;$ $y = 3 - x + 2x + 2 - 2x - x + 1 = -2x + 6.$

$-1 \leqslant x < 0;$ $y = 3 - x - 2x - 2 - 2x - x + 1 = -6x + 2.$

$0 \leqslant x < 3;$ $y = 3 - x - 2x - 2 + 2x - x + 1 = -2x + 2.$

$x \geqslant 3;$ $y = x - 3 - 2x - 2 + 2x - x + 1 = -4.$

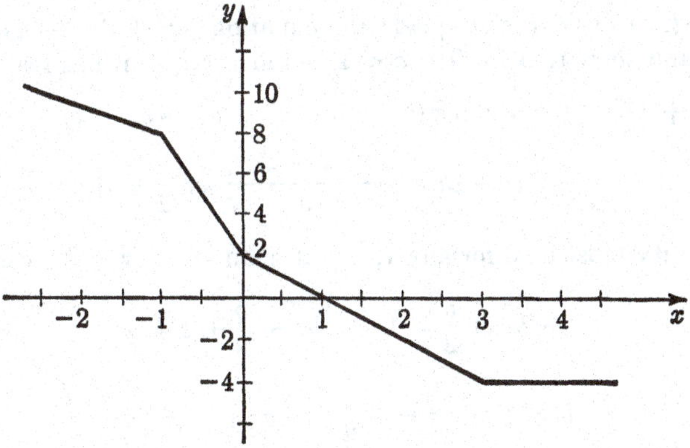

FIGURE 77

The graph is given in Figure 77. The function is not rational.

(b) $-3 \leqslant x < -2, y = 0;$ $-2 \leqslant x < 0, y = (x + 1)^2 - 1;$
$0 \leqslant x < 2, y = (x - 1)^2 - 1;$ $2 \leqslant x \leqslant 3, y = 0.$ Not rational.

54. $f(x + 1) = f[(x + 2) - 1] = 2(x + 2)^2 - 3(x + 2) + 1$
$= 2x^2 + 5x + 3.$

55. $$f(y) = \frac{\dfrac{x + 2}{3x - 1} + 2}{3\dfrac{x + 2}{3x - 1} - 1} = \frac{7x}{7} = x.$$

56. $$f\left(\frac{1}{x}\right) = \frac{\left(\dfrac{1}{x} - 1\right)^2}{\left(\dfrac{1}{x} + 1\right)^2} = \frac{(1 - x)^2}{(1 + x)^2} = f(x).$$

57. The inequality holds only when $f(x)$ and $g(x)$ are of different signs We thus have to solve the inequality $f(x)g(x) < 0$, i.e., $(x - 3)(4 - x) < 0$. By drawing the parabola (Fig. 78) we find $x < 3$, $x > 4$.

Remark. In general,

$$|x_1 + x_2 + \ldots + x_n| \leqslant |x_1| + |x_2| + \ldots + |x_n|$$

with equality holding only when all x_k ($k = 1, 2, \ldots n$) are of the same sign.

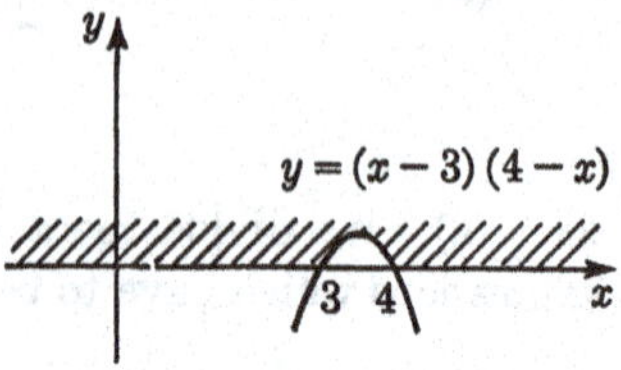

FIGURE 78

58. If one of the two numbers is $(a/2) + x$, the other is $(a/2) - x$. Their product, $(a^2/4) - x^2$, will be clearly maximal when $x^2 = 0$, i.e., $x = 0$. Hence each of the two numbers equals $a/2$.

59. *Hint.* Consider the quadratic equation $(x^2 + 4x + 3c)y = x^2 + 2x + c$, and show that for $0 < c < 1$ and arbitrary y it has real solutions.

61. $f(x + 1) - f(x) = a \cos (bx + b + c) - a(\cos bx + c)$

$$= -2a \sin \frac{2bx + b + 2c}{2} \sin \frac{b}{2} = \sin x.$$

The equality must hold for any x, i.e., it is an identity. We conclude

$$-2a \sin \frac{b}{2} = 1; \qquad bx + \frac{b}{2} + c = x.$$

Hence $$b = 1; \qquad c = -\frac{1}{2}; \qquad a = -\frac{1}{2 \sin \frac{1}{2}}.$$

63. Multiply both sides by $2 \sin \dfrac{x}{2}$:

$$2 \sin x \sin \frac{x}{2} + 2 \sin 2x \sin \frac{x}{2} + 2 \sin 3x \sin \frac{x}{2} + \ldots + 2 \sin nx \sin \frac{x}{2}$$

$$= \left(\cos\frac{x}{2} - \cos\frac{3x}{2}\right) + \left(\cos\frac{3x}{2} - \cos\frac{5x}{2}\right) + \left(\cos\frac{5x}{2} - \cos\frac{7x}{2}\right) + \dots$$

$$+ \left(\frac{\cos n - \frac{1}{2}}{2}x - \cos\frac{n + \frac{1}{2}}{2}x\right)$$

$$= \cos\frac{x}{2} - \cos\frac{n+\frac{1}{2}}{2}x = 2\sin\frac{n+1}{2}x\sin\frac{nx}{2}.$$

65. $\left(\frac{e^x + e^{-x}}{2}\right)^2 - \left(\frac{e^x - e^{-x}}{2}\right)^2 = \frac{1}{4}(e^{2x} + 2 + e^{-2x} - e^{2x} + 2 - e^{-2x}) = 1.$

66. $1 - \tanh^2 x = 1 - \frac{\sinh^2 x}{\cosh^2 x} = \frac{\cosh^2 x - \sinh^2 x}{\cosh^2 x} = \frac{1}{\cosh^2 x}.$

67. $\cosh x \pm \sinh x = \frac{e^x + e^{-x}}{2} \pm \frac{e^x - e^{-x}}{2} = e^{\pm x};$

$\cosh nx \pm \sinh nx = e^{\pm nx}$; $(e^{\pm x})^n = e^{\pm nx}$.

68. From the given equation there follows
$\sec t + \tan t = e^x$; $\sec t \pm \sqrt{\sec^2 t - 1} = e^x$;

$$\sec^2 t - 1 = e^{2x} - 2e^x \sec t + \sec^2 t;\ \sec t = \frac{e^{2x}+1}{2e^x} = \frac{e^x + e^{-x}}{2} = \cosh x.$$

69. $\frac{\sinh 2x}{\cosh 2x + 1} = \frac{2\sinh x\cosh x}{\cosh^2 x + \sinh^2 x + 1} = \frac{2\sinh x \cosh x}{2\cosh^2 x} = \tanh x.$

We used here the hyperbolic identities

$$\sinh 2x = 2\sinh x\cosh x, \qquad \cosh 2x = \cosh^2 x + \sinh^2 x,$$
$$1 + \sinh^2 x = \cosh^2 x.$$

70. $\frac{\cosh 2x + \cosh 4y}{\sinh 2x + \sinh 4y} = \frac{2\cosh(x+2y)\cosh(x-2y)}{2\sinh(x+2y)\cosh(x-2y)} = \coth(x+2y).$

We used $\sinh x \pm \sinh y = 2\sinh\frac{x \pm y}{2}\cosh\frac{x \mp y}{2}$;

$$\cosh x + \cosh y = 2\cosh\frac{x+y}{2}\cosh\frac{x-y}{2}.$$

These, as well as the following two identities, can be proved directly from the definition of the hyperbolic functions:

$$\cosh x - \cosh y = 2\sinh\frac{x+y}{2}\sinh\frac{x-y}{2};$$

$$\tanh x \pm \tan y = \frac{\sinh(x \pm y)}{\cosh x\cosh y}.$$

72. For $x \geqslant 0$, $x = \sqrt{1-y}$;
for $x < 0$, $x = -\sqrt{1-y}$.

73. $x^2 = y^3 - 1$; for $x \geqslant 0$, $x = \sqrt{y^3 - 1}$;

for $x < 0$, $x = -\sqrt{y^3 - 1}$.

74. $cxy - ay = ax - b$, $x = \dfrac{ay - b}{cy - a}$. In the form $y = f(x)$, $y = \dfrac{ax - b}{cx - a}$.

Here the given function and its inverse are identical.

75. We use the formula $(u + v)^3 = u^3 + v^3 + 3uv(u + v)$:

$$y^3 = x + \sqrt{1 + x^2} + x - \sqrt{1 + x^2} + 3\sqrt[3]{x^2 - 1 - x^2},\ y^3 = 2x - 3y;$$

$x = \dfrac{y^3 + 3y}{2}$ or, in the usual notation, $y = \dfrac{x^3 + 3x}{2}$. We remark that the given y is a solution of the cubic equation

$$y^3 + 3y - 2x = 0.$$

76. $x + 1 = \log_{10} y$; in the notation $y = f(x)$,

$$y = \log_{10} x - 1 = \log_{10} \frac{x}{10}.$$

77. $\ln (x + 2) = y - 1$; $x = e^{y-1} - 2$ or $y = e^{x-1} - 2$.

78. $x^y = 2$; $x = 2^{1/y}$ or $y = 2^{1/x}$.

79. $2^x(1 - y) = y$; $x = \log_2 \dfrac{y}{1 - y}$ or $y = \log_2 \dfrac{x}{1 - x}$.

80. *Answer.* $(LB - 2KC - 2AM)^2 - (L^2 - 4KM)(B^2 - 4AC) = 0.$

82. $\dfrac{1}{2}(y - 1) = \sin \dfrac{x - 1}{x + 1}$, $\dfrac{x - 1}{x + 1} = \arcsin \dfrac{y - 1}{2}$;

$$x = \frac{1 + \arcsin \dfrac{y - 1}{2}}{1 - \arcsin \dfrac{y - 1}{2}} \quad \text{or} \quad y = \frac{1 + \arcsin \dfrac{x - 1}{2}}{1 - \arcsin \dfrac{x - 1}{2}}.$$

83. Here $\quad -1 \leqslant x \leqslant 1; \quad 0 \leqslant \sqrt{1 - x^2} \leqslant 1;$

i.e., $0 \leqslant \arcsin \sqrt{1 - x^2} \leqslant \pi/2$ and $0 \leqslant 4 \arcsin \sqrt{1 - x^2} \leqslant 2\pi$. We now solve the given equation for x:

$$\sin \frac{y}{4} = \sqrt{1 - x^2}; \qquad x^2 = 1 - \sin^2 \frac{y}{4} = \cos^2 \frac{y}{4}; \qquad x = \pm\cos \frac{y}{4}.$$

We have $0 \leqslant y \leqslant 2\pi$, i.e., $0 \leqslant y/4 \leqslant \pi/2$ and $\cos (y/4) \geqslant 0$ in this interval. We conclude:

In the interval $x \geqslant 0$ the inverse function is $x = \cos (y/4)$;

In the interval $x < 0$ the inverse function is $x = -\cos (y/4)$.

85. Put $\alpha = \arcsin x$, i.e., $\sin\alpha = x$. Now $-\pi/2 \leqslant \alpha \leqslant \pi/2$ and $\cos\alpha = \sqrt{1-\sin^2\alpha} = \sqrt{1-x^2}$. Put $\beta = \arccos\sqrt{1-x^2}$. Then $\cos\alpha = \cos\beta$. If $0 \leqslant x \leqslant 1$, then $0 \leqslant \alpha \leqslant \pi/2$ and $0 \leqslant \beta \leqslant \pi/2$. $\cos\alpha = \cos\beta$ implies in this case $\alpha = \beta$. But if $-1 \leqslant x < 0$, then $-\pi/2 \leqslant \alpha < 0$ and $0 < \beta \leqslant \pi/2$. $\cos\alpha = \cos\beta$ implies in this case $\alpha = -\beta$.

86. *Hint.* See the solution of Problem 85.

87. Set $\arctan x = \alpha$, $\arctan y = \beta$, i.e.,

$$\tan\alpha = x, \qquad \tan\beta = y,$$

and

$$\tan(\alpha+\beta) = \frac{\tan\alpha+\tan\beta}{1-\tan\alpha\tan\beta} = \frac{x+y}{1-xy}.$$

Now set

$$\gamma = \arctan\frac{x+y}{1-xy}, \quad \text{i.e.,} \quad \tan\gamma = \frac{x+y}{1-xy} = \tan(\alpha+\beta).$$

This equality implies $\gamma = \pi n + (\alpha+\beta)$. γ must fulfill $-\pi/2 < \gamma < \pi/2$. If $xy < 0$ the signs of α and β are different. One of these numbers lies in interval $[0, \pi/2)$ and the second in $(-\pi/2, 0]$. Their sum is clearly in the interval $(-\pi/2, \pi/2)$. In this case $n = 0$ and $\gamma = \alpha + \beta$. Similar reasoning shows that the given equation is true in all other cases.

88. We have to show that

$$\arccos\left(\frac{x}{2} + \frac{1}{2}\sqrt{3-3x^2}\right) = \frac{\pi}{3} - \arccos x, \qquad \frac{1}{2} \leqslant x \leqslant 1.$$

Now $\cos\left(\frac{\pi}{3} - \arccos x\right) = \cos\frac{\pi}{3}\cos(\arccos x) + \sin\frac{\pi}{3}\sin(\arccos x)$.

$\cos(\arccos x) = x$; and if $\frac{1}{2} \leqslant x \leqslant 1$, $0 \leqslant \arccos x \leqslant \pi/3$, i.e., $\sin(\arccos x) = \sqrt{1-x^2}$; we conclude

$$\cos\left(\frac{\pi}{3} - \arccos x\right) = \frac{1}{2}x + \frac{1}{2}\sqrt{3}\sqrt{1-x^2},$$

and this is equal to the value of cosine on the left-hand side. If $\frac{1}{2} \leqslant x \leqslant 1$, both sides are within the interval $[0, \pi/2]$ and the equality is established.

89. For $x < 0$ the left side is positive, whereas the other side is negative.

91. $y = \operatorname{argtanh} x$; $\tanh y = x$; $x = \dfrac{e^y - e^{-y}}{e^y + e^{-y}} = \dfrac{e^{2y}-1}{e^{2y}+1}$;

$xe^{2y} + x = e^{2y} - 1$; $e^{2y} = \dfrac{1+x}{1-x}$;

$2y = \ln\dfrac{1+x}{1-x}$; $y = \dfrac{1}{2}\ln\dfrac{1+x}{1-x}$.

92. $x = \ln(\sec t + \tan t) = \ln(\sec t + \sqrt{\sec^2 t - 1}) = \operatorname{argcosh} \sec t$; $\sec t = \cosh x$.

93. $\tanh x = \sin t$;

$$x = \operatorname{argtanh} \sin t = \frac{1}{2} \ln \frac{1 + \sin t}{1 - \sin t}$$

$$= \frac{1}{2} \ln \frac{1 + \cos\left(\frac{\pi}{2} - t\right)}{1 - \cos\left(\frac{\pi}{2} - t\right)} = \frac{1}{2} \ln \frac{2\cos^2\left(\frac{\pi}{4} - \frac{t}{2}\right)}{2\sin^2\left(\frac{\pi}{4} - \frac{t}{2}\right)} = \frac{1}{2} \ln \cot^2\left(\frac{\pi}{4} - \frac{t}{2}\right)$$

$$= \frac{1}{2} \cdot 2 \ln \tan\left[\frac{\pi}{2} - \left(\frac{\pi}{4} - \frac{t}{2}\right)\right] = \ln \tan\left(\frac{\pi}{4} + \frac{t}{2}\right).$$

94. $y = \sqrt{z + 1} = \sqrt{\tan^2 x + 1} = |\sec x|$.

95. $u = \sqrt{1 + v^2} = \sqrt{1 + \ln^2 y} = \sqrt{1 + \ln^2 \sin x}$.
The function is defined only if $\sin x > 0$.

96. $y = u^3$; $u = \sin z$; $z = 2x + 1$.

97. $y = 5^u$; $u = z^2$; $z = 3x + 1$.

98. (a) $f\left[g\left(\frac{\pi}{12}\right)\right] = f\left(\sin\frac{\pi}{6}\right) = f\left(\frac{1}{2}\right) = \left(\frac{1}{2}\right)^3 - \frac{1}{2} = -\frac{3}{8}$.

(b) $f(0) = 0$ and $f\{f[f(0)]\} = 0$.

100. $f(x + T) = \sin(x + T)^2 = \sin(x^2 + 2Tx + T^2)$.
This can be equal to $\sin x^2$ only when $2Tx + T^2 = 2\pi n$. But here T depends on x and the function is not periodic.

102. $T_1 = \dfrac{2\pi}{\pi/3} = 6$; $T_2 = \dfrac{2\pi}{\pi/4} = 8$. The least period of y is 24.

103. π is the least period of $|\sin x|$ and also of $|\cos x|$. But

$$\left|\sin\left(x + \frac{\pi}{2}\right)\right| = |\cos x| \quad \text{and} \quad \left|\cos\left(x + \frac{\pi}{2}\right)\right| = |-\sin x| = |\sin x|.$$

Hence also $\pi/2$ is a period of the sum. Indeed,

$$\left|\sin\left(x + \frac{\pi}{2}\right)\right| + \left|\cos\left(x + \frac{\pi}{2}\right)\right| = |\cos x| + |\sin x|.$$

To show that $\pi/2$ is the least period, put $x = 0$ in $|\sin(x + T)| + |\cos(x + T)| = |\sin x| + |\cos x|$. Suppose $0 < T < \pi/2$. We have

$$|\sin T| + |\cos T| = \sin T + \cos T = 1.$$

$$\cos T + \sin T = \cos T - \cos(T + \pi/2) = 2\sin(T + \pi/4)\sin(\pi/4).$$

So
$$\sqrt{2}\sin\left(T+\frac{\pi}{4}\right)=1; \qquad T+\frac{\pi}{4}=\pi n+(-1)^n\frac{\pi}{4};$$
$$T=\pi n+[(-1)^n-1]\frac{\pi}{4};$$

and the least positive T satisfying this is $T=\frac{\pi}{2}$.

104. Here there are two possibilities,
$$f(x)=x^2\cos 2\pi x \quad \text{and} \quad f(x)=x^2+\sin \pi x.$$

110. *Answer.* $\delta < \epsilon/6$; $\delta < \sqrt{(x_0-1)^2+\epsilon}-|x_0-1|$.

112. This is of the form 0/0. Let us divide the numerator by $x-1$:
$$\begin{array}{l|l}
3x^4-4x^3+1 & x-1 \\ \cline{2-2}
\underline{3x^4-3x^3} & 3x^3-x^2-x-1 \\
\quad -x^3+1 & \\
\quad \underline{-x^3+x^2} & \\
\qquad -x^2+1 & \\
\qquad \underline{-x^2+x} & \\
\qquad\quad -x+1 & \\
\qquad\quad \underline{-x+1} &
\end{array}$$

Now
$$\lim_{x\to 1}\frac{3x^4-4x^3+1}{(x-1)^2}=\lim_{x\to 1}\frac{3x^3-x^2-x-1}{x-1}.$$

This is again an indeterminate form 0/0. Dividing again we obtain
$$\lim_{x\to 1}\frac{3x^3-x^2-x-1}{x-1}=\lim_{x\to 1}\frac{(x-1)(3x^2+2x+1)}{x-1}=\lim_{x\to 1}(3x^2+2x+1).$$

Finally
$$\lim_{x\to 1}\frac{3x^4-4x^3+1}{(x-1)^2}=\lim_{x\to 1}(3x^2+2x+1)=6.$$

We remark that the result can be obtained by dividing directly by $(x-1)^2$.

113. $$\lim_{x\to 1}\frac{x^m-1}{x^n-1}=\lim_{x\to 1}\frac{(x-1)(x^{m-1}+x^{m-2}+\ldots+1)}{(x-1)(x^{n-1}+x^{n-2}+\ldots+1)}=\frac{m}{n}.$$

115. Put $1+x=u^3$, i.e., $\lim\limits_{x\to 0} u=1$ and $x=u^3-1$.
$$\lim_{x\to 0}\frac{\sqrt[3]{1+x}-1}{x}=\lim_{u\to 1}\frac{u-1}{u^3-1}=\lim_{u\to 1}\frac{1}{1+u+u^2}=\frac{1}{3}.$$

116. Put $x=u^{15}$. Then $\lim\limits_{x\to -1} u=-1$.
$$\lim_{x\to -1}\frac{1+\sqrt[3]{x}}{1+\sqrt[5]{x}}=\lim_{u\to -1}\frac{1+u^5}{1+u^3}=\lim_{u\to -1}\frac{(1+u)(1-u+u^2-u^3+u^4)}{(1+u)(1-u+u^2)}=\frac{5}{3}.$$

117.

$$\lim_{x\to 2}\frac{\sqrt{2+x}-\sqrt{3x-2}}{\sqrt{4x+1}-\sqrt{5x-1}}$$

$$=\lim_{x\to 2}\frac{(\sqrt{2+x}-\sqrt{3x-2})(\sqrt{2+x}+\sqrt{3x-2})(\sqrt{4x+1}+\sqrt{5x-1})}{(\sqrt{4x+1}-\sqrt{5x-1})(\sqrt{4x+1}+\sqrt{5x-1})(\sqrt{2+x}+\sqrt{3x-2})}$$

$$=\lim_{x\to 2}\frac{(2+x-3x+2)(3+3)}{(4x+1-5x+1)(2+2)}=\lim_{x\to 2}\frac{4-2x}{2-x}\cdot\frac{3}{2}=\frac{3}{2}\lim_{x\to 2}\frac{2(2-x)}{2-x}=3.$$

118. *Answer.* $\frac{1}{2}$. (*Hint:* Multiply numerator and denominator by $\sqrt{1+x+x^2}+1$.)

119. $$\lim_{x\to 0}\frac{\sqrt[3]{1+3x}-\sqrt{1-2x}}{x+x^2}=\lim_{x\to 0}\frac{\sqrt[3]{(1+3x)^2}-(1-2x)}{(x+x^2)(\sqrt[3]{1+3x}+\sqrt{1-2x})}$$

$$=\frac{1}{2}\lim_{x\to 0}\frac{(1+3x)^2-(1-2x)^3}{(x+x^2)[\sqrt[3]{(1+3x)^4}+\sqrt[3]{(1+3x)^2}(1-2x)+(1-2x)^2]}$$

$$=\frac{1}{6}\lim_{x\to 0}\frac{8x^3-3x^2+12x}{x+x^2}=\frac{1}{6}\lim_{x\to 0}\frac{8x^2-3x+12}{x+1}=2.$$

122. Put $x=\pi+u$; $\lim_{x\to\pi}u=0$. Now

$$\lim_{x\to\pi}\frac{\sin mx}{\sin nx}=\lim_{u\to 0}\frac{\sin(m\pi+mu)}{\sin(n\pi+nu)}=\lim_{u\to 0}\frac{(-1)^m\sin mu}{(-1)^n\sin nu}$$

$$=(-1)^{m-n}\lim_{u\to 0}\frac{m\dfrac{\sin mu}{mu}}{n\dfrac{\sin nu}{nu}}=(-1)^{m-n}\frac{m}{n}.$$

123. $$\lim_{x\to 0}\frac{1-\cos x}{x^2}=\lim_{x\to 0}\frac{2\sin^2\dfrac{x}{2}}{x^2}=\lim_{x\to 0}\frac{2\left(\sin\dfrac{x}{2}\right)^2}{4\left(\dfrac{x}{2}\right)^2}=\frac{1}{2}.$$

124. $$\lim_{x\to 0}\frac{\tan x-\sin x}{x^3}=\lim_{x\to 0}\frac{\sin x}{x\cos x}\cdot\frac{1-\cos x}{x^2}=1\cdot\frac{1}{2}=\frac{1}{2}.$$

125. $$\lim_{x\to 0}\frac{\sin(a+x)+\sin(a-x)-2\sin a}{x^2}$$

$$=\lim_{x\to 0}\frac{2\sin a\cos x-2\sin a}{x^2}$$

$$=2\sin a\lim_{x\to 0}\frac{\cos x-1}{x^2}=-\sin a.$$

126. $\lim_{x\to\infty}(\sin\sqrt{x+1}-\sin\sqrt{x})$

$$= \lim_{x\to\infty} 2\sin\frac{\sqrt{x+1}-\sqrt{x}}{2}\cos\frac{\sqrt{x+1}+\sqrt{x}}{2}$$

$$= \lim_{x\to\infty} 2\sin\frac{1}{2(\sqrt{x+1}+\sqrt{x})}\cos\frac{\sqrt{x+1}+\sqrt{x}}{2} = 0;$$

Because $\lim_{u\to\infty}\sin(1/u) = 0$ and $|\cos u| \leqslant 1$.

127. $$\lim_{x\to 0}\frac{\cos x-\sqrt[3]{\cos x}}{\sin^2 x} = \lim_{x\to 0}\frac{\cos^3 x-\cos x}{\sin^2 x(\cos^2 x+\cos x\sqrt[3]{\cos x}+\sqrt[3]{\cos^2 x})}$$

$$= \frac{1}{3}\lim_{x\to 0}\frac{\cos^2 x-1}{\sin^2 x} = -\frac{1}{3}.$$

128. $$\lim_{x\to 0}\frac{\sqrt{1+x\sin x}-\cos x}{\sin^2\frac{x}{2}}$$

$$= \lim_{x\to 0}\frac{1+x\sin x-\cos^2 x}{\left(\sin^2\frac{x}{2}\right)(\sqrt{1+x\sin x}+\cos x)}$$

$$= \frac{1}{2}\lim_{x\to 0}\frac{\sin x\,(x+\sin x)}{\sin^2\frac{x}{2}} = \frac{1}{2}\lim_{x\to 0}\frac{2\sin\frac{x}{2}\cos\frac{x}{2}(\sin x+x)}{\sin^2\frac{x}{2}}$$

$$= \lim_{x\to 0}\left(\frac{\sin x}{\sin\frac{x}{2}}+\frac{x}{\sin\frac{x}{2}}\right) = \lim_{x\to 0}\left(2\cos\frac{x}{2}+2\frac{\frac{x}{2}}{\sin\frac{x}{2}}\right) = 2+2 = 4.$$

129. $$\lim_{x\to 0}\frac{1-\cos 2x+\tan^2 x}{x\sin x} = \lim_{x\to 0}\frac{2\sin^2 x+\tan^2 x}{x\sin x}$$

$$= \lim_{x\to 0}\frac{\sin x}{x}\left(2+\frac{1}{\cos^2 x}\right) = 3.$$

130. $$\lim_{x\to 0}\frac{1-\cos(1-\cos x)}{x^4}$$

$$= \lim_{x\to 0}\frac{2\sin^2\left(\frac{1-\cos x}{2}\right)}{x^4}$$

$$= \lim_{x\to 0}\frac{2\sin^2\left(\sin^2\frac{x}{2}\right)}{x^4} = 2\lim_{x\to 0}\left(\frac{\sin\sin^2\frac{x}{2}}{\sin^2\frac{x}{2}}\right)^2\cdot\frac{\sin^4\frac{x}{2}}{x^4}$$

$$= 2 \lim_{x\to 0} \left(\frac{\sin \frac{x}{2}}{\frac{x}{2}} \right)^4 \cdot \frac{1}{16} = \frac{1}{8}.$$

132. Here we shall use again the method of Problem 131.

$$\lim_{x\to 0} (1 + \tan x - 1) \cot x = \lim_{x\to 0} 1 = 1;$$

the limit is $e^1 = e$.

133. $\lim\limits_{m\to\infty} \left(\cos \frac{x}{m} - 1\right) m = \lim\limits_{m\to\infty} \frac{-2 \sin^2 (x/2m)}{1/m}.$

Put $1/2m = \alpha$; $\lim\limits_{m\to\infty} \alpha = 0$. We have

$$-2 \lim_{\alpha\to 0} \frac{\sin^2 \alpha x}{2\alpha} = -\lim_{\alpha\to 0} \frac{x \sin \alpha x}{\alpha x} \cdot \sin \alpha x = 0;$$

the limit is $e^0 = 1$.

134. $\lim\limits_{x\to\infty} \left(1 + \frac{1}{x^2} - 1\right) x = \lim\limits_{x\to\infty} \frac{1}{x} = 0;\ e^0 = 1.$

135. $\lim\limits_{x\to\pi/4} (\tan x - 1) \tan 2x = \lim\limits_{x\to\pi/4} \frac{2 \tan x}{1 - \tan^2 x} (\tan x - 1)$

$$= -2 \lim_{x\to\pi/4} \frac{\tan x}{1 + \tan x} = -1;$$

the limit is e^{-1}.

142. The function $f(x) = x^5 - 3x - 1$ is continuous at any x, and

$$f(1) = 1 - 3 - 1 = -3 < 0, \qquad f(2) = 32 - 6 - 1 = 25 > 0.$$

By continuity $f(x)$ must obtain the value 0 at least once between $x = 1$ and $x = 2$.

143. For $|x| > 1$ we obtain

$$f(x) = \lim_{n\to\infty} \frac{x^{2n} - 1}{x^{2n} + 1} = \lim_{n\to\infty} \frac{1 - \frac{1}{x^{2n}}}{1 + \frac{1}{x^{2n}}} = 1.$$

For $|x| < 1$, $f(x) = \lim\limits_{n\to\infty} \frac{x^{2n} - 1}{x^{2n} + 1} = \frac{-1}{1} = -1$

(because $\lim\limits_{m\to\infty} x^m = 0$ if $|x| < 1$). Now we find

$$f(1) = \frac{1 - 1}{1 + 1} = 0; \qquad f(-1) = \frac{1 - 1}{1 + 1} = 0;$$

$$\lim_{x\to 1+} f(x) = 1, \qquad \lim_{x\to 1-} f(x) = -1, \qquad f(1) = 0.$$

The function is not continuous at $x = 1$. The same holds at $x = -1$, but at all other points $f(x)$ is continuous. The graph of the function is sketched in Figure 79.

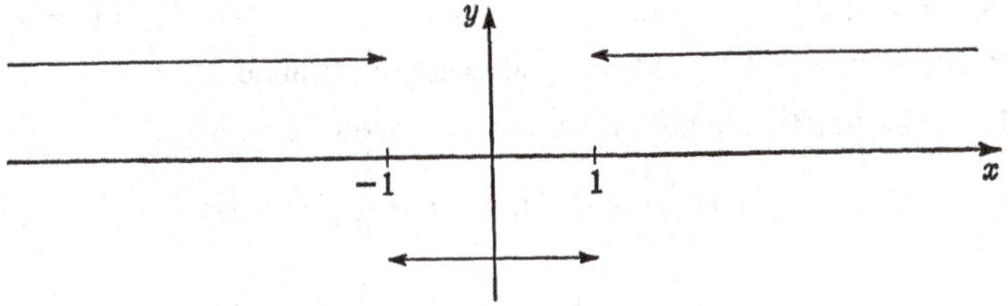

FIGURE 79

144. The function is continuous at any point except 0 by the theorem of continuity of the elementary functions. At $x = 0$ it is not defined. Moreover, $\lim_{x \to 0} \sin(1/x)$ does not exist ($1/x$ increases to infinity and $\sin(1/x)$ oscillates between -1 and $+1$). Thus we cannot extend the definition of the function to make it continuous at $x = 0$.

145. $\lim_{x \to 0} x \sin(1/x) = 0$ since $|\sin(1/x)| \leqslant 1$ and the first factor tends to zero). The function $f(x) = x \sin(1/x)$, $f(0) = 0$ is continuous everywhere.

146.
$$\lim_{x \to 0-} f(x) = \lim_{x \to 0-} e^{1/x} = 0 \qquad (1/x \text{ tends to } -\infty);$$
$$\lim_{x \to 0+} f(x) = \lim_{x \to 0+} e^{1/x} = \infty \qquad (1/x \text{ tends to } \infty).$$

Defining $f(0) = 0$ we can make the resulting function continuous from the left, but not from the right; i.e., the new function is still noncontinuous.

147. In the intervals $0 \leqslant x < 1$ and $1 < x \leqslant 2$ the function is clearly continuous. At $x = 1$ we have $f(1) = 3 - 1 = 2$. Now

$$\lim_{x \to 1-} f(x) = \lim_{x \to 1} 2x = 2; \quad \text{also,} \quad \lim_{x \to 1+} f(x) = 2.$$

Hence the above function is continuous everywhere in the given interval (see Fig. 80).

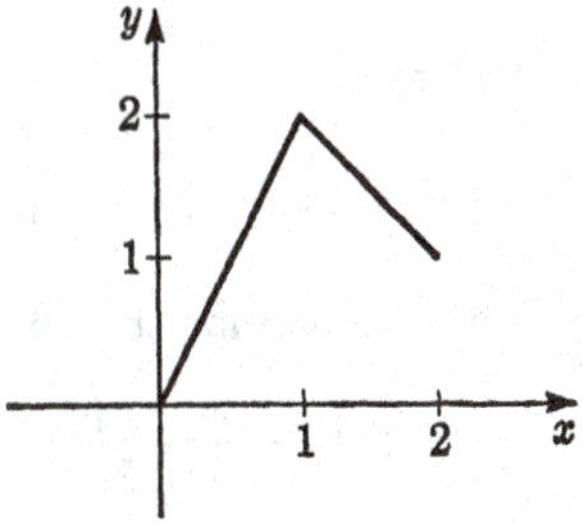

FIGURE 80

148. $\lim_{x\to 0} x \ln x^2 = \lim_{x\to 0} 2x \ln |x| = 0$ (cf. Prob. 139).
$a = 0$ makes the function continuous.

149. $x = \pm 2$. **150.** $x = \pm\sqrt{3}$. **151.** $x = 0, x = \pm 1$.

152. $x = 2, x = 4$. **153.** Everywhere continuous.

154. The function is not defined at the following points:

$$3x = \frac{\pi}{2}(2n+1), \qquad x = \frac{\pi}{6}(2n+1);$$

$$2x = \frac{\pi}{2}(2n+1), \qquad x = \frac{\pi}{4}(2n+1),$$

$$2x = \pi n, \qquad x = \frac{\pi}{2}n.$$

These are clearly points of discontinuity. At some of them the function can be made continuous by extending the definition; for example,

$$\lim_{x\to 0} \frac{\tan 3x}{\tan 2x} = \frac{3}{2} \quad \text{and we can define} \quad f(0) = \frac{3}{2};$$

$$\lim_{x\to \pi/4} \frac{\tan 3x}{\tan 2x} = 0 \quad \text{and we can define} \quad f\left(\frac{\pi}{4}\right) = 0, \text{ etc.}$$

155. $\sin x = 0$ at $x = \pi n$. $\lim_{x\to 0} \frac{x}{\sin x} = 1$, and for $n = 0$, i.e., $x = 0$, we can define $f(0) = 1$. The new function is not continuous at $x = \pi n, n \neq 0$.

156. At $x = 0$ the function is not defined.

157. $x^2(x-1) = 0$; $x = 0$, $x = 1$. At $x = 1$ the numerator vanishes too. Let us find

$$\lim_{x\to 1} \frac{\cos \frac{\pi x}{2}}{x^2(x-1)} = \lim_{x\to 1} \frac{\sin\left(\frac{\pi}{2} - \frac{\pi x}{2}\right)}{x-1} = -\lim_{(1-x)\to 0} \frac{\sin\left[\frac{\pi}{2}(1-x)\right]}{(1-x)} = -\frac{\pi}{2}.$$

We can define $f(1) = -\pi/2$ and the new function will be continuous at this point. The discontinuity at $x = 0$ cannot be removed.

158. $f(-1) = 6$; $\lim_{x\to -1} f(x) = 6$.

$$|f(x) - f(x_0)| = |x^2 - 5x - 6| = |x+1|\,|x-6| < \epsilon,$$

i.e., $|x+1| < \frac{\epsilon}{|x-6|}$. We can assume $|x-6| < 8$, and

$$|x+1| < \frac{\epsilon}{8} = \frac{1}{800};$$

$\delta = \frac{1}{800}$ is clearly not the largest possible.

159. Let n be any integer, then $f(n) = n - n = 0$. When $n < x < n + 1$ we have $f(x) = x - n$. The graph (Fig. 81) illustrates the behavior of the function. At integer values of x the function is continuous from the right.

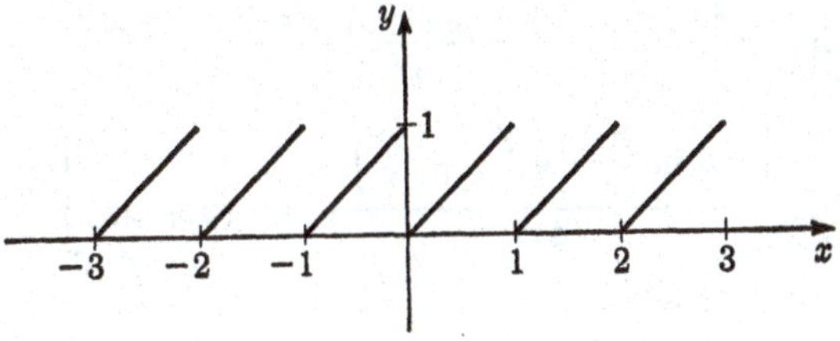

FIGURE 81

160. The graph of the function is given in Figure 82.

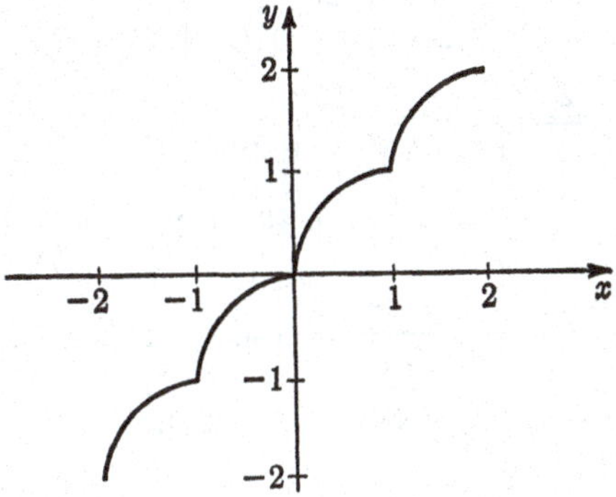

FIGURE 82

162. $y' = \lim_{\Delta x \to 0} \frac{a^{x+\Delta x} - a^x}{\Delta x} = a^x \lim_{\Delta x \to 0} \frac{a^{\Delta x} - 1}{\Delta x} = a^x \ln a.$

The limit was evaluated in Problem 138. For $a = e$, we have $\ln e = 1$, i.e., $(e^x)' = e^x$.

163. $y' = \lim_{\Delta x \to 0} \frac{\tan (x + \Delta x) - \tan x}{\Delta x}$

$$= \lim_{\Delta x \to 0} \frac{\sin \Delta x}{\Delta x \cos (x + \Delta x) \cos x} = \frac{1}{\cos^2 x}.$$

We used here the trigonometric identity

$$\tan \alpha - \tan \beta = \frac{\sin (\alpha - \beta)}{\cos \alpha \cos \beta}$$

and the known limit $\lim\limits_{\Delta x \to 0} (\sin \Delta x)/\Delta x = 1$.

164.
$$y' = \lim_{\Delta x \to 0} \frac{(x + \Delta x)\sin(x + \Delta x) - x \sin x}{\Delta x}$$
$$= \lim_{\Delta x \to 0} \frac{x[\sin(x + \Delta x) - \sin x] + \Delta x \sin(x + \Delta x)}{\Delta x}$$
$$= \lim_{\Delta x \to 0} \left[\frac{x\, 2 \sin \dfrac{\Delta x}{2} \cos\left(x + \dfrac{\Delta x}{2}\right)}{\Delta x} + \sin(x + \Delta x)\right]$$
$$= x \cos x + \sin x.$$

177.
$$y' = -12x^{-4} + 20x^3 + 35x^{-6} - 8ax^{-9}$$
$$= -\frac{12}{x^4} + 20x^3 + \frac{35}{x^6} - \frac{8a}{x^9}.$$

178.
$$y' = [(2x - 1)^{-1/3} + 5(x^2 + 2)^{-3/4}]'$$
$$= -\tfrac{1}{3}(2x - 1)^{-4/3} \cdot 2 + 5 \cdot (-\tfrac{3}{4}) \cdot (x^2 + 2)^{-7/4} \cdot 2x$$
$$= -\frac{2}{3\sqrt[3]{(2x - 1)^4}} - \frac{15x}{2\sqrt[4]{(x^2 + 2)^7}}.$$

179. $y = (1 + x^2)^{-1/3}$; $y' = -\dfrac{1}{3}(1 + x^2)^{-4/3} \cdot 2x = -\dfrac{2x}{3\sqrt[3]{(1 + x^2)^4}}$.

180. $y' = \left(\dfrac{e^x - e^{-x}}{2}\right)' = \dfrac{e^x + e^{-x}}{2} = \cosh x$. This is formula (14).

181. $y' = 3(\sin x + \cos 2x)^2(\cos x - 2 \sin 2x)$.

182. $y' = \dfrac{1}{\sqrt{1 - x}} \cdot \dfrac{1}{2\sqrt{x}} = \dfrac{1}{2\sqrt{x - x^2}}$.

183. $y' = \sinh \dfrac{1}{x} \cdot \left(-\dfrac{1}{x^2}\right) = -\dfrac{1}{x^2} \sinh \dfrac{1}{x}$.

184.
$$y' = \frac{1}{2\sqrt{x + 1}} - \frac{1}{x + \sqrt{x + 1}}\left(1 + \frac{1}{2\sqrt{x + 1}}\right)$$
$$= \frac{1}{2\sqrt{x + 1}} - \frac{1}{x + \sqrt{x + 1}} \cdot \frac{2\sqrt{x + 1} + 1}{2\sqrt{x + 1}}$$
$$= \frac{x + \sqrt{x + 1} - 2\sqrt{x + 1} - 1}{2\sqrt{x + 1}(x + \sqrt{x + 1})} = \frac{x - \sqrt{x + 1} - 1}{2\sqrt{x + 1}(x + \sqrt{x + 1})}.$$

185.
$$y' = \frac{1}{2\sqrt{\arctan\left(\sinh \dfrac{x}{3}\right)}} \cdot \frac{1}{1 + \sinh^2 \dfrac{x}{3}} \cdot \cosh \frac{x}{3} \cdot \frac{1}{3}$$

$$= \frac{1}{6\sqrt{\arctan\left(\sinh\frac{x}{3}\right)}} \cdot \frac{1}{\cosh^2\frac{x}{3}} \cdot \cosh\frac{x}{3}$$

$$= \frac{1}{6\cosh\frac{x}{3}\sqrt{\arctan\left(\sinh\frac{x}{3}\right)}}.$$

186. $$y' = -\frac{1}{x^2} + 2\left(-\frac{1}{2}\right)x^{-3/2} + 3\left(-\frac{1}{3}\right)x^{-4/3}$$

$$= -\frac{1}{x^2} - \frac{1}{x\sqrt{x}} - \frac{1}{x\sqrt[3]{x}}.$$

187. $$y' = \frac{1}{2}\left(a^2\frac{1}{\sqrt{1-(x^2/a^2)}} \cdot \frac{1}{a} + \sqrt{a^2-x^2} + x\frac{1}{2\sqrt{a^2-x^2}}(-2x)\right)$$

$$= \frac{1}{2}\left(\frac{a^2}{\sqrt{a^2-x^2}} + \sqrt{a^2-x^2} - \frac{x^2}{\sqrt{a^2-x^2}}\right) = \frac{1}{2}\frac{a^2 + a^2 - x^2 - x^2}{\sqrt{a^2-x^2}}$$

$$= \sqrt{a^2-x^2}.$$

188. $$y' = \frac{2x \cdot x\sqrt{1-x^2} - \left(\sqrt{1-x^2} + x\frac{1}{2\sqrt{1-x^2}}(-2x)\right)(1+x^2)}{x^2(1-x^2)}$$

$$= \frac{2x^2(1-x^2) - (1+x^2)(1-x^2-x^2)}{x^2(1-x^2)\sqrt{1-x^2}}$$

$$= \frac{2x^2 - 2x^4 - 1 + x^4 + x^2 + x^4}{x^2(1-x^2)\sqrt{1-x^2}} = \frac{3x^2-1}{x^2(1-x^2)\sqrt{1-x^2}}.$$

189. $$y' = \frac{1}{2}\frac{1}{1+x^2} + \frac{1}{4}\frac{x^2+1}{(x+1)^2}\left[\frac{2(x+1)(x^2+1) - (x+1)^2 \cdot 2x}{(x^2+1)^2}\right]$$

$$= \frac{1}{2(1+x^2)} + \frac{x^2+1-x^2-x}{2(x+1)(x^2+1)} = \frac{x+1+1-x}{2(1+x)(1+x^2)}$$

$$= \frac{1}{(1+x)(1+x^2)}.$$

190. $$y' = \frac{\frac{2x+1}{2\sqrt{x^2+x-1}} \cdot x - \sqrt{x^2+x-1}}{x^2}$$

$$+ \frac{1}{2}\frac{1}{\sqrt{1-\frac{(x-2)^2}{5x^2}}} \cdot \frac{x-x+2}{\sqrt{5}x^2}$$

$$= \frac{2x^2+x-2x^2-2x+2}{2x^2\sqrt{x^2+x-1}} + \frac{1}{x\sqrt{5x^2-x^2+4x-4}}$$

$$= \frac{2-x}{2x^2\sqrt{x^2+x-1}} + \frac{1}{2x\sqrt{x^2+x-1}}$$

$$= \frac{2-x+x}{2x^2\sqrt{x^2+x-1}} = \frac{1}{x^2\sqrt{x^2+x-1}}.$$

191.

$$y' = \frac{1}{\sqrt{a^2-b^2}} \cdot \frac{1}{\sqrt{1-\left(\frac{a\sin x+b}{a+b\sin x}\right)^2}}$$

$$\cdot \frac{a\cos x(a+b\sin x)-(a\sin x+b)b\cos x}{(a+b\sin x)^2}$$

$$= \frac{1}{\sqrt{a^2-b^2}}$$

$$\cdot \frac{a^2\cos x + ab\sin x\cos x - ab\sin x\cos x - b^2\cos x}{\sqrt{(a^2+2ab\sin x+b^2\sin^2 x-a^2\sin^2 x-2ab\sin x-b^2)}(a+b\sin x)}$$

$$= \frac{(a^2-b^2)\cos x}{\sqrt{a^2-b^2}\sqrt{(a^2-b^2)\cos^2 x}(a+b\sin x)} = \frac{1}{a+b\sin x}.$$

192. $y' = -1 + \frac{1}{2}\frac{1}{1+e^{2x}} \cdot e^{2x}.2 - \left(-e^{-x}\operatorname{arccot} e^x + e^{-x} \cdot \frac{-1}{1+e^{2x}} \cdot e^x\right)$

$$= -1 + \frac{e^{2x}}{1+e^{2x}} + \frac{1}{1+e^{2x}} + e^{-x}\operatorname{arccot} e^x$$

$$= \frac{-1-e^{2x}+e^{2x}+1}{1+e^{2x}} + e^{-x}\operatorname{arccot} e^x = \frac{\operatorname{arccot} e^x}{e^x}.$$

193. $y' = 3\sinh^2 x\cosh x.$

194. $y' = \sinh(\sinh x)\cosh x.$

195. $y' = \frac{1}{\cosh^2 \ln x} \cdot \frac{1}{x}.$

196. $y' = \frac{3}{4\sqrt[4]{1+\tanh^2 x}} \cdot \frac{2\tanh x}{\cosh^2 x} = \frac{3\tanh x}{2\cosh^2 x\sqrt[4]{1+\tanh^2 x}}$

197. $y' = \frac{1}{2\cosh^2 x} + \frac{\sqrt{2}}{8} \cdot \frac{1-\sqrt{2}\tanh x}{1+\sqrt{2}\tanh x}$

$$\cdot \frac{\sqrt{2}\frac{1}{\cosh^2 x}(1-\sqrt{2}\tanh x) - \left(-\sqrt{2}\frac{1}{\cosh^2 x}\right)(1+\sqrt{2}\tanh x)}{(1-\sqrt{2}\tanh x)^2}$$

$$= \frac{1}{2\cosh^2 x} + \frac{1}{4} \cdot \frac{2}{(1-2\tanh^2 x)\cosh^2 x}$$

$$= \frac{1 - 2\tanh^2 x + 1}{2\cosh^2 x\,(1 - 2\tanh^2 x)} = \frac{1 - \tanh^2 x}{\cosh^2 x\,(1 - 2\tanh^2 x)}$$

$$= \frac{\cosh^2 x - \sinh^2 x}{\cosh^2 x\,(\cosh^2 x - 2\sinh^2 x)} = \frac{1}{\cosh^2 x\,(1 - \sinh^2 x)}$$

$$= \frac{1}{(1 + \sinh^2 x)(1 - \sinh^2 x)} = \frac{1}{1 - \sinh^4 x}.$$

199. $y' = x^x\left(\frac{1}{x}x + 1\cdot\ln x\right) = x^x(1 + \ln x).$

200. $y' = x^{\sin x}\left(\frac{1}{x}\sin x + \cos x \ln x\right).$

201. $\ln y = x^x \ln x;$

$$\frac{y'}{y} = \ln x\cdot x^x(1 + \ln x) + x^x\cdot\frac{1}{x};$$

$$y' = x^{x^x}\cdot x^x\left[(1 + \ln x)\ln x + \frac{1}{x}\right].$$

Here we used the result of Problem 199.

202. $\ln y = \frac{2}{x}\ln(x + 1);$

$$\frac{y'}{y} = -\frac{2}{x^2}\ln(x + 1) + \frac{2}{x(x + 1)};$$

$$y' = \frac{2\sqrt[x]{(x + 1)^2}}{x}\left[\frac{1}{x + 1} - \frac{\ln(x + 1)}{x}\right].$$

203. This exercise too is easier to solve by logarithmic differentiation.

$$\ln y = 3\ln(x + 1) + \frac{1}{4}\ln(x - 2) - \frac{2}{5}\ln(x - 3);$$

$$\frac{y'}{y} = \frac{3}{x + 1} + \frac{1}{4(x - 2)} - \frac{2}{5}\cdot\frac{1}{x - 3} = \frac{57x^2 - 302x + 361}{20(x + 1)(x - 2)(x - 3)};$$

$$y' = \frac{(x + 1)^2\sqrt[4]{x - 2}}{\sqrt[5]{(x - 3)^2}}\cdot\frac{57x^2 - 302x + 361}{20(x - 2)(x - 3)}.$$

204. $\ln y = \frac{1}{2}\ln x + \frac{1}{2}\ln\sin x + \frac{1}{4}\ln(1 - e^x);$

$$\frac{y'}{y} = \frac{1}{2x} + \frac{\cos x}{2\sin x} + \frac{-e^x}{4(1 - e^x)};$$

$$y' = \frac{1}{2}\sqrt{x\sin x\sqrt{1 - e^x}}\left[\frac{1}{x} + \cot x - \frac{e^x}{2(1 - e^x)}\right].$$

205. We have to find y' and then substitute it in the given equation.

The statement that y is a solution means, of course, that an identity will then be obtained.

$$y' = x + \frac{1}{2}\sqrt{x^2+1} + \frac{1}{2}x \cdot \frac{2x}{2\sqrt{x^2+1}} + \frac{1}{2}\frac{1}{x+\sqrt{x^2+1}}\left(1 + \frac{2x}{2\sqrt{x^2+1}}\right)$$

$$= x + \frac{1}{2}\sqrt{x^2+1} + \frac{x^2}{2\sqrt{x^2+1}} + \frac{1}{2}\cdot\frac{1}{\sqrt{x^2+1}}$$

$$= x + \frac{1}{2}\sqrt{x^2+1} + \frac{x^2+1}{2\sqrt{x^2+1}} = x + \sqrt{x^2+1};$$

$$2y = x^2 + x\sqrt{x^2+1} + \ln(x + \sqrt{x^2+1}),$$

$$xy' + \ln y' = x^2 + x\sqrt{x^2+1} + \ln(x + \sqrt{x^2+1}).$$

The required identity is indeed obtained.

206. $\dfrac{du}{dv} = \dfrac{1}{2}\cdot\dfrac{1-v}{1+v}\cdot\dfrac{1-v+1+v}{(1-v)^2} = \dfrac{1}{1-v^2}$; let us express v as a function of u:

$$2u = \ln\frac{1+v}{1-v}; \qquad \frac{1+v}{1-v} = e^{2u};$$

$$v = \frac{e^{2u}-1}{e^{2u}+1} = \frac{e^u - e^{-u}}{e^u + e^{-u}} = \frac{\sinh u}{\cosh u} = \tanh u;$$

$$\frac{dv}{du} = \frac{1}{\cosh^2 u} = \frac{4}{e^{2u} + 2 + e^{-2u}} = \frac{4}{\dfrac{1+v}{1-v} + 2 + \dfrac{1-v}{1+v}}$$

$$= \frac{4(1-v^2)}{1 + 2v + v^2 + 2 - 2v^2 + 1 - 2v + v^2} = 1 - v^2;$$

then

$$\frac{du}{dv}\cdot\frac{dv}{du} = \frac{1}{1-v^2}\cdot(1-v^2) = 1.$$

207. By differentiation of the given identity we obtain

$$1 + 2x + 3x^2 + \ldots + nx^{n-1} = \frac{(n+1)x^n(x-1) - (x^{n+1}-1)}{(x-1)^2}$$

$$= \frac{nx^{n+1} - (n+1)x^n + 1}{(x-1)^2}.$$

208. By differentiating,

$$-\sin x - 3\sin 3x - \ldots - (2n-1)\sin(2n-1)x$$

$$= \frac{2n\cos 2nx \sin x - \sin 2nx \cos x}{2\sin^2 x}$$

$$= \frac{2n[\sin (2n+1)x - \sin (2n-1)x] - [\sin (2n+1)x + \sin (2n-1)x]}{4 \sin^2 x}$$

$$= \frac{(2n-1) \sin (2n+1)x - (2n+1) \sin (2n-1)x}{4 \sin^2 x}.$$

Now $\sin x + 3 \sin 3x + \ldots + (2n-1) \sin (2n-1)x$

$$= \frac{1}{4 \sin^2 x} [(2n+1) \sin (2n-1)x - (2n-1) \sin (2n+1)x].$$

209. $f'(x) = \frac{1}{2}\sqrt{\frac{x-1}{x+1}} \cdot \frac{x-1-x-1}{(x-1)^2} = -\sqrt{\frac{x-1}{x+1}} \cdot \frac{1}{(x-1)^2};$

$f'(2) = -\sqrt{\frac{1}{3}} \cdot \frac{1}{1} = -\frac{\sqrt{3}}{3}.$

210. $\frac{2x}{a^2} + \frac{2yy'}{b^2} = 0;\ y' = -\frac{x}{a^2} \cdot \frac{b^2}{y} = -\frac{b^2x}{a^2y}.$

211. $\frac{1}{2\sqrt{x}} + \frac{1}{2\sqrt{y}} y' = 0;\ y' = -\sqrt{\frac{y}{x}}.$

212. $3x^2 + 3y^2y' - 3ay - 3axy' = 0;\ y' = \frac{x^2 - ay}{ax - y^2}.$

213. $2yy' \cos x - y^2 \sin x = 3a^2 \cos 3x;$

$$y' = \frac{3a^2 \cos 3x + y^2 \sin x}{2y \cos x}.$$

214. $y \ln x = x \ln y;\ y' \ln x + \frac{y}{x} = \ln y + x\frac{y'}{y};$

$$y' = \frac{\ln y - \frac{y}{x}}{\ln x - \frac{x}{y}}.$$

215. $y' = e^y + xe^yy';\ y' = \frac{e^y}{1 - xe^y} = \frac{e^y}{1-(y-1)} = \frac{e^y}{2-y}.$

216. $y' = 1 + \frac{1}{1+y^2} y';\ y' = \frac{1+y^2}{y^2}.$

217. $\sin y + x \cos y \cdot y' + \sin y \cdot y' - 2 \sin 2y \cdot y' = 0;$

$$y' = \frac{-\sin y}{x \cos y + \sin y - 2 \sin 2y}.$$

218. $y + xy' - \frac{y'}{y} = 0;\ y' = \frac{y^2}{1 - xy}.$

After substitution,

$$y^2 + (xy - 1)\frac{y^2}{1 - xy} = y^2 - y^2 = 0.$$

219. $\frac{1}{2} \cdot \frac{2x + 2yy'}{x^2 + y^2} = \frac{1}{1 + (y^2/x^2)} \cdot \frac{y'x - y}{x^2}; \frac{x + yy'}{x^2 + y^2} = \frac{xy' - y}{x^2 + y^2};$
$x + yy' = xy' - y;$

$$y' = \frac{x + y}{x - y}.$$

220. $2(x - 1) + 2(y + 3)y' = 0.$ Substitute now $x = 2; y = 1$:

$$2(2 - 1) + 2(1 + 3)y' = 0; \qquad y' = \frac{-2}{8} = -\frac{1}{4}.$$

221. $\frac{dx}{dt} = -a \sin t, \frac{dy}{dt} = b \cos t;$

$$\frac{dy}{dx} = \frac{dy/dt}{dx/dt} = \frac{b \cos t}{-a \sin t} = -\frac{b}{a}\cot t.$$

222. $\frac{dx}{d\varphi} = a(1 - \cos \varphi), \frac{dy}{d\varphi} = a \sin \varphi;$

$$\frac{dy}{dx} = \frac{a \sin \varphi}{a(1 - \cos \varphi)} = \frac{2 \sin \frac{\varphi}{2} \cos \frac{\varphi}{2}}{2 \sin^2 \frac{\varphi}{2}} = \cot \frac{\varphi}{2}.$$

223. $\frac{dx}{dt} = -\frac{1}{t^2}, \frac{dy}{dt} = \frac{1}{t^2}; \frac{dy}{dx} = -1.$

224. $\frac{dx}{dt} = \frac{2t}{1 + t^2}, \frac{dy}{dt} = 1 - \frac{1}{1 + t^2} = \frac{t^2}{1 + t^2}; \frac{dy}{dx} = \frac{t^2}{2t} = \frac{t}{2}.$

225. $\frac{dx}{dt} = e^t \sin t + e^t \cos t; \frac{dy}{dt} = e^t \cos t - e^t \sin t;$

$$\frac{dy}{dx} = \frac{\cos t - \sin t}{\cos t + \sin t} = \frac{1 - \tan t}{1 + \tan t} = \tan\left(\frac{\pi}{4} - t\right).$$

226. $x = \frac{1}{t^3} + \frac{1}{t^2}; \frac{dx}{dt} = -\frac{3}{t^4} - \frac{2}{t^3} = -\frac{3 + 2t}{t^4};$

$$\frac{dy}{dt} = -\frac{3}{t^3} - \frac{2}{t^2} = -\frac{3 + 2t}{t^3}; \frac{dy}{dx} = t.$$

By substituting in the given equation,

$$\frac{1 + t}{t^3} \cdot t^3 = 1 + t; \qquad 1 + t = 1 + t.$$

227. $$\frac{dx}{dt} = -\frac{1}{2}\cdot\frac{2t}{(1+t^2)^{3/2}} - \frac{1}{1+\sqrt{1+t^2}}\cdot\frac{2t}{2(1+t^2)^{1/2}} + \frac{1}{t}$$
$$= -\frac{t}{(1+t^2)^{3/2}} - \frac{t^2 - \sqrt{1+t^2} - 1 - t^2}{t(1+\sqrt{1+t^2})\sqrt{1+t^2}}$$
$$= -\frac{t}{(1+t^2)^{3/2}} + \frac{1}{t\sqrt{1+t^2}} = \frac{-t^2+1+t^2}{t(1+t^2)^{3/2}} = \frac{1}{t(1+t^2)^{3/2}};$$

$$\frac{dy}{dt} = \frac{\sqrt{1+t^2} - \frac{t}{2}\cdot\frac{2t}{\sqrt{1+t^2}}}{1+t^2} = \frac{1+t^2-t^2}{(1+t^2)^{3/2}}\cdot = \frac{1}{(1+t^2)^{3/2}};$$

$$\frac{dy}{dx} = t.$$

After substitution,

$$\frac{t}{\sqrt{1+t^2}}\cdot\sqrt{1+t^2} = t, \qquad t = t.$$

228. First we find the value of t corresponding to the given point:

$$1 = 2\cos t; \qquad \cos t = \frac{1}{2}; \qquad t = \pm\frac{\pi}{3};$$

$$y = \sin t = \sin\left(-\frac{\pi}{3}\right) = -\frac{\sqrt{3}}{2};$$

the corresponding t is $t = -\pi/3$. Now we differentiate,

$$\frac{dx}{dt} = -2\sin t, \qquad \frac{dy}{dt} = \cos t; \qquad \frac{dy}{dx} = \frac{\cos t}{-2\sin t} = -\frac{1}{2}\cot t;$$

$$y'\left(1, -\frac{\sqrt{3}}{2}\right) = -\frac{1}{2}\cot\left(-\frac{\pi}{3}\right) = -\frac{1}{2}\left(-\frac{\sqrt{3}}{3}\right) = \frac{\sqrt{3}}{6}.$$

235. $y' = 3x^2 - 2x + 1;\ y'' = 6x - 2;\ y''' = 6;\ y^{iv} = 0.$

236. $y' = \dfrac{1}{x^2+1},\ y'' = -\dfrac{2x}{(x^2+1)^2}.$

237. $y' = \frac{3}{5}x^{-2/5},\ y'' = -\frac{6}{25}x^{-7/5},\ y''' = \frac{42}{125}x^{-12/5}.$

238.
$$\begin{aligned} y' &= 5x^4\ln x + x^4;\ y'' = 20x^3\ln x + 5x^3 + 4x^3 = 20x^3\ln x + 9x^3; \\ y''' &= 60x^2\ln x + 20x^2 + 27x^2 = 60x^2\ln x + 47x^2; \\ y^{iv} &= 120x\ln x + 60x + 94x = 120x\ln x + 154x; \\ y^{v} &= 120\ln x + 120 + 154 = 120\ln x + 274;\ y^{vi} = 120/x. \end{aligned}$$

239. $$y = \frac{x^3}{x-1} = \frac{x^3-1+1}{x-1} = x^2 + x + 1 + \frac{1}{x-1};$$

$$y' = 2x + 1 - \frac{1}{(x-1)^2};\ y'' = 2 + \frac{2}{(x-1)^3};\ y''' = -\frac{6}{(x-1)^4}.$$

240. $y' = 2x\,e^{2x} + 2x^2\,e^{2x} = 2e^{2x}(x + x^2);$
$y'' = 4e^{2x}(x + x^2) + 2e^{2x}(1 + 2x) = 2e^{2x}(2x^2 + 4x + 1);$
$y''' = 4e^{2x}(2x^2 + 4x + 1) + 2e^{2x}(4x + 4) = 4e^{2x}(2x^2 + 6x + 3);$
$y^{iv} = 8e^{2x}(2x^2 + 6x + 3) + 4e^{2x}(4x + 6) = 16e^{2x}(x^2 + 4x + 3).$

We shall see later that this exercise can be solved in a simpler way using a special formula derived by Leibniz.

241. $y' = 2x \arctan x + \dfrac{1 + x^2}{1 + x^2} = 2x \arctan x + 1;$

$y'' = 2 \arctan x + \dfrac{2x}{1 + x^2}.$

242. $y' = e^x \cos x + e^x \sin x;$
$y'' = -e^x \sin x + e^x \cos x + e^x \cos x + e^x \sin x = 2e^x \cos x.$

We substitute this in the given differential equation:

$$2e^x \cos x - 2e^x \cos x - 2e^x \sin x + 2e^x \sin x = 0.$$

243. $\dfrac{dy}{dt} = A\omega \cos (\omega t + \omega_0) - B\omega \sin (\omega t + \omega_0);$

$\dfrac{d^2y}{dt^2} = -A\omega^2 \sin (\omega t + \omega_0) - B\omega^2 \cos (\omega t + \omega_0).$

Substitution turns the differential equation into an identity.

244. $2b^2x + 2a^2yy' = 0;\ y' = -\dfrac{b^2x}{a^2y};$

$$y'' = -\frac{b^2}{a^2} \cdot \frac{y - xy'}{y^2} = -\frac{b^2}{a^2} \cdot \frac{y + x\dfrac{b^2x}{a^2y}}{y^2} = -\frac{b^2}{a^4y^3}(a^2y^2 + b^2x^2)$$

$$= -\frac{b^2 \cdot a^2b^2}{a^4y^3} = -\frac{b^4}{a^2y^3}.$$

245. Here it seems preferable to pass first to the inverse function,

$$x + y = \arctan y;$$

$1 + y' = y' \dfrac{1}{1 + y^2},$

$$y' = \frac{1}{\dfrac{1}{1 + y^2} - 1} = \frac{1 + y^2}{-y^2} = -\frac{1}{y^2} - 1;$$

$$y'' = \frac{2y'}{y^3} = \frac{2}{y^3}\left(-\frac{1}{y^2} - 1\right) = -\frac{2}{y^5} - \frac{2}{y^3};$$

$$y''' = \frac{10}{y^6} y' + \frac{6}{y^4} y' = -\frac{10 + 16y^2 + 6y^4}{y^8}.$$

246. We first take logarithms of both sides:

$$x + y = \ln x + \ln y;$$

$$1 + y' = \frac{1}{x} + \frac{y'}{y}, \qquad y' = \frac{(x-1)y}{(1-y)x}.$$

To find y'' we shall differentiate once more the equality $1 + y' = \dfrac{1}{x} + \dfrac{y'}{y}$:

$$y'' = -\frac{1}{x^2} + \frac{y''}{y} - \frac{y'^2}{y^2}, \qquad y''\left(1 - \frac{1}{y}\right) = -\frac{1}{x^2} - \frac{(x-1)^2}{(1-y)^2x^2};$$

$$y'' = \frac{(1-y)^2 + (x-1)^2}{(1-y)^2x^2} \cdot \frac{y}{1-y} = \frac{y[(x-1)^2 + (y-1)^2]}{x^2(1-y)^3}.$$

247. $e^y y' + xy' + y = 0.$ We differentiate once more:

$$e^y(y')^2 + e^y y'' + y' + xy'' + y' = 0.$$

We shall not express every one of the derivatives in terms of x and y, but rather substitute $x = 0$ directly first into the original expression to find $y(0)$ and then into the equation derived above to find $y'(0)$ and $y''(0)$.

$$e^y + xy = e, \qquad x = 0 \Rightarrow e^y = e, \qquad y = 1.$$

$$e^y y' + xy' + y = 0 \Rightarrow ey' + 1 = 0, \qquad y' = -1/e.$$

$$e^y(y')^2 + e^y y'' + 2y' + xy'' = 0 \Rightarrow e\frac{1}{e^2} + ey'' - \frac{2}{e} = 0, \qquad y'' = \frac{1}{e^2}.$$

248. (1,1) of course satisfies the given equation. Now

$$2x + 2y + 2xy' + 2yy' - 4 + 2y' = 0.$$

Dividing by 2,

$$x + y + xy' + yy' - 2 + y' = 0.$$

y' at (1,1) can be now found from this equation:

$$1 + 1 + y' + y' - 2 + y' = 0, \qquad y' = 0.$$

Now we differentiate once more:

$$1 + y' + y' + xy'' + y'^2 + yy'' + y'' = 0,$$

and substitute $x = 1$, $y = 1$, $y' = 0$:

$$1 + y'' + y'' + y'' = 0, \qquad y'' = -\tfrac{1}{3}.$$

By another differentiation,

$$2y'' + y'' + xy''' + 2y'y'' + y'y'' + yy''' + y''' = 0,$$

or

$$3y'' + xy''' + 3y'y'' + yy''' + y''' = 0.$$

To find y''' at (1, 1) we substitute $x = 1$, $y = 1$, $y' = 0$, $y'' = -\frac{1}{3}$:

$$-1 + y''' + y''' + y''' = 0, \qquad y''' = \tfrac{1}{3}.$$

249. $\frac{dx}{dt} = 2at, \frac{dy}{dt} = 3bt^2; \frac{dy}{dx} = \frac{3bt^2}{2at} = \frac{3b}{2a}t;$

$$\frac{d^2y}{dx^2} = \frac{d\left(\frac{dy}{dx}\right)}{dx} = \frac{\frac{d\left(\frac{dy}{dx}\right)}{dt}}{\frac{dx}{dt}} = \frac{\frac{3b}{2a}}{2at} = \frac{3b}{4a^2t}.$$

250. $\frac{dy}{dx} = \frac{a\cos t}{-a\sin t} = -\cot t; \frac{d^2y}{dx^2} = \frac{1/\sin^2 t}{-a\sin t} = -\frac{1}{a\sin^3 t};$

$$\frac{d^3y}{dx^3} = \frac{d(d^2y/dx^2)}{dx} = \frac{\frac{d(d^2y/dx^2)}{dt}}{\frac{dx}{dt}} = \frac{\frac{3\cos t}{a\sin^4 t}}{-a\sin t} = \frac{3\cos t}{a^2\sin^5 t}.$$

251. $\frac{dy}{dx} = \frac{a(\sin t + t\cos t)}{a(\cos t - t\sin t)} = \frac{\sin t + t\cos t}{\cos t - t\sin t};$

$$\frac{d^2y}{dx^2} = \frac{(\cos t + \cos t - t\sin t)(\cos t - t\sin t) - (\sin t + t\cos t)(-\sin t - \sin t - t\cos t)}{a(\cos t - t\sin t)^3}$$

$$= \frac{2 + t^2}{a(\cos t - t\sin t)^3}.$$

252. $\frac{dy}{dx} = \frac{e^t\cos t - e^t\sin t}{e^t\sin t + e^t\cos t} = \frac{\cos t - \sin t}{\sin t + \cos t} = \frac{1 - \tan t}{1 + \tan t} = \tan\left(\frac{\pi}{4} - t\right);$

$$\frac{d^2y}{dx^2} = \frac{-\frac{1}{\cos^2\left(\frac{\pi}{4} - t\right)}}{e^t(\sin t + \cos t)} = -\frac{1}{e^t\cos^2\left(\frac{\pi}{4} - t\right)\sqrt{2}\cos\left(\frac{\pi}{4} - t\right)}$$

$$= -\frac{1}{\sqrt{2}\,e^t\cos^3\left(\frac{\pi}{4} - t\right)}.$$

Now we substitute in the given equation:

$$-\frac{1}{\sqrt{2}e^t\cos^3\left(\frac{\pi}{4} - t\right)}e^{2t}(\sin t + \cos t)^2 = 2\left[e^t\sin t\tan\left(\frac{\pi}{4} - t\right) - e^t\cos t\right];$$

$$-\frac{2\cos^2\left(\frac{\pi}{4} - t\right)}{\sqrt{2}\cos^3\left(\frac{\pi}{4} - t\right)} = \frac{2}{\cos\left(\frac{\pi}{4} - t\right)}\left[\sin t\sin\left(\frac{\pi}{4} - t\right) - \cos t\cos\left(\frac{\pi}{4} - t\right)\right];$$

$$-\frac{1}{\sqrt{2}} = -\cos\left(\frac{\pi}{4} - t + t\right) = -\cos\frac{\pi}{4} = -\frac{1}{\sqrt{2}}.$$

254. $y = x^{1/2}$;

$$y^{(n)} = (x^{1/2})^{(n)} = \tfrac{1}{2}(-\tfrac{1}{2})(-\tfrac{3}{2}) \ldots (\tfrac{1}{2} - n + 1)x^{1/2-n}$$

$$= \frac{(-1)^{n-1}\,1\cdot 3\cdots(2n-3)}{2^n}\,x^{1/2-n} = \frac{(-1)^{n-1}(2n-3)!!\sqrt{x}}{2^n x^n}.$$

We used here the symbolic notation $(2n+1)!! = 1\cdot 3\cdot 5\cdots(2n+1)$. Analogously, $(2n)!! = 2\cdot 4\cdot 6\cdots 2n$.

255. $y^{(n)} = \rho(\rho-1)\cdots(\rho-n+1)a^n(ax+b)^{\rho-n}$.

This was obtained directly from the solution of Problem 253.

256. By Problem 255 we have

$$y^{(n)} = \frac{(-1)^n\,4\cdot 5\cdots(n+3)\cdot 2^n}{(2x+3)^{n+4}} = \frac{(-1)^n(n+3)!2^{n-1}}{3(2x+3)^{n+4}}.$$

257. By Problem 255 we have

$$y^{(n)} = \frac{(-1)^n n!(-1)^n}{(1-x)^{n+1}} = \frac{n!}{(1-x)^{n+1}}.$$

Here $a = -1$ and $a^n = (-1)^n$. Additionally, $(-1)^n\cdot(-1)^n = (-1)^{2n} = 1$.

258. The first derivative is $y' = \dfrac{a}{ax+b}$. Now by Problem 255,

$$y^{(n)} = [\ln(ax+b)]^{(n)} = \left(\frac{a}{ax+b}\right)^{(n-1)} = \frac{(-1)^{n-1}a^n(n-1)!}{(ax+b)^n} \qquad (0! = 1).$$

259. $y = \ln(x+1) - \ln(3-2x)$;

$$y^{(n)} = (n-1)![(-1)^{n-1}(x+1)^{-n} + 2^n(3-2x)^{-n}].$$

We used the solution of Problem 258.

260. It can be seen that

$$\frac{1}{x^2-1} = \frac{1}{2}\left(\frac{1}{x-1} - \frac{1}{x+1}\right).$$

This operation is called decomposition into partial fractions and will be discussed in more detail in connection with integration. Here we shall employ the method without further application:

$$\frac{1}{x^2-1} = \frac{A}{x-1} + \frac{B}{x+1}; \qquad A(x+1) + B(x-1) = 1;$$

$$\left.\begin{aligned} A + B &= 0 \\ A - B &= 1 \end{aligned}\right\} \Rightarrow A = \tfrac{1}{2}, \qquad B = -\tfrac{1}{2}.$$

We have thus $$\frac{1}{x^2-1} = \frac{1}{2}\left(\frac{1}{x-1} - \frac{1}{x+1}\right).$$

Using Problem 255 we obtain

$$\left(\frac{1}{x^2 - 1}\right)^{(n)} = \frac{1}{2} \cdot (-1)^n \cdot n! \left[\frac{1}{(x-1)^{n+1}} - \frac{1}{(x+1)^{n+1}}\right].$$

261. First we factor the denominator: $x^2 - 3x + 2 = (x - 1)(x - 2)$. Then decomposing into partial fractions,

$$\frac{1}{x^2 - 3x + 2} = \frac{A}{x - 1} + \frac{B}{x - 2}; \qquad A(x - 2) + B(x - 1) = 1.$$

$x = 2$, $B = 1$; $x = 1$, $-A = 1$, $A = -1$; i.e.,

$$y = \frac{1}{x - 2} - \frac{1}{x - 1} \quad \text{and} \quad y^{(n)} = (-1)^n n! \left[\frac{1}{(x-2)^{n+1}} - \frac{1}{(x-1)^{n+1}}\right].$$

262. By dividing the numerator by the denominator we obtain

$$y = -1 + \frac{2}{1 - x}.$$

Now by Problem 257, $\qquad y^{(n)} = \dfrac{2n!}{(1 - x)^{n+1}}.$

263. $y' = a^{px} p \ln a$; $y'' = a^{px} p^2 (\ln a)^2$.

By induction it is easy to show

$$y^{(n)} = a^{px} p^n (\ln a)^n.$$

In particular, $(e^x)^{(n)} = e^x$.

264.
$$y' = \cos x = \sin\left(x + \frac{\pi}{2}\right);$$

$$y'' = \cos\left(x + \frac{\pi}{2}\right) = \sin\left(x + \frac{\pi}{2} + \frac{\pi}{2}\right) = \sin\left(x + 2\frac{\pi}{2}\right);$$

$$\vdots$$

$$y^{(n)} = (\sin x)^{(n)} = \sin\left(x + n\frac{\pi}{2}\right).$$

Analogously we obtain $(\cos x)^{(n)} = \cos\left(x + n\dfrac{\pi}{2}\right)$.

265. As in Problem 264, $y^{(n)} = 3^n \cos\left(3x + n\dfrac{\pi}{2}\right)$.

266. $y' = 2 \sin x \cos x = \sin 2x$; $y^{(n)} = 2^{n-1} \sin\left[2x + (n - 1)\dfrac{\pi}{2}\right]$.

We used the solution of Problem 264.

267.
$$\begin{aligned} y' &= 3x^2 \ln x + x^2 = x^2(3 \ln x + 1); \\ y'' &= 2x(3 \ln x + 1) + 3x = x(6 \ln x + 5); \\ y''' &= 6 \ln x + 5 + 6 = 6 \ln x + 11; \; y^{iv} = 6/x. \end{aligned}$$

Using the solution of Problem 253 we obtain

$$(x^3 \ln x)^{(n)} = (6/x)^{(n-4)} = 6(-1)^{n-4}(n-4)!x^{3-n} = 6(-1)^n(n-4)!x^{3-n}.$$

This formula holds for $n \geqslant 4$. y', y'', y''' were found above.

268. $\sin^4 x + \cos^4 x = (\sin^2 x + \cos^2 x)^2 - 2\sin^2 x \cos^2 x = 1 - \frac{1}{2}\sin^2 2x$

$$= 1 - \tfrac{1}{4}(1 - \cos 4x) = \tfrac{3}{4} + \tfrac{1}{4}\cos 4x.$$

$$(\sin^4 x + \cos^4 x)^{(n)} = (\tfrac{3}{4} + \tfrac{1}{4}\cos 4x)^{(n)} = \tfrac{1}{4}\cdot 4^n \cos(4x + n(\pi/2))$$
$$= 4^{n-1}\cos(4x + n(\pi/2)).$$

269. First we divide the numerator by the denominator:

$$\begin{array}{l|l} x^4 + 1 & x^3 - x \\ \underline{x^4 - x^2} & x \\ \quad\;\; x^2 + 1 & \end{array} \qquad y = x + \frac{x^2+1}{x(x-1)(x+1)}.$$

Then decomposing into partial fractions,

$$\frac{x^2+1}{x(x-1)(x+1)} = \frac{A}{x} + \frac{B}{x-1} + \frac{C}{x+1},$$
$$A(x+1)(x-1) + Bx(x+1) + Cx(x-1) - x^2 + 1;$$

$$\begin{array}{lll} x = 0, & -A = 1, & A = -1; \\ x = 1, & 2B = 2, & B = 1; \\ x = -1, & 2C = 2, & C = 1. \end{array}$$

Resuming, $$y = x - \frac{1}{x} + \frac{1}{x+1} + \frac{1}{x-1}$$

and $$y^{(n)} = (-1)^n n!\left[\frac{1}{(x+1)^{n+1}} + \frac{1}{(x-1)^{n+1}} - \frac{1}{x^{n+1}}\right] \qquad (n > 1).$$

For $n = 1$ we have

$$y' = 1 + \frac{1}{x^2} - \frac{1}{(x+1)^2} - \frac{1}{(x-1)^2}.$$

270. $y' = ae^{ax}\sin bx + be^{ax}\cos bx = e^{ax}(a\sin bx + b\cos bx)$.
Let us bring the expression in the brackets into the form $A\sin(bx + \varphi)$ (see Problem 62):

$$a\sin bx + b\cos bx = \sqrt{a^2+b^2}\sin(bx + \varphi);$$

here $\tan\varphi = b/a$ and $\sin\varphi = b/\sqrt{a^2+b^2}$. We thus have

$$y' = e^{ax}\sqrt{a^2+b^2}\sin(bx+\varphi).$$

$$y'' = \sqrt{a^2+b^2}\,[ae^{ax}\sin(bx+\varphi) + be^{ax}\cos(bx+\varphi)]$$
$$= \sqrt{a^2+b^2}\,e^{ax}[a\sin(bx+\varphi) + b\cos(bx+\varphi)]$$
$$= (a^2+b^2)e^{ax}\sin(bx+2\varphi).$$

By induction we obtain

$$y^{(n)} = (\sqrt{a^2 + b^2})^n e^{ax} \sin (bx + n\varphi).$$

272. We put $u = \cos ax$; $v = x^2$.

$$y^{(30)} = a^{30} \cos\left(ax + 30 \cdot \frac{\pi}{2}\right) \cdot x^2 + 30a^{29} \cos\left(ax + 29 \cdot \frac{\pi}{2}\right) \cdot 2x$$

$$+ \frac{30 \cdot 29}{1 \cdot 2} a^{28} \cos\left(ax + 28 \cdot \frac{\pi}{2}\right) \cdot 2$$

$$= a^{28}[(870 - a^2x^2) \cos ax - 60\, ax \sin ax].$$

We used here the trigonometric identities

$\cos (ax + 30(\pi/2)) = \cos (ax + 14\pi + \pi) = \cos (ax + \pi) = -\cos ax$, etc.

273. $y^{iv} = e^x \sin x + 4e^x \sin\left(x + \frac{\pi}{2}\right) + 6e^x \sin (x + \pi)$

$$+ 4e^x \sin\left(x + \frac{3\pi}{2}\right) + e^x \sin (x + 2\pi)$$

$$= e^x(\sin x + 4 \cos x - 6 \sin x - 4 \cos x + \sin x) = -4e^x \sin x.$$

We can also use the solution of Problem 270, with $a = 1$, $b = 1$, $\varphi = \pi/4$.

$$y^{iv} = (\sqrt{2})^4 e^x \sin\left(x + 4 \cdot \frac{\pi}{4}\right) = 4e^x \sin (x + \pi) = -4e^x \sin x.$$

274. Put $\ln x = u$, $x^3 = v$.

$$y^{(n)} = \frac{(-1)^{n-1}(n-1)!}{x^n} x^3 + n \frac{(-1)^{n-2}(n-2)!}{x^{n-1}} \cdot 3x^2$$

$$+ \frac{n(n-1)}{1 \cdot 2} \cdot \frac{(-1)^{n-3}(n-3)!}{x^{n-2}} \cdot 6x$$

$$+ \frac{n(n-1)(n-2)}{1 \cdot 2 \cdot 3} \cdot \frac{(-1)^{n-4}(n-4)!}{x^{n-3}} \cdot 6$$

$$= \frac{(-1)^n}{x^{n-3}} [-(n-1)! + 3n(n-2)! - 3n(n-1)(n-3)!$$

$$+ n(n-1)(n-2)(n-4)!]$$

$$= (-1)^n(n-4)!x^{3-n}[-(n-1)(n-2)(n-3) + 3n(n-2)(n-3)$$

$$- 3n(n-1)(n-3) + n(n-1)(n-2)]$$

$$= (-1)^n(n-4)!x^{3-n} \cdot 6.$$

Compare the solution of the same exercise in Problem 267.

276. *Hint.* Show that the derivative is

$$\frac{\sqrt{a^2 + b^2}}{a + b \cos x} - \frac{\sqrt{a^2 + b^2}}{a + b \cos x} = 0.$$

277. By ordinary differentiation $f'(x) = (x \cos x - \sin x)/x^2$ provided $x \neq 0$. To prove $f'(0) = 0$ we write

$$f'(0) = \lim_{h \to 0} \frac{f(h) - f(0)}{h} = \lim_{h \to 0} \frac{\dfrac{\sin h}{h} - 1}{h} = \lim_{h \to 0} \frac{\sin h - h}{h^2}.$$

In a circle, center C, with unit radius, let a small central angle intercept an arc of length h $(h > 0)$. Perpendiculars to AC through B and A intercept lengths of $\sin h$ and $\tan h$, respectively. We have, from the figure, $\sin h < h < \tan h$. Then

$$0 < h - \sin h < \tan h - \sin h, \quad 0 < \frac{h - \sin h}{h^2} < \frac{\tan h - \sin h}{h^2}.$$

This last fraction can be written

$$\frac{1}{\cos h} \cdot \frac{h}{1} \cdot \frac{\sin h}{h} \cdot \frac{1 - \cos h}{h^2}$$

which has for its limit (as $h \to 0$)$1 \cdot 0 \cdot 1 \cdot \frac{1}{2} = 0$. Therefore

$$0 \leqslant \lim_{h \to 0} \frac{h - \sin h}{h^2} \leqslant 0$$

which proves $f'(0) = 0$. (We used the result of Problem 123.)

$f'(x)$ is evidently continuous everywhere except possibly at $x = 0$. To prove continuity at $x = 0$ we must show

$$\lim_{h \to 0} \frac{h \cos h - \sin h}{h^2} = 0,$$

but this follows from $\sin h < h < \tan h$, above. We have

$$\sin h \cos h < h \cos h < \sin h,$$
$$\sin h \cos h - \sin h < h \cos h - \sin h < 0,$$
$$\frac{\sin h \cos h - \sin h}{h^2} < \frac{h \cos h - \sin h}{h^2} < 0,$$

and note that the first fraction has for its limit

$$\lim_{h \to 0} \sin h \cdot \lim_{h \to 0} \frac{\cos h - 1}{h^2} = 0 \cdot (-\tfrac{1}{2}) = 0,$$

so that
$$0 \leqslant \lim_{h \to 0} \frac{h \cos h - \sin h}{h^2} \leqslant 0.$$

The same argument holds when $h < 0$ except for reversal of all inequality signs.

279. *Hint.* Consider $(\log y)'$.

280. *Hint.* Let $f(x) = (x - a)^m \cdot g(x)$, where $m \geqslant 1$ and $g(x)$ is not divisible by $(x - a)$. Then

$$f'(x) = m(x - a)^{m-1}g(x) + (x - a)^m g'(x)$$

which is divisible by $(x - a)^{m-1}$ but not by $(x - a)^m$.

281. $y' = \frac{x + y}{x - y}$; $y'' = 2\frac{x^2 + y^2}{(x - y)^3}$.

282. *Hint.* Show that $x^2y'' + xy' + m^2y = 0$, and differentiate this n times using Leibniz's formula.

284. $\frac{dy}{dx} = t$, $\frac{d^2y}{dx^2} = \frac{1}{f''(t)}$.

286. $y(-1) = -1 + 4 + 7 - 10 = 0$; $y(2) = 8 + 16 - 14 - 10 = 0$; $y' = 3x^2 + 8x - 7$.

We shall find the points for which $y' = 0$.

$$3x^2 + 8x - 7 = 0, \qquad x_{1,2} = \frac{-4 \pm \sqrt{16 + 21}}{3} \sim \frac{4 \pm 6}{3},$$

$$x_1 \sim \tfrac{10}{3} = 3\tfrac{1}{3}, \qquad x_2 \sim -\tfrac{2}{3}.$$

The second point is between -1 and 2, and Rolle's theorem is fulfilled.

287. Rolle's theorem does not apply. At $x = 0$ the function is not defined and certainly not continuous.

288. $f(0) = 0$; $f(x_0) = 0$. By Rolle's theorem there is a point c, $0 < c < x_0$, for which $f'(c) = 0$. The second equation is exactly $f'(x) = 0$, thus c is its solution.

289. $f(0) = f(1) = f(2) = f(3) = f(4) = 0$. By Rolle's theorem, between every two points at which $f(x) = 0$ there exists at least one point c with $f'(c) = 0$. Thus $f'(x) = 0$ has four solutions which lie in the intervals $(0, 1)$, $(1, 2)$, $(2, 3)$, $(3, 4)$. It cannot have more solutions because $f'(x)$ is a polynomial of degree 4.

290. $f'(x) = nx^{n-1} + p$; $\qquad nx^{n-1} + p = 0$, $x^{n-1} = -p/n$.

For an even n this equation has exactly one real solution ($n - 1$ is odd!) and by Rolle's theorem there can be at most two points for which $f(x) = 0$, since between every two such points there must be a c such that $f'(c) = 0$.

If n is odd, $n - 1$ is even and there can be no more than two real solutions of the equation $x^{n-1} = -p/n$. By Rolle's theorem there can be at most three points with $f(x) = 0$ in this case.

292. $f'(x) = 1/2\sqrt{x}$.

$$f(4) - f(1) = (4 - 1)f'(c);\ f(4) = \sqrt{4} = 2,\ f(1) = \sqrt{1} = 1;$$

$$2 - 1 = 3\frac{1}{2\sqrt{c}};\ \sqrt{c} = \frac{2}{3};\ c = \frac{9}{4}.$$

293. $b^3 - a^3 = 3(b - a)c^2,\ c = \sqrt{\dfrac{b^2 + ab + a^2}{3}}.$

294. $\dfrac{\pi}{4} - 0 = 1 \cdot \dfrac{1}{1 + c^2},\ c = \sqrt{\dfrac{4}{\pi} - 1}.$

295. $\dfrac{\pi}{2} - 0 = 1 \dfrac{1}{\sqrt{1 - c^2}},\ c = \sqrt{1 - \dfrac{4}{\pi^2}}.$

296. $\ln 2 - \ln 1 = 1 \cdot \dfrac{1}{c},\ c = \dfrac{1}{\ln 2}.$

297. Cauchy's formula is $\dfrac{f(b) - f(a)}{g(b) - g(a)} = \dfrac{f'(c)}{g'(c)}.$ In our case,

$$\frac{1 - 0}{0 - 1} = \frac{\cos c}{-\sin c}; \qquad -\tan c = -1; \qquad c = \frac{\pi}{4}.$$

298. $\dfrac{4^2 - 1^2}{\sqrt{4} - \sqrt{1}} = \dfrac{2c}{1/2\sqrt{c}},\ 15 = 4c\sqrt{c},\ c = \sqrt[3]{\tfrac{225}{16}} \approx 2.4.$

299. Suppose $f(x) = x^n$. We shall write for this function Lagrange's formula in the interval $[b, a]$:

$$a^n - b^n = (a - b)nc^{n-1}, \qquad b < c < a$$

All numbers a, b, and c are positive; consequently,

$$b^{n-1} < c^{n-1} < a^{n-1}.$$

From this, $\quad a^n - b^n = (a - b)nc^{n-1} > (a - b)nb^{n-1},$
and $\quad a^n - b^n = (a - b)nc^{n-1} < (a - b)na^{n-1}.$

Both inequalities are thus proved.

300. Consider the function $f(x) = \ln x$. By Lagrange's formula,

$$\ln a - \ln b = (a - b)\frac{1}{c}, \qquad \ln\frac{a}{b} = \frac{a - b}{c}.$$

But $b < c < a$, and the given inequalities follow directly.

301. Consider $f(x) = \log_{10} x$. $f'(x) = \dfrac{\log_{10} e}{x}$, and

$$\log_{10} 11 \approx \log_{10} 10 + \frac{\log_{10} e}{10.5} \cdot 1 = 1 + \frac{0.43429}{10.5} \approx 1.0414.$$

302. Consider the function $f(x) = e^x$ in the interval $[0,x]$. Lagrange's formula gives

$$e^x - e^0 = (x - 0)e^c, \qquad 0 < c < x.$$

Thus $e^x - 1 = xe^c$. But $e^c > e^0 = 1$, i.e., $xe^c > x$, whence $e^x - 1 = xe^c > x$. This is the required inequality, $e^x > 1 + x$.

306. $f(4) = 256 - 5\cdot 64 + 16 - 3\cdot 4 + 4 = -56,$
$f'(4) = 4\cdot 4^3 - 15\cdot 4^2 + 2\cdot 4 - 3 = 21,$
$f''(4) = 4\cdot 3\cdot 4^2 - 15\cdot 2\cdot 4 + 2 = 74,$
$f'''(4) = 4\cdot 3\cdot 2\cdot 4 - 15\cdot 2 = 66,$
$f^{iv}(4) = 4\cdot 3\cdot 2 = 24;$

$$f(x) = f(4) + f'(4)(x-4) + \frac{f''(4)}{2!}(x-4)^2 + \frac{f'''(4)}{3!}(x-4)^3 + \frac{f^{iv}(4)}{4!}(x-4)^4$$
$$= -56 + 21(x-4) + 37(x-4)^2 + 11(x-4)^3 + (x-4)^4.$$

307. $f(0) = 1;$
$f'(x) = 3(x^2 - 3x + 1)^2(2x - 3),\ f'(0) = -9;$
$f''(x) = 6(x^2 - 3x + 1)(2x - 3)^2 + 6(x^2 - 3x + 1)^2,\ f''(0) = 60.$

In the same way,

$$f'''(0) = -270; \quad f^{iv}(0) = 720; \quad f^{v}(0) = -1080; \quad f^{vi}(0) = 720.$$

We obtain

$$f(x) = 1 - 9x + \tfrac{60}{2}x^2 - \tfrac{270}{6}x^3 + \tfrac{720}{24}x^4 - \tfrac{1080}{120}x^5 + \tfrac{720}{720}x^6$$
$$= 1 - 9x + 30x^2 - 45x^3 + 30x^4 - 9x^5 + x^6.$$

Use of the binomial theorem will lead to a quicker solution in this case.

308. $f(x) \approx 1 + 60(x - 1) + 2570(x - 1)^2;$
$f(1.005) \approx 1 + 60\cdot 0.005 + 2570\cdot 0.005^2 \approx 1.364.$

310. In Problem 309 we found

$$R_{n+1} = \frac{e^{\theta x}}{(n+1)!}x^{n+1}.$$

The greatest possible error will occur, in the present case, when $x = \frac{1}{2}$ and $\theta = 1$. Here $n = 3$:

$$R_4 < \frac{e^{1/2}}{4!}\left(\frac{1}{2}\right)^4 < \frac{\sqrt{3}}{24\cdot 16} < 0.01.$$

311. We found in Problem 264,

$$f^{(n)}(x) = \sin\left(x + \frac{n\pi}{2}\right), \quad f^{(n)}(0) = \sin\frac{n\pi}{2}.$$

For an even n, $f^{(n)}(0) = 0$. For $n = 2k - 1$, $\sin\dfrac{(2k-1)\pi}{2} = (-1)^{k-1}$.

The expansion is:

$$\sin x = x - \frac{x^3}{3!} + \frac{x^5}{5!} - \ldots + (-1)^{k-1}\frac{x^{2k-1}}{(2k-1)!} + R_{2k+1}.$$

We use the fact that the $2k^{\text{th}}$ term is 0 and thus the remainder is the $(2k+1)^{\text{th}}$ term:

$$R_{2k+1} = \frac{\sin[\theta x + (2k+1)(\pi/2)]}{(2k+1)!} x^{2k+1} = (-1)^k \cos\theta x \frac{x^{2k+1}}{(2k+1)!}.$$

As $|\cos\alpha| \leqslant 1$,

$$|R_{2k+1}| \leqslant \frac{|x|^{2k+1}}{(2k+1)!}.$$

For small values of x we obtain the approximate formulas

$$\sin x \sim x; \qquad \sin x \sim x - \frac{x^3}{6}; \qquad \sin x \sim x - \frac{x^3}{6} + \frac{x^5}{210}.$$

Additional terms improve the approximation.

313. $\cos bx = 1 - \dfrac{(bx)^2}{2!} + \dfrac{(bx)^4}{4!} - \ldots + (-1)^k \dfrac{(bx)^{2k}}{(2k)!} + R_{2k+2};$

$$R_{2k+2} = (-1)^{k+1} \cos\theta bx \frac{(bx)^{2k+2}}{(2k+2)!}; \; |R_{2k+2}| \leqslant \frac{|bx|^{2k+2}}{(2k+2)!}.$$

For $b = 1$ we obtain

$$\cos x = 1 - \frac{x^2}{2!} + \frac{x^4}{4!} - \ldots + (-1)^k \frac{x^{2k}}{(2k)!} + R_{2k+2}.$$

On assuming, e.g., $\cos x \sim 1 - \dfrac{x^2}{2}$, the error is not greater than $\dfrac{x^4}{24}$.

315. $\dfrac{x^2 \cdot 2}{2!} - \dfrac{x^4 \cdot 2^3}{4!} + \dfrac{x^6 \cdot 2^5}{6!} - \ldots.$

316. $f(0) = \ln 1 = 0$; as in Problem 258, $f^{(k)}(x) = \dfrac{(-1)^{k-1}(k-1)!}{(1+x)^k}$;

$f^{(k)}(0) = (-1)^{k-1}(k-1)!$.

The expansion is

$$\ln(1+x) = x - \frac{x^2}{2} + \frac{x^3}{3} - \ldots + (-1)^{n-1} \frac{x^n}{n} + R_{n+1}.$$

The remainder term in this case will be dealt with in Chapter XI.

317. As in Problem 255,

$$f^{(k)}(x) = m(m-1)\cdots(m-k+1)(1+x)^{m-k};$$

$$f(0) = 1, \qquad f^{(k)}(0) = m(m-1)\cdots(m-k+1).$$

This gives the expansion

$$(1+x)^m = 1 + mx + \frac{m(m-1)}{1\cdot 2} x^2 + \ldots$$

$$+ \frac{m(m-1)\cdots(m-n+1)}{n!} x^n + R_{n+1}.$$

The estimation of R_{n+1} will be postponed to Chapter XI.

We remark that the above expansion is analogous to Newton's binomial formula; moreover, for natural m we get here Newton's expansion.

Approximate formulas for $-1 < x < 1$ obtained from the above expansion will be given:

$$\frac{1}{1+x} = 1 - x + x^2 - x^3 + x^4 - \dots;$$

$$\frac{1}{(1+x)^2} = 1 - 2x + 3x^2 - \dots;$$

$$\sqrt{1+x} = 1 + \tfrac{1}{2}x - \tfrac{1}{8}x^2 + \dots.$$

For x very small in comparison with 1, these formulas give a good approximation with a few terms.

318. $y = 2 - (x-2) + (x-2)^2 - (x-2)^3 + \dots.$

320. $$\cosh x = 1 + \frac{x^2}{2!} + \frac{x^4}{4!} + \dots + \frac{x^{2n}}{(2n)!} + \frac{x^{2n+2}}{(2n+2)!}\cosh \theta x, \quad 0 < \theta < 1.$$

$$\sinh x = x + \frac{x^3}{3!} + \frac{x^5}{5!} + \dots + \frac{x^{2n-1}}{(2n-1)!} + \frac{x^{2n+1}}{(2n+1)!}\cosh \theta x.$$

321. $$f(x) = x^2 + \frac{x^3}{1!} + \frac{x^4}{2!} + \dots + \frac{x^{n+1}}{(n-1)!} + \frac{x^{n+2}}{n!}e^{\theta x}.$$

322. $$f'(x) = \frac{1}{\cos^2 x}, \; f''(x) = \frac{2\sin x}{\cos^3 x},$$

$$f'''(x) = 2\,\frac{1 + 2\sin^2 x}{\cos^4 x}, \; f^{iv}(x) = 8\sin x\,\frac{2 + \sin^2 x}{\cos^5 x}, \dots.$$

$$f(0) = 0;\; f'(0) = 1;\; f''(0) = 0;\; f'''(0) = 2;\; f^{iv}(0) = 0.$$

Hence the first terms of the expansion are

$$\tan x = x + \frac{x^3}{3} + \dots.$$

In this case we have not found the explicit form of the general term of the expansion.

It should be remarked that the same result can be obtained by dividing the expansion of $\sin x$ by that of $\cos x$:

$$\begin{array}{l|l} x - \dfrac{x^3}{3!} + \dfrac{x^5}{5!} - \dfrac{x^7}{7!} + \dots & 1 - \dfrac{x^2}{2!} + \dfrac{x^4}{4!} - \dfrac{x^6}{6!} + \dots \\ \cline{2-2} x - \dfrac{x^3}{2!} + \dfrac{x^5}{4!} - \dfrac{x^7}{6!} & x + \dfrac{x^3}{3} + \dfrac{2}{15}x^5 \dots \\ \hline \dfrac{x^3}{3} - \dfrac{x^5}{30} + \dfrac{6x^7}{7!} & \end{array}$$

$$\frac{\dfrac{x^3}{3} - \dfrac{x^5}{6} + \dfrac{x^7}{72}}{\dfrac{2x^5}{15} \cdots}$$

More details about this method are found in Chapter XI.

323. $f'(x) = -\sin\sin x \cos x; f''(x) = -\cos\sin x \cos^2 x + \sin\sin x \sin x;$

$$f'''(x) = \sin\sin x \cos^3 x + 3\cos\sin x \cos x \sin x + \sin\sin x \cos x;$$

$$f^{iv}(x) = \cos\sin x \cos^4 x - 6\sin\sin x \cos^2 x \sin x - 3\cos\sin x \sin^2 x + 4\cos\sin x \cos^2 x - \sin\sin x \sin x;$$

$$f(0) = 1; f'(0) = 0; f''(0) = -1; f'''(0) = 0; f^{iv}(0) = 5;$$

$$\cos\sin x = 1 - \tfrac{1}{2}x^2 + \tfrac{5}{24}x^4 + \cdots.$$

A technique which is often more efficient, especially in cases similar to those dealt with in this problem, will be introduced in Chapter XI.

326. $\lim\limits_{x\to 0} \dfrac{\sin x - x}{\sinh x} = \lim\limits_{x\to 0} \dfrac{\cos x - 1}{\cosh x} = 0.$

327. $$\lim_{x\to 0} \frac{\sqrt{1+x} - \sqrt[3]{1-x}}{2x} = \lim_{x\to 0} \frac{\dfrac{1}{2\sqrt{1+x}} + \dfrac{1}{3\sqrt[3]{(1-x)^2}}}{2} = \frac{\frac{1}{2} + \frac{1}{3}}{2} = \frac{5}{12}.$$

328. $$\lim_{x\to 0} \frac{e^{\sin x} - e^x}{\sin x - x} = \lim_{x\to 0} \frac{e^{\sin x}\cos x - e^x}{\cos x - 1}$$

$$= \lim_{x\to 0} \frac{e^{\sin x}\cos^2 x - e^{\sin x}\sin x - e^x}{-\sin x}$$

$$= \lim_{x\to 0} \frac{e^{\sin x}\cos^3 x - 3e^{\sin x}\sin x \cos x - e^{\sin x}\cos x - e^x}{-\cos x}$$

$$= 1.$$

Here we applied l'Hôpital's rule three times.

329. $\lim\limits_{x\to\pi} \dfrac{\sin 3x}{\tan 5x} = \lim\limits_{x\to\pi} \dfrac{3\cos 3x}{5(1/\cos^2 5x)} = -\dfrac{3}{5}.$

330. $$\lim_{x\to 0} \frac{\ln(1-x) + x^2}{(1+x)^m - 1 + x^2} = \lim_{x\to 0} \frac{-\dfrac{1}{1-x} + 2x}{m(1+x)^{m-1} + 2x} = -\frac{1}{m}.$$

331. $\lim\limits_{x\to\infty} \dfrac{x^\alpha}{a^x} = \lim\limits_{x\to\infty} \dfrac{\alpha x^{\alpha-1}}{a^x \ln a}.$

If $\alpha < 1$, the limit is directly seen to be zero. If $\alpha > 1$ we proceed to differentiate until a negative power of x appears in the numerator. In the denominator there remains always a^x, i.e., the limit is zero also in this case.

332. $$\lim_{x\to\infty} \frac{\ln x}{x^\alpha} = \lim_{x\to\infty} \frac{1}{x\alpha x^{\alpha-1}} = \frac{1}{\alpha} \lim_{x\to\infty} \frac{1}{x^\alpha} = 0.$$

333. $$\lim_{x\to\pi/4} \frac{\sin x - \cos x}{1 - \tan^2 x} = \lim_{x\to\pi/4} \frac{\cos x + \sin x}{\dfrac{-2\tan x}{\cos^2 x}} = \frac{\dfrac{\sqrt{2}}{2} + \dfrac{\sqrt{2}}{2}}{\dfrac{-2}{\frac{1}{2}}} = -\frac{\sqrt{2}}{4}.$$

334. $$\lim_{h\to 0} \frac{f(x+h) - 2f(x) + f(x-h)}{h^2} = \lim_{h\to 0} \frac{f'(x+h) - f'(x-h)}{2h}$$
$$= \lim_{h\to 0} \frac{f''(x+h) + f''(x-h)}{2}$$
$$= f''(x).$$

The same limit can also be evaluated directly by using the definition of the derivative.

335. $$\lim_{x\to 0} \frac{\tan ax - a\sin x}{x(1 - \cos ax)} = \lim_{x\to 0} \frac{\dfrac{a}{\cos^2 ax} - a\cos x}{1 - \cos ax + ax\sin ax}$$
$$= \lim_{x\to 0} \frac{\dfrac{2a^2\sin ax}{\cos^3 ax} + a\sin x}{2a\sin ax + a^2x\cos ax}$$
$$= \lim_{x\to 0} \frac{2\left(\dfrac{a^3\cos ax}{\cos^3 ax} + \dfrac{3a^3\sin^2 ax}{\cos^4 ax}\right) + a\cos x}{3a^2\cos ax - a^2x\sin ax}$$
$$= \frac{2a^3 + a}{3a^2} = \frac{2a^2 + 1}{3a}.$$

336. $$\lim_{x\to 0} \frac{\pi/x}{\cot(\pi x/2)} = \pi \lim_{x\to 0} \frac{\sin(\pi x/2)}{\cos(\pi x/2)\cdot x} = \pi \lim_{x\to 0} \frac{\sin(\pi x/2)}{x}$$
$$= \pi \lim_{x\to 0} \frac{\pi}{2} \cos\frac{\pi x}{2} = \frac{\pi^2}{2}.$$

337. $$\lim_{x\to 0} \frac{\ln(\sin ax)}{\ln \sin x} = \lim_{x\to 0} \frac{a\dfrac{\cos ax}{\sin ax}}{\dfrac{\cos x}{\sin x}} = a \lim_{x\to 0} \frac{\sin x}{\sin ax}$$
$$= a \lim_{x\to 0} \frac{\cos x}{a\cos ax} = 1.$$

338. $\lim_{x\to 0} x^\alpha \ln x = \lim_{x\to 0} \frac{\ln x}{x^{-\alpha}} = \lim_{x\to 0} \frac{1/x}{-\alpha x^{-\alpha-1}} = \lim_{x\to 0} \frac{x^\alpha}{-\alpha} = 0.$

339. $$\lim_{x\to 1-} \ln x \ln (1-x) = \lim_{x\to 1-} \frac{\ln (1-x)}{1/\ln x} = \lim_{x\to 1-} \frac{-1/(1-x)}{-1/(x \ln^2 x)}$$

$$= \lim_{x\to 1} \frac{x \ln^2 x}{1-x} = \lim_{x\to 1} \frac{\ln^2 x}{1-x} = \lim_{x\to 1} \frac{2 \ln x}{-x} = 0.$$

340. This is an indeterminate form $\infty - \infty$. We proceed as follows:

$$\lim_{x\to 1} \left(\frac{x}{x-1} - \frac{1}{\ln x}\right) = \lim_{x\to 1} \frac{x \ln x - x + 1}{(x-1) \ln x} = \lim_{x\to 1} \frac{\ln x + 1 - 1}{\frac{x-1}{x} + \ln x}$$

$$= \lim_{x\to 1} \frac{\ln x}{1 - \frac{1}{x} + \ln x} = \lim_{x\to 1} \frac{\frac{1}{x}}{\frac{1}{x^2} + \frac{1}{x}} = \frac{1}{2}.$$

341. $$\lim_{x\to 0} \left(\frac{1}{x} - \frac{1}{\cosh x - 1}\right) = \lim_{x\to 0} \frac{\cosh x - 1 - x}{x(\cosh x - 1)}$$

$$= \lim_{x\to 0} \frac{\sinh x - 1}{\cosh x - 1 + x \sinh x} = -\infty.$$

342. $$\lim_{x\to\infty} \left[x - x^2 \ln \left(1 + \frac{1}{x}\right)\right] = \lim_{x\to\infty} x \left[1 - x \ln \left(1 + \frac{1}{x}\right)\right].$$

We first evaluate the limit of the second term inside the brackets, to see if the whole expression is an indeterminate form.

$$\lim_{x\to\infty} x \ln \left(1 + \frac{1}{x}\right) = \lim_{x\to\infty} \ln \left(1 + \frac{1}{x}\right)^x = \ln e = 1.$$

The expression in the brackets thus tends to 0, i.e., the given expression is of the form $\infty \cdot 0$. We proceed as follows:

$$\lim_{x\to\infty} \frac{1 - x \ln \left(1 + \frac{1}{x}\right)}{1/x} = \lim_{x\to\infty} \frac{1 - x \ln (1 + x) + x \ln x}{1/x}$$

$$= \lim_{x\to\infty} \frac{-\ln (1 + x) - \frac{x}{1 + x} + \ln x + 1}{-1/x^2}$$

$$= \lim_{x\to\infty} \frac{-\frac{1}{1 + x} - \frac{1 + x - x}{(1 + x)^2} + \frac{1}{x}}{2/x^3}$$

$$= \lim_{x\to\infty} \frac{x^3(-x - x^2 - x + 1 + 2x + x^2)}{2(1 + x)^2 x} = \lim_{x\to\infty} \frac{x^2}{2(1 + x)^2} = \frac{1}{2}.$$

343. This is an indeterminate form 1^∞. We put

$$y = x^{\frac{1}{x-1}}, \qquad \ln y = \frac{\ln x}{1-x};$$

$$\lim_{x\to1} \ln y = \lim_{x\to1} \frac{1/x}{-1} = -1; \qquad \lim_{x\to1} y = e^{-1} = \frac{1}{e}.$$

344. This is an indeterminate form ∞^0.

$$y = (\tan x)^{2x-\pi}, \qquad \ln y = (2x-\pi)\ln\tan x.$$

$$\lim_{x\to\pi/2} \ln y = \lim_{x\to\pi/2} \frac{\ln\tan x}{1/(2x-\pi)} = \lim_{x\to\pi/2} \frac{\dfrac{1}{\tan x}\cdot\dfrac{1}{\cos^2 x}}{-2/(2x-\pi)^2} = -\lim_{x\to\pi/2} \frac{(2x-\pi)^2}{2\sin x\cos x}$$

$$= -\lim_{x\to\pi/2} \frac{(2x-\pi)^2}{\sin 2x} = -\lim_{x\to\pi/2} \frac{2\cdot 2(2x-\pi)}{2\cos 2x} = 0;$$

$$\lim_{x\to\pi/2} y = e^0 = 1.$$

345. This is an indeterminate form 0^0.

$$y = x^{\frac{1}{\ln(e^x-1)}}, \qquad \ln y = \frac{\ln x}{\ln(e^x-1)};$$

$$\lim_{x\to0} \ln y = \lim_{x\to0} \frac{1/x}{\dfrac{e^x}{e^x-1}} = \lim_{x\to0} \frac{e^x-1}{xe^x} = \lim_{x\to0} \frac{e^x}{e^x+xe^x} = 1;$$

$$\lim_{x\to0} y = e^1 = e.$$

346. $y = \left(\dfrac{2}{\pi}\arccos x\right)^{1/x}$, $\ln y = \dfrac{\ln\dfrac{2}{\pi} + \ln\arccos x}{x}$;

$$\lim_{x\to0} \ln y = \lim_{x\to0} \frac{\dfrac{-1}{\arccos x\cdot\sqrt{1-x^2}}}{1} = -\frac{2}{\pi}; \qquad \lim_{x\to0} y = e^{-2/\pi}.$$

347. 1.

348. 1.

349. $e^{2/\pi}$.

350. e.

351. See Figure 83.

$$S_1 = \frac{1}{2}R^2\alpha - \frac{1}{2}R^2\sin\alpha;$$

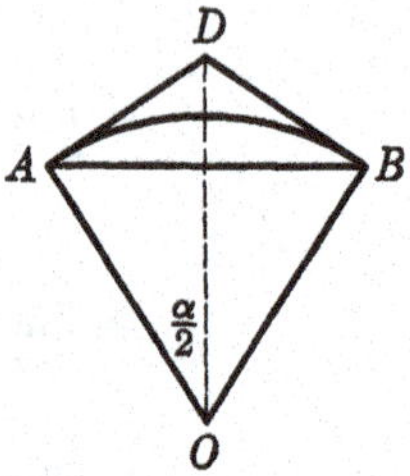

FIGURE 83

$$S_2 = 2 \cdot \frac{1}{2} AO \cdot OD \sin\frac{\alpha}{2} - \frac{1}{2} R^2\alpha = \frac{R^2 \sin\frac{\alpha}{2}}{\cos\frac{\alpha}{2}} - \frac{1}{2} R^2\alpha;$$

$$\frac{S_1}{S_2} = \frac{\frac{1}{2}(\alpha - \sin\alpha)}{\tan\frac{\alpha}{2} - \frac{1}{2}\alpha};$$

$$\lim_{\alpha\to 0} \frac{S_1}{S_2} = \frac{1}{2} \lim_{\alpha\to 0} \frac{1 - \cos\alpha}{\frac{1}{2} \cdot \frac{1}{\cos^2(\alpha/2)} - \frac{1}{2}}$$

$$= \lim_{\alpha\to 0} \frac{1 - \cos\alpha}{1 - \cos^2\frac{\alpha}{2}} = \lim_{\alpha\to 0} \frac{\sin\alpha}{2\sin\frac{\alpha}{2}\cos\frac{\alpha}{2} \cdot \frac{1}{2}} = 2.$$

352. $S(4) = 64 + \frac{3}{4} = 64.75$, $S(6) = 216 + \frac{1}{2} = 216.5$;

$$v_{\text{av}} = \frac{S(6) - S(4)}{6 - 4} = 75.875.$$

$$v_{t=4} = \frac{ds}{dt}(t = 4) = \left(3t^2 - \frac{3}{t^2}\right)_{t=4} = 3 \cdot 16 - \frac{3}{16} = 47\tfrac{13}{16}.$$

353. (a) $\gamma_{AN(\text{av})} = 8\ g/2$ cm $= 4$ g/cm. (γ is density.)

(b) Let $x = AM$. Then $m_{AM} = kx^2$, with k a coefficient of proportionality. We find k from the given mass of AN:

$$8\ g = k \cdot 2^2\ \text{cm}^2, \qquad k = 2\ \text{g/cm}^2.$$

Now

$$m_{AB} = k \cdot L^2 = 2\ \text{g/cm}^2 \cdot 400\ \text{cm}^2 = 800\ \text{g},$$
$$\gamma_{AB(\text{av})} = 800\ \text{g}/20\ \text{cm} = 40\ \text{g/cm}.$$

(c) The density at $M(x)$ is given by

$$\gamma_M = \lim_{\Delta x\to 0} \frac{m(x + \Delta x) - m(x)}{\Delta x} = [m(x)]' = (kx^2)' = 2kx = 4x\ \text{g/cm}.$$

354. Denote the specific heat by c.

$$c = \frac{dQ}{dt} = 1 + 0.00004\,t + 0.0000009\,t^2.$$

For $t = 100$°C. we obtain

$$c = 1 + 0.004 + 0.009 = 1.013.$$

355. y and x are functions of the time t. Differentiating both sides of the given equation with respect to t,

$$2y\frac{dy}{dt} = 12\frac{dx}{dt},$$

$$\frac{dy}{dt} = \frac{6}{y}\frac{dx}{dt} = \frac{6}{\sqrt{12x}}\frac{dx}{dt} = \frac{6}{\sqrt{12 \cdot 3}} \cdot 2 = 2\,\frac{\text{units}}{\text{sec}}.$$

356. $32x\frac{dx}{dt} + 18y\frac{dy}{dt} = 0$. We require $\frac{dy}{dt} = -\frac{dx}{dt}$, i.e.,

$$32x = 18y \quad \text{or} \quad 16x = 9y.$$

Using the equation of the ellipse we find for the points in question

$$(3, \tfrac{16}{3}) \quad \text{and} \quad (-3, -\tfrac{16}{3}).$$

357. $V = \frac{4}{3}\pi r^3, \qquad \frac{dV}{dt} = 4\pi r^2 \frac{dr}{dt} = 4\pi r^2 v.$

$S = 4\pi r^2, \qquad \frac{dS}{dt} = 8\pi r \frac{dr}{dt} = 8\pi r v.$

We remark that both rates depend on the instantaneous value of the radius.

358. $y = \sin x, \frac{dy}{dt} = \cos x \frac{dx}{dt}$. We require

$$\frac{dy}{dt} = \frac{1}{2}\left|\frac{dx}{dt}\right|, \quad \text{i.e., } \cos x = \pm\frac{1}{2},$$

whence $\quad x = \pm\frac{\pi}{3} + 2\pi n \quad$ and $\quad x = \pm\frac{2\pi}{3} + 2\pi n.$

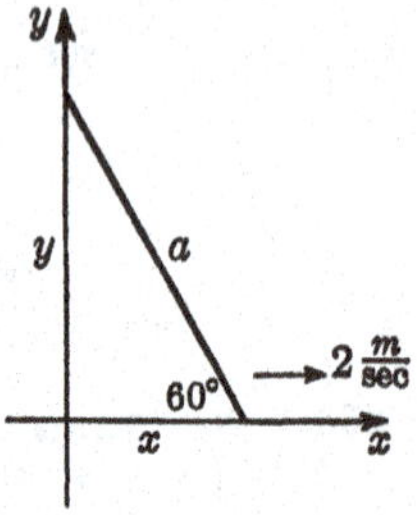

FIGURE 84

359. Denote the length of the ladder by a (Fig. 84). Then

$$y = \sqrt{a^2 - x^2},$$

$$\frac{dy}{dt} = \frac{-x}{\sqrt{a^2 - x^2}}\frac{dx}{dt} = -\frac{x}{y}\frac{dx}{dt} = -\cot\frac{\pi}{3} \cdot 2\,\frac{\text{m}}{\text{sec}} = -\frac{2\sqrt{3}}{3}\,\frac{\text{m}}{\text{sec}}.$$

The minus sign appears here because the top of the ladder moves downward.

360. See the figure (Fig. 14, Sec. 6.1). Put

$$\alpha = \measuredangle MOA, \qquad MA = x, \qquad \tan\alpha = x/a;$$

then

$$\frac{1}{\cos^2\alpha}\frac{d\alpha}{dt} = \frac{1}{a}\frac{dx}{dt} = \frac{v}{a}, \qquad \frac{d\alpha}{dt} = \frac{v\cos^2\alpha}{a}.$$

The angular velocity is a function of α.

362. $f'(x) = \frac{2}{1 + x^2},$

$$g'(x) = \frac{1}{1 + \frac{4x^2}{(1 - x^2)^2}} \cdot \frac{2 - 2x^2 + 4x^2}{(1 - x^2)^2} = \frac{2(1 + x^2)}{(1 + x^2)^2} = \frac{2}{1 + x^2}$$

(because $(1 - x^2)^2 + 4x^2 = (1 + x^2)^2$).

$$f'(x) = g'(x), \quad \text{i.e.,} \quad f(x) = g(x) + C.$$

At the points $x = \pm 1$, $g(x)$ is not defined. The values of C can be different for the three intervals in which these points divide the number axis. Let us find these values. For $-\infty < x < -1$, we substitute $x = -\sqrt{3}$:

$$2 \arctan(-\sqrt{3}) = -\frac{2\pi}{3}; \qquad \arctan \frac{-2\sqrt{3}}{1-3} = \arctan \sqrt{3} = \frac{\pi}{3}.$$

We have
$$-\frac{2\pi}{3} = \frac{\pi}{3} + C, \quad \text{i.e.,} \quad C = -\pi.$$

For $-1 < x < 1$, $f(0) = 0$, $g(0) = 0$, $C = 0$;
and for $1 < x < \infty$, $f(\sqrt{3}) = \frac{2\pi}{3}$, $g(\sqrt{3}) = \arctan \frac{2\sqrt{3}}{1-3} = -\frac{\pi}{3}$, $C = \pi$.

364. $f'(x) = a - a \cos ax = a(1 - \cos ax) = 2a \sin^2 \frac{ax}{2} \geqslant 0.$

The function increases for every x. Here the derivative vanishes for infinitely many x's, which, however, form a set of isolated points. Theorem (4) applies.

365. $f'(x) = \dfrac{1 + x^2 - 2x^2}{(1 + x^2)^2} = \dfrac{1 - x^2}{(1 + x^2)^2}.$

We have to solve the inequality

$$\frac{1 - x^2}{(1 + x^2)^2} > 0.$$

It is equivalent to $1 - x^2 > 0$, i.e., $|x| < 1$. In this interval the function increases.

366. $f'(x) = 3x^2 - 4x^3$;
$3x^2 - 4x^3 < 0, x^2(3 - 4x) < 0, 3 - 4x < 0, x > \frac{3}{4}.$

368. $f'(x) = \cos x - \sin x.$
$f'(0) = 1 > 0; f'(1) = \cos 1 - \sin 1 < 0.$

At $x = 0$ the function increases, and at $x = 1$ it decreases.

370. $y' = 5(x - 2)^4(2x + 1)^4 + 4(x - 2)^5 \cdot 2(2x + 1)^3$
$= (x - 2)^4(2x + 1)^2[5(2x + 1)^2 + 8(x - 2)(2x + 1)]$;

The first two factors are always $\geqslant 0$. The third one is

$$(2x + 1)(10x + 5 + 8x - 16) = (2x + 1)(18x - 11).$$

The solutions of the inequality $(2x + 1)(18x - 11) > 0$ are $-\infty < x < -\frac{1}{2}$

and $\frac{11}{18} < x < \infty$. In these intervals the function increases. For $-\frac{1}{2} < x < \frac{11}{18}$ it decreases. The behavior of $f(x)$ at $x = -\frac{1}{2}$ and $x = \frac{11}{18}$ will be studied in the next section (Prob. 382).

371. $y' = 1 - e^x$; $1 - e^x > 0$, $e^x < 1$, $x < 0$.

The function increases for $-\infty < x < 0$ and decreases for $0 < x < \infty$. The behavior of the function at $x = 0$ will be studied in the next section (Prob. 383).

372. The domain of definition of the function is given by

$$ax - x^2 \geqslant 0, \qquad x(a - x) \geqslant 0, \qquad 0 \leqslant x \leqslant a.$$

Consequently for $y = \sqrt{ax^3 - x^4}$,

$$y' = \frac{3ax^2 - 4x^3}{2\sqrt{ax^3 - x^4}} = \frac{3ax - 4a^2}{2\sqrt{ax - x^2}},$$

$$3ax - 4x^2 > 0, \qquad x(3a - 4x) > 0, \qquad 0 < x < 3a/4.$$

In this interval the function increases. It decreases for $3a/4 < x < a$. What happens at $x = 3a/4$ will be clarified in Problem 384.

373. $y' = 2xe^{-x} - x^2e^{-x} = e^{-x}x(2 - x)$.
$x(2 - x) > 0$; $0 < x < 2$ is the interval of increasing. $-\infty < x < 0$ and $2 < x < \infty$ are intervals of decreasing. The behavior at $x = 0$ and $x = 2$ will be discussed in Problem 396.

374. $0 < x < e$, decreases; $x > e$, increases.

376. Put $f(x) = x - \dfrac{x^2}{2}$, $g(x) = \ln(1 + x)$, $h(x) = x$. Then

$$f(0) = g(0) = h(0) = 0;$$
$$f'(x) = 1 - x; \qquad g'(x) = \frac{1}{1 + x}; \qquad h'(x) = 1.$$

Clearly for $x > 0$, $g'(x) < h'(x)$, i.e., the second part of the inequality is proved. To prove the first part we compute

$$1 - x - \frac{1}{1 + x} = \frac{1 - x^2 - 1}{1 + x} = \frac{-x^2}{1 + x} < 0,$$

i.e., $f'(x) < g'(x)$.

377. Put $f(x) = \tan x$, $g(x) = x + \frac{1}{3}x^3$. Then

$$f(0) = g(0) = 0; \qquad f'(x) = \frac{1}{\cos^2 x}, \quad g'(x) = 1 + x^2;$$

$$\frac{1}{\cos^2 x} = \sec^2 x = \tan^2 x + 1 > 1 + x^2$$

(because $\tan x > x$ for $0 < x < \pi/2$, as can be proved by the same method: $\tan 0 = 0$; $1/\cos^2 x > 1$ in $0 < x < \pi/2$).

378. $2\sqrt{1} = 2;\ 3 - \frac{1}{1} = 2.$

$$(2\sqrt{x})' = \frac{1}{\sqrt{x}};\ \left(3 - \frac{1}{x}\right)' = \frac{1}{x^2}.$$

For $x > 1$, $\frac{1}{\sqrt{x}} > \frac{1}{x^2}$ and the inequality follows.

389. $y' = 6 - 2x;\quad 6 - 2x = 0,\ x = 3.$
$y'' = -2 < 0$; Max (3,9).

390. $y' = 4(x - b)^3;\quad 4(x - b)^3 = 0,\ x = b.$
$y'' = 12(x - b)^2,\ y''(b) = 0;\quad y''' = 24(x - b),\ y'''(b) = 0;$
$y^{iv} = 24 > 0$; Min (b,a).

391. $y'''(b) = 6 \neq 0.$ No extremum.

392. $y' = 1 - \frac{1}{x^2};\quad 1 - \frac{1}{x^2} = 0,\ x = \pm 1;$

$y'' = \frac{2}{x^3};\ y''(1) > 0$; Min (1,2).

$y''(-1) < 0$; Max $(-1,-2)$.

The derivative is undefined at $x = 0$, but so is the function itself.

393. $y' = \frac{2x(x^4 + 4) - x^2 \cdot 4x^3}{(x^4 + 4)^2} = \frac{8x - 2x^5}{(x^4 + 4)^2};$

$8x - 2x^5 = 0;\quad x_1 = 0,\ x_{2,3} = \pm\sqrt{2}.$

We shall differentiate only the numerator (see the solution of Problem 384):

$f'(x) = 8 - 10x^4;$
$f'(0) = 8 > 0,$ Min (0,0);
$f'(\pm\sqrt{2}) = 8 - 40 = -32 < 0,$ Max $(\pm\sqrt{2},\frac{1}{4})$.

394. $y' = \ln x + 1;\ \ln x + 1 = 0,\ x = e^{-1};$

$$y'' = \frac{1}{x},\ y''\left(\frac{1}{e}\right) = e > 0,\ \text{Min}\left(\frac{1}{e}, -\frac{1}{e}\right).$$

395. $y' = x^x(1 + \ln x);\ y' = 0,\ x = \frac{1}{e};$

$$f'\left(\frac{1}{e} -\right) < 0,\ f'\left(\frac{1}{e} +\right) > 0,\ \text{Min}\left(\frac{1}{e}, \left(\frac{1}{e}\right)^{1/e}\right).$$

396. Compare Problem 373.

$$y' = e^{-x}x(2 - x); \qquad y' = 0, \quad x = 0, \quad x = 2.$$

		Sign of		
x	y	$y'(x-)$	$y'(x+)$	*Result*
0	0	$-$	$+$	Min
2	$4/e^2$	$+$	$-$	Max

397. $y' = \sinh x$; $\sinh x = 0$, $x = 0$;
$y'' = \cosh x$; $\cosh 0 = 1 > 0$, Min (0,1).

398. $y' = -\sin x + \sinh x$; $\sinh x - \sin x = 0$;

The only solution of this equation is $x = 0$. (Indeed, $\sinh 0 = \sin 0 = 0$; $(\sinh x)' = \cosh x$; $(\sin x)' = \cos x$, and $\cosh x > \cos x$ for $x > 0$. Thus $\sinh x > \sin x$ for $x > 0$.)

$$\begin{aligned} y'' &= -\cos x + \cosh x, & y''(0) &= 0; \\ y''' &= \sin x + \sinh x, & y'''(0) &= 0; \\ y^{iv} &= \cos x + \cosh x, & y^{iv}(0) &= 2 > 0, \qquad \text{Min } (0,2). \end{aligned}$$

399. $y' = \dfrac{1}{x} - \dfrac{1}{1 + x^2} = \dfrac{x^2 - x + 1}{x(1 + x^2)}$.

$x^2 - x + 1 = 0$ has no real solutions. The function is not defined at $x = 0$. There is no extremum.

400. $y' = e^x(\cos x - \sin x)$; $\cos x - \sin x = 0$, $\tan x = 1$,

$$x = \frac{\pi}{4} + \pi n \ (n = 0, \pm 1, \pm 2, \ldots).$$

$$y'' = e^x(\cos x - \sin x - \sin x - \cos x) = -2e^x \sin x.$$

$$y''\left(\frac{\pi}{4} + 2k\pi\right) = -2e^{\frac{\pi}{4}+2k\pi} \sin \frac{\pi}{4} < 0,$$

$$\text{Max}\left(\frac{\pi}{4} + 2k\pi, \frac{\sqrt{2}}{2}\, e^{\frac{\pi}{4}+2k\pi}\right);$$

$$y''\left(\frac{\pi}{4} + (2k+1)\pi\right) = -2e^{\frac{\pi}{4}+(2k+1)\pi}\left(-\sin \frac{\pi}{4}\right) > 0,$$

$$\text{Min}\left(\frac{\pi}{4} + (2k+1)\pi, -\frac{\sqrt{2}}{2}\, e^{\frac{\pi}{4}+(2k+1)\pi}\right).$$

401. $y' = 3\cos 3x - 3\cos x$; $\cos 3x - \cos x = 0$, $x = n\pi/2$.

The expression $x = n\pi/2$ gives all the solutions of the trigonometric equation. Now

$$y'' = 3\sin x - 9\sin 3x;$$

the function is periodic with 2π as the least period, and we can limit ourselves to the interval $[0,2\pi]$, or $[-0.1,2\pi - 0.1)$ to include the point $x = 0$ inside the interval.

$$y''(0) = 0; \qquad y''\left(\frac{\pi}{2}\right) = 3 + 9 = 12 > 0, \qquad \text{Min}\left(\frac{\pi}{2},-4\right);$$

$$y''(\pi) = 0; \qquad y''\left(\frac{3\pi}{2}\right) = -3 - 9 = -12 < 0, \qquad \text{Max}\left(\frac{3\pi}{2},4\right).$$

To decide about the remaining two points ($x = 0$ and $x = \pi$) we differentiate once more:

$$y''' = 3(\cos x - 9\cos 3x); \qquad y'''(0) \neq 0, \quad y'''(\pi) \neq 0.$$

These are not extrema. The periodicity determines the behavior outside the above interval.

402. $y' = 3x^2\sqrt[3]{x-1} + \dfrac{x^3}{3\sqrt[3]{(x-1)^2}} = x^2\dfrac{9x - 9 + x}{3\sqrt[3]{(x-1)^2}} = \dfrac{x^2(10x - 9)}{3\sqrt[3]{(x-1)^2}}.$

We have to check the points $x = 0$, $x = \frac{9}{10}$, and $x = 1$. When passing $x = 0$ and $x = 1$ the derivative does not change sign and there is no extremum. Close to the left of $x = \frac{9}{10}$ it is negative and close to the right, positive; therefore

$$\left(\frac{9}{10}, -\frac{9^3}{10^3}\cdot\frac{1}{10^{1/3}}\right)$$

is a minimum point of the given function.

405. First we find the domain of definition of this function:

$$x(6 - x) \geqslant 0; 0 \leqslant x \leqslant 6.$$

Now we look for the extrema of the given function:

$$f'(x) = \frac{6 - 2x}{2\sqrt{x(6 - x)}} = \frac{3 - x}{\sqrt{x(6 - x)}}.$$

The derivative of the numerator is $-1 < 0$. The denominator is positive. We have thus Max (3,3). The greatest value of the function is 3, and the least is at the ends of the interval, $x = 0$, $x = 6$, and equals 0.

407. $2x + 2y + 2xy' + 2yy' - 4 + 2y' = 0;$

$$y' = \frac{2 - x - y}{x + y + 1}; \; y' = 0 \text{ gives } x + y = 2.$$

Let us write the given equation in the form

$$(x + y)^2 - 6x + 2(x + y) - 2 = 0.$$

After substitution of $x + y = 2$ we obtain

$$4 - 6x + 4 - 2 = 0, \qquad 6x = 6, \qquad x = 1;$$

$$y = 2 - 1 = 1.$$

The denominator of y' is not zero at this point. We find now the second derivative:

$$2(1 + y' + y' + xy'' + y'^2 + yy'' + y'') = 0.$$

Substituting $x = 1$, $y = 1$, $y' = 0$,

$$1 + y'' + y'' + y'' = 0, \qquad y'' = -\tfrac{1}{3} < 0, \qquad \text{Max } (1,1).$$

When $x + y + 1 = 0$ the first derivative does not exist. We substitute $x + y = -1$ and obtain

$$1 - 6x - 2 - 2 = 0, \qquad x = -\tfrac{1}{2}, \qquad y = -\tfrac{1}{2}.$$

We can find y from the equation

$$(x + y)^2 + 2(x + y) - 6x - 2 = 0;$$

$$x + y = -1 \pm \sqrt{1 + 6x + 2} = -1 \pm \sqrt{6x + 3};$$

$$y = -1 - x \pm \sqrt{6x + 3}.$$

We see that for $x < -\frac{1}{2}$, y is not real, i.e., the two functions are undefined there. Consequently $(-\frac{1}{2},-\frac{1}{2})$ cannot be an extremum. The tangent to the curve representing the given equation is vertical at this point, and the whole curve lies to one side of this tangent (see Fig. 85).

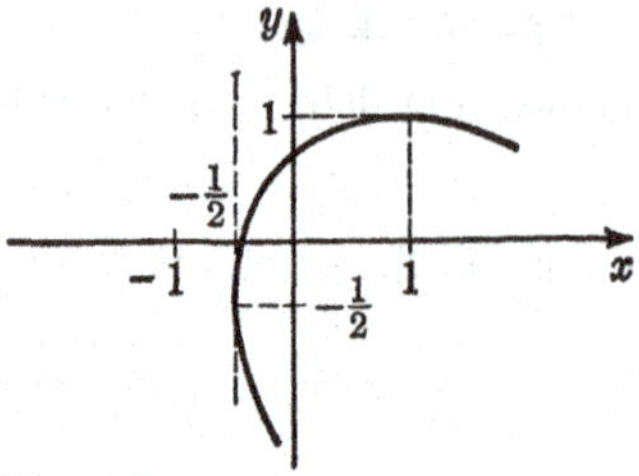

FIGURE 85

408. Max $(a\sqrt[3]{2}, a\sqrt[3]{4})$.

409. The curve representing the given equation is clearly symmetric with regard to both axes and to the line $y = x$ (see remark (b) in Section 6.6). Hence it suffices to investigate its behavior for $y \geqslant x \geqslant 0$. We differentiate:

$$4x^3 + 4y^3y' = 2x + 2yy', \qquad y' = \frac{x(2x^2 - 1)}{y(1 - 2y^2)};$$

$$y' = 0; \qquad x_1 = 0, \quad x_2 = \sqrt{1/2};$$

(we are concerned only with nonnegative x);

$$x_1 = 0; \qquad y^4 = y^2, \qquad y = 0, \qquad y = 1;$$

($y = -1$ is left aside for the present). The point (0,0) is an isolated point of the graph and cannot be an extremum. Really, if $-1 < x < 1$ and $-1 < y < 1$, then $x^4 + y^4 < x^2 + y^2$ for every point except (0,0). Now we investigate (0,1). First we differentiate once more after dividing by 2:

$$6x^2 + 6y^2(y')^2 + 2y^3y'' = 1 + (y')^2 + yy''.$$

We substitute $x = 0$, $y = 1$, $y' = 0$;

$$2y'' = 1 + y'', \quad y'' = 1 > 0, \qquad \text{Min } (0,1).$$

Now we find the y's corresponding to $x = \sqrt{\frac{1}{2}}$:

$$\tfrac{1}{4} + y^4 = \tfrac{1}{2} + y^2, \qquad 4y^4 - 4y^2 - 1 = 0,$$

$$y^2 = \frac{2 \pm \sqrt{4 + 4}}{4} = \frac{1 \pm \sqrt{2}}{2};$$

only the $+$ sign can be taken, and

$$y^2 = \frac{1 + \sqrt{2}}{2}, \qquad y = \sqrt{\frac{1 + \sqrt{2}}{2}}.$$

(Once more, the negative y is not being investigated now.) We substitute in the expression for y''

$$x = \sqrt{\frac{1}{2}}, \quad y = \sqrt{\frac{1 + \sqrt{2}}{2}}, \qquad y' = 0:$$

$$6 \cdot \frac{1}{2} + 2\left(\sqrt{\frac{1 + \sqrt{2}}{2}}\right)^3 y'' = 1 + \sqrt{\frac{1 + \sqrt{2}}{2}}\, y'';$$

$$y'' < 0, \qquad \text{Max}\left(\frac{1}{\sqrt{2}}, \sqrt{\frac{1 + \sqrt{2}}{2}}\right).$$

y' does not exist at $y = 0$ and $y = 1/\sqrt{2}$. The corresponding values of x are greater than these values of y and we do not need to discuss these points separately. By symmetry we obtain

$$\text{Min } (0,1), \qquad \text{Max}\left(\pm\frac{1}{\sqrt{2}}, \sqrt{\frac{1 + \sqrt{2}}{2}}\right),$$

$$\text{Max}\,(0,-1),\qquad \text{Min}\left(\pm\frac{1}{\sqrt{2}},-\sqrt{\frac{1+\sqrt{2}}{2}}\right).$$

The graph is given in Figure 86.

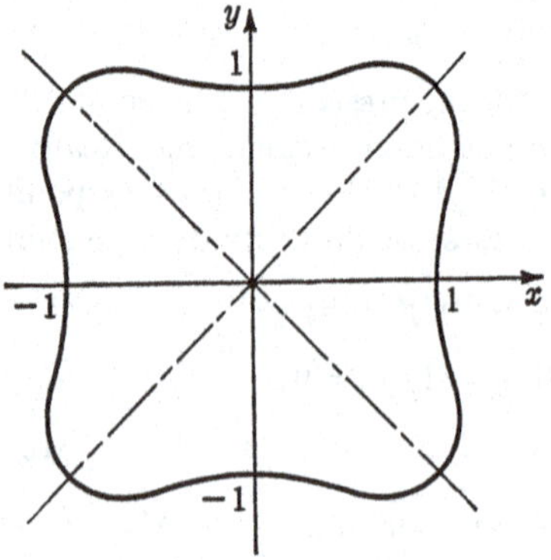

FIGURE 86

411. $\dfrac{dy}{dx} = \dfrac{e^{-t} - te^{-t}}{e^t + te^t} = e^{-2t}\dfrac{1-t}{1+t};\ y' = 0,\ t = 1;$

y' does not exist for $t = -1$.

$$\frac{dx}{dt} = e^t(1+t);\qquad \frac{dx}{dt} > 0 \quad\text{for}\quad t > 1,$$

i.e., x increases monotonically together with t for positive t. Now

$$y'(t = 1-) > 0,\qquad y'(t = 1+) < 0;\qquad \text{Max}\left(e,\frac{1}{e}\right).$$

(This conclusion is also obtained easily by consideration of d^2y/dx^2.)

$$\frac{dx}{dt}(t = -1) = 0,\qquad \frac{dx}{dt}(t = -1-) < 0,\qquad \frac{dx}{dt}(t = -1+) > 0,$$

i.e., x reaches its minimum $-1/e$ for $t = -1$; consequently there are no points of the graph to the left of $1/e$ and so this x cannot be an extremum.

416. See Figure 87. The volume of the inscribed cone is $V = \frac{1}{3}\pi r^2 h$. Now

$$\frac{r}{R} = \frac{H-h}{H},\quad\text{i.e.,}\quad h = H - \frac{r}{R}H;$$

$$V = \frac{1}{3}\pi r^2 H\left(1 - \frac{r}{R}\right) = \frac{1}{3}\pi H\left(r^2 - \frac{r^3}{R}\right);$$

$$\frac{dV}{dr} = \frac{1}{3}\pi H\left(2r - \frac{3r^2}{R}\right); \qquad \frac{dV}{dr} = 0, \qquad r_1 = 0, \quad r_2 = \frac{2}{3}R.$$

The second solution is clearly the required one. The altitude is equal to

$$h = H(1 - \tfrac{2}{3}) = \tfrac{1}{3}H.$$

417. See Figure 88. Take α as the independent variable. The length S of the segment is

$$S = \frac{y_0}{\sin\alpha} + \frac{x_0}{\cos\alpha} \qquad \left(0 < \alpha < \frac{\pi}{2}\right).$$

$$S' = \frac{dS}{d\alpha} = -\frac{y_0\cos\alpha}{\sin^2\alpha} + \frac{x_0\sin\alpha}{\cos^2\alpha} = y_0\frac{\sin\alpha}{\cos^2\alpha}\left(\frac{x_0}{y_0} - \cot^3\alpha\right).$$

S' does not exist for $\alpha = 0$ and $\alpha = \pi/2$. (The line is parallel to one of the axes.) The required solution is given by

$$\cot\alpha = \sqrt[3]{x_0/y_0}.$$

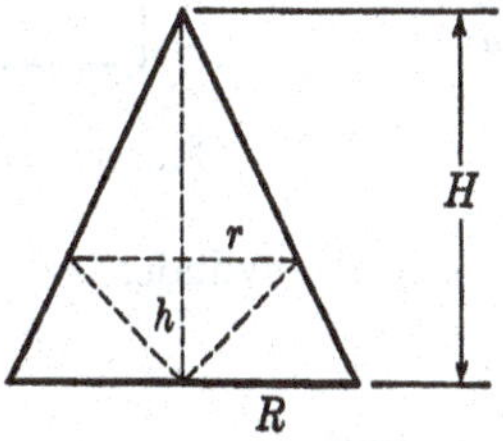

FIGURE 87

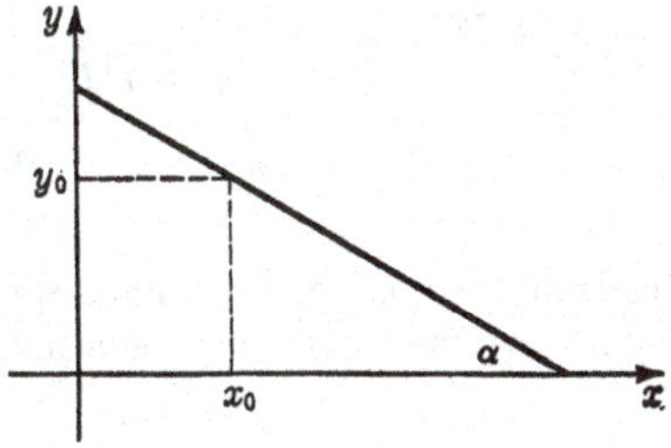

FIGURE 88

418. See Figure 89. Put $AB = z$. Then

$$2p\cdot OA = AB^2, \qquad OA = \frac{z^2}{2p}, \qquad BE = \frac{m^2}{2p} - \frac{z^2}{2p}.$$

$$S = 2z\cdot\frac{1}{2p}(m^2 - z^2) = \frac{1}{p}(m^2 z - z^3);$$

$$S' = \frac{1}{p}(m^2 - 3z^2); \qquad S' = 0, \qquad z = \frac{m}{\sqrt{3}};$$

$$S'' = -\frac{6z}{p}; \qquad S''(m/\sqrt{3}) < 0;$$

$$S_{\max} = \frac{m}{\sqrt{3}p}\left(m^2 - \frac{m^2}{3}\right) = \frac{2m^3}{3\sqrt{3}p}.$$

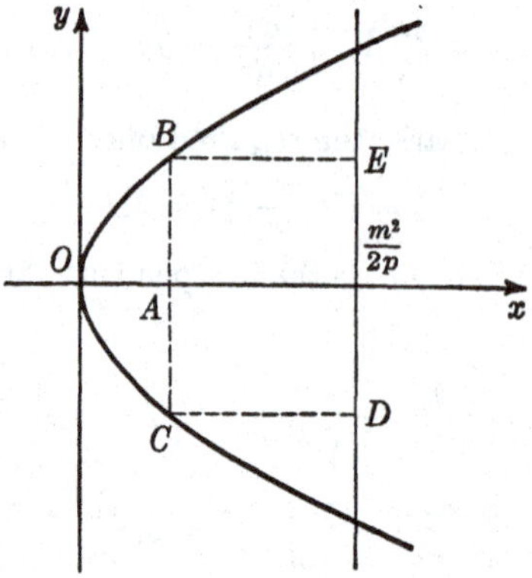

FIGURE 89

419. See Figure 90. The argument will be r.

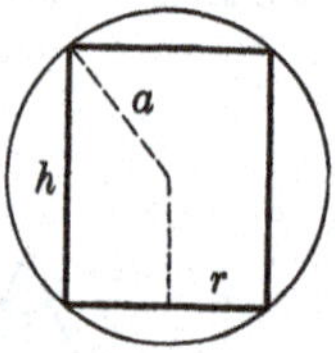

FIGURE 90

$$S = 2\pi rh; \quad \frac{h}{2} = \sqrt{a^2 - r^2}; \quad S = 4\pi r\sqrt{a^2 - r^2}.$$

$$S' = 4\pi\left(\sqrt{a^2 - r^2} - \frac{r^2}{\sqrt{a^2 - r^2}}\right) = 4\pi\,\frac{a^2 - 2r^2}{\sqrt{a^2 - r^2}}.$$

$$S' = 0; \quad a^2 - 2r^2 = 0, \quad r = \frac{a}{\sqrt{2}};$$

r varies from 0 to a. At both ends the volume of the cylinder is 0. The single point we find must then be a maximum.

$$S_{\max} = 4\pi\,\frac{a}{\sqrt{2}}\sqrt{a^2 - \frac{a^2}{2}} = 2\pi a^2.$$

Observe that this area is exactly half of the surface of the sphere.

420. Let the time t be the argument. The situation of the points at t is given by

$$x_1 = a - v_1t,\ y_1 = 0 \quad \text{and} \quad x_2 = 0,\ y_2 = b - v_2t.$$

The distance between the points is

$$S = \sqrt{(x_1 - x_2)^2 + (y_1 - y_2)^2} = \sqrt{(a - v_1t)^2 + (b - v_2t)^2}.$$

Now $S > 0$ and it will attain its maximum together with

$$S^2 = (a - v_1t)^2 + (b - v_2t)^2.$$

$$\frac{dS^2}{dt} = -2v_1(a - v_1t) - 2v_2(b - v_2t); \quad \frac{dS^2}{dt} = 0, \quad t = \frac{av_1 + bv_2}{v_1^2 + v_2^2}.$$

$$\frac{d^2S^2}{dt^2} = 2v_1^2 + 2v_2^2 > 0,$$

i.e., this is a minimum and because there exists only one extremum, it must be also the least value (the function is continuous).

$$S^2_{\min} = a^2 + b^2 - 2(av_1 + bv_2)\frac{av_1 + bv_2}{v_1^2 + v_2^2} + (v_1^2 + v_2^2)\frac{(av_1 + bv_2)^2}{(v_1^2 + v_2^2)^2}$$

$$= \frac{a^2v_1^2 + a^2v_2^2 + b^2v_1^2 + b^2v_2^2 - a^2v_1^2 - 2abv_1v_2 - b^2v_2^2}{v_1^2 + v_2^2}$$

$$= \frac{(av_2 - bv_1)^2}{v_1^2 + v_2^2}.$$

421. Take the angle x as the argument.

$$H = k + \frac{h}{2} = k + b \sin x.$$

Now we shall connect k with x. We have

$$MO = (k + h) \cot \beta; \qquad ON = k \cot \alpha.$$

Consequently,

$$(k + 2b \sin x) \cot \beta + k \cot \alpha = 2b \cos x; \qquad k = 2b\,\frac{\cos x - \sin x \cot \beta}{\cot \alpha + \cot \beta}.$$

$$H = b\left(2\,\frac{\cos x - \sin x \cot \beta}{\cot \alpha + \cot \beta} + \sin x\right)$$

$$= \frac{b}{\cot \alpha + \cot \beta}\,[2 \cos x + \sin x\,(\cot \alpha - \cot \beta)].$$

$$H' = \frac{b}{\cot \alpha + \cot \beta}\,[-2 \sin x + \cos x\,(\cot \alpha - \cot \beta)];$$

$$H' = 0, \ \tan x_0 = \tfrac{1}{2}(\cot \alpha - \cot \beta).$$

$$H'' = \frac{b}{\cot \alpha + \cot \beta}\,[-2 \cos x - \sin x\,(\cot \alpha - \cot \beta)];$$

$$H''(x_0) = \frac{-b \cos x_0}{\cot \alpha + \cot \beta}\,[2 + \tan x_0\,(\cot \alpha - \cot \beta)]$$

$$= \frac{-b \cos x_0}{\cot \alpha + \cot \beta}\left[2 + \frac{1}{2}\,(\cot \alpha - \cot \beta)^2\right] < 0,$$

because $\alpha < \pi/2$, $\beta < \pi/2$, i.e., $\cot \alpha$ and $\cot \beta$ are positive and $-\alpha \leqslant x_0 \leqslant \beta$, i.e., also $\cos x_0 > 0$. Thus $x_0 = \arctan \frac{1}{2}\,(\cot \alpha - \cot \beta)$ is the required angle.

422. See Figure 91. Given $AC = R$, $AD = H$, then $L = \sqrt{H^2 + R^2}$; put $OB = r$ and $AE = h$. The volume of the water displaced by the sphere is equal to the volume V of the spherical segment inside the funnel. From solid geometry we know that

$$V = \frac{\pi}{3} h^2(3r - h).$$

Now $\Delta OBD \sim \Delta ADC$, i.e., $\frac{r}{R} = \frac{OD}{L} = \frac{H + r - h}{L}$,

whence $r = \frac{R(H - h)}{L - R}$.

Then

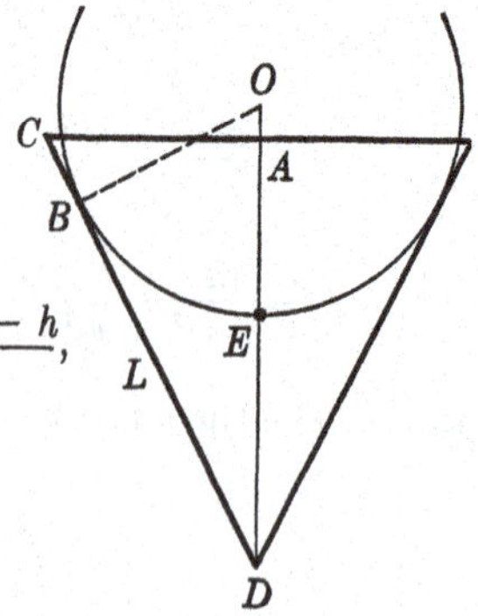

FIGURE 91

$$V = \frac{\pi}{3} h^2 \left[\frac{3R(H - h)}{L - R} - h\right] = \frac{\pi}{3(L - R)} (3RHh^2 - 2Rh^3 - Lh^3)$$

$$= \frac{\pi}{3(L - R)} [3RHh^2 - h^3(2R + L)];$$

$$\frac{dV}{dh} = \frac{\pi}{3(L - R)} [6RHh - 3h^2(2R + L)]; \quad \frac{dV}{dh} = 0,\ h = \frac{2RH}{2R + L}$$

($h = 0$ is clearly not a solution for us).

$$\frac{d^2V}{dh^2} = \frac{\pi}{L - R} [2RH - 2h(2R + L)];$$

$$\frac{d^2V}{dh^2}\left(h = \frac{2RH}{2R + L}\right) = \frac{\pi}{L - R} (2RH - 4RH) = \frac{-2\pi RH}{L - R} < 0 \quad (L > R).$$

We find the r_0 that makes the maximum:

$$r_0 = \frac{R\left(H - \dfrac{2RH}{2R + L}\right)}{L - R} = \frac{RHL}{(L - R)(2R + L)}.$$

Note. The formula for V can also be derived using the method of Section 10.3.

423. $S = \sum_{k=1}^{n} (A_1 - x_k)^2;$

$$\frac{dS}{dA_1} = 2 \sum_{k=1}^{n} (A_1 - x_k) = 2\left(A_1 n - \sum_{k=1}^{n} x_k\right);$$

$$\frac{dS}{dA_1} = 0,\ A_1 = \frac{1}{n} \sum_{k=1}^{n} x_k; \quad \frac{d^2S}{dA_1^2} = 2n > 0.$$

A_1 so obtained is the arithmetic mean of the n measurements.

424. $h = L/\sqrt{3}, r = L\sqrt{2/3}$.

425. The nearest point to M on the parabola is $(p/\sqrt[3]{2}, p\sqrt[3]{2})$.

426. $\left|\dfrac{ax_0 + by_0 + c}{\sqrt{a^2 + b^2}}\right|$.

427. $r = \sqrt[3]{\dfrac{V}{2\pi}}; h = \sqrt[3]{\dfrac{4V}{\pi}}$.

428. The required position is defined by

$$\cos \alpha = \frac{b + \sqrt{b^2 + 128a^2}}{16a},$$

where α is the angle between the rod and the horizontal line.

429. $m_2 = \sqrt{m_1 m_3}$.

433. $y' = 3x^2 + a$, $y'' = 6x$; for $x < 0$, $y'' < 0$ and this is the interval in which the curve is concave downward.

434. $x = \frac{3}{2}; \frac{3}{2}(\frac{9}{4} - y^2) + y^2 = 0; \frac{1}{2}y^2 = \frac{27}{8}; y = 3\sqrt{3}/2 \ (y > 0)$.
Let us differentiate $x^3 - xy^2 + y^2 = 0$;

$$3x^2 - y^2 - 2xyy' + 2yy' = 0.$$

We substitute $x = \dfrac{3}{2}, y = \dfrac{3\sqrt{3}}{2}$,

$$3 \cdot \frac{9}{4} - \frac{27}{4} - 2 \cdot \frac{3}{2} \cdot \frac{3\sqrt{3}}{2} y' + 2 \cdot \frac{3\sqrt{3}}{2} y' = 0; \qquad y' = 0.$$

We differentiate again:

$$6x - 2yy' - 2yy' - 2x(y')^2 - 2xyy'' + 2(y')^2 + 2yy'' = 0,$$

and substitute $x = \frac{3}{2}, y = 3\sqrt{3}/2, y' = 0$:

$$9 - 2 \cdot \frac{3}{2} \cdot \frac{3\sqrt{3}}{2} y'' + 2 \cdot \frac{3\sqrt{3}}{2} y'' = 0; \qquad y'' = \frac{9}{3\sqrt{3}/2} = 2\sqrt{3} > 0$$

At $\left(\dfrac{3}{2}, \dfrac{3\sqrt{3}}{2}\right)$ the curve is concave upward.

435. $y' = 2x \ln x + x; y'' = 2 \ln x + 2 + 1 = 2 \ln x + 3$.
$y''(1) = 3 > 0$: concave upward;
$y''(1/e^2) = -4 + 3 = -1 < 0$: concave downward.

436. $x < \frac{5}{3}$, concave downward; $x > \frac{5}{3}$, concave upward; $(\frac{5}{3}, -\frac{250}{27})$, an inflection point.

437. Concave upward for any x.

438. Concave upward for any x.

439. $(1,-1)$ is an inflection point; $x < 1$, concave downward; $x > 1$, concave upward.

440. $$y' = \frac{3x^2(x^2 + 3a^2) - x^3 \cdot 2x}{(x^2 + 3a^2)^2} = \frac{x^4 + 9a^2x^2}{(x^2 + 3a^2)^2};$$

$$y'' = \frac{(4x^3 + 18a^2x)(x^2 + 3a^2) - (x^4 + 9a^2x^2) \cdot 2 \cdot 2x}{(x^2 + 3a^2)^3} = \frac{54a^4x - 6a^2x^3}{(x^2 + 3a^2)^3}.$$

$$y'' = 0,\ 6a^2x(9a^2 - x^2) = 0,\ x_1 = 0,\ x_2 = 3a,\ x_3 = -3a.$$

The denominator of y'' is always positive. The numerator factors into $6a^2x(3a - x)(3a + x)$. Hence for

$x < -3a$,	$f''(x) > 0$,	concave upward;
$-3a < x < 0$,	$f''(x) < 0$,	" downward;
$0 < x < 3a$,	$f''(x) > 0$,	" upward;
$x > 3a$,	$f''(x) < 0$,	" downward.

$(0,0)$, $(-3a,-9a/4)$, $(3a,9a/4)$ are inflection points.

441. $y' = -\frac{1}{3}(x - b)^{-2/3};\ y'' = \frac{2}{9}(x - b)^{-5/3}.$

For	$x > b$,	$y'' > 0$,	concave upward;
	$x < b$,	$y'' < 0$,	concave downward.

Consequently (b,a) is an inflection point. The second derivative does not exist there.

442. $x < -1$, concave downward;
$-1 < x < 1$, " upward;
$x > 1$, " downward.

$(\pm 1, \ln 2)$ are inflection points.

443. $$y' = e^{\arctan x} \cdot \frac{1}{1 + x^2};\ y'' = e^{\arctan x} \cdot \frac{1 - 2x}{(1 + x^2)^2};$$

$x < \frac{1}{2}$, concave upward;
$x > \frac{1}{2}$, concave downward.

$(\frac{1}{2}, e^{\arctan 1/2})$ is the inflection point.

444. $y' = x\cos x + \sin x;$
$y'' = \cos x - x\sin x + \cos x = 2\cos x - x\sin x;$
$y'' = 0,\ 2\cos x - x\sin x = 0,\ x = 2\cot x.$

This equation can be solved by graphical or numerical methods. We are interested in showing that its solutions satisfy $y^2(4 + x^2) = 4x^2$. Let us substitute $x = 2 \cot x$; $y = x \sin x = 2 \cos x$.

$$4 \cos^2 x(4 + 4 \cot^2 x) = 4 \cdot 4 \cot^2 x,$$

$$16 \cos^2 x \csc^2 x = 16 \cot^2 x,$$

and this is a known identity.

445. $y' = e^x + 3ax^2$; $y'' = e^x + 6ax$; $e^x + 6ax = 0$; $e^x = -6ax$. Let us draw the curves $y_1 = e^x$ and $y_2 = -6ax$ (Fig. 92). For $a > 0$ the

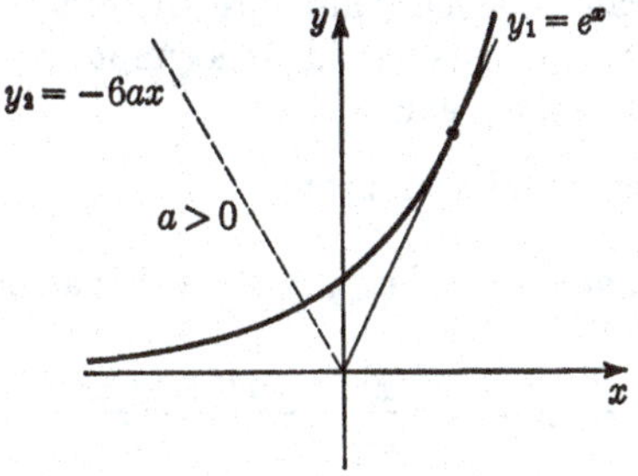

FIGURE 92

straight line intersects the curve $y_1 = e^x$. For $a < 0$ the limit straight line $y = -6ax$ intersecting the curve is the tangent (through the origin). Let us find this tangent. If the point of contact is (x_1,e^{x_1}), then the slope of the tangent is $y_1'(x_1) = e^{x_1}$ and its equation is $y = e^{x_1}x$. Now (x_1,e^{x_1}) is a point on the tangent, i.e., $e^{x_1} = e^{x_1}x_1$; consequently $x_1 = 1$ and the slope $e^1 = e = -6a$, i.e., $a = -e/6$. This is the limit value of a. For $a \leqslant -e/6$ there is an intersection between the two lines. Inflection points exist, thus, for $a > 0$ and $a \leqslant -e/6$.

448. See Problem 446. $g' = e^t$, $g'' = e^t$; $f' = \cos t$, $f'' = -\sin t$.

$$\frac{g'f'' - f'g''}{g'} = \frac{-e^t \sin t - e^t \cos t}{e^t} = -(\sin t + \cos t)$$

$$= -\sqrt{2} \sin\left(t + \frac{\pi}{4}\right).$$

$$\sin\left(t + \frac{\pi}{4}\right) = 0;\ t + \frac{\pi}{4} = \pi n,\ t = \pi n - \frac{\pi}{4}\ (n = 0, \pm 1, \pm 2, \ldots).$$

These are the inflection points.

455. There are no asymptotes parallel to the axes because $\lim\limits_{x\to\pm\infty} y = \infty$ and the function is continuous for every x.

$$\lim_{x\to\infty} \frac{y}{x} = \lim_{x\to\infty} \arctan x = \frac{\pi}{2};$$

$$\lim_{x\to\infty} \left(x \arctan x - \frac{\pi}{2} x \right) = \lim_{x\to\infty} \frac{\arctan x - \dfrac{\pi}{2}}{\dfrac{1}{x}} = \lim_{x\to\infty} \frac{\dfrac{1}{1+x^2}}{-\dfrac{1}{x^2}} = -1;$$

$$y = \frac{\pi}{2} x - 1 \quad \text{is an asymptote.}$$

By investigating $x \to -\infty$ we find that also $y = -(\pi/2)x - 1$ is an asymptote. This can be deduced also directly from the fact that the given function is even, i.e., the curve (and its asymptotes, of course) are symmetric with respect to the y axis.

456. $x = 0$ is a horizontal asymptote.

$y = 2x + 1 + \dfrac{1}{x^3}$; $\lim_{x\to\infty} \dfrac{1}{x^3} = 0$; i.e., $y = 2x + 1$ is an oblique asymptote.

458. Substitute $y = kx + n$: $\dfrac{x^2}{a^2} - \dfrac{k^2x^2 + 2knx + n^2}{b^2} = 1;$

$$\frac{1}{a^2} - \frac{k^2}{b^2} = 0,\ k = \pm\frac{b}{a};\ -\frac{2kn}{b^2} = 0,\ n = 0.$$

The given curve (called a hyperbola) has two asymptotes: $y = \dfrac{b}{a} x$ and

$y = -\dfrac{b}{a} x.$

459. $x = -1;\ x = 5;\ y = 0.$

460. $x = -1;\ y = \frac{1}{2}x - 1.$

461. $y = kx + b;\ k^2x^3 + 2kbx^2 + b^2x + kx^3 + bx^2 - a^3 = 0;$

$$\left.\begin{aligned} k^2 + k &= 0 \\ 2kb + b &= 0 \end{aligned}\right\} \Rightarrow \begin{aligned} &k_1 = 0,\ k_2 = -1, \\ &\text{in both cases } b = 0. \end{aligned}$$

$y = 0$ and $y = -x$ are asymptotes. When $x \to 0$, y tends to ∞, and so $x = 0$ is also an asymptote.

462. $\lim_{x\to\infty} y = \infty$; $\lim_{x\to 0} y = \lim_{x\to 0} \dfrac{\ln\left(e + \dfrac{1}{x}\right)}{\dfrac{1}{x}} = \lim_{x\to 0} \dfrac{-\dfrac{1}{x^2}}{-\dfrac{1}{x^2}\left(e + \dfrac{1}{x}\right)} = 0.$

Consequently, the straight line $x = 0$ is not an asymptote.

$$\lim_{x\to(-1/e)-} y = +\infty, \quad \text{i.e.,} \quad x = -1/e \text{ is a vertical asymptote.}$$

Now $\lim_{x\to\infty} \frac{y}{x} = \lim_{x\to\infty} \ln\left(e + \frac{1}{x}\right) = 1;$

$$\lim_{x\to\infty}\left[x \ln\left(e + \frac{1}{x}\right) - x\right] = \lim_{x\to\infty} \frac{\ln\left(e + \frac{1}{x}\right) - 1}{\frac{1}{x}}$$

$$= \lim_{x\to\infty} \frac{-\frac{1}{x^2}}{\left(e + \frac{1}{x}\right)\left(-\frac{1}{x^2}\right)} = \frac{1}{e};$$

$y = x + \frac{1}{e}$ is an oblique asymptote. For $x \to -\infty$ we obtain the same asymptote.

463. $x = 0$, $y = x$.

464. $y = 2x + \frac{\pi}{2}$, $y = 2x - \frac{\pi}{2}$.

467. $\lim_{t\to\pm1} x = \infty$ (or $-\infty$), $\lim_{t\to\pm1} y = \infty$ (or $-\infty$);

$$\lim_{t\to\pm1} \frac{y}{x} = \lim_{t\to\pm1} \frac{t}{2} = \pm\frac{1}{2};$$

$$\lim_{t\to\pm1}\left(y \mp \frac{1}{2}x\right) = \lim_{t\to\pm1}\left(\frac{t^2}{1-t^2} \mp \frac{t}{1-t^2}\right) = \lim_{t\to\pm1} \frac{t(t \mp 1)}{1-t^2}$$

$$= \lim_{t\to\pm1} \frac{\mp t}{1 \pm t} = -\frac{1}{2};$$

There are two asymptotes, $y = \pm\frac{x}{2} - \frac{1}{2}$.

468. $y = -x - a$.

469. y becomes infinite as t tends to 0; $\lim_{t\to0} x = 2$; $x = 2$ is an asymptote.

$\lim_{t\to\pm2} y = \infty$ (or $-\infty$), $\lim_{t\to\pm2} x = \infty$ (or $-\infty$);

$$\lim_{t\to2} \frac{y}{x} = \lim_{t\to2} \frac{3}{t(t-8)} = \frac{1}{4},$$

$$\lim_{t\to-2} \frac{y}{x} = \lim_{t\to-2} \frac{3}{t(t-8)} = \frac{3}{20},$$

$$\lim_{t\to 2}\left(y + \frac{1}{4}x\right) = \lim_{t\to 2}\left[\frac{3}{t(t^2-4)} + \frac{t-8}{4(t^2-4)}\right] = \lim_{t\to 2}\frac{12+t^2-8t}{4t(t^2-4)}$$

$$= \lim_{t\to 2}\frac{(t-2)(t-6)}{4t(t-2)(t+2)} = -\frac{1}{8}.$$

$$\lim_{t\to -2}\left(y - \frac{3}{20}x\right) = \lim_{t\to -2}\left[\frac{3}{t(t^2-4)} - \frac{3}{20}\cdot\frac{t-8}{t^2-4}\right] = 3\lim_{t\to -2}\frac{20-t^2+8t}{20t(t^2-4)}$$

$$= -3\lim_{t\to -2}\frac{(t+2)(t-10)}{20t(t-2)(t+2)} = \frac{9}{40}.$$

The asymptotes are $y = -\frac{1}{4}x - \frac{1}{8}$, $y = \frac{3}{20}x + \frac{9}{40}$.

471. The function is defined everywhere.

$$f(-x) = \frac{(-x)^3 - 9(-x)}{10} = -\frac{x^3 - 9x}{10} = -f(x),$$

i.e., the function is odd and its graph is symmetric with respect to the origin. The intersections with the axes are

$$y = 0, \qquad x^3 - 9x = 0; \qquad x = 0, \qquad x = \pm 3,$$

i.e., (0,0), (3,0), (−3,0). There are no asymptotes.

$y' = \frac{1}{10}(3x^2 - 9)$; $y' = 0$, $x = \pm\sqrt{3}$;

$y'' = \frac{3x}{5}$, $y''(\sqrt{3}) > 0$, Min $\left(\sqrt{3}, -\frac{3\sqrt{3}}{5}\right)$.

$y''(-\sqrt{3}) < 0$, Max $\left(-\sqrt{3}, \frac{3\sqrt{3}}{5}\right)$.

$y'' = 0$, $x = 0$; $y''' = \frac{3}{5} \neq 0$, i.e., (0,0) is an inflection point. Two additional points are (1,−0.8), (2,−1). The graph of this cubic is given in Figure 93.

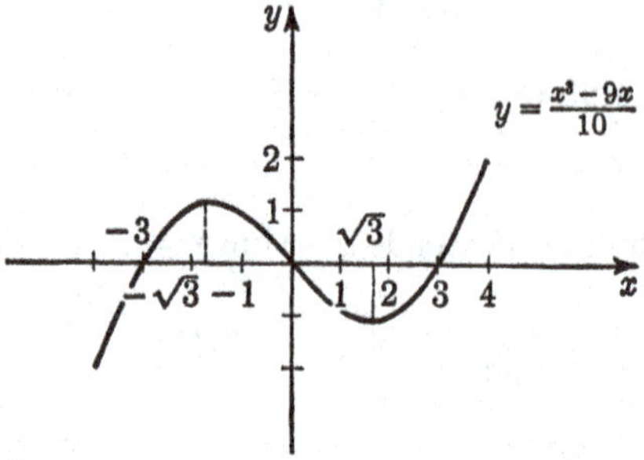

FIGURE 93

472. The function is defined for all x. There is no symmetry with respect to the axes. Points of intersection with the axes are (0,0), (1,0), (2,0), (3,0). There are no asymptotes.

$y = \frac{1}{24}(x^2 - x)(x^2 - 5x + 6) = \frac{1}{24}(x^4 - 6x^3 + 11x^2 - 6x);$
$y' = \frac{1}{12}(2x^3 - 9x^2 + 11x - 3),\ y'' = \frac{1}{12}(6x^2 - 18x + 11).$
$y' = 0,\ 2x^3 - 9x^2 + 11x - 3 = (2x - 3)(x^2 - 3x + 1) = 0;$

$$x_1 = \frac{3}{2}, \qquad x_{2,3} = \frac{3}{2} \pm \sqrt{\frac{9}{4} - 1} = \frac{3 \pm \sqrt{5}}{2}.$$

$y''(\frac{3}{2}) = \frac{1}{12}(6\cdot\frac{9}{4} - 18\cdot\frac{3}{2} + 11) < 0$, Max $(\frac{3}{2}, \frac{3}{128})$.
By continuity of the function we clearly have minimum points at $x = \dfrac{3 \pm \sqrt{5}}{2}$. We find the inflection points:

$$y'' = 0, \qquad 6x^2 - 18x + 11 = 0;$$

$$x_{1,2} = \frac{9 \pm \sqrt{81 - 66}}{6} = \frac{9 \pm \sqrt{15}}{6} \sim \frac{9 \pm 4}{6};$$

$$x_1 \sim \frac{13}{6}, \qquad x_2 \sim \frac{5}{6}.$$

For $x \to \pm\infty$, $y \to \infty$. We find in addition the points $x = -1$, $y = 1$; $x = 4$, $y = 1$. The graph is given in Figure 94.

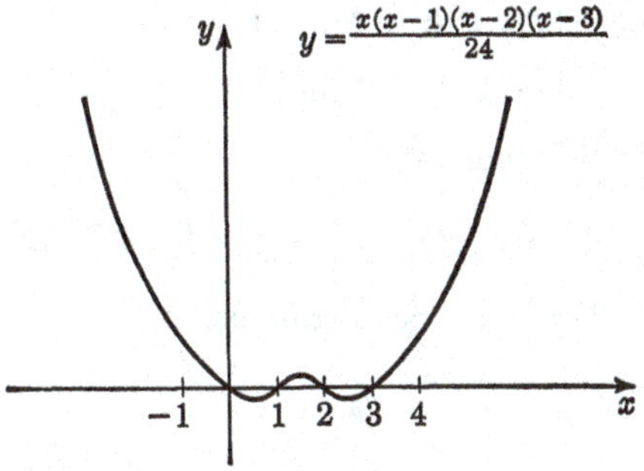

FIGURE 94

473. Defined for every x. The y axis is an axis of symmetry (the function is even). We have $x = 0$, $y = 1$, and

$$\lim_{x \to \pm\infty} y = 0, \quad \text{i.e.,} \quad y = 0 \text{ is an asymptote.}$$

$$y' = \frac{-2x}{(1 + x^2)^2};\ y' = 0,\ x = 0;$$

$$y'' = -2\,\frac{1 + x^2 - x\cdot 2\cdot 2x}{(1 + x^2)^3} = \frac{6x^2 - 2}{(1 + x^2)^3};\ y''(0) < 0,\ \text{Max } (0,1);$$

$y'' = 0$, $x = \pm\dfrac{\sqrt{3}}{3}$, i.e., $\left(\pm\dfrac{\sqrt{3}}{3}, \dfrac{3}{4}\right)$ are inflection points. In addition, we find $x = 1$, $y = \frac{1}{2}$; $x = 2$, $y = \frac{1}{5}$. See Figure 95.

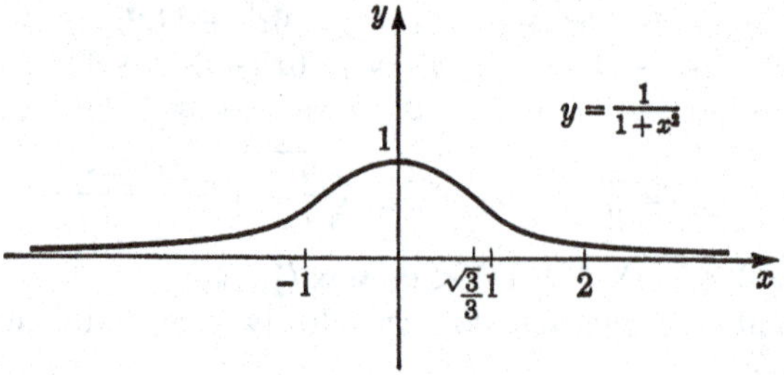

FIGURE 95

474. The curve is called Newton's serpentine. It is defined everywhere and is odd, i.e., symmetric with respect to the origin. At $x = 0$, $y = 0$.

$$\lim_{x\to\pm\infty} y = 0, \quad \text{i.e.,} \quad y = 0 \text{ is an asymptote.}$$

$$y' = \frac{1 + x^2 - x\cdot 2x}{(1 + x^2)^2} = \frac{1 - x^2}{(1 + x^2)^2},$$

$$y'' = \frac{-2x(1 + x^2) - (1 - x^2)\cdot 2\cdot 2x}{(1 + x^2)^3} = \frac{2x^3 - 6x}{(1 + x^2)^3}.$$

$y' = 0$, $x_{1,2} = \pm 1$; $y''(1) < 0$, Max $(1, \frac{1}{2})$.

$y''(-1) > 0$, Min $(-1, -\frac{1}{2})$.

(This is also clear by symmetry.)

$y'' = 0$; $x_1 = 0$, $x_{2,3} = \pm\sqrt{3}$, i.e., $(0,0)$, $\left(\pm\sqrt{3}, \pm\frac{\sqrt{3}}{4}\right)$ are inflection points. For $x = 2$, $y = \frac{2}{5}$. See Figure 96.

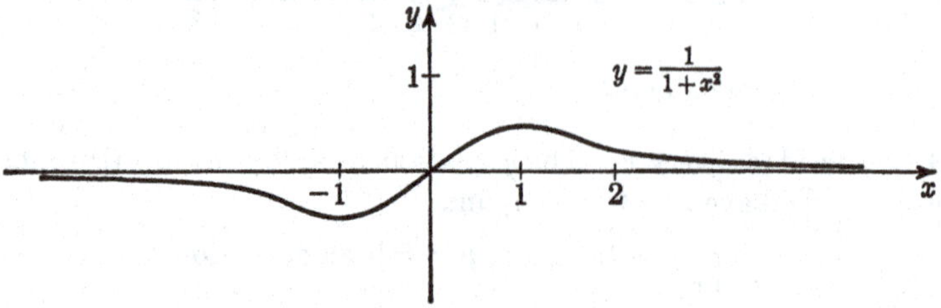

FIGURE 96

475. Defined everywhere except at $x = \pm 1$. $x = \pm 1$ are asymptotes. $\lim_{x\to\pm\infty} y = 0$, i.e., $y = 0$ is also an asymptote. $x = 0, y = 1; f(-x) = f(x)$, i.e., the curve is symmetric with respect to the y axis.

$$y' = \frac{2x}{(1 - x^2)^2}, \quad y'' = \frac{2(1 - x^2) - 2x\cdot 2(-2x)}{(1 - x^2)^3} = \frac{6x^2 + 2}{(1 - x^2)^3}.$$

$y' = 0$, $x = 0$; $y''(0) > 0$, Min (0,1).

$y'' \neq 0$ and there are no inflection points. Further, $x = \pm\frac{1}{2}$, $y = \frac{4}{3}$; $x = \pm 2$, $y = -\frac{1}{3}$. See Figure 97.

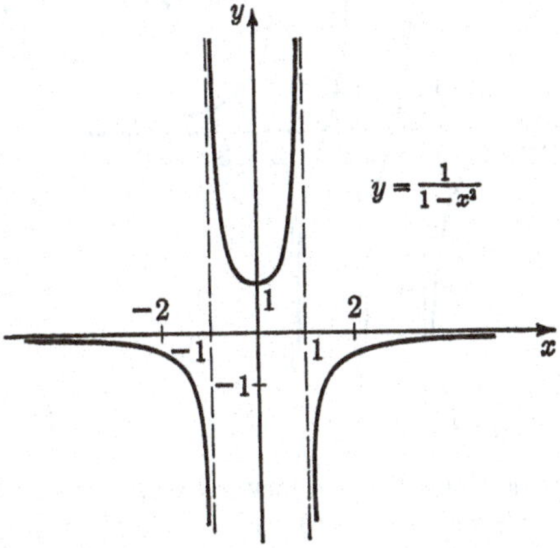

FIGURE 97

476. The domain of definition is $x \neq 0$, $x \neq 1$, $x \neq 2$. $x = 0$, $x = 1$, $x = 2$ are asymptotes. $\lim_{x \to \pm\infty} y = 0$ and $y = 0$ is also an asymptote. Intersections with the axes are:

$$y = 0; \quad x^2 - 3x + 2 + x^2 - 2x + x^2 - x = 0, \quad 3x^2 - 6x + 2 = 0,$$
$$x_1 \sim 1.6, \qquad x_2 \sim 0.4.$$

$$y' = -\frac{1}{x^2} - \frac{1}{(x-1)^2} - \frac{1}{(x-2)^2};$$

$y' < 0$ for every x and the function decreases everywhere.

$$y'' = 2\left[\frac{1}{x^3} + \frac{1}{(x-1)^3} + \frac{1}{(x-2)^3}\right];$$

the equation $y'' = 0$ is of sixth degree, and we shall find intervals in which the solutions of this equation are located. For $x > 2$, $y'' > 0$ and for $x < 0$, $y'' < 0$; i.e., all real solutions of the equation $y'' = 0$ are between 0 and 2. Near to 0 from the right $y'' > 0$ (because $1/x^3 \to +\infty$ when $x \to 0$ from the right). Near to 1 from the left $y'' < 0$ (because $1/(x-1)^3$ tends to $-\infty$ when $x \to 1$ from the left). It follows that between 0 and 1 there is an inflection point. Similarly it can be shown that between 1 and 2 there is an inflection point. Additional points of the graph are $x = -1$, $y = -1\frac{5}{6}$; $x = 3$, $y = 1\frac{5}{6}$. See Figure 98.

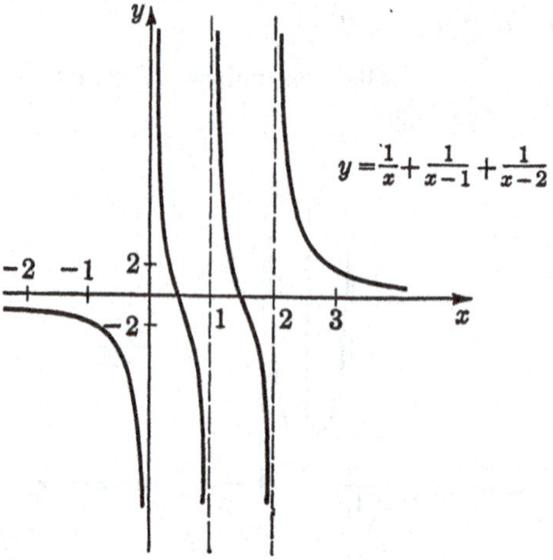

FIGURE 98

477. The function is defined for every x except $x = \pm 1$. $x = \pm 1$ are asymptotes. $\lim\limits_{x\to\pm\infty} 2x/(x^2 - 1) = 0$, i.e., $y = x$ is an oblique asymptote. (0,0) is a center of symmetry and also a point of intersection with the axes.

$$y' = 1 + \frac{2x^2 - 2 - 4x^2}{(x^2 - 1)^2} = 1 + \frac{-2x^2 - 2}{(x^2 - 1)^2},$$

$$y'' = -2\,\frac{2x(x^2 - 1) - (x^2 + 1)\cdot 2\cdot 2x}{(x^2 - 1)^3} = 4\,\frac{x^3 + 3x}{(x^2 - 1)^3}.$$

$y' = 0$, $(x^2 - 1)^2 - 2x^2 - 2 = 0$, $x^4 - 4x^2 - 1 = 0$; $x_{1,2} \sim \pm 2.12$.
$y''(2.12) > 0$, Min $(\sim 2.12, \sim 3.3)$.
$y''(-2.12) < 0$, Max $(\sim -2.12, \sim -3.3)$.
$y'' = 0$, $x = 0$, (0,0) is an inflection point.
For $-1 < x < 1$, $y' < 0$, i.e., the function decreases.
This gives a clear idea of the position of the curve with respect to the asymptotes. Additional points are $x = \frac{1}{2}$, $y = -\frac{5}{6}$; $x = 2$, $y = 3\frac{1}{3}$; $x = 3$, $y = 3\frac{3}{4}$. See Figure 99.

487. The graph of $y = +\sqrt{x - 2}$ is symmetric to the graph of $y = -\sqrt{x - 2}$ with respect to the x axis. The domain of either function is $x \geqslant 2$. The union of the two graphs is the graph of $y^2 = x - 2$ which is a parabola. It is investigated in analytic geometry. The graph is given in Figure 100.

488. The domain of either function is $-5 \leqslant x \leqslant 5$. Here also we can square and obtain the equation

$$\frac{x^2}{4} + y^2 = \frac{25}{4} \quad \text{or} \quad \frac{x^2}{25} + \frac{y^2}{\frac{25}{4}} = 1.$$

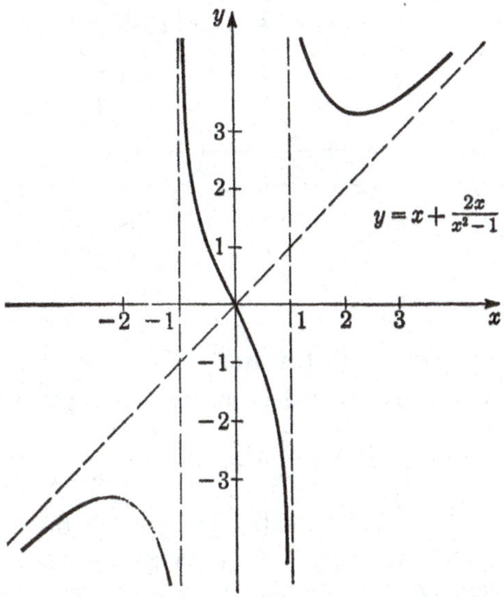

FIGURE 99

This is an ellipse with semiaxes 5 and 2.5. Its graph is given in Figure 101. This curve is also investigated in analytic geometry.

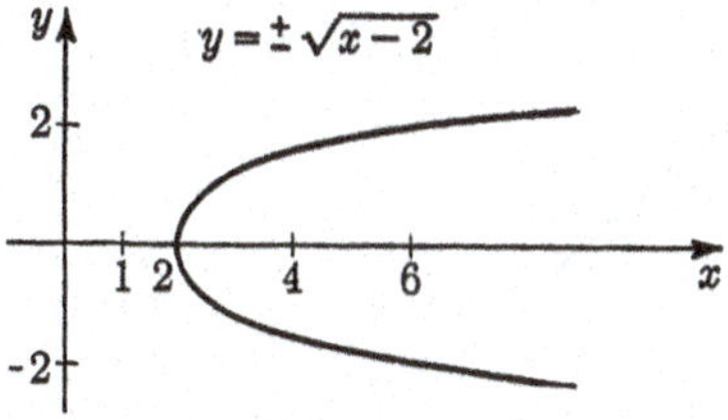

FIGURE 100

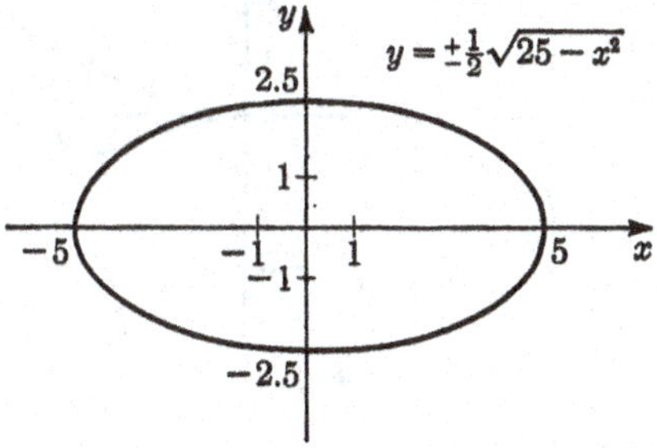

FIGURE 101

489. $|x| \geqslant 1$. The straight lines $y = \pm\frac{1}{2}x$ are asymptotes (cf. Prob. 458). After squaring we obtain

$$x^2 - \frac{y^2}{\frac{1}{4}} = 1.$$

This is a hyperbola, which is also investigated in analytic geometry. See Figure 102.

490. $x \geqslant 0$. The curve is symmetric with respect to the x axis. (0,0) is

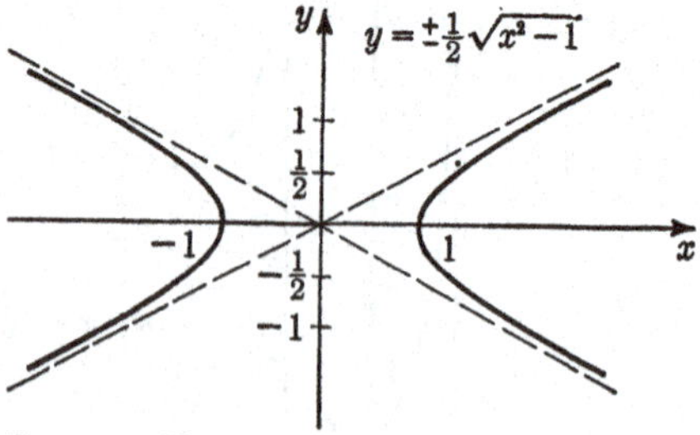

FIGURE 102

the point of intersection with the axes. There are no asymptotes. We investigate the upper branch of the curve, i.e., $y = \frac{1}{2}x\sqrt{x}$.

$$y' = \frac{3}{2} \cdot \frac{1}{2} x^{1/2} = \frac{3}{4}\sqrt{x}; \qquad y'' = \frac{3}{8} \cdot \frac{1}{\sqrt{x}}.$$

$y' = 0$; $x = 0$. This is not an extremum because the curve exists only for $x \geqslant 0$. Anyway, we know that the tangent to the curve at (0,0) is horizontal. There are no inflection points. Further, we have $x = 1$, $y = \pm\frac{1}{2}$; $x = 4$, $y = \pm 4$. See Figure 103.

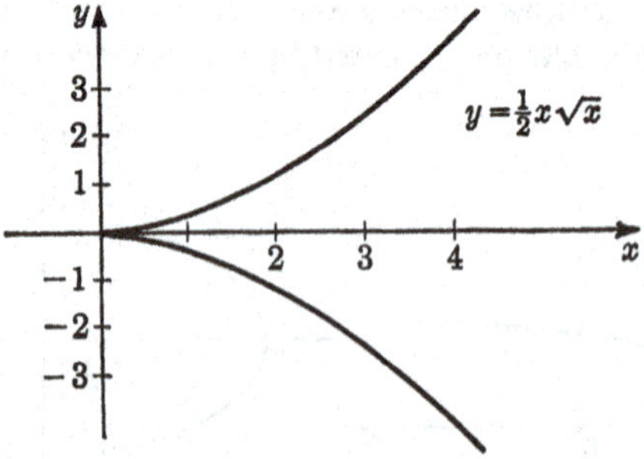

FIGURE 103

491. The domain of definition will be obtained by solving the inequality $x/(4 - x) \geqslant 0$. Its solutions are $0 \leqslant x < 4$. The curve is symmetric with respect to Ox. $\lim\limits_{x \to 4-} y = \infty$ (or $-\infty$), i.e., $x = 4$ is an asymptote.

For $y = x\sqrt{\dfrac{x}{4 - x}}$ we have

$$y' = \sqrt{\frac{x}{4-x}} + \frac{x(4 - x + x)}{2\sqrt{\dfrac{x}{4-x}}\,(4-x)^2} = \frac{4x + 8x - 2x^2}{2\sqrt{\dfrac{x}{4-x}}\,(4-x)^2} = \frac{6x - x^2}{\sqrt{x}\,(4-x)^{3/2}}.$$

$y' = 0$; $x_1 = 0$. The second solution is outside the domain of definition of the function. The tangent at (0,0) is horizontal. Additional points are $x = 1, y = \pm\sqrt{\frac{1}{3}}; x = 2, y = \pm 2; x = 3, y = \pm 3\sqrt{3}$. The curve, known as the cissoid of Diocles, is given in Figure 104.

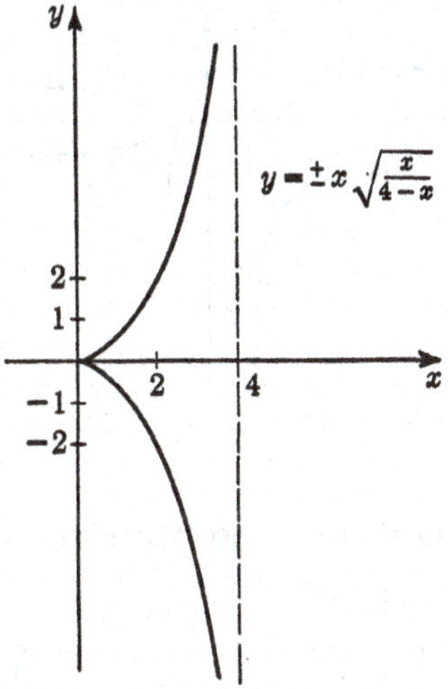

FIGURE 104

492. $(x - 1)/(x + 1) \geqslant 0 \Rightarrow x < -1,\ x \geqslant 1$. This is the domain of definition. There is one intersection with the axes: $x = 1, y = 0$. $x = 0$ is not in the domain of definition. $\lim\limits_{x\to\pm\infty} y = 1$ (or -1); $y = \pm 1$ are two asymptotes of the curve. Another asymptote is $x = -1$. For the upper part of the curve:

$$y' = \frac{1}{2}\sqrt{\frac{x+1}{x-1}} \cdot \frac{x + 1 - x + 1}{(x+1)^2} = \sqrt{\frac{x+1}{x-1}} \cdot \frac{1}{(x+1)^2}.$$

There are no extrema. $y' > 0$ and the upper part of the curve increases for every x in the domain of definition. Additional points are $x = 2$, $y = \pm\sqrt{\frac{1}{3}}$; $x = -2$, $y = \pm\sqrt{3}$; $x = -3$, $y = \pm\sqrt{2}$. See Figure 105.

493. We require $x^2 + x + 1 \geqslant 0$ and $x^2 - x + 1 \geqslant 0$. These inequalities hold for all x, and consequently the domain of definition of the function is $-\infty < x < \infty$. There is one intersection with the axes, at $x = 0$, $y = 1$. Investigating symmetry,

$$2f(-x) = \sqrt{x^2 - x + 1} + \sqrt{x^2 + x + 1} = 2f(x),$$

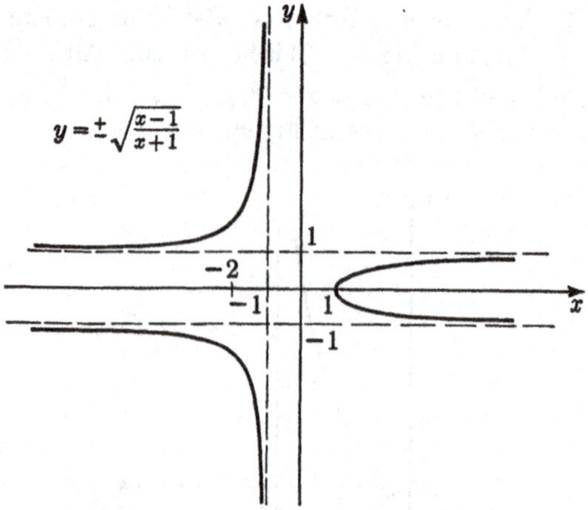

FIGURE 105

i.e., Oy is an axis of symmetry. Seeking asymptotes, $\lim_{x\to\infty} y = \infty$. We try to find oblique asymptotes:

$$\lim_{x\to\infty} \frac{y}{x} = \lim_{x\to\infty} \frac{1}{2}\left(\sqrt{1 + \frac{1}{x} + \frac{1}{x^2}} + \sqrt{1 - \frac{1}{x} + \frac{1}{x^2}}\right) = 1;$$

$$\lim_{x\to\infty} (y - x) = \lim_{x\to\infty} \left[\frac{1}{2}\sqrt{x^2 + x + 1} + \frac{1}{2}\sqrt{x^2 - x + 1} - x\right]$$

$$= \frac{1}{2}\lim_{x\to\infty} [\sqrt{x^2 + x + 1} - x + \sqrt{x^2 - x + 1} - x]$$

$$= \frac{1}{2}\lim_{x\to\infty} \left[\frac{x^2 + x + 1 - x^2}{\sqrt{x^2 + x + 1} + x} + \frac{x^2 - x + 1 - x^2}{\sqrt{x^2 - x + 1} + x}\right]$$

$$= \frac{1}{2}\left(\frac{1}{2} - \frac{1}{2}\right) = 0;$$

$y = x$ is an asymptote, and by symmetry also $y = -x$ is an asymptote. Looking for extrema,

$$2y' = \frac{2x + 1}{2\sqrt{x^2 + x + 1}} + \frac{2x - 1}{2\sqrt{x^2 - x + 1}};$$

$$y' = 0, \qquad \frac{2x + 1}{\sqrt{x^2 + x + 1}} = \frac{1 - 2x}{\sqrt{x^2 - x + 1}},$$

$$(4x^2 + 4x + 1)(x^2 - x + 1) = (4x^2 - 4x + 1)(x^2 + x + 1),$$
$$4x^4 - 4x^3 + 4x^2 + 4x^3 - 4x^2 + 4x + x^2 - x + 1$$
$$= 4x^4 + 4x^3 + 4x^2 - 4x^3 - 4x^2 - 4x + x^2 + x + 1,$$
$$3x = -3x, \qquad x = 0.$$

This is the only solution, and clearly (0,1) is a minimum point. The graph of the curve is given in Figure 106.

$2y = \sqrt{x^2+x+1} + \sqrt{x^2-x+1}$

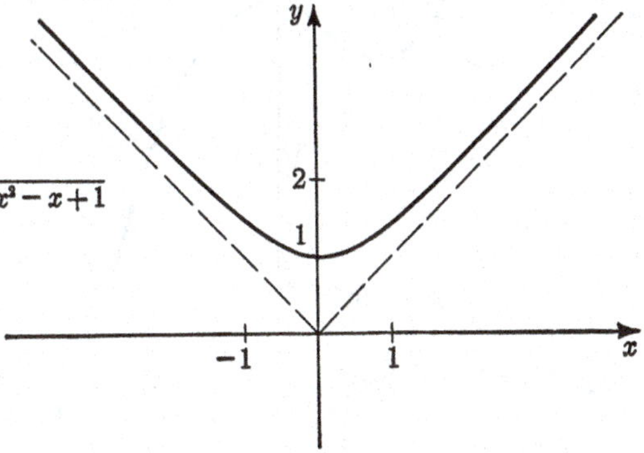

FIGURE 106

494. The domain of definition is $x \geqslant -1$. Intersections with the axes are $(-1,0)$, $(0,0)$. Ox is an axis of symmetry. There are no asymptotes. For the positive sign we have

$$y' = 2x\sqrt{x+1} + \frac{x^2}{2\sqrt{x+1}} = \frac{4x^2 + 4x + x^2}{2\sqrt{x+1}} = \frac{5x^2 + 4x}{2\sqrt{x+1}};$$

$$y'' = \frac{(10x + 4)\sqrt{x+1} - (5x^2 + 4x)(1/2\sqrt{x+1})}{2(x+1)} = \frac{15x^2 + 24x + 8}{4(x+1)^{3/2}}.$$

$y' = 0$; $x_1 = 0$, $x_2 = -4/5$;
$y''(0) > 0$, Min (0,0); $y''(-4/5) < 0$, Max $(-4/5, 16/25\sqrt{5})$.
$y'' = 0$, $15x^2 + 24x + 8 = 0$, $x_1 \sim -17/15$, $x_2 \sim -7/15$.

The first solution is outside the domain of definition, and the second gives an inflection point $(\sim -7/15, \sim 0.2)$. Further, $\lim_{x\to\infty} y = \infty$. An additional point is $x = 1$, $y = \sqrt{2}$. See Figure 107.

495. The function is defined everywhere. (0,0) and $(-1,0)$ are points of intersection with the axes. There are no asymptotes and no symmetry.

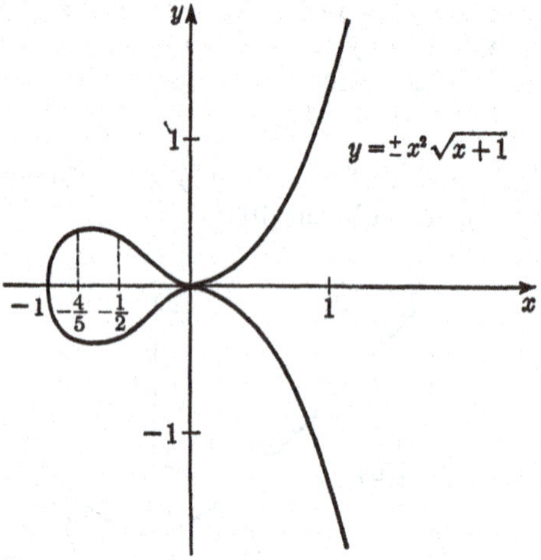

FIGURE 107

$$y' = 3(x+1)^2\sqrt[3]{x^2} + \frac{2}{3}(x+1)^3\frac{1}{\sqrt[3]{x}} = \frac{(x+1)^2(11x+2)}{3\sqrt[3]{x}};$$

$$y'' = \frac{[2(x+1)(11x+2)+(x+1)^2\cdot 11]\sqrt[3]{x} - (x+1)^2(11x+2)\cdot\frac{1}{3}\cdot\frac{1}{\sqrt[3]{x^2}}}{3\sqrt[3]{x^2}}$$

$$= \frac{(x+1)(66x^2+12x+33x^2+33x-11x^2-11x-2x-2)}{9\sqrt[3]{x^4}}$$

$$= \frac{2(x+1)(44x^2+16x-1)}{9\sqrt[3]{x^4}}.$$

$y' = 0;\ x_1 = -1,\ x_2 = -\frac{2}{11}$.

For $x < -\frac{2}{11}$, $y' > 0$, i.e., the function increases in this interval and $x = -1$ is not an extremum point. Now $y'(-\frac{2}{11}+) < 0$, i.e., Max $(-\frac{2}{11}, \sim 0.7)$. The point $x = 0$ must also be investigated.

$y'(0-) < 0$, $y'(0+) > 0$; Min $(0,0)$.
$y'' = 0;\ x_1 = -1;\ 44x^2 + 16x - 1 = 0,\ x_2 \sim 0.057,\ x_3 \sim -0.4$.

There are three inflection points:

$$(-1,0), \qquad (\sim -0.4, \sim 0.11), \qquad (\sim 0.057, \sim 0.14).$$

Finally, $\lim\limits_{x\to\pm\infty} y = \pm\infty$. The graph is given in Figure 108.

496. The domain of definition is $x \leqslant 1$, $x \neq -1$. $\lim\limits_{x\to -1\pm} y = \pm\infty$, i.e., $x = -1$ is an asymptote. Intersections with the axes are $(1,0)$, $(0,0)$.

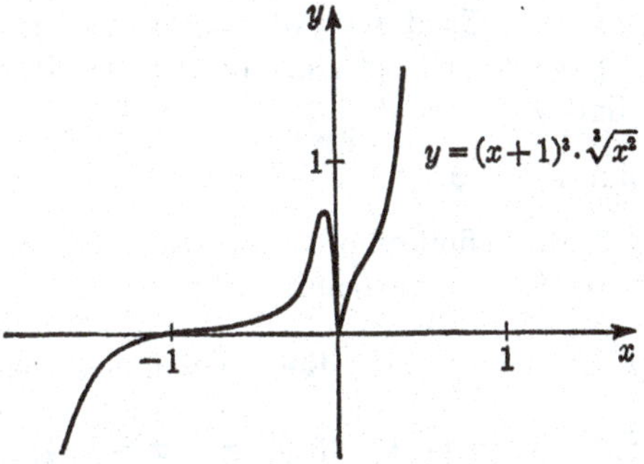

FIGURE 108

$$y' = \frac{\sqrt{1-x}}{1+x} - \frac{x}{2(1+x)\sqrt{1-x}} - \frac{x\sqrt{1-x}}{(1+x)^2}$$

$$= \frac{2 - 2x^2 - x - x^2 - 2x + 2x^2}{2(1+x)^2\sqrt{1-x}} = \frac{-x^2 - 3x + 2}{2(1+x)^2\sqrt{1-x}}.$$

$y' = 0$; $x^2 + 3x - 2 = 0$; $x_1 \sim -3.5$, $x_2 \sim 0.5$.

The derivative of the numerator of y' is $-2x - 3$. This expression is positive at x_1 and negative at x_2, and because the denominator of y' is always positive we have Min $(\sim -3.5, \sim 3)$ and Max $(\sim 0.5, \sim 0.25)$. We find now

$$\lim_{x\to-\infty} y = \infty \quad \text{and} \quad \lim_{x\to 1-} y' = -\infty,$$

i.e., the tangent tends to a vertical position. See Figure 109.

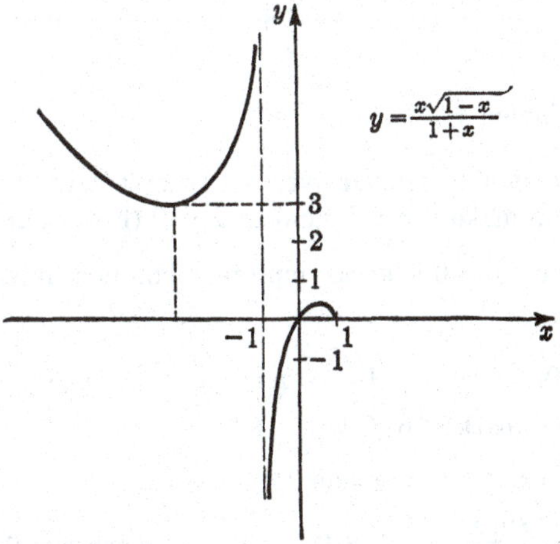

FIGURE 109

500. The function is defined everywhere except at $x = 0$. There is no symmetry and there are no points of intersection with the axes. $y > 0$ for every x. $\lim_{x\to\pm\infty} y = 1$, and $\lim_{x\to 0+} y = \infty$; i.e., $y = 1$ and $x = 0$ are asymptotes. $\lim_{x\to 0-} e^{1/x} = 0$.

Now, by adding to the definition of the function $y(0) = 0$, we can make it continuous from the left at this point. Then

$y' = -\frac{1}{x^2}e^{1/x}$; $y' < 0$ for all x, i.e., the function decreases everywhere.

$y'' = \frac{1}{x^4}e^{1/x} + \frac{2}{x^3}e^{1/x} = \frac{e^{1/x}}{x^4}(1 + 2x)$; $y'' = 0$, $x = -\frac{1}{2}$;

$(-\frac{1}{2}, e^{-2})$ is an inflection point.

For $x < -\frac{1}{2}$, $y'' < 0$, and the curve is concave downward; for $-\frac{1}{2} < x < 0$ and for $x > 0$, $y'' > 0$, and the curve is concave upward. $\lim_{x\to 0-} y' = 0$, i.e., the tangent tends to a horizontal position when the point on the curve tends from the left to the origin. More points are $x = 1$, $y = e$; $x = 2$, $y = \sim 1.65$. See Figure 110.

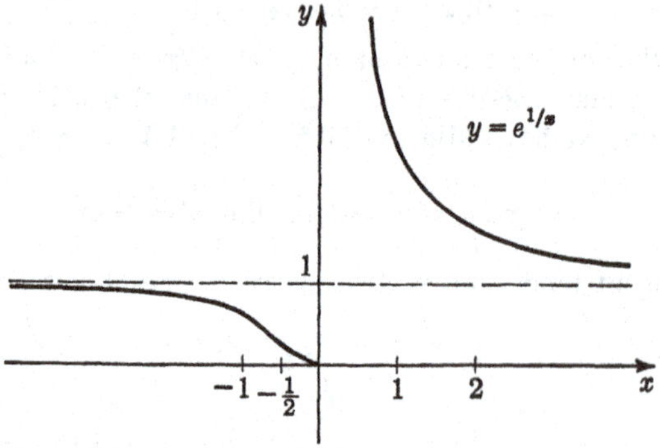

FIGURE 110

501. The function is defined everywhere except at $x = 0$. By adding $f(0) = 1$ we can make it continuous at $x = 0$ (because $\lim_{x\to 0} (\sin x)/x = 1$). $\lim_{x\to\pm\infty} y = 0$, i.e., $y = 0$ is an asymptote. Points of intersection with the axes are

$$y = 0, \quad \sin x = 0; \quad x = \pi n, \quad n = \pm 1, \pm 2, \ldots$$

(by the above remark also (0,1)).

$f(-x) = f(x)$, i.e., Oy is a symmetry axis.

$$y' = \frac{x\cos x - \sin x}{x^2}; \quad y' = 0, \quad x\cos x - \sin x = 0; \quad \tan x = x.$$

This equation can be solved graphically (see Fig. 111). The points of intersection (which are the extrema of the investigated curve) are very close

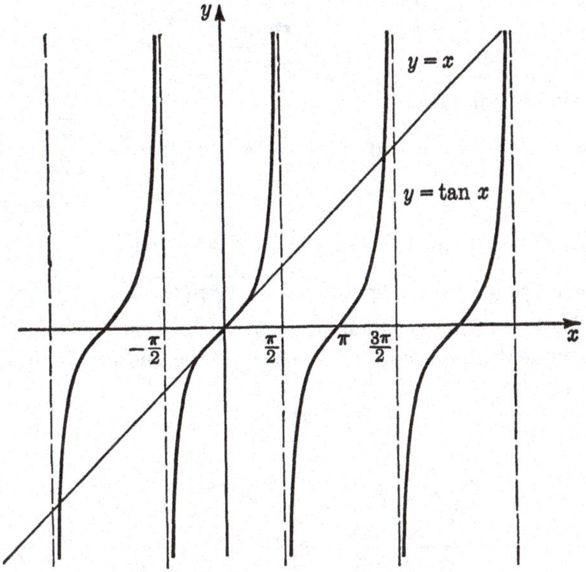

FIGURE 111

to $x = \pm(2n + 1)(\pi/2)$ (and $x = 0$, of course, but this point is outside the domain of definition). At $x = 0$ we find

$$\lim_{x\to 0} y' = \lim_{x\to 0} \frac{x \cos x - \sin x}{x^2} = \lim_{x\to 0} \frac{\cos x - x \sin x - \cos x}{2x} = 0,$$

i.e., the tangent tends to a horizontal position when $x \to 0$. The inequality $|\sin x| < |x|$ shows that the curve decreases at both sides of $x = 0$. To find the inflection points we would need to solve another transcendental equation, and this will not be done here. The graph is sketched in Figure 112.

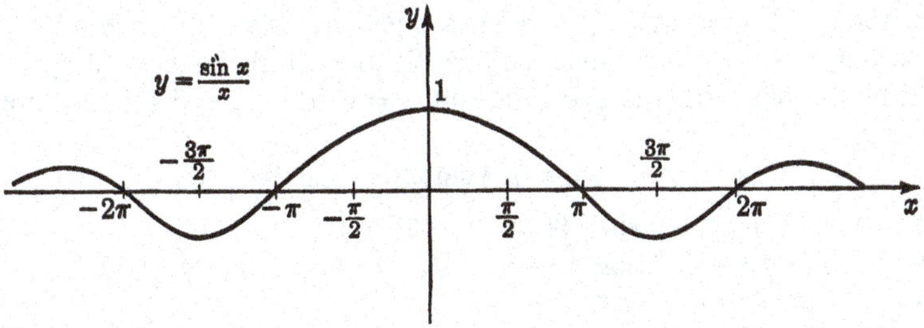

FIGURE 112

502. The function is defined everywhere. There are no asymptotes. Oy is an axis of symmetry. Intersections with the axes are

$$y = 0, \qquad y = 0;$$
$$x = 0; \qquad x^2 = \pi n,\ n = 0, 1, \ldots,$$
$$x = \pm\sqrt{\pi n}.$$

The function is not periodic.

$$y' = 2x \cos x^2;\ y' = 0,\ x = 0,\ x^2 = \frac{\pi}{2}(2n+1),\ x = \pm\sqrt{\frac{\pi}{2}(2n+1)},$$
$$(n = 0, 1, 2, \ldots).$$

$y'' = 2 \cos x^2 - 4x^2 \sin x^2;\ y''(0) = 2 > 0;$ Min $(0,0)$.

The other extrema are maxima at $(\pm\sqrt{(\pi/2)(4n+1)}, 1)$ and minima at $(\pm\sqrt{(\pi/2)(4n+3)}, -1)$. The inflection points can be found by solving the transcendental equation $2x^2 = \cot x^2$. See the graph in Figure 113.

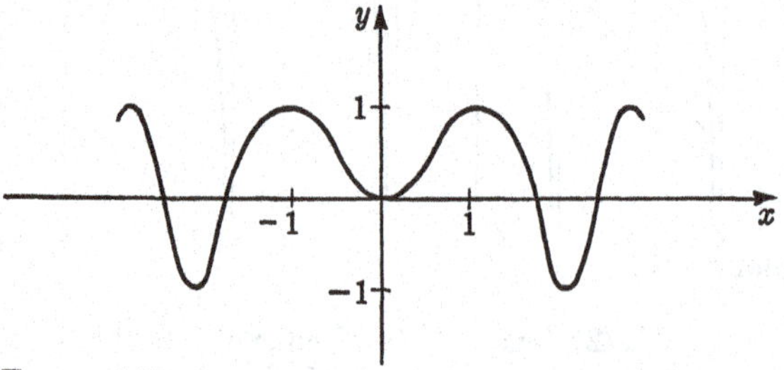

FIGURE 113

503. This curve can be drawn by superposing three functions. We shall investigate the function as a whole. It is defined everywhere. $f(-x) = -f(x)$, i.e., (0,0) is a center of symmetry. There are no asymptotes. Now

$$f(x + 2\pi) = \sin(x + 2\pi) + \tfrac{1}{2}\sin(2x + 4\pi) + \tfrac{1}{3}\sin(3x + 6\pi)$$
$$= \sin x + \tfrac{1}{2}\sin 2x + \tfrac{1}{3}\sin 3x = f(x),$$

i.e., the function is periodic and 2π is a period. This, together with the symmetry, enables us to limit the investigation to the interval $[0,\pi]$.

Intersections with the axes occur at $x = 0$, $y = 0$, and at other solutions of

$$y = 0, \qquad \sin x + \tfrac{1}{2}\sin 2x + \tfrac{1}{3}\sin 3x = 0,$$

$\sin x + \sin x \cos x + \frac{1}{3}\sin x\,(4\cos^2 x - 1) = 0,$
$\sin x\,(3 + 3\cos x + 4\cos^2 x - 1) = 0;$
$\sin x = 0;\ x = 0,\ x = \pi;$
$3 + 3\cos x + 4\cos^2 x - 1 = 0,\ 4\cos^2 x + 3\cos x + 2 = 0.$

The quadratic equation for $\cos x$ has no real solution. Thus intercepts are $(0,0)$ and $(\pi,0)$, and other values by symmetry.

$$y' = \cos x + \cos 2x + \cos 3x; \qquad y'' = -\sin x - 2\sin 2x - 3\sin 3x.$$
$$y' = 0; \qquad \cos x + \cos 2x + \cos 3x = 0;$$
$$2\cos 2x \cos x + \cos 2x = 0, \qquad \cos 2x = 0, \qquad \cos x = -\tfrac{1}{2};$$

In the interval $[0,\pi]$ we obtain $x = \dfrac{\pi}{4}, x = \dfrac{3\pi}{4}, x = \dfrac{2\pi}{3}.$

$$y''\left(\frac{\pi}{4}\right) = -\frac{\sqrt{2}}{2} - 2 - 3\frac{\sqrt{2}}{2} < 0; \qquad \text{Max}\left(\frac{\pi}{4}, \frac{3+4\sqrt{2}}{6}\right).$$

Now we have Min $\left(\dfrac{2\pi}{3}, \sim 0.43\right)$, Max $\left(\dfrac{3\pi}{4}, \sim 0.44\right)$.

To find the inflection points one can solve the trigonometric equation $y'' = 0$. The graph of one period is given in Figure 114.

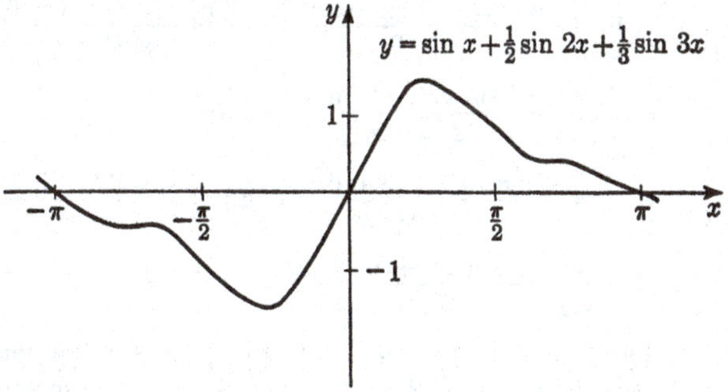

FIGURE 114

504. The function is symmetric with respect to Oy.

$$f(x+\pi) = \frac{\tan(3x+3\pi)}{\tan(x+\pi)} = \frac{\tan 3x}{\tan x} = f(x),$$

i.e., π is a period of the function, and we can limit ourselves to the interval $[0,\pi/2]$. At $x = 0$, $x = \pi/6$, and $x = \pi/2$ the function is not defined. We have

$$\lim_{x\to\pi/6-} y = \infty, \qquad \lim_{x\to\pi/6+} y = -\infty;$$

$x = \pi/6$ is an asymptote.

$$\lim_{x\to 0} y = \lim_{x\to 0}\frac{\tan 3x}{\tan x} = \lim_{x\to 0}\frac{3\cos^2 x}{\cos^2 3x} = 3.$$

$$\lim_{x\to\pi/2} y = \lim_{x\to\pi/2}\frac{\tan 3x}{\tan x} = \lim_{x\to\pi/2}\frac{3\cos^2 x}{\cos^2 3x} = \lim_{x\to\pi/2}\frac{3\sin 2x}{3\sin 6x}$$
$$= \lim_{x\to\pi/2}\frac{2\cos 2x}{6\cos 6x} = \frac{1}{3}.$$

We can extend the definition of the function by putting $f(0) = 3$ and $f(\pi/2) = 1/3$, and it will be continuous at these points. The intercepts are $x = 0$, $y = 3$ (by the extended definition), and

$$y = 0, \qquad \tan 3x = 0, \qquad 3x = \pi, \qquad x = \pi/3$$

(we limit ourselves to $[0,\pi/2]$).

$$y' = \frac{\tan x \cdot 3 \cdot \dfrac{1}{\cos^2 3x} - \tan 3x \cdot \dfrac{1}{\cos^2 x}}{\tan^2 x} = \frac{3 \sin x \cos x - \sin 3x \cos 3x}{\sin^2 x \cos^2 3x}.$$

$$y' = 0; \qquad 3 \sin 2x - \sin 6x = 0, \qquad \sin 6x - \sin 2x - 2 \sin 2x = 0,$$

$$2 \sin 2x \cos 4x - 2 \sin 2x = 0,$$

$$\sin 2x = 0, \qquad x = 0, \qquad x = \pi/2; \qquad \cos 4x = 1, \qquad x = 0, \qquad x = \pi/2.$$

At both points the above expression for y' is not defined.

$$\lim_{x \to 0} y' = \lim_{x \to 0} \frac{\frac{3}{2} \sin 2x - \frac{1}{2} \sin 6x}{\sin^2 x \cos^2 3x} = \lim_{x \to 0} \frac{3 \cos 2x - 3 \cos 6x}{\sin 2x}$$

$$= 3 \lim_{x \to 0} \frac{-2 \sin 2x + 6 \sin 6x}{2 \cos 2x} = 0;$$

$$\lim_{x \to \pi/2} y' = \lim_{x \to \pi/2} \frac{\frac{3}{2} \sin 2x - \frac{1}{2} \sin 6x}{\sin^2 x \cos^2 3x} = \lim_{x \to \pi/2} \frac{3 \cos 2x - 3 \cos 6x}{-3 \sin 6x}$$

$$= \lim_{x \to \pi/2} \frac{-2 \sin 2x + 6 \sin 6x}{-6 \cos 6x} = 0.$$

The limit of the tangent to the curve at both points is horizontal. For $0 < x < \pi/6$ and $\pi/6 < x < \pi/2$, $y' > 0$, i.e., the function increases there. See Figure 115.

505. The function is defined for $x > 0$.

$$\lim_{x \to 0+} y = \lim_{x \to 0+} \frac{\ln x}{1/x} = \lim_{x \to 0+} \frac{1/x}{-1/x^2} = 0; \qquad \lim_{x \to \infty} y = \infty.$$

There is no symmetry nor asymptotes.

$$y' = \ln x + 1, \qquad y'' = 1/x;$$

$y'' > 0$ for the whole domain of definition and the curve is everywhere concave upward.

$$y' = 0; \qquad \ln x + 1 = 0, \qquad x = 1/e; \qquad \text{Min } (1/e, -1/e).$$

(1,0) is the point of intersection with the axes. See Figure 116.

506. The domain of definition is $x > 0$;

$$\lim_{x \to 0+} y = \lim_{x \to 0+} \frac{1/x}{-2/x^3} = 0;$$

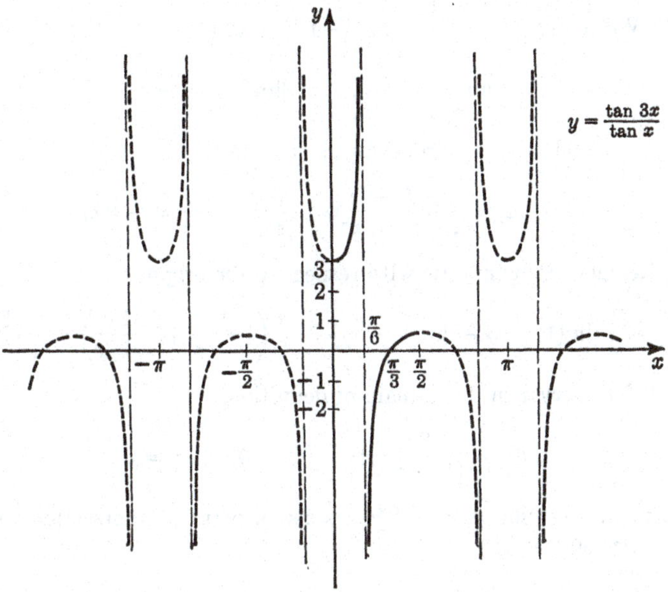

FIGURE 115

there is no symmetry nor asymptotes. (1,0) is a point of intersection with the axes.

$$y' = 2x \ln x + x; \qquad y'' = 2 \ln x + 3.$$

$y' = 0$, $x = 0$ (although $x = 0$ is outside the domain of definition, $\lim_{x \to 0+} y' = 0$, and the tangent tends to horizontal position when $x \to 0+$); $2 \ln x + 1 = 0$, $x = e^{-1/2}$;

$$y''(e^{-1/2}) = -1 + 3 > 0, \qquad \text{Min}\left(e^{-1/2}, -\frac{1}{2e}\right).$$

$$y'' = 0; \qquad 2 \ln x + 3 = 0, \qquad x = e^{-3/2};$$

$\left(e^{-3/2}, -\dfrac{3}{2e^3}\right)$ is an inflection point. See Figure 117.

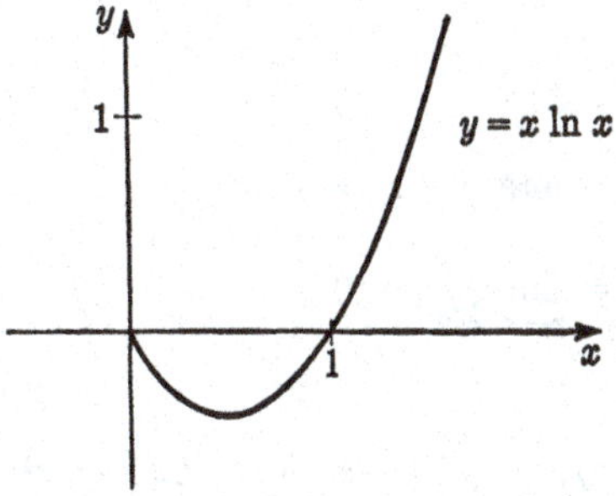

FIGURE 116

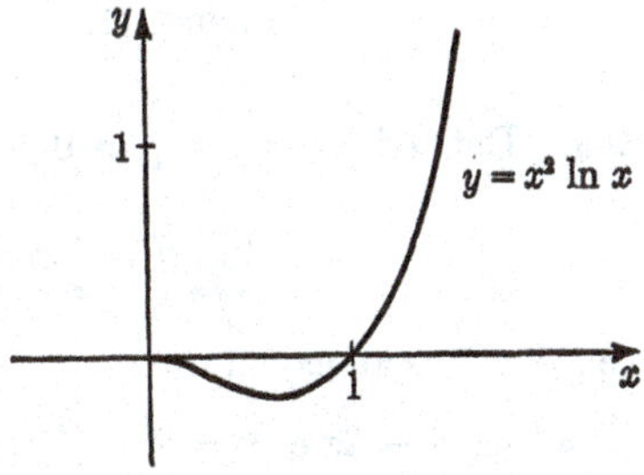

FIGURE 117

507. We require $\frac{1+x}{1-x} > 0$, i.e., $-1 < x < 1$.

$$\lim_{x\to 1-} y = +\infty, \qquad \lim_{x\to -1+} y = -\infty;$$

i.e., $x = 1$ and $x = -1$ are asymptotes. Now

$$f(-x) = \frac{1}{2}\ln\frac{1-x}{1+x} = -\frac{1}{2}\ln\frac{1+x}{1-x} = -f(x)$$

and the curve is symmetric with respect to the origin.

$$y' = \frac{1}{2}[\ln(1+x) - \ln(1-x)]' = \frac{1}{2}\left(\frac{1}{1+x} + \frac{1}{1-x}\right) = \frac{1}{1-x^2};$$

$y' > 0$ for every x in the domain of definition.

$$y'' = \frac{2x}{(1-x^2)^2}; \qquad y'' = 0, \qquad x = 0;$$

(0,0) is an inflection point. This is also a point of intersection with the axes. See Figure 118.

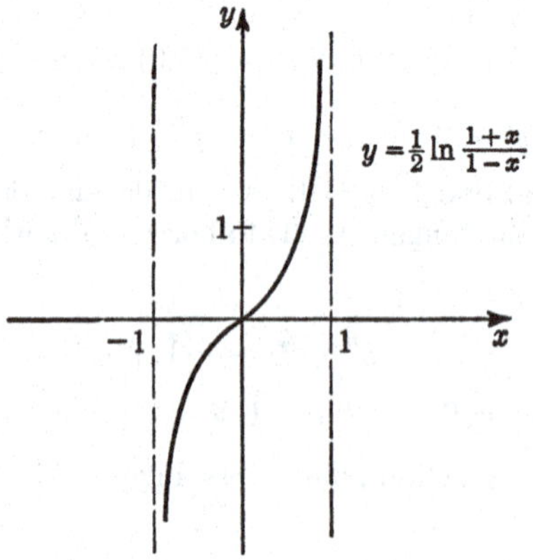

FIGURE 118

508. Defined for any x; $y > 0$ everywhere; $f(-x) = f(x)$.

$$\lim_{x\to\infty} y = \lim_{x\to\infty}\frac{x^2}{e^{x^2}} = \lim_{x\to\infty}\frac{2x}{2xe^{x^2}} = 0,$$

i.e., $y = 0$ is an asymptote.

$$y' = 2xe^{-x^2} - 2x^3e^{-x^2} = 2xe^{-x^2}(1-x^2); \qquad y'' = 2e^{-x^2}(1 - 5x^2 + 2x^4).$$

$$y' = 0; \qquad x_1 = 0, \qquad x_{2,3} = \pm 1;$$

$$y''(0) = 2 > 0, \qquad \text{Min } (0,0);$$

$$y''(\pm 1) = \frac{2}{e}(1 - 5 + 2) < 0, \qquad \text{Max}\left(\pm 1, \frac{1}{e}\right).$$

$$y'' = 0; \qquad x_{1,2,3,4} = \pm\tfrac{1}{2}\sqrt{5 \pm \sqrt{17}}.$$

There are 4 inflection points. See Figure 119.

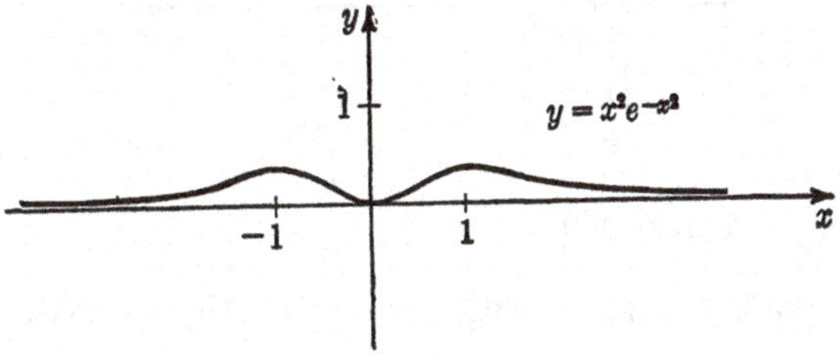

FIGURE 119

509. The domain is $x > 0$;

$$\lim_{x \to 0+} y = \lim_{x \to 0+} \frac{\ln x}{x} = -\infty,$$

i.e., $x = 0$ is an asymptote.

$$\lim_{x \to \infty} \frac{\ln x}{x} = \lim_{x \to \infty} \frac{1}{x} = 0,$$

i.e., $y = x$ is an oblique asymptote. The curve intersects this asymptote.

$$x + \frac{\ln x}{x} = x, \qquad \ln x = 0, \qquad x = 1, \qquad y = 1.$$

Intersection with the x axis occurs at

$$y = 0; \qquad x + \frac{\ln x}{x} = 0, \qquad x^2 + \ln x = 0.$$

This is a transcendental equation with a solution between 0 and 1.

$$y' = 1 + \frac{1 - \ln x}{x^2}; \qquad y'' = \frac{-\frac{1}{x} \cdot x - 2(1 - \ln x)}{x^3} = \frac{2 \ln x - 3}{x^3}.$$

$y' = 0$, $x^2 + 1 = \ln x$. This equation has no real solutions. Indeed, for $x > 1$, $x^2 + 1 > \ln x$ clearly, and for $0 < x < 1$ the left-hand side is positive and $\ln x$ is negative. There are no extrema.

$$y'' = 0; \qquad 2 \ln x = 3, \qquad x = e^{3/2};$$

$(e^{3/2}, e^{3/2} + \frac{3}{2}e^{-3/2})$ is an inflection point. See Figure 120.

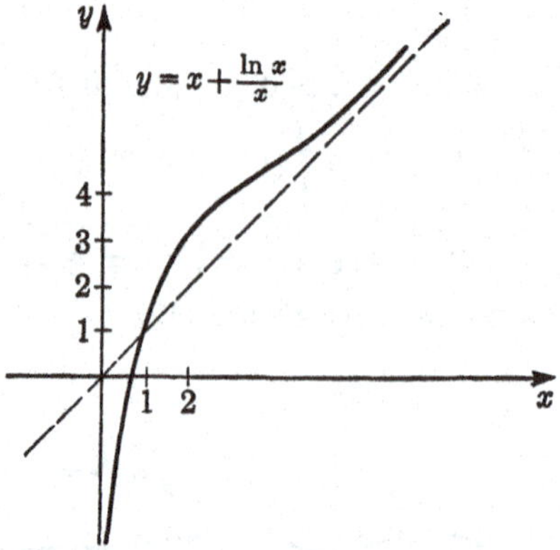

FIGURE 120

516. The curve is symmetric with respect to both coordinate axes. Let us solve the equation for y: $y = \pm x^2\sqrt{1 - x^2}$. The domain of definition is given by $1 - x^2 \geqslant 0$ or $|x| \leqslant 1$. Intersection with the axes are $x = 0$, $y = 0$, and $x = \pm 1$, $y = 0$. Now $y^2 = x^4 - x^6$, and x^4 and x^6 are no more than 1, i.e., y^2 is less than 1, i.e. the whole curve lies inside a square of side 2 and with center at the origin. Thus there are clearly no asymptotes. Now, to find extrema let us investigate the upper half of the curve: $y = x^2\sqrt{1 - x^2}$.

$$y' = 2x\sqrt{1 - x^2} - \frac{x^3}{\sqrt{1 - x^2}} = \frac{2x - 2x^3 - x^3}{\sqrt{1 - x^2}} = \frac{2x - 3x^3}{\sqrt{1 - x^2}}.$$

$$y' = 0; \qquad x_1 = 0, \qquad x_{2,3} = \pm\sqrt{\tfrac{2}{3}}.$$

The derivative of the numerator of y' is $2 - 9x^2$. It is positive at $x = 0$ and negative at the two other points. We have Min (0,0), Max $\left(\pm\sqrt{\frac{2}{3}}, \frac{2}{3\sqrt{3}}\right)$.

$$y'' = \frac{(2 - 9x^2)\sqrt{1 - x^2} + (2x - 3x^3)\dfrac{x}{\sqrt{1 - x^2}}}{1 - x^2}$$

$$= \frac{2 - 9x^2 - 2x^2 + 9x^4 + 2x^2 - 3x^4}{(1 - x^2)^{3/2}}$$

$$= \frac{6x^4 - 9x^2 + 2}{(1 - x^2)^{3/2}}.$$

$y'' = 0$, $x^2 = (9 \pm \sqrt{33})/12$. The solution corresponding to the $+$ sign is outside the domain of definition. We have inflection points at $x = \pm\sqrt{(9 - \sqrt{33})/12} \approx \pm 0.5$. To sketch the graph let us note further that $\lim_{x \to 1-} y' = -\infty$, i.e., the tangent to the curve tends to a vertical position when $x \to 1-$. The graph is given in Figure 121. We remark that the main part of the discussion here was made in the explicit form. When this is possible, it is in general more convenient.

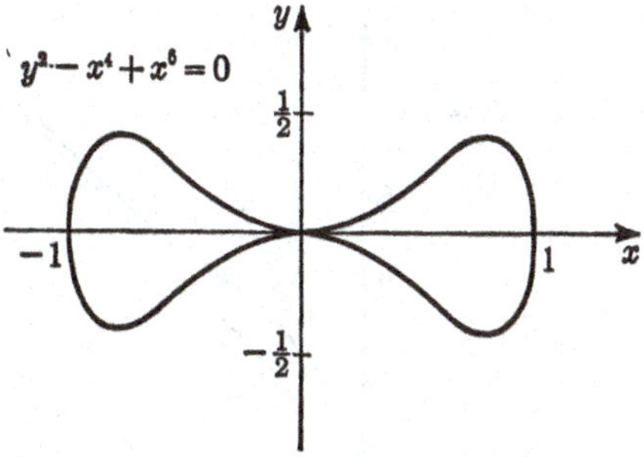

FIGURE 121

517. The curve is symmetric with respect to the origin, because only odd powers of the variables appear. Finding intersection with the axes:

$$x = 0; \qquad y^3 + y = 0, \qquad y = 0,$$

($y^2 + 1 = 0$ has no solutions). $y = 0$ gives $x^3 + 2x = 0$, i.e., $x = 0$. There is, consequently, one point of intersection with the axes, (0,0). To find the asymptotes let us substitute $y = ax + b$:

$$a^3x^3 + 3a^2bx^2 + 3ab^2x + b^3 - x^3 + ax + b - 2x = 0;$$

$$a^3 - 1 = 0, \qquad a = 1; \qquad 3a^2b = 0, \qquad b = 0.$$

$y = x$ is an asymptote. Seeking extrema,

$$3y^2y' - 3x^2 + y' - 2 = 0, \qquad y' = \frac{3x^2 + 2}{3y^2 + 1}.$$

There are no extrema. Let us differentiate again:

$$6y(y')^2 + 3y^2y'' - 6x + y'' = 0, \qquad y'' = 6\,\frac{x - y(y')^2}{3y^2 + 1}.$$

$$y'' = 0; \qquad x = y(y')^2 = y\left(\frac{3x^2 + 2}{3y^2 + 1}\right)^2.$$

To find inflection points we have to solve the system

$$\begin{cases} y(3x^2 + 2)^2 = x(3y^2 + 1)^2, \\ y^3 - x^3 + y - 2x = 0. \end{cases}$$

One solution is (0,0). Other points are difficult to obtain and we shall not look for them here. Let us find the points of intersection of the curve with the asymptote:

$$y = x; \qquad y^3 - y^3 + y - 2y = 0, \qquad y = 0,$$

(0,0) is such a point. The direction of the curve at (0,0) is given by $y'(0,0) = 2$; to find additional points we must solve a cubic equation. We obtain, e.g., (3,~3.11). The schematic graph is given in Figure 122.

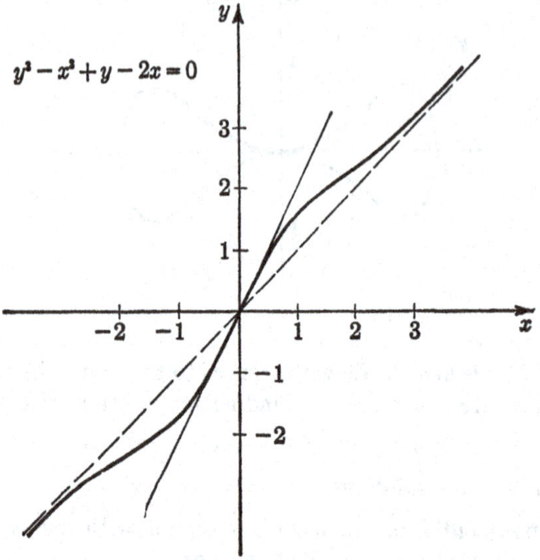

FIGURE 122

521. The variables x and y are defined for any t except $t = -1$. One point of intersection with the axes is

$$t = 0; \qquad x = 0, \qquad y = 0.$$

In Problem 468 we found that $y = -x - 1$ is an asymptote of the given curve (substitute there $a = 1$). Let us now substitute $t = t_1$ and $t = 1/t_1$ in the given equations:

$$x(t = t_1) = \frac{3t_1}{1 + t_1^3}, \qquad y(t = t_1) = \frac{3t_1^2}{1 + t_1^3};$$

$$x\left(t = \frac{1}{t_1}\right) = \frac{\dfrac{3}{t_1}}{1 + \dfrac{1}{t_1^3}} = \frac{3t_1^2}{1 + t_1^3}, \qquad y\left(t = \frac{1}{t_1}\right) = \frac{3\,\dfrac{1}{t_1^2}}{1 + \dfrac{1}{t_1^3}} = \frac{3t_1}{1 + t_1^3}.$$

Consequently

$$x\left(\frac{1}{t_1}\right) = y(t_1) \quad \text{and} \quad y\left(\frac{1}{t_1}\right) = x(t_1);$$

i.e., together with any (a,b) also (b,a) belongs to the curve, hence it is symmetric with respect to the straight line $y = x$. The only value for which we cannot find the reciprocal is $t = 0$, but the corresponding point (0,0) is on the axis of symmetry. The curve intersects its axis of symmetry:

$$\frac{3t}{1+t^3} = \frac{3t^2}{1+t^3}; \qquad t^2 = t, \quad t = 0 \quad \text{and} \quad t = 1.$$

One intersection point (0,0) we know already; the second is $(\frac{3}{2},\frac{3}{2})$. Now the extrema:

$$\frac{dy}{dt} = \frac{6t(1+t^3) - 3t^2\cdot 3t^2}{(1+t^3)^2} = \frac{6t - 3t^4}{(1+t^3)^2},$$

$$\frac{dx}{dt} = \frac{3(1+t^3) - 3t\cdot 3t^2}{(1+t^3)^2} = \frac{3 - 6t^3}{(1+t^3)^2},$$

$$\frac{dy}{dx} = \frac{2t - t^4}{1 - 2t^3};$$

$$\frac{d^2y}{dx^2} = \frac{(1+t^3)^2}{3-6t^3}\cdot\frac{(2-4t^3)(1-2t^3) - (2t-t^4)(-6t^2)}{(1-2t^3)^2}$$

$$= \frac{(1+t^3)^2(2t^6+4t^3+2)}{3(1-2t^3)^3} = \frac{2(1+t^3)^4}{3(1-2t^3)^3}.$$

$y' = 0; \qquad 2t - t^4 = 0,\ t_1 = 0,\ t_2 = \sqrt[3]{2};$

$$\frac{d^2y}{dx^2}(t = 0) = \frac{2}{3} > 0, \qquad \text{Min } (0,0).$$

$$\frac{d^2y}{dx^2}(t = \sqrt[3]{2}) < 0, \qquad \text{Max } (\sqrt[3]{2},\sqrt[3]{4}).$$

$y'' = 0$; $1 + t^3 = 0$; $t = -1$; this point is outside the domain of definition and cannot be an inflection point.

The value $t = \sqrt[3]{1/2}$, for which both dy/dx and d^2y/dx^2 do not exist, corresponds to a point symmetric to the maximum obtained above. Hence there is a maximum of x, or in other words the tangent there is vertical and the curve passes to the left of the tangent. An analogous situation is to be found at (0,0). We come to this point for a second time when $t \to \infty$ (then $x \to 0$ and $y \to 0$). (We include in the graph also the point obtainable by letting t tend to ∞, if such a point exists.) (0,0) is a "double point" and by symmetry, together with the x axis the y axis is also a tangent to the curve at the origin. The curve, known as the folium of Descartes, is given in Figure 123.

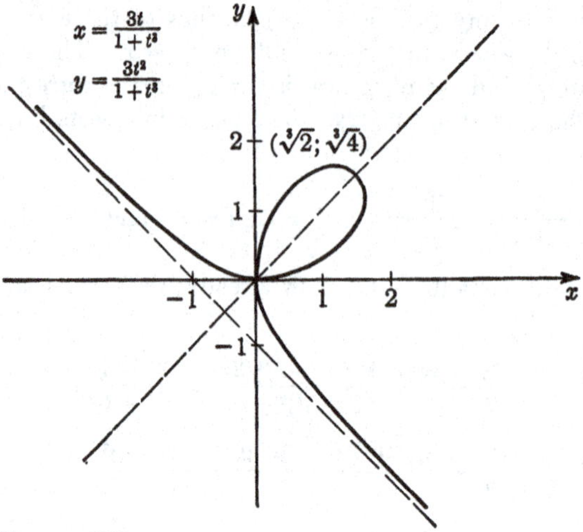

FIGURE 123

524. Any θ is permitted. $r(\theta) = r(2\pi - \theta)$, i.e., the x axis is an axis of symmetry. We restrict ourselves to $0 \leqslant \theta \leqslant \pi$.

$$\theta = 0, r = 2a; \qquad \theta = \pi/2, r = a; \qquad \theta = \pi, r = 0.$$

More points can be found in the same way. Let us find the position of the vertical and horizontal tangents to the curve:

$$x = r\cos\theta = a(1 + \cos\theta)\cos\theta,$$

$$\frac{dx}{d\theta} = a(-\sin\theta\cos\theta - \sin\theta - \sin\theta\cos\theta) = -a\sin\theta\,(2\cos\theta + 1).$$

$$\frac{dx}{d\theta} = 0; \text{ in our interval } \theta = 0, \theta = \pi, \theta = \frac{2\pi}{3}.$$

$$y = r\sin\theta = a(1 + \cos\theta)\sin\theta,$$

$$\frac{dy}{d\theta} = a(-\sin^2\theta + \cos\theta + \cos^2\theta) = a(\cos 2\theta + \cos\theta) = 2a\cos\frac{3\theta}{2}\cos\frac{\theta}{2}.$$

$$\frac{dy}{d\theta} = 0; \text{ in the interval } 0 \leqslant \theta \leqslant \pi, \text{ we obtain } \theta = \frac{\pi}{3}, \theta = \pi.$$

The behavior at $\theta = 0$, $\theta = 2\pi/3$, and $\theta = \pi/3$ is clear. At ($\theta = 0$, $r = 2a$) and ($\theta = 2\pi/3$, $r = a/2$) the tangents are vertical. At ($\theta = \pi/3$, $r = \frac{3}{2}a$) the tangent is horizontal. The point ($\theta = \pi$, $r = 0$) must be investigated further:

$$\lim_{\theta\to\pi}\frac{dy}{dx} = \lim_{\theta\to\pi}\frac{2a\cos\frac{3\theta}{2}\cos\frac{\theta}{2}}{-a\cdot 2\sin\frac{\theta}{2}\cos\frac{\theta}{2}(2\cos\theta+1)}$$

$$= -\lim_{\theta\to\pi}\frac{\cos\frac{3\theta}{2}}{\sin\frac{\theta}{2}(2\cos\theta+1)} = 0;$$

consequently the limit of the slope of the tangent is zero when $\theta \to \pi$. The graph of the curve is shown in Figure 124.

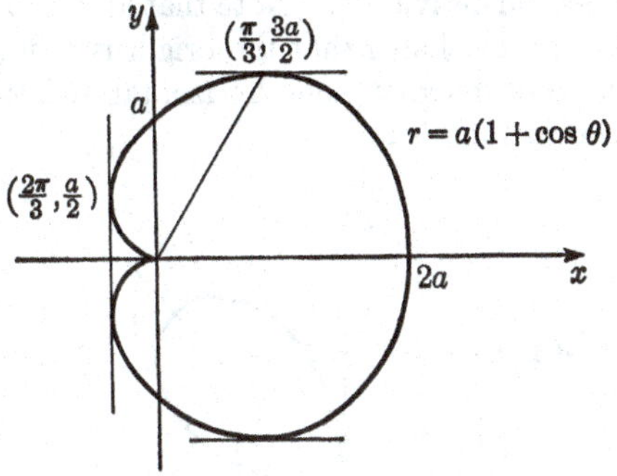

FIGURE 124

525. Before we pass to polar coordinates, let us remark that the curve is symmetric with respect to $y = x$, and with respect to the origin. Now

$$x = r\cos\theta, \qquad y = r\sin\theta.$$

$$r^4(\cos^4\theta + \sin^4\theta) = 2r^2\sin\theta\cos\theta;$$

we divide by r^2 (remarking that $r = 0$ is a point of the graph):

$$r^2(1 - 2\sin^2\theta\cos^2\theta) = \sin 2\theta;$$

$$r^2 = \frac{\sin 2\theta}{1 - \frac{1}{2}\sin^2 2\theta} = \frac{2\sin 2\theta}{1 + \cos^2 2\theta}.$$

The curve exists for $\sin 2\theta \geqslant 0$, i.e., $0 \leqslant \theta \leqslant \pi/2$ and $\pi \leqslant \theta \leqslant 3\pi/2$. By symmetry to the origin we can restrict ourselves to $0 \leqslant \theta \leqslant \pi/2$ and, using symmetry to $y = x$, to $0 \leqslant \theta \leqslant \pi/4$ only. Here

$$r = \sqrt{\frac{2\sin 2\theta}{1 + \cos^2 2\theta}}.$$

When θ increases from 0 to $\pi/4$, r also increases. $r_{\max} = \sqrt{2}$ is obtained for $\theta = \pi/4$.

The maxima of x and y are easier to compute in the original form of the equation:

$$x^4 + y^4 = 2xy; \qquad 4x^3 + 4y^3y' = 2y + 2xy'.$$

$$y' = \frac{2x^3 - y}{x - 2y^3}; \qquad y' = 0, \qquad y = 2x^3.$$

We substitute $y = 2x^3$ in the given equation and obtain

$$x^4 + 16x^{12} = 4x^4; \qquad x_1 = 0, \qquad x_2 = \sqrt[8]{\tfrac{3}{16}}.$$

(We look for $x \geqslant 0$ only.) $x = \sqrt[8]{\frac{3}{16}}$ is a maximum point, as can be proved by computing the second derivative. Note that at $x = 0$ the denominator of y' vanishes. It can be shown that the origin is a double point of the curve and there are two tangents, one horizontal and the other vertical. The graph is shown in Figure 125.

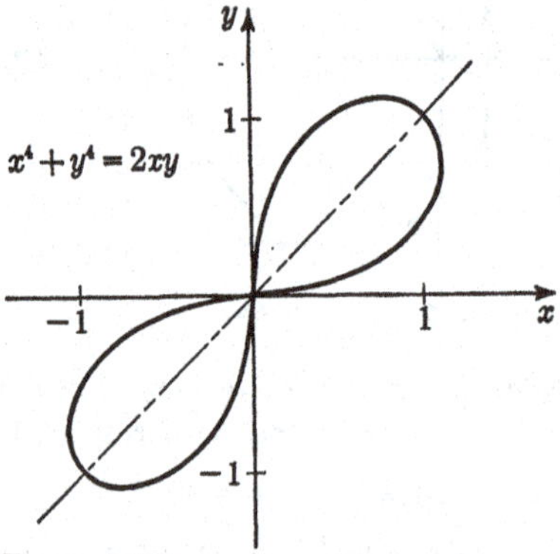

FIGURE 125

528. We have $-1 \leqslant t \leqslant 1$. It is clear that $\theta \geqslant -\pi/2$. Now

$$\frac{d\theta}{dt} = \frac{1}{\sqrt{1 - t^2}} - \frac{t}{\sqrt{1 - t^2}} = \frac{1 - t}{\sqrt{1 - t^2}} > 0$$

for $|t| < 1$, and θ increases when t increases from -1 to $+1$. Consequently, $-\pi/2 \leqslant \theta \leqslant \pi/2$.

$$\frac{dr}{dt} = \frac{-t}{\sqrt{1 - t^2}},$$

i.e., r increases for negative t and decreases for positive t. $r_{\max}(t = 0) = 1$; the corresponding θ is also equal to 1; and the curve can be drawn after additional points have been computed by substituting various values of t in the given equation (see Fig. 126.):

$$t = -1; \qquad \theta = -\frac{\pi}{2}, \qquad r = 0.$$

$$t = \pm\frac{1}{2}; \qquad \theta = \pm\frac{\pi}{6} + \frac{\sqrt{3}}{2}, \quad r = \frac{\sqrt{3}}{2}.$$

$$t = \pm\frac{\sqrt{3}}{2}; \qquad \theta = \pm\frac{\pi}{3} + \frac{1}{2}, \qquad r = \frac{1}{2}.$$

$$t = 1; \qquad \theta = \frac{\pi}{2}, \qquad r = 0.$$

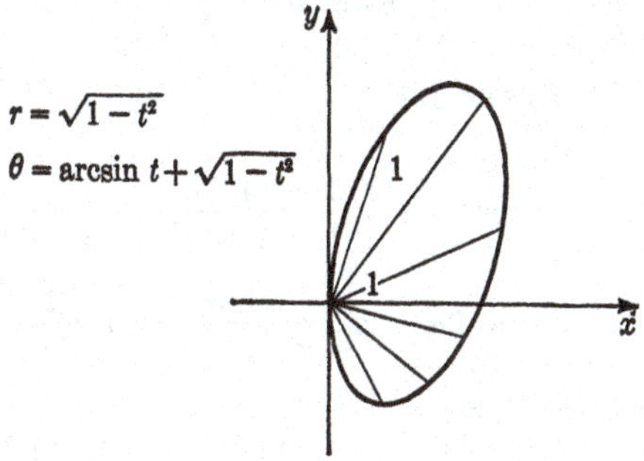

FIGURE 126

531. We obtain $x - y = 2$. This is a straight line. We shall determine whether the range of the parametric equations consists of the whole line.

$$x = (t - 1)^2 + 2;$$

the minimal x obtainable by the given equation will be $x_{\min} = 2$. We see consequently that the given equations represent the part of the straight line $y = x - 2$, which lies in the half plane determined by $x \geqslant 2$. The same conclusion can be obtained by investigating y.

532. We have immediately

$$\frac{x}{a} + \frac{y}{b} = 1.$$

This is a straight line. But from the given equations it follows directly that $0 \leqslant x \leqslant a$, $0 \leqslant y \leqslant b$, i.e., it is the segment of the above line between the points $(0,b)$ and $(a,0)$.

534. See Figure 127. $l = 2a \cos \alpha$;

$$x = l \cos \alpha - a = a(2 \cos^2 \alpha - 1)$$
$$= a \cos 2\alpha.$$
$$y = l \sin \alpha = 2a \cos \alpha \sin \alpha = a \sin 2\alpha.$$

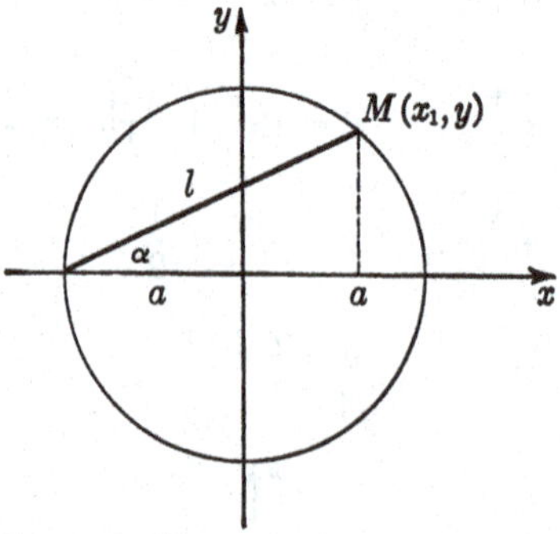

FIGURE 127

Now $$x = a\cos 2\alpha = a\frac{1-\tan^2\alpha}{1+\tan^2\alpha} = a\frac{1-t^2}{1+t^2},$$

$$y = a\sin 2\alpha = a\frac{2\tan\alpha}{1+\tan^2\alpha} = a\frac{2t}{1+t^2}.$$

535. We put $y = tx$. Then

$$x^3 + t^3x^3 = 3atx^2,$$

i.e., $$x = \frac{3at}{1+t^3} \quad \text{and} \quad y = \frac{3at^2}{1+t^3}.$$

(By dividing by x^2 we have not lost the value $x = 0$, as this can be obtained also for $t = 0$.) The resulting equations have been investigated in Problem 521. This method of parametrization is often useful when the given implicit equation is algebraic.

536. It is well known that the condition

$$\begin{vmatrix} x_1 & y_1 & 1 \\ x_2 & y_2 & 1 \\ x_3 & y_3 & 1 \end{vmatrix} = 0$$

ensures that the corresponding three points lie on the same straight line.

$$\begin{vmatrix} \dfrac{3at_1}{1+t_1^3} & \dfrac{3at_1^2}{1+t_1^3} & 1 \\ \dfrac{3at_2}{1+t_2^3} & \dfrac{3at_2^2}{1+t_2^3} & 1 \\ \dfrac{3at_3}{1+t_3^3} & \dfrac{3at_3^2}{1+t_3^3} & 1 \end{vmatrix} = \frac{1}{(1+t_1^3)(1+t_2^3)(1+t_3^3)} \begin{vmatrix} t_1 & t_1^2 & t_1^3+1 \\ t_2 & t_2^2 & t_2^3+1 \\ t_3 & t_3^2 & t_3^3+1 \end{vmatrix}.$$

The determinant equals

$$\begin{vmatrix} t_1 & t_1^2 & 1 \\ t_2 & t_2^2 & 1 \\ t_3 & t_3^2 & 1 \end{vmatrix} + \begin{vmatrix} t_1 & t_1^2 & t_1^3 \\ t_2 & t_2^2 & t_2^3 \\ t_3 & t_3^2 & t_3^3 \end{vmatrix} = (1 + t_1t_2t_3) \begin{vmatrix} 1 & t_1 & t_1^2 \\ 1 & t_2 & t_2^2 \\ 1 & t_3 & t_3^2 \end{vmatrix}.$$

The last determinant equals $(t_1 - t_2)(t_2 - t_3)(t_3 - t_1)$ and is clearly different from zero (all t_i are different). Thus the original determinant is zero if and only if $t_1t_2t_3 + 1 = 0$.

537. We put

$$\frac{x^2 + y^2}{a(x - y)} = t.$$

By substitution in the equation of the lemniscate we obtain

$$t = \frac{a(x + y)}{x^2 + y^2}.$$

Multiplying we obtain

$$t^2 = \frac{x + y}{x - y}, \quad \text{i.e.,} \quad y = x\frac{t^2 - 1}{t^2 + 1}.$$

We substitute this in the first equation:

$$x^2 + x^2\left(\frac{t^2 - 1}{t^2 + 1}\right)^2 = at\left(x - x\frac{t^2 - 1}{t^2 + 1}\right).$$

We divide by x (remembering that (0,0) is on the curve), and obtain

$$x = at\frac{\dfrac{t^2 + 1 - t^2 + 1}{t^2 + 1}}{\dfrac{t^4 + 2t^2 + 1 + t^4 - 2t^2 + 1}{(t^2 + 1)^2}} = at\frac{t^2 + 1}{t^4 + 1}.$$

And now
$$y = x\frac{t^2 - 1}{t^2 + 1} = at\frac{t^2 - 1}{t^4 + 1}.$$

538. We put $y = kx$:

$$x^{kx} = (kx)^x, \qquad x^k = kx;$$

$k = 1$ satisfies this equation, i.e., $y = x$ $(x > 0)$ is a part of the graph. Now

$$x^{k-1} = k, \qquad x = k^{1/(k-1)}.$$

For $k \neq 1$ ($k = 1$ was discussed already) we can denote $1/(k - 1)$ by t. Then $k = 1 + (1/t)$. We have for x,

$$x = \left(1 + \frac{1}{t}\right)^t,$$

and for y,

$$y = kx = \left(1 + \frac{1}{t}\right)\left(1 + \frac{1}{t}\right)^t = \left(1 + \frac{1}{t}\right)^{t+1}.$$

540. $\frac{2x}{a^2} + \frac{2yy'}{b^2} = 0, \qquad y' = -\frac{b^2x}{a^2y};$

$$y - y_0 = -\frac{b^2x_0}{a^2y_0}(x - x_0); \qquad \frac{yy_0}{b^2} - \frac{y_0^2}{b^2} + \frac{xx_0}{a^2} - \frac{x_0^2}{a^2} = 0.$$

The equation of the tangent is

$$\frac{xx_0}{a^2} + \frac{yy_0}{b^2} = 1.$$

It holds also for the vertices of the ellipse.

541. $\frac{dy}{dx} = \frac{a \sin t}{a(1 - \cos t)} = \frac{2 \sin \frac{t}{2} \cos \frac{t}{2}}{2 \sin^2 \frac{t}{2}} = \cot \frac{t}{2}.$

The equation of the tangent at (x_0, y_0) is

$$y - y_0 = \cot \frac{t_0}{2}(x - x_0).$$

One can substitute here

$$x_0 = a(t_0 - \sin t_0), \qquad y_0 = a(1 - \cos t_0).$$

Now

$$n = |y_0|\sqrt{1 + (y_0')^2} = a(1 - \cos t_0)\sqrt{1 + \cot^2 \frac{t_0}{2}}$$

$$= a \cdot 2 \sin^2 \frac{t_0}{2} \csc \frac{t_0}{2} = 2a \sin \frac{t_0}{2}, \qquad 0 < t_0 < 2\pi.$$

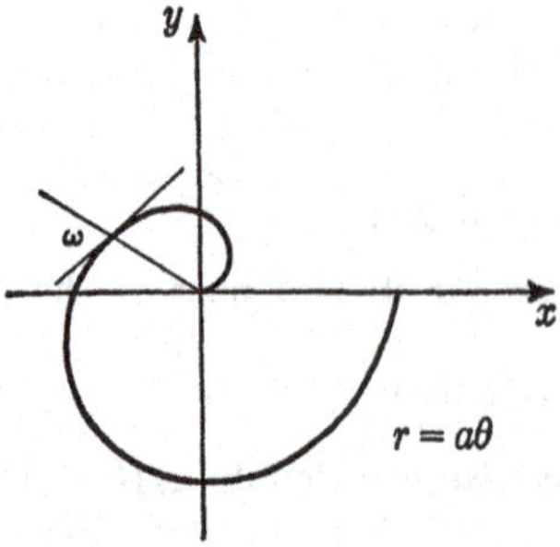

FIGURE 128

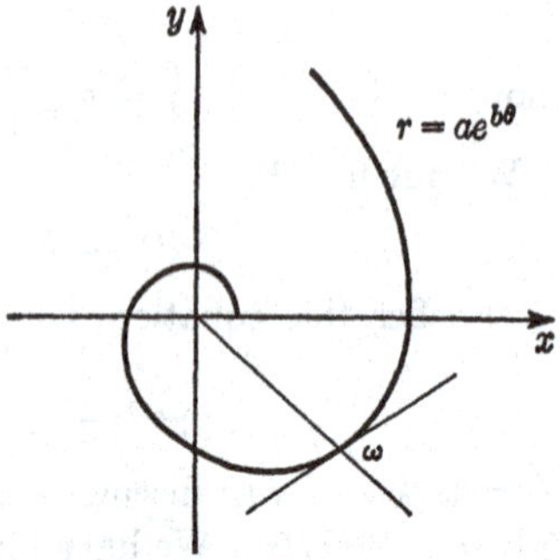

FIGURE 129

542. The graph of the curve is sketched in Figure 128. Now $r' = a$. It follows that sbn_p is constant and equal to a.

$$\tan \omega = \frac{r}{r'} = \frac{a\theta}{a} = \theta;$$

if we permit θ to increase above 2π, we see that as $\theta \to \infty$, $\omega \to \pi/2$, i.e., for large θ the curve approximates a large circle.

543. See Figure 129.

$$r' = abe^{b\theta},$$

$$\tan \omega = \frac{r}{r'} = \frac{ae^{b\theta}}{abe^{b\theta}} = \frac{1}{b}.$$

The radius vector intersects the curve at a constant angle equal to arctan $(1/b)$.

544. $y' = -\dfrac{16a^3x}{(4a^2 + x^2)^2};$

$$y'(2a) = -\frac{32a^4}{64a^4} = -\frac{1}{2};$$

$$y(2a) = \frac{8a^3}{8a^2} = a.$$

The equation of the tangent is $y - a = -\frac{1}{2}(x - 2a)$ or

$$2y + x - 4a = 0.$$

The equation of the normal is

$$y - a = 2(x - 2a), \quad \text{i.e.,} \quad y - 2x + 3a = 0.$$

545. The slope of the given straight line is 1. We must find a point on the curve where the slope of the normal is 1, i.e., where the slope of the tangent is -1.

$$y' = \ln x + 1; \qquad \ln x_0 + 1 = -1, \qquad \ln x_0 = -2,$$

$$x_0 = \frac{1}{e^2}, \qquad y_0 = -\frac{2}{e^2}.$$

The equation of the normal is

$$y + \frac{2}{e^2} = x - \frac{1}{e^2} \quad \text{or} \quad e^2(x - y) - 3 = 0.$$

547. The point of intersection is $(\pi/4, \sqrt{2}/2)$. The required angle is $\varphi = \arctan 2\sqrt{2}$.

548. We find the equation of the tangent at a point (x_0, y_0) on the curve:

$$\tfrac{2}{3}x^{-1/3} + \tfrac{2}{3}y^{-1/3}y' = 0; \qquad y' = -y^{1/3}x^{-1/3}.$$

The equation of the tangent at (x_0, y_0) is

$$y - y_0 = -x_0^{-1/3}y_0^{1/3}(x - x_0).$$

Now we find the intercepts,

$$x = 0, \qquad y_1 = y_0^{1/3}(y_0^{2/3} + x_0^{2/3}) = a^{2/3}y_0^{1/3};$$
$$y = 0, \qquad x_1 = x_0^{1/3}(x_0^{2/3} + y_0^{2/3}) = a^{2/3}x_0^{1/3}.$$

The required segment is

$$d = \sqrt{x_1^2 + y_1^2} = a^{2/3}\sqrt{x_0^{2/3} + y_0^{2/3}} = a^{2/3}\sqrt{a^{2/3}} = a.$$

549. We find the slope of the tangent:

$$y = a \cosh \frac{x}{a}, \qquad y' = \sinh \frac{x}{a}, \qquad y'(x_0) = \sinh \frac{x_0}{a}.$$

Now the slope of MP can be obtained by geometric considerations:

$$\measuredangle MQN = \measuredangle MNP;$$

$$\tan \measuredangle MQN = \tan \measuredangle MNP = \frac{MP}{PN}$$

$$= \frac{\sqrt{MN^2 - PN^2}}{PN} = \frac{\sqrt{a^2 \cosh^2 \frac{x_0}{a} - a^2}}{a}$$

$$= \sqrt{\cosh^2 \frac{x_0}{a} - 1} = \sinh \frac{x_0}{a}.$$

We see that $\tan \measuredangle MQN = y'(x_0)$, and this proves the construction.

551. In this case it seems to be preferable to use the implicit equation of the second curve.

$$\cos t = \tfrac{3}{5}x,\ \sin t = \tfrac{4}{5}y; \qquad (\tfrac{3}{5}x)^2 + (\tfrac{4}{5}y)^2 = 1.$$

We obtain the ellipse $9x^2 + 16y^2 = 25$. The points of intersection of the curves are $(\pm 1,1)$. By symmetry we can restrict ourselves to one of them, $(1,1)$.

$$y_1' = 2x, \qquad y_1'(1) = 2;$$
$$18x + 32yy_2' = 0, \qquad y_2'(1,1) = -\tfrac{9}{16};$$
$$\tan \alpha = \left|\frac{2 + \frac{9}{16}}{1 - \frac{9}{8}}\right| = \frac{41}{2}, \qquad \alpha = \arctan \frac{41}{2}.$$

552. To find the points of intersection let us note that all points on the first curve satisfy $x^2 + y^2 = a^2$. Now

$$a^2\left[\frac{t^4}{(1 + t^2)^2} + \frac{3t^2}{(1 + t^2)^2}\right] = a^2,$$

$$t^4 + 3t^2 = t^4 + 2t^2 + 1, \qquad t^2 = 1, \quad t = \pm 1;$$

$$\cos \varphi = \frac{t^2}{1 + t^2} = \frac{1}{2}, \qquad \sin \varphi = \frac{t\sqrt{3}}{1 + t^2} = \pm\frac{\sqrt{3}}{2}.$$

We have $t = \pm 1$, $\varphi = \pm\frac{\pi}{3}$. Now

$$\frac{dy_1}{dx} = \frac{a\cos\varphi}{-a\sin\varphi} = -\cot\varphi; \qquad y_1'\left(\varphi = \pm\frac{\pi}{3}\right) = \mp\frac{\sqrt{3}}{3}.$$

$$\frac{dy_2}{dx} = \frac{\dfrac{a\sqrt{3}(1 + t^2 - t\cdot 2t)}{(1+t^2)^2}}{\dfrac{a[2t\cdot(1+t^2) - t^2\cdot 2t]}{(1+t^2)^2}} = \frac{\sqrt{3}(1-t^2)}{2t}; \qquad y_2'(t = \pm 1) = 0;$$

$$\tan\alpha_1 = \frac{\sqrt{3}}{3},\ \alpha_1 = \frac{\pi}{6}; \qquad \tan\alpha_2 = -\frac{\sqrt{3}}{3},\ \alpha_2 = -\frac{\pi}{6}\left(\text{or } \frac{5\pi}{6}\right).$$

An additional point of intersection is obtained if we include in the second curve the limit point when $t \to \infty$, i.e., $(a,0)$. This point is clearly on the circle (for $\varphi = 0$). It can be seen that at this point both curves have vertical tangents and the angle between them is 0.

553. Let $t = t_0$ ($x = x_0$, $y = y_0$) be an arbitrary point on the involute.

$$\frac{dy}{dx} = \frac{a(\cos t - \cos t + t\sin t)}{a(-\sin t + \sin t + t\cos t)} = \tan t; \qquad y_1'(t_0) = \tan t_0.$$

The equation of the normal to the involute at (x_0,y_0) is

$$y - y_0 = -\frac{1}{\tan t_0}(x - x_0).$$

We shall prove that this straight line is a tangent to the circle. We use the normal equation of the straight line and substitute there (0,0), the coordinates of the center of the circle. This gives the distance from the center to the straight line and if it is a, i.e., equal to the radius, everything will be proved. The computation follows:

$$\left|\frac{x_0 + y_0\tan t_0}{\sqrt{1+\tan^2 t_0}}\right| = |\cos t_0\,[a(\cos t_0 + t_0\sin t_0) + a\tan t_0\,(\sin t_0 - t_0\cos t_0)]|$$
$$= a|\cos^2 t_0 + t_0\sin t_0\cos t_0 + \sin^2 t_0 - t_0\sin t_0\cos t_0| = a.$$

554. $\tan\omega = \dfrac{r}{r'}$; $r' = \dfrac{dr}{d\theta} = \dfrac{3at^2}{2bt} = \dfrac{3at}{2b}$.

$$\tan\omega = \frac{at^3}{3at/2b} = \frac{2bt^2}{3} = \frac{2}{3}\theta.$$

555. $$\tan\omega_1 = \frac{r_1}{r_1'} = \frac{a(1+\cos\theta)}{-a\sin\theta} = -\frac{2\cos^2\dfrac{\theta}{2}}{2\sin\dfrac{\theta}{2}\cos\dfrac{\theta}{2}} = -\cot\frac{\theta}{2},$$

$$\tan\omega_2 = \frac{r_2}{r_2'} = \frac{a(1-\cos\theta)}{a\sin\theta} = \frac{2\sin^2\dfrac{\theta}{2}}{2\sin\dfrac{\theta}{2}\cos\dfrac{\theta}{2}} = \tan\frac{\theta}{2}.$$

The θ of the point of intersection is the same for each curve, and consequently

$$\tan \omega_2 = \tan \frac{\theta}{2} = \frac{1}{\cot (\theta/2)} = -\frac{1}{\tan \omega_1} = -\cot \omega_1 = \tan \left(\frac{\pi}{2} + \omega_1\right).$$

This shows that the curves intersect at a right angle.

556. The equation of the straight line is $y = kx$. The points of intersection are

$$kx^2 = a^2, \qquad x = \pm\frac{a}{\sqrt{k}}, \qquad y = \pm a\sqrt{k}$$

(we assume $k > 0$). The derivative for the hyperbola at these points is

$$y = \frac{a^2}{x}, \qquad y' = -\frac{a^2}{x^2}, \qquad y'\left(\pm\frac{a}{\sqrt{k}}\right) = -k.$$

We require $k(-k) = -1$, i.e., $k^2 = 1$ and $k = 1$ ($k > 0$). The solution is $y = x$.

557. The corresponding value of x can be obtained from the equation

$$x^3 + 9 + 2x - 6 = 0 \quad \text{or} \quad x^3 + 2x + 3 = 0.$$

Now $$x^3 + 2x + 3 = (x + 1)(x^2 - x + 3).$$

The only solution of $x^3 + 2x + 3 = 0$ is $x = -1$, because $x^2 - x + 3 = 0$ has no real solution. Now

$$3x^2 + 2yy' + 2 = 0.$$

We substitute $(-1,3)$: $3 + 6y' + 2 = 0$, $y' = -\frac{5}{6}$. The equation of the normal is

$$y - 3 = \tfrac{6}{5}(x + 1) \quad \text{or} \quad 6x - 5y + 21 = 0.$$

558. $y' = 1$; $3x^2 - 3y^2y' = 6x$.

We have to solve the system

$$\begin{cases} x^2 - y^2 = 2x \\ x^3 - y^3 = 3x^2. \end{cases}$$

One solution is $(0,0)$. Now

$$\begin{cases} (x - y)(x + y) = 2x \\ (x - y)(x^2 + xy + y^2) = 3x^2; \end{cases}$$

if $x = y$ we have again $(0,0)$. For $x \neq y$ we obtain

$$\frac{2x}{x + y}(x^2 + xy + y^2) = 3x^2;$$

and (because now $x \neq 0$)

$$2x^2 + 2xy + 2y^2 = 3x^2 + 3xy \quad \text{or} \quad x^2 + xy - 2y^2 = 0,$$

i.e., $(x + 2y)(x - y) = 0$, and consequently $x = -2y$. Now

$$4y^2 - y^2 = -4y, \qquad 3y^2 = -4y.$$

The new solution is $y_2 = -\frac{4}{3}$, $x_2 = \frac{8}{3}$. Hence $y + \frac{4}{3} = x - \frac{8}{3}$, and the two tangents are

$$y = x \quad \text{and} \quad y = x - 4.$$

559. $2b^2x + 2a^2yy' = 0,\ y_1' = -\dfrac{b^2x}{a^2y};$

$$b^2 \ln y = a^2 \ln x + \ln m,\ \frac{b^2y'}{y} = \frac{a^2}{x},\ y_2' = \frac{a^2y}{b^2x}.$$

$$y_1' \cdot y_2' = -1.$$

560. We shall take the parabola in the form $y = ax^2$. Let $P(x_1,y_1)$ be a point of the required locus. First we shall find a normal to the parabola through P. Let (x_0,y_0) be the point of intersection of the normal with the parabola. $y'(x_0) = 2ax_0$, and the equation of the normal is

$$y - y_0 = -\frac{1}{2ax_0}(x - x_0).$$

(x_1,y_1) satisfies this equation, i.e.,

$$2ax_0y_1 - 2ax_0y_0 + x_1 - x_0 = 0$$

or (after substitution of $y_0 = ax_0^2$)

$$2a^2x_0^3 - x_0(2ay_1 - 1) - x_1 = 0.$$

Let x_0' and x_0'' be the two solutions of this equation which give, as required,

$$-\frac{1}{2ax_0'}\left(-\frac{1}{2ax_0''}\right) = -1, \quad \text{i.e.,} \quad x_0'x_0'' = -\frac{1}{4a^2}.$$

Denote the third solution of the cubic equation by x_0'''. It is known from algebra that $x_0' \cdot x_0'' \cdot x_0''' = x_1/2a^2$ and consequently

$$-\frac{1}{4a^2}x_0''' = \frac{x_1}{2a^2} \quad \text{or} \quad x_0''' = -2x_1.$$

The given equation must have $x_0 + 2x_1$ as a factor. Let us divide:

$$\begin{array}{l|l}
2a^2x_0^3 - x_0(2ay_1 - 1) - x_1 & x_0 + 2x_1 \\
\underline{2a^2x_0^3 + 4a^2x_1x_0^2} & 2a^2x_0^2 - 4a^2x_1x_0 + 8a^2x_1^2 - 2ay_1 + 1 \\
\quad - 4a^2x_1x_0^2 - x_0(2ay_1 - 1) - x_1 & \\
\quad \underline{- 4a^2x_1x_0^2 - 8a^2x_1^2x_0} & \\
\qquad x_0(8a^2x_1^2 - 2ay_1 + 1) - x_1 & \\
\qquad \underline{x_0(8a^2x_1^2 - 2ay_1 + 1) + 16a^2x_1^3 - 4ax_1y_1 + 2x_1} & \\
\qquad\qquad - 16a^2x_1^3 + 4ax_1y_1 - 3x_1 &
\end{array}$$

The remainder must be equal to 0. This gives the required connection on x_1 and y_1:

$$16a^2x_1^3 - 4ax_1y_1 + 3x_1 = 0, \qquad 16a^2x_1^2 - 4ay_1 + 3 = 0.$$

The locus is a parabola $\qquad y = \dfrac{3}{4a} + 4ax^2.$

562. $\pi/2.$

563. $\arctan \dfrac{\sqrt{17}}{3}.$

564. $\left(\sqrt[6]{32}, \dfrac{1}{2\sqrt[3]{4}}\right).$

566. $t = 2\sqrt{5}$, $n = \sqrt{5}$, $\alpha = \arctan 3.$

570. $y_1(1) = A \sin(a + b)$; $y_2(1) = 1.$

$y_1' = Aa \cos(ax + b)$, $y_1'(1) = Aa \cos(a + b)$; $y_2' = \dfrac{1}{2\sqrt{x}}$, $y_2'(1) = \dfrac{1}{2}.$

$y_1'' = -Aa^2 \sin(ax + b)$, $y_1''(1) = -Aa^2 \sin(a + b)$;

$$y_2'' = -\frac{1}{4x\sqrt{x}}, \quad y_2''(1) = -\frac{1}{4}.$$

We have to satisfy the three equations

$$A \sin(a + b) = 1, \qquad Aa \cos(a + b) = \tfrac{1}{2}, \qquad -Aa^2 \sin(a + b) = -\tfrac{1}{4}.$$

By dividing the third equation by the first, $a^2 = \frac{1}{4}$, $a = \frac{1}{2}$ $(a > 0)$. Dividing the second equation by the first, we obtain

$$a \cot(a + b) = \tfrac{1}{2}, \quad \text{i.e.,} \quad \cot(a + b) = 1.$$

Take $a + b = \dfrac{\pi}{4}$, i.e., $b = \dfrac{\pi}{4} - \dfrac{1}{2}.$ Now

$$A = \frac{1}{\sin(a + b)} = \frac{1}{\sin(\pi/4)} = \sqrt{2}.$$

$$y_1 = \sqrt{2} \sin\left(\frac{x}{2} + \frac{\pi - 2}{4}\right).$$

571. First let us find the corresponding value of y:

$$\sqrt{3} + \sqrt{y} = 2\sqrt{3}; \qquad \sqrt{y} = \sqrt{3},\ y = 3.$$

It is readily seen that (3,3) is on the circle. We now find the consecutive derivatives at this point, first for the equation of the circle:

$$2x + 2yy' - 18 - 18y' = 0, \quad 6 + 6y' - 18 - 18y' = 0, \quad y' = -1.$$

We divide by 2 and differentiate once more,

$$1 + (y')^2 + yy'' - 9y'' = 0.$$

We substitute $y = 3$, $y' = -1$:

$$1 + 1 + 3y'' - 9y'' = 0, \qquad y'' = \tfrac{1}{3}.$$

Now $3y'y'' + yy''' - 9y''' = 0$.

We substitute $y = 3$, $y' = -1$, $y'' = \frac{1}{3}$:

$$-1 + 3y''' - 9y''' = 0, \qquad y''' = -\tfrac{1}{6}.$$

Now we find the explicit equation and then the derivatives of the parabola:

$$y = 12 - 4\sqrt{3}\cdot\sqrt{x} + x.$$

$$y' = -\frac{2\sqrt{3}}{\sqrt{x}} + 1, \qquad y'(3) = -1,$$

$$y'' = \frac{\sqrt{3}}{x\sqrt{x}}, \qquad y''(3) = \tfrac{1}{3},$$

$$y''' = -\frac{\sqrt{3}\cdot 3}{2x^2\sqrt{x}}, \qquad y'''(3) = -\tfrac{1}{6}.$$

The order of contact is at least 3. We leave for the reader to show that the order is exactly 3, i.e., that the fourth derivatives differ at (3,3).

572. The corresponding y and t can be found from the equations of the curves:

$$3(\pi a - \pi a)^2 + 4(y + a)^2 = 36a^2, \qquad y + a = \pm 3a,$$

$$y_1 = 2a, \qquad y_2 = -4a.$$

But $a(1 - \cos t) \geqslant 0$, i.e., only the first solution, $y = 2a$, can serve. Now

$$2a = a(1 - \cos t), \qquad \pi a = a(t - \sin t),$$

i.e.,

$$\cos t = -1 \quad \text{and} \quad t - \sin t = \pi.$$

The only solution is $t = \pi$. Now, the derivatives of the equation of the ellipse are

$6(x - \pi a) + 8(y + a)y' = 0$, $3(x - \pi a) + 4(y + a)y' = 0$;

$$3 + 4(y')^2 + 4(y + a)y'' = 0;$$

$8y'y'' + 4y'y'' + 4(y + a)y''' = 0$,

$$3y'y'' + (y + a)y''' = 0;$$

$$3(y'')^2 + 3y'y''' + y'y''' + (y + a)y^{iv} = 0.$$

The values of the derivatives at $(\pi a, 2a)$ are

$$\begin{array}{ll} 3(\pi a - \pi a) + 4(2a + a)y' = 0, & y' = 0; \\ 3 + 4\cdot 3ay'' = 0, & y'' = -1/4a; \\ 3ay''' = 0, & y''' = 0; \\ \dfrac{3}{16a^2} + 3ay^{iv} = 0, & y^{iv} = -\dfrac{1}{16a^3}. \end{array}$$

Now for the cycloid,

$$y' = \frac{a \sin t}{a(1 - \cos t)} = \frac{2 \sin (t/2) \cos (t/2)}{2 \sin^2 (t/2)} = \cot \frac{t}{2},$$

$$y'' = -\frac{1}{2 \sin^2 (t/2)} \cdot \frac{1}{a(1 - \cos t)} = -\frac{1}{4a} \sin^{-4} \frac{t}{2},$$

$$y''' = \frac{1}{2a} \sin^{-2} \frac{t}{2} \cdot \frac{1}{4a} \cdot 4 \sin^{-5} \frac{t}{2} \cdot \cos \frac{t}{2} \cdot \frac{1}{2} = \frac{1}{4a^2} \sin^{-7} \frac{t}{2} \cos \frac{t}{2},$$

$$y^{iv} = \frac{1}{4a^2} \left(-7 \sin^{-8} \frac{t}{2} \cos^2 \frac{t}{2} \cdot \frac{1}{2} - \sin^{-7} \frac{t}{2} \sin \frac{t}{2} \cdot \frac{1}{2} \right) \frac{1}{2a \sin^2 (t/2)}.$$

For the cycloid at $t = \pi$,

$$y'(t = \pi) = \cot \frac{\pi}{2} = 0,$$

$$y''(t = \pi) = -\frac{1}{4a \sin^4 (\pi/2)} = -\frac{1}{4a},$$

$$y'''(t = \pi) = \frac{1}{4a^2} \sin^{-7} \frac{\pi}{2} \cos \frac{\pi}{2} = 0,$$

$$y^{iv}(t = \pi) = \frac{1}{16a^3 \sin^2 (\pi/2)} \left(-7 \sin^{-8} \frac{\pi}{2} \cos^2 \frac{\pi}{2} - \sin^{-6} \frac{\pi}{2} \right) = -\frac{1}{16a^3}.$$

The corresponding derivatives are equal at the given point.

573. $y_1(0) = a;$

$$y_1' = \sinh (x/a), \qquad y_1'(0) = 0;$$

$$y_1'' = \frac{1}{a} \cosh \frac{x}{a}, \qquad y_1''(0) = \frac{1}{a};$$

$$y_1''' = \frac{1}{a^2} \sinh \frac{x}{a}, \qquad y_1'''(0) = 0;$$

$$y_1^{iv} = \frac{1}{a^3} \cosh \frac{x}{a}, \qquad y_1^{iv}(0) = \frac{1}{a^3}.$$

For the parabola,

$$y_2 = m_0x^4 + m_1x^3 + m_2x^2 + m_3x + m_4; \qquad y_2(0) = m_4,\ m_4 = a.$$

$$y_2' = 4m_0x^3 + 3m_1x^2 + 2m_2x + m_3; \qquad y_2'(0) = m_3,\ m_3 = 0.$$

$$y_2'' = 12m_0x^2 + 6m_1x + 2m_2; \qquad y_2''(0) = 2m_2,\ 2m_2 = \frac{1}{a},\ m_2 = \frac{1}{2a}.$$

$$y_2''' = 24m_0x + 6m_1; \qquad y_2'''(0) = 6m_1,\ 6m_1 = 0,\ m_1 = 0.$$

$$y_2^{iv} = 24m_0; \qquad 24m_0 = \frac{1}{a^3},\ m_0 = \frac{1}{24a^3}.$$

The equation of the parabola is

$$y = \frac{1}{24a^3}x^4 + \frac{1}{2a}x^2 + a.$$

It can be seen that $y_1^{v}(0) = 0$, and $y_2^{v}(0) = 0$, but that $y_1^{vi}(0) \neq 0$ while $y_2^{vi}(0) = 0$, i.e., the order of contact is 5. We could not require this in advance, since in a parabola of fourth degree 5 coefficients are free, whereas to have contact of order 5, 6 conditions must be satisfied.

575. Let us use the formulas of this section: $y(0) = 1$, $y'(0) = 1$, $y''(0) = 1$.

$$a = x - y'\frac{1 + (y')^2}{y''} = -2, \qquad b = y + \frac{1 + (y')^2}{y''} = 3:$$

$$R^2 = \frac{[1 + (y')^2]^3}{(y'')^2} = 8.$$

The equation of the osculating circle is $(x + 2)^2 + (y - 3)^2 = 8$.

576. $x^3 + xy^2 - 2ey^2 = 0;$

$$3x^2 + y^2 + 2xyy' - 4eyy' = 0, \qquad 3e^2 + e^2 + 2c^2y' - 4e^2y' = 0,$$

$$y'(e,e) = 2;$$

$$6x + 2yy' + 2yy' + 2x(y')^2 + 2xyy'' - 4e(y')^2 - 4eyy'' = 0,$$

$$6e + 4e\cdot 2 + 2e\cdot 4 + 2e^2y'' - 4e\cdot 4 - 4e^2y'' = 0, \qquad y''(e,e) = \frac{3}{e}.$$

By the formulas we obtain

$$a = e - 2\frac{1+4}{3/e} = -\frac{7}{3}e, \qquad b = e + \frac{1+4}{3/e} = \frac{8}{3}e,$$

$$R^2 = \frac{5^3}{(3/e)^2} = \frac{125e^2}{9}.$$

The circle is $$\left(x + \frac{7}{3}e\right)^2 + \left(y - \frac{8}{3}e\right)^2 = \frac{125e^2}{9}.$$

577. $y + xy' = 0,\ 2 + 2y' = 0,\ y' = -1;$

$y' + y' + xy'' = 0,\ -2 + 2y'' = 0,\ y'' = 1.$

By the formula,

$$R = \left|\frac{(1 + y'^2)^{3/2}}{y''}\right| = \left|\frac{(1+1)^{3/2}}{1}\right| = 2\sqrt{2}.$$

The curvature will be $k = 1/R = \sqrt{2}/4$.

578. $y' = 1/x,\ y'(1) = 1;\ y'' = -1/x^2,\ y''(1) = -1;$

$R = 2\sqrt{2},\ k = \sqrt{2}/4.$

579. $\dfrac{2x}{a^2} - \dfrac{2yy'}{b^2} = 0,\ y' = \dfrac{b^2x}{a^2y};$

$$y'' = \frac{b^2}{a^2}\cdot\frac{y - xy'}{y^2} = \frac{b^2}{a^2}\cdot\frac{y - (b^2x^2/a^2y)}{y^2} = \frac{b^2}{a^2}\cdot\frac{a^2y^2 - b^2x^2}{a^2y^3}$$

$$= -\frac{b^2}{a^2}\cdot\frac{a^2b^2}{a^2y^3} = -\frac{b^4}{a^2y^3}.$$

$$R = \left|\frac{\left(1 + \dfrac{b^4x^2}{a^4y^2}\right)^{3/2}}{-\dfrac{b^4}{a^2y^3}}\right| = \frac{(a^4y^2 + b^4x^2)^{3/2}}{a^4b^4}.$$

580. $\frac{2}{3}x^{-1/3} + \frac{2}{3}y^{-1/3}y' = 0,\ y' = -x^{-1/3}y^{1/3};$

$$y'' = \tfrac{1}{3}x^{-4/3}y^{1/3} - \tfrac{1}{3}x^{-1/3}y^{-2/3}y' = \tfrac{1}{3}x^{-4/3}y^{1/3} + \tfrac{1}{3}x^{-2/3}y^{-1/3}$$

$$= \tfrac{1}{3}x^{-4/3}y^{-1/3}(y^{2/3} + x^{2/3}) = \tfrac{1}{3}a^{2/3}x^{-4/3}y^{-1/3}.$$

$$R = \left|\frac{(1 + y^{2/3}x^{-2/3})^{3/2}}{\frac{1}{3}a^{2/3}x^{-4/3}y^{-1/3}}\right| = \left|\frac{3a}{a^{2/3}x\cdot x^{-4/3}y^{-1/3}}\right| = 3a^{1/3}|x|^{1/3}|y|^{1/3}.$$

581. We use the parametric form of the formula for R:

$$x'_{t=1} = 6,\ x''_{t=1} = 6; \qquad y'_{t=1} = 0,\ y''_{t=1} = -6.$$

$$R = \left|\frac{[(x_t')^2 + (y_t')^2]^{3/2}}{y_t''x_t' - y_t'x_t''}\right| = \left|\frac{216}{-36}\right| = 6.$$

582. $x_t' = -2a\sin t + 2a\sin 2t, \qquad x_t'' = -2a\cos t + 4a\cos 2t;$

$y_t' = 2a\cos t - 2a\cos 2t, \qquad y_t'' = -2a\sin t + 4a\sin 2t;$

$$(x_t')^2 + (y_t')^2$$
$$= 4a^2(\sin^2 t - 2\sin t\sin 2t + \sin^2 2t + \cos^2 t - 2\cos t\cos 2t + \cos^2 2t)$$
$$= 4a^2(2 - 2\cos t) = 16a^2\sin^2(t/2);$$

$$y_t''x_t' - y_t'x_t''$$
$$= 4a^2(\sin^2 t - 3\sin t\sin 2t + 2\sin^2 2t + \cos^2 t - 3\cos t\cos 2t + 2\cos^2 2t)$$
$$= 4a^2(3 - 3\cos t) = 24a^2\sin^2(t/2);$$

$$R = \left|\frac{(16a^2\sin^2 t/2)^{3/2}}{24a^2\sin^2(t/2)}\right| = \frac{8a}{3}\left|\sin\frac{t}{2}\right|.$$

583. $r' = a^\theta\ln a,\ r'(0) = \ln a; \qquad r'' = a^\theta\ln^2 a,\ r''(0) = \ln^2 a;$

$$R = \left|\frac{[r^2 + (r')^2]^{3/2}}{r^2 + 2(r')^2 - rr''}\right| = \left|\frac{(1 + \ln^2 a)^{3/2}}{1 + 2\ln^2 a - \ln^2 a}\right| = \sqrt{1 + \ln^2 a}.$$

584. $r' = ka\theta^{k-1},\ r'' = k(k-1)a\theta^{k-2};$

$$R = \left|\frac{(a^2\theta^{2k} + k^2a^2\theta^{2k-2})^{3/2}}{a^2\theta^{2k} + 2a^2k^2\theta^{2k-2} - k(k-1)a^2\theta^{2k-2}}\right| = \frac{a\theta^{k-1}(\theta^2 + k^2)^{3/2}}{\theta^2 + k^2 + k}.$$

585. $2rr' = -2a^2 \sin 2\theta,\ rr' = -a^2 \sin 2\theta;$

$(r')^2 + rr'' = -2a^2 \cos 2\theta = -2r^2,\ -rr'' = (r')^2 + 2r^2;$

$$r^2 + (r')^2 = a^2 \cos 2\theta + \frac{a^4 \sin^2 2\theta}{r^2} = \frac{a^4 \cos^2 2\theta + a^4 \sin^2 2\theta}{r^2} = \frac{a^4}{r^2};$$

$r^2 + 2(r')^2 - rr'' = r^2 + 2(r')^2 + (r')^2 + 2r^2 = 3(r^2 + r'^2);$

$$R = \frac{[(r')^2 + r^2]^{3/2}}{3[r^2 + (r')^2]} = \frac{1}{3}\sqrt{r^2 + (r')^2} = \frac{1}{3}\sqrt{\frac{a^4}{r^2}} = \frac{a^2}{3r}.$$

586. $y = \ln x,\ y' = 1/x,\ y'' = -1/x^2;$

$$R = \left|\frac{\left(1 + \frac{1}{x^2}\right)^{3/2}}{-\frac{1}{x^2}}\right| = \frac{(1 + x^2)^{3/2}}{x}, \qquad (x > 0).$$

$$\frac{dR}{dx} = \frac{\frac{3}{2}(1 + x^2)^{1/2}\cdot 2x\cdot x - (1 + x^2)^{3/2}}{x^2} = \frac{(1 + x^2)^{1/2}}{x^2}(2x^2 - 1);$$

$$\frac{dR}{dx} = 0,\ 2x^2 - 1 = 0,\ x = \sqrt{\tfrac{1}{2}}\ (x > 0).$$

The vertex is at $(\sqrt{2}/2, -\frac{1}{2}\ln 2)$. It is easy to show that R attains at this point a minimum equal to

$$\frac{(\frac{3}{2})^{3/2}}{(\frac{1}{2})^{1/2}} = \frac{3\sqrt{3}}{2}.$$

587. $k = \left|\frac{y''}{(1 + y'^2)^{3/2}}\right| = \left|\frac{y''}{(1 + \tan^2 \alpha)^{3/2}}\right| = \left|\frac{y''}{\sec^3 \alpha}\right| = |y'' \cos^3 \alpha|.$

588. We have to choose a, b, and c so that the order of contact of $y = x^3$ and $y = ax^2 + bx + c$ at (1,1) will be 2.

$$y_1 = x^3,\ y_1' = 3x^2,\ y_1'' = 6x; \qquad y_1(1) = 1,\ y_1'(1) = 3,\ y_1'' = 6.$$
$$a + b + c = 1,\ 2a + b = 3,\ 2a = 6; \qquad a = 3,\ b = -3,\ c = 1.$$

The required parabola (see Fig. 130) is

$$y = 3x^2 - 3x + 1.$$

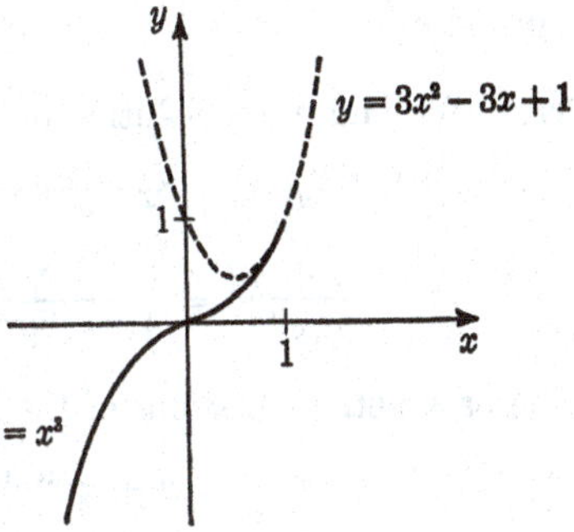

FIGURE 130

589. The equation of the circle is

$$x^2 + (y_1 - 5)^2 = 25$$

or

$$x^2 - 10y_1 + y_1^2 = 0.$$

Let the equation of the parabola be

$$y_2 = ax^5 + bx^4 + cx^3 + dx^2 + ex + f.$$

The parabola must pass through $M(0,0)$, i.e., $f = 0$. At $M(0,0)$, $y_2'(0) = 0$, i.e., $y_2'(0) = e = 0$. Now

$$2x - 10y_1' + 2y_1y_1' = 0,$$

$$1 - 5y_1'' + (y_1')^2 + y_1y_1'' = 0.$$

We substitute $x = 0$, $y_1 = 0$, $y_1' = 0$, $y_1'' = \frac{1}{5}$:

$$y_2'' = 20ax^3 + 12bx^2 + 6cx + 2d, \qquad y_2''(0) = 2d, \qquad d = 0.1.$$

We turn to $B(1,3)$:

$$y_2(1) = a + b + c + 0.1 = 3, \text{ i.e., } a + b + c = 2.9.$$

The slope of the segment BC is $\dfrac{66 - 3}{11 - 1} = 6.3.$

$$y_2'(1) = 5a + 4b + 3c + 0.2 = 6.3, \text{ i.e., } 5a + 4b + 3c = 6.1.$$

The second derivative at B must be zero, i.e.,

$$y_2''(1) = 20a + 12b + 6c + 0.2 = 0, \text{ or } 10a + 6b + 3c = -0.1.$$

The solution of the system for a, b, c is $a = -1$, $b = -0.6$, $c = 4.5$, i.e., the required parabola is

$$y = -x^5 - 0.6x^4 + 4.5x^3 + 0.1x^2.$$

590. $(\frac{1}{4},\frac{1}{4})$.

592. $\left(\dfrac{a}{2}, 0\right)$.

593. $\left(x - \dfrac{\pi}{2}\right)^2 + y^2 = 1.$

594. $\dfrac{(a^2 + b^2)^{3/2}}{2\sqrt{2}ab}$.

595. $\frac{3}{8}$.

597. We find in the solution of Problem 580,

$y' = -x^{-1/3}y^{1/3}$, $y'' = \frac{1}{3}c^{2/3}x^{-4/3}y^{-1/3}$,

$$\frac{1 + y'^2}{y''} = \frac{1 + y^{2/3}x^{-2/3}}{\frac{1}{3}c^{2/3}x^{-4/3}y^{-1/3}} = 3 \cdot \frac{x^{2/3} + y^{2/3}}{c^{2/3}x^{-2/3}y^{-1/3}} = 3x^{2/3}y^{1/3}.$$

The parametric equations of the evolute are therefore

$$a = x + x^{-1/3}y^{1/3} \cdot 3x^{2/3}y^{1/3} = x + 3x^{1/3}y^{2/3},$$

$$b = y + 3x^{2/3}y^{1/3}.$$

Let us eliminate x and y:

$$a + b = x + 3x^{2/3}y^{1/3} + 3x^{1/3}y^{2/3} + y = (x^{1/3} + y^{1/3})^3,$$

$$a - b = x - 3x^{2/3}y^{1/3} + 3x^{1/3}y^{2/3} - y = (x^{1/3} - y^{1/3})^3,$$

$$(a + b)^{2/3} + (a - b)^{2/3} = (x^{1/3} + y^{1/3})^2 + (x^{1/3} - y^{1/3})^2$$

$$= 2(x^{2/3} + y^{2/3}) = 2c^{2/3}.$$

Now we shall replace a and b by x and y. The implicit equation of the evolute is

$$(x + y)^{2/3} + (x - y)^{2/3} = 2c^{2/3}.$$

598. $x' = 3,\ x'' = 0,\ y' = 2t,\ y'' = 2;$

$$x = 3t - \frac{2t(9 + 4t^2)}{6} = -\frac{4}{3}t^3, \qquad y = t^2 - 6 + \frac{3(9 + 4t^2)}{6} = 3t^2 - \frac{3}{2}.$$

We shall eliminate the parameter:

$$t^3 = -\frac{3x}{4},\ t^2 = \left(-\frac{3x}{4}\right)^{2/3} = \left(\frac{3x}{4}\right)^{2/3}.$$

The equation of the evolute is

$$y = 3\left(\frac{3x}{4}\right)^{2/3} - \frac{3}{2}.$$

599. $$x' = -a\left(\frac{1}{\tan(t/2)\cos^2(t/2)} \cdot \frac{1}{2} - \sin t\right) = -a\left(\frac{1}{\sin t} - \sin t\right)$$

$$= -a\frac{\cos^2 t}{\sin t} = -a \cos t \cot t,$$

$$x'' = -a\left(-\sin t \cot t - \frac{\cos t}{\sin^2 t}\right) = a \cos t\left(1 + \frac{1}{\sin^2 t}\right);$$

$$y' = a \cos t, \qquad y'' = -a \sin t.$$

$$(x')^2 + (y')^2 = a^2(\cos^2 t \cot^2 t + \cos^2 t) = a^2 \cos^2 t \operatorname{cosec}^2 t = a^2 \cot^2 t,$$

$$x'y'' - x''y' = -a \cos t \cot t \cdot (-a \sin t) - a \cos t\left(1 + \frac{1}{\sin^2 t}\right) \cdot a \cos t$$

$$= a^2(\cos^2 t - \cos^2 t - \cot^2 t) = -a^2 \cot^2 t,$$

$$\frac{x'^2 + y'^2}{x'y'' - x''y'} = \frac{a^2 \cot^2 t}{-a^2 \cot^2 t} = -1.$$

The parametric equations of the evolute are

$$x = -a[\ln \tan(t/2) + \cos t] + a \cos t = -a \ln \tan(t/2),$$

$$y = a \sin t + a \cos t \cot t = \frac{a}{\sin t}.$$

From the first equation, $\tan (t/2) = e^{-x/a}$. Now

$$y = \frac{a}{\sin t} = a \frac{1 + \tan^2 (t/2)}{2 \tan (t/2)} = a \frac{1 + e^{2(-x/a)}}{2e^{-x/a}}$$

$$= \frac{a}{2} (e^{x/a} + e^{-x/a}) = a \cosh \frac{x}{a}.$$

601. $r = a^\theta$, $r' = a^\theta \ln a$, $r'' = a^\theta \ln^2 a$;

$$\frac{r^2 + (r')^2}{r^2 + 2(r')^2 - rr''} = \frac{a^{2\theta}(1 + \ln^2 a)}{a^{2\theta}(1 + 2 \ln^2 a - \ln^2 a)} = 1.$$

The equations of the evolute are

$$x = r \cos \theta - r' \sin \theta - r \cos \theta = -r' \sin \theta,$$
$$y = r \sin \theta + r' \cos \theta - r \sin \theta = r' \cos \theta.$$

We transform to polar coordinates:

$$r^2 = x^2 + y^2 = (r')^2 = a^{2\theta} \ln^2 a,$$

$$r = a^\theta \ln a = a^\theta a^{\frac{\ln \ln a}{\ln a}} = a^{\theta + \frac{\ln \ln a}{\ln a}}.$$

For $a = e$ we have

$$\frac{\ln \ln e}{\ln e} = \frac{\ln 1}{1} = 0.$$

Hence, in the case $r = e^\theta$ the evolute of the logarithmic spiral will be the curve itself.

602. $x = t + \sin t$, $y = -1 + \cos t$.

605. $f(x) = x^3 - 1.8x^2 - 10x + 17, f'(x) = 3x^2 - 3.6x - 10,$

$$f''(x) = 6x - 3.6.$$

$f(1.6) = 0.48; f(1.7) = 1.7^3 - 1.8 \cdot 1.7^2 < 0.$

$f'(x) < 0$ and $f''(x) > 0$ in this interval. We begin with 1.6:

x	$f(x)$	$f'(x)$	$-f(x)/f'(x)$
1.6	0.48	−8.08	0.059
1.66	0.013		

$$f(1.665) = -0.025$$

The solution is clearly 1.66.

606. $f(x) = x^4 + 96x - 80, f'(x) = 4x^3 + 96, f''(x) = 12x^2;$
$f(0) = -80$, $f(1) = 17$. (0,1) is clearly an interval of isolation. For $x > 0, f'(x) > 0$, i.e., there is only one positive solution

x	$f(x)$	$f'(x)$	$-f(x)/f'(x)$
1	17	100	−0.17
0.83	0.1546	98.3	−0.00157

$$f(0.825) = -0.337 < 0$$

Therefore 0.83 is the required solution.

607. $f(x) = \sin x - x + 2, f'(x) = \cos x - 1, f''(x) = -\sin x;$
$f(\pi/2) \approx 1.43, f(\pi) \approx -1.14;$
$(\pi/2, \pi)$ is an interval of isolation.

x	$f(x)$	$f'(x)$	$-f(x)/f'(x)$
π	~ -1.14	−2	−0.57
2.57	−0.02903	−1.84104	−0.01577
2.55423	−0.0001		

The solution is 2.5542 (since $f(2.55415) > 0$).

It is easy to see by sketching a graph that the above equation has no more solutions. We remark that the first computations can be performed with less accuracy than is required at the end, but care must be taken not to pass ξ.

608. 2.45.

609. 1.557.

610. 0.0570 and 1.4678.

611. 2.49.

613. For $\Delta x = -0.1$,

$$\Delta y = f(1) - f(1.1) = \sin 1 - \sin 1.1$$
$$= 0.84147 - 0.89121 = -0.04974.$$
$$dy = f'(1.1)\, dx = \cos 1.1 \cdot (-0.1)$$
$$= 0.45360 \cdot (-0.1) = -0.04536.$$

For $\Delta x = -0.01$,

$$\Delta y = \sin 1.09 - \sin 1.1 = 0.88663 - 0.89121 = -0.00458.$$
$$dy = \cos 1.1 \cdot (-0.01) = 0.45360 \cdot (-0.01) = -0.004536.$$

615. $dy = 4(5 - 2x + x^5)^3(-2 + 5x^4)\, dx.$

616. $dy = \left(\frac{1}{2\sqrt{\arcsin x}} \cdot \frac{1}{\sqrt{1-x^2}} + \frac{2 \arctan x}{1+x^2}\right) dx.$

617. $dy = \left(5^{-1/x^2} \cdot \ln 5 \cdot \frac{2}{x^3} - \frac{4}{x^3} - 10x\right) dx.$

618. $$dy = \frac{\frac{1}{x}(x - x\ln x) - (1 + \ln x)\left(1 - \ln x - \frac{x}{x}\right)}{(x - x\ln x)^2}\, dx$$

$$= \frac{1 - \ln x + \ln x + \ln^2 x}{x^2(1-\ln x)^2}\, dx = \frac{1 + \ln^2 x}{x^2(1 - \ln x)^2}\, dx;$$

$$2x^2\, dy = 2\frac{1 + \ln^2 x}{(1 - \ln x)^2}\, dx;$$

$$x^2y^2 + 1 = x^2 \frac{(1 + \ln x)^2}{x^2(1 - \ln x)^2} + 1$$

$$= \frac{1 + 2\ln x + \ln^2 x + 1 - 2\ln x + \ln^2 x}{(1 - \ln x)^2}$$

$$= 2\frac{1 + \ln^2 x}{(1 - \ln x)^2}.$$

622. $$dS = -2\cos z \sin z \cdot \frac{2t}{4}\, dt = -\sin 2z \frac{t}{2}\, dt$$

$$= -\sin\frac{t^2 - 1}{2} \cdot \frac{t}{2}\, dt.$$

623. In this case it seems to be more convenient to express y directly as a function of x and then compute the differential:

$$y = e^u = e^{\frac{1}{2}\ln v} = \sqrt{v} = \sqrt{2x^2 - 3x + 1},$$

$$dy = \frac{4x - 3}{2\sqrt{2x^2 - 3x + 1}}\, dx.$$

626. $y = \sin x;\ \Delta y \approx dy = \cos x\, dx.$

$$dx = 3' = \frac{3\pi}{60 \cdot 180} = \frac{\pi}{3600};\ x = \pi/3;\ \cos\frac{\pi}{3} = \frac{1}{2};$$

$$\sin 60°3' = \sin 60° + \Delta y \approx \sin 60° + dy$$

$$= \frac{\sqrt{3}}{2} + \frac{1}{2} \cdot \frac{\pi}{3600} \approx 0.8665.$$

627. $y = e^x,\ dy = e^x\, dx;\ x = 0,\ dx = 0.1;$

$$e^{0.1} \approx e^0 + dy = 1 + e^0 \cdot 0.1 = 1.1.$$

630. $V = \frac{\pi}{6} D^3,\ dV = \frac{\pi}{2} D^2\, dD,$

$$\left|\frac{dV}{V}\right| = \frac{(\pi/2)D^2|dD|}{(\pi/6)D^3} = 3\frac{|dD|}{D} = 3\alpha.$$

631. $dR_x = R\dfrac{a - x + x}{(a - x)^2}\,dx = \dfrac{Ra\,dx}{(a - x)^2},$

$$\left|\frac{dR_x}{R_x}\right| = \frac{Ra}{(a - x)^2}\,|dx| \cdot \frac{a - x}{Rx} = \frac{a|dx|}{x(a - x)}.$$

For a constant $|dx|$, dR_x/R_x attains its minimal value when the denominator $x(a - x)$ is maximal. We obtain $x = a/2$, i.e., the relative error of R_x will be minimal when x bisects AB.

633. $\frac{2}{3}\%$.

634. $\left|\dfrac{9}{\pi} - \dfrac{1}{2}\cot\dfrac{\pi}{18}\right| \cdot \dfrac{\pi}{180} \cdot 100\%.$

636. $dy = \frac{2}{3}x^{-1/3}\,dx,\ d^2y = -\frac{2}{9}x^{-4/3}\,dx^2.$

637. $y = x^5 + x^4 - 2x^3 - 2x^2 + x + 1,$
$d^4y = y^{iv}\,dx^4 = (5\cdot4\cdot3\cdot2x + 4\cdot3\cdot2\cdot1)\,dx^4 = (120x + 24)\,dx^4.$

638. $d^3y = -4\sin 2x\,dx^3.$

639. $r^2 = a^2\tan^3\theta,\ r = a\tan^{3/2}\theta,$

$$dr = \tfrac{3}{2}a\tan^{1/2}\theta\cos^{-2}\theta\,d\theta,$$

$$d^2r = \tfrac{3}{2}a(\tfrac{1}{2}\tan^{-1/2}\theta\cos^{-4}\theta + \tan^{1/2}\theta\cdot2\cos^{-3}\theta\sin\theta)\,d\theta^2$$

$$= \frac{3}{4}a\frac{1}{\sqrt{\tan\theta}} \cdot \frac{1 + 4\sin^2\theta}{\cos^4\theta}\,d\theta^2.$$

640. (a) $dy = \dfrac{1 + x^2}{1 - x^2} \cdot \dfrac{-2x(1 + x^2) - (1 - x^2)2x}{(1 + x^2)^2}\,dx = \dfrac{-4x\,dx}{1 - x^4},$

$$d^2y = \frac{-4(1 + 3x^4)}{(1 - x^4)^2}\,dx^2 - \frac{4x}{1 - x^4}\,d^2x.$$

(b) $y = \ln\dfrac{1 - x^2}{1 + x^2} = \ln\dfrac{1 - \tan^2 t}{1 + \tan^2 t} = \ln\,[(1 - \tan^2 t)\cos^2 t]$

$$= \ln(\cos^2 t - \sin^2 t) = \ln\cos 2t,$$

$$dy = -\frac{1}{\cos 2t} \cdot 2\sin 2t\,dt = -2\tan 2t\,dt,$$

$$d^2y = -\frac{4}{\cos^2 2t}\,dt^2.$$

We can obtain the same expression by substitution in d^2y obtained in (a).

641. (a) $dy = \cos z\,dz,\ d^2y = -\sin z\,dz^2 + \cos z\,d^2z.$

(b) $y = \sin a^x$, $dy = \cos a^x \cdot \ln a \cdot a^x \, dx$,

$$d^2y = (-\sin a^x \cdot \ln^2 a \cdot a^{2x} + \cos a^x \cdot \ln^2 a \cdot a^x) \, dx^2 + \cos a^x \ln a \cdot a^x \, d^2x$$

$$= a^x \ln a[(\ln a \cos a^x - \ln a \cdot a^x \cdot \sin a^x) \, dx^2 + \cos a^x \, d^2x].$$

(c) Here it seems more convenient to use the above expression instead of making a direct computation:

$$x = t^3, \qquad dx = 3t^2 \, dt, \qquad d^2x = 6t \, dt^2;$$

$$d^2y = a^{t^3} \ln a[(\ln a \cos a^{t^3} - \ln a \cdot a^{t^3} \sin a^{t^3})9t^4 \, dt^2 + \cos a^{t^3} \cdot 6t \, dt^2]$$

$$= 3ta^{t^3} \ln a \, [\cos a^{t^3} (2 + 3t^3 \ln a) - 3t^3 a^{t^3} \ln a \sin a^{t^3}] \, dt^2.$$

643. $I = \int (x^6 + 4x^3 + 4) \, dx = \frac{x^7}{7} + x^4 + 4x + C.$

644. $I = \int \left(x^2 + 2x - 1 + \frac{3}{x}\right) dx = \frac{x^3}{3} + x^2 - x + 3 \ln |x| + C.$

649. $I = -\frac{1}{5}e^{-5x} + C.$

651. $I = \int (e^{3x} + e^{2x} - e^{-2x}) \, dx = \frac{1}{3}e^{3x} + \frac{1}{2}e^{2x} + \frac{1}{2}e^{-2x} + C.$

652. $I = \int 3 \, dx - 2 \int \left(\frac{3}{2}\right)^x dx = 3x - 2 \cdot \left(\frac{3}{2}\right)^x \frac{1}{\ln \frac{3}{2}} + C.$

653. $I = \frac{1}{a} \sinh ax + C.$

654. $I = \frac{1}{a} \tanh ax + C.$

659. $I = -\frac{1}{4} \cot (4x + 3) + C.$

661. $I = \int \left(\frac{x^4 - 1}{x - 1} + \frac{1}{x - 1}\right) dx = \int \left(x^3 + x^2 + x + 1 + \frac{1}{x - 1}\right) dx$

$$= \frac{x^4}{4} + \frac{x^3}{3} + \frac{x^2}{2} + x + \ln |x - 1| + C.$$

663. $I = \int \frac{\sin x}{\cos x} \, dx = -\int \frac{d(\cos x)}{\cos x} = -\ln |\cos x| + C.$

664. $I = \int \frac{x}{9x^2 + 1} \, dx + \int \frac{1}{(3x)^2 + 1} \, dx$

$$= \frac{1}{18} \int \frac{d(9x^2 + 1)}{9x^2 + 1} + \frac{1}{3} \int \frac{d(3x)}{(3x)^2 + 1}$$

$$= \frac{1}{18} \ln (9x^2 + 1) + \frac{1}{3} \arctan 3x + C.$$

665. $I = \int \frac{d(e^x + 1)}{e^x + 1} = \ln(e^x + 1) + C.$

666. $I = \int -\tfrac{1}{2}d(1 - x^2)\cdot(1 - x^2)^{1/2} = -\tfrac{1}{2}\cdot\tfrac{2}{3}(1 - x^2)^{3/2} + C$

$= -\tfrac{1}{3}(1 - x^2)\sqrt{1 - x^2} + C.$

667. $I = \int (x^3 + 2)^{2/3}\cdot\tfrac{1}{3}d(x^3 + 2) = \tfrac{1}{5}(x^3 + 2)^{5/3} + C.$

668. $I = \int \frac{1}{2}\frac{d(3x^2 - 5x + 6)}{\sqrt{3x^2 - 5x + 6}} = \sqrt{3x^2 - 5x + 6} + C.$

669. $I = \int \sin^7 x\, d(\sin x) = \frac{\sin^8 x}{8} + C.$

670. $\int (\arctan x)^2\, d(\arctan x) = \tfrac{1}{3}(\arctan x)^3 + C.$

671. $I = \int -e^{-x^3}\cdot\tfrac{1}{3}d(-x^3) = -\tfrac{1}{3}e^{-x^3} + C.$

672. $I = \int \frac{1 + x^2}{1 + x^2}\,dx + \int \frac{2x}{1 + x^2}\,dx = x + \ln(1 + x^2) + C.$

673. $I = \int \frac{d(\arcsin x)}{(\arcsin x)^3} = -\frac{1}{2}\frac{1}{(\arcsin x)^2} + C.$

674. $I = \int \frac{1}{2}\frac{d(x^2)}{\sqrt{a^2 - (x^2)^2}} = \frac{1}{2a}\int \frac{d(x^2)}{\sqrt{1 - (x^2/a)^2}} = \frac{1}{2}\int \frac{d(x^2/a)}{\sqrt{1 - (x^2/a)^2}}$

$= \frac{1}{2}\arcsin\frac{x^2}{a} + C.$

675. $I = \int 2\cos^4 x \sin x\, dx = -\tfrac{2}{5}\cos^5 x + C.$

676. $I = \int \frac{dx}{4\sinh^2 x} = -\frac{1}{4}\coth x + C.$

677. $I = \int \frac{dx}{1 + \cos\left(x - \frac{\pi}{2}\right)} = \int \frac{dx}{2\cos^2\left(\frac{x}{2} - \frac{\pi}{4}\right)} = \tan\left(\frac{x}{2} - \frac{\pi}{4}\right) + C.$

678. $I = \int \left(\frac{2}{1 - \sin x} - 1\right) dx = \int \left(\frac{2}{1 - \cos\left(x - \frac{\pi}{2}\right)} - 1\right) dx$

$= \int \left(\frac{1}{\sin^2\left(\frac{x}{2} - \frac{\pi}{4}\right)} - 1\right) dx = -2\cot\left(\frac{x}{2} - \frac{\pi}{4}\right) - x + C$

$= 2\tan\left(\frac{x}{2} + \frac{\pi}{4}\right) - x + C.$

Alternately:

$$I = \int \frac{(1 + \sin x)^2}{1 - \sin^2 x}\,dx = \int \frac{1 + 2\sin x + \sin^2 x}{\cos^2 x}\,dx$$

$$= \int \frac{2 + 2\sin x - \cos^2 x}{\cos^2 x}\,dx = 2\int \frac{dx}{\cos^2 x} - 2\int \frac{d(\cos x)}{\cos^2 x} - \int dx$$

$$= 2\tan x + 2\cdot\frac{1}{\cos x} - x + C.$$

679. $I = \int \cosh 2ax\,dx = \frac{1}{2a}\sinh 2ax + C.$

680. $I = \frac{1}{2}\int (1 + \cos 2ax)\,dx = \frac{1}{2}\left(x + \frac{1}{2a}\sin 2ax\right) + C.$

681. $I = \ln|\ln x| + C.$

693. $\ln\cos x = t,\ -\frac{\sin x}{\cos x}\,dx = dt.$

$$I = -\int t\,dt = -\frac{t^2}{2} + C = -\frac{1}{2}(\ln\cos x)^2 + C.$$

694. $I = \int \frac{dx}{2\sin\frac{x}{2}\cos\frac{x}{2}} = \frac{1}{2}\int \frac{dx}{\tan\frac{x}{2}\cos^2\frac{x}{2}} = \ln\left|\tan\frac{x}{2}\right| + C.$

695. $I = \frac{1}{2}\int \frac{d\left(\frac{\pi}{2} + 2x\right)}{\sin\left(\frac{\pi}{2} + 2x\right)} = \frac{1}{2}\ln\left|\tan\left(\frac{\pi}{4} + x\right)\right| + C.$

(See Problem 694.)

696. $\frac{1}{x^2} = t,\ -\frac{2}{x^3}\,dx = dt.$

$$I = -\frac{1}{2}\int \sinh t\,dt = -\frac{1}{2}\cosh t + C = -\frac{1}{2}\cosh\frac{1}{x^2} + C.$$

697. $I = \int \frac{\sinh x}{2\cosh x}\,dx = \frac{1}{2}\ln\cosh x + C.$

698. $x = a\tan t,\ dx = \frac{a}{\cos^2 t}\,dt.$

$$I = \int \frac{a\,dt}{\cos^2 t\,(a^2\tan^2 t + a^2)^2} = \frac{1}{a^3}\int \frac{dt}{\cos^2 t\sec^4 t}$$

$$= \frac{1}{a^3}\int \cos^2 t\,dt = \frac{1}{a^3}\cdot\frac{1}{2}\int (1 + \cos 2t)\,dt = \frac{1}{2a^3}\left(t + \frac{1}{2}\sin 2t\right) + C$$

$$= \frac{1}{2a^3}\left(\arctan\frac{x}{a} + \frac{ax}{a^2 + x^2}\right) + C$$

(because $x = a \tan t$, i.e., $\tan t = x/a$ and $t = \arctan (x/a)$). We have also

$$\sin 2t = \frac{2\tan t}{1 + \tan^2 t} = \frac{2x/a}{1 + (x^2/a^2)} = \frac{2ax}{a^2 + x^2}.$$

699. $x = a \sin z$, $dx = a \cos z\,dz$.

$$I = \int \frac{a^2 \sin^2 z \cdot a \cos z\,dz}{a \cos z} = a^2 \int \sin^2 z\,dz$$

$$= \frac{1}{2}a^2 \int (1 - \cos 2z)\,dz = \frac{1}{2}a^2\left(z - \frac{1}{2}\sin 2z\right) + C$$

$$= \frac{1}{2}a^2\left(\arcsin\frac{x}{a} - \frac{x}{a}\sqrt{1 - \frac{x^2}{a^2}}\right) + C$$

$$= \frac{1}{2}a^2 \arcsin\frac{x}{a} - \frac{1}{2}x\sqrt{a^2 - x^2} + C.$$

700. We use here the substitution $x^2 - a^2 = u^2$, $2x\,dx = 2u\,du$.

$$I = \int \frac{u \cdot u\,du}{a^2 + u^2} = \int \left(1 - \frac{a^2}{a^2 + u^2}\right) du$$

$$= u - a \arctan\frac{u}{a} + C = \sqrt{x^2 - a^2} - a \arctan\frac{\sqrt{x^2 - a^2}}{a} + C.$$

701. $x = a \sinh t$, $dx = a \cosh t\,dt$.

$$I = \int \frac{a \cosh t\,dt}{a \cosh t} = \int dt = t + C = \operatorname{arg\,sinh}\frac{x}{a} + C$$

$$= \ln\left(\frac{x}{a} + \sqrt{1 + \frac{x^2}{a^2}}\right) + C = \ln (x + \sqrt{a^2 + x^2}) + C_1.$$

$-\ln a$ was included in the constant.

702. $x = a \cosh t$, $dx = a \sinh t\,dt$.

$$I = \int a^2 \sinh^2 t\,dt = \frac{a^2}{2}\int (-1 + \cosh 2t)\,dt$$

$$= \frac{a^2}{2}\left(-t + \frac{1}{2}\sinh 2t\right) + C = \frac{a^2}{2}\left(-\operatorname{arg\,cosh}\frac{x}{a} + \frac{x}{a}\sqrt{\frac{x^2}{a^2} - 1}\right) + C$$

$$= \frac{1}{2}x\sqrt{x^2 - a^2} - \frac{a^2}{2}\operatorname{arg\,cosh}\frac{x}{a} + C.$$

703. $x - 2 = t^2,\ dx = 2t\,dt,\ x = 2 + t^2.$

$$I = \int \frac{(t^2 + 2 + 1)2t\,dt}{(t^2 + 2)t} = 2\int \left(1 + \frac{1}{2 + t^2}\right) dt$$

$$= 2t + \sqrt{2}\arctan\frac{t}{\sqrt{2}} + C = 2\sqrt{x - 2} + \sqrt{2}\arctan\frac{\sqrt{x - 2}}{\sqrt{2}} + C.$$

704. $\sqrt{x} = t,\ \dfrac{dx}{2\sqrt{x}} = dt.$

$$I = 2\int \cos t\,dt = 2\sin t + C = 2\sin\sqrt{x} + C.$$

705. $\sin x = t,\ \cos x\,dx = dt.$

$$I = \int \frac{t^2\,dt}{1 + t^2} = t - \arctan t + C = \sin x - \arctan \sin x + C.$$

706. $I = \displaystyle\int \frac{e^{x/2} - e^{-x/2}}{e^{x/2} + e^{-x/2}}\,dx = \int \frac{\sinh (x/2)}{\cosh (x/2)}\,dx = 2\ln\cosh (x/2) + C.$

707. $\ln\tan x = t,\ dt = \dfrac{1}{\tan x}\cdot\dfrac{1}{\cos^2 x}\,dx = \dfrac{1}{\sin x\cos x}\,dx.$

$$I = \int t\,dt = \frac{t^2}{2} + C = \frac{1}{2}(\ln\tan x)^2 + C.$$

708. $a^3 - x^3 = t^2,\ -3x^2\,dx = 2t\,dt,\ x^3 = a^3 - t^2.$

$$I = \int \frac{(a^3 - t^2)(-\frac{2}{3}t)\,dt}{t} = -\frac{2}{3}\left(a^3 t - \frac{t^3}{3}\right) + C$$

$$= -\frac{2}{3}\left[a^3\sqrt{a^3 - x^3} - \frac{1}{3}\sqrt{(a^3 - x^3)^3}\right] + C$$

$$= -\frac{2}{9}\sqrt{a^3 - x^3}(2a^3 + x^3) + C.$$

709. $x = a\sin t,\ dx = a\cos t\,dt.$

$$I = \int \frac{a\cos t\cdot a\cos t\,dt}{a^2\sin^2 t} = \int \left(\frac{1}{\sin^2 t} - 1\right) dt = -\cot t - t + C$$

$$= -\frac{\sqrt{1 - (x^2/a^2)}}{x/a} - \arcsin\frac{x}{a} + C = -\frac{\sqrt{a^2 - x^2}}{x} - \arcsin\frac{x}{a} + C.$$

710. $x = a\cosh t,\ dx = a\sinh t\,dt.$

$$I = \int \frac{a\sinh t}{a^3\sinh^3 t}\,dt = \frac{1}{a^2}\int \frac{1}{\sinh^2 t}\,dt = -\frac{1}{a^2}\coth t + C$$

$$= -\frac{1}{a^2}\frac{\cosh t}{\sinh t} + C = -\frac{1}{a^2}\frac{x/a}{\sqrt{(x^2/a^2) - 1}} + C = -\frac{1}{a^2}\frac{x}{\sqrt{x^2 - a^2}} + C.$$

711. $e^x - 1 = t^2$, $e^x\,dx = 2t\,dt$, $e^x = 1 + t^2$.

$$I = \int t \cdot \frac{2t\,dt}{1 + t^2} = 2\int \frac{t^2 + 1 - 1}{1 + t^2}\,dt = 2t - 2\arctan t + C$$

$$= 2\sqrt{e^x - 1} - 2\arctan\sqrt{e^x - 1} + C.$$

712. $x = \frac{1}{t}$, $dx = -\frac{1}{t^2}\,dt$.

$$I = \int \sqrt{\frac{(1/t) - 1}{(1/t) + 1}}\,t^2\left(-\frac{1}{t^2}\right)dt = -\int \sqrt{\frac{1 - t}{1 + t}}\,dt = -\int \frac{1 - t}{\sqrt{1 - t^2}}\,dt.$$

Now we substitute once more,

$$t = \cos z, \qquad dt = -\sin z\,dz.$$

$$I = \int \frac{1 - \cos z}{\sin z}\sin z\,dz = \int (1 - \cos z)\,dz = z - \sin z + C$$

$$= \arccos t - \sqrt{1 - t^2} + C = \arccos\frac{1}{x} - \sqrt{1 - \frac{1}{x^2}} + C$$

$$= \arccos\frac{1}{x} - \frac{\sqrt{x^2 - 1}}{x} + C.$$

713. $1 + xe^x = t$, $dt = (e^x + xe^x)\,dx = e^x(1 + x)\,dx$.

$$I = \int \frac{dt}{t(t - 1)} = \int\left(-\frac{dt}{t} + \frac{dt}{t - 1}\right) = \ln\left|\frac{t - 1}{t}\right| + C = \ln\left|\frac{xe^x}{1 + xe^x}\right| + C.$$

714. First we change the form of the integrand:

$$I = \int \frac{\ln\left(1 + \frac{1}{x}\right)dx}{\left(1 + \frac{1}{x}\right)x^2}.$$

Now we substitute $\ln\left(1 + \frac{1}{x}\right) = t$,

$$\frac{1}{1 + \frac{1}{x}}\left(-\frac{1}{x^2}\right)dx = dt.$$

$$I = \int -t\,dt = -\frac{t^2}{2} + C = -\frac{1}{2}\ln^2\left(1 + \frac{1}{x}\right) + C.$$

717. $I = \frac{1}{2}\int x\sin 2x\,dx = -\frac{1}{4}x\cos 2x + \frac{1}{8}\sin 2x + C.$

We used the solution of Problem 715.

718. $I = \frac{1}{2}\int x^2(1 + \cos 2x)\,dx = \frac{x^3}{6} + \frac{1}{2}\int x^2\cos 2x\,dx.$

We put $u = x^2$, $v''' \, dx = \cos 2x \, dx$. Then

$$v'' = \tfrac{1}{2} \sin 2x, \qquad v' = -\tfrac{1}{4} \cos 2x, \qquad v = -\tfrac{1}{8} \sin 2x.$$

By formula (A) we have

$$I = \frac{x^3}{6} + \frac{1}{2}\left(\frac{1}{2} x^2 \sin 2x + \frac{1}{2} x \cos 2x - \frac{1}{4} \sin 2x\right) + C.$$

The same result can be obtained by two consecutive integrations by parts.

719. We can solve this integral by three consecutive integrations by parts, but it is simpler to use formula (A):

$$u = x^3, \qquad v^{iv} \, dx = e^x \, dx;$$

$$v''' = e^x, \qquad v'' = e^x, \qquad v' = e^x, \qquad v = e^x;$$

$$u' = 3x^2, \qquad u'' = 6x, \qquad u''' = 6.$$

Now $$I = e^x(x^3 - 3x^2 + 6x - 6) + C.$$

720. $u = x^2$, $du = 2x \, dx$, $dv = a^x \, dx$, $v = \dfrac{a^x}{\ln a}$.

$$I = \frac{a^x}{\ln a} x^2 - \frac{2}{\ln a} \int x a^x \, dx.$$

Once more: $u = x$, $du = dx$, $dv = a^x \, dx$, $v = \dfrac{a^x}{\ln a}$.

$$I = \frac{a^x}{\ln a} \cdot x^2 - \frac{2}{\ln a}\left(x \frac{a^x}{\ln a} - \int \frac{a^x}{\ln a} \, dx\right)$$

$$= \frac{a^x}{\ln a}\left(x^2 - \frac{2x}{\ln a} + \frac{2}{\ln^2 a}\right) + C.$$

723. $u = \ln x$, $du = \dfrac{1}{x} \, dx$, $dv = \sqrt{x} \, dx$, $v = \dfrac{2}{3} x^{3/2}$.

$$I = \frac{2}{3} x^{3/2} \ln x - \frac{2}{3} \int x^{3/2} \cdot \frac{1}{x} \, dx = \frac{2}{3} x^{3/2} \ln x - \frac{2}{3} \int x^{1/2} \, dx$$

$$= \frac{2}{3} x^{3/2} \ln x - \frac{4}{9} x^{3/2} + C = \frac{2}{3} x^{3/2}\left(\ln x - \frac{2}{3}\right) + C.$$

724. $u = \ln^2 x$, $dv = dx$, $du = 2 \ln x \cdot \dfrac{1}{x} \, dx$, $v = x$.

$$I = x \ln^2 x - 2 \int \ln x \, dx = x \ln^2 x - 2(x \ln x - x) + C.$$

We used Problem 722.

726. $u = \arcsin x$, $du = \dfrac{1}{\sqrt{1 - x^2}} \, dx$, $dv = dx$, $v = x$.

$$I = x \arcsin x - \int \frac{x\,dx}{\sqrt{1 - x^2}} = x \arcsin x + \sqrt{1 - x^2} + C.$$

727. $u = \arctan x,\ du = \dfrac{1}{1 + x^2}\,dx,\ dv = dx,\ v = x.$

$$I = x \arctan x - \int \frac{x}{1 + x^2}\,dx = x \arctan x - \frac{1}{2} \ln (1 + x^2) + C.$$

728. $I = x \sinh x - \int \sinh x\,dx = x \sinh x - \cosh x + C.$

729. $I = \dfrac{x^2}{2} \arctan x - \dfrac{1}{2} \displaystyle\int \frac{x^2}{1 + x^2}\,dx$

$$= \frac{x^2}{2} \arctan x - \frac{1}{2} x + \frac{1}{2} \arctan x + C.$$

We used here $\qquad \dfrac{x^2}{1 + x^2} = 1 - \dfrac{1}{1 + x^2}.$

730. $u = \ln (x + 1),\ du = \dfrac{dx}{x + 1},\ dv = (x^2 + x)\,dx,\ v = \dfrac{x^3}{3} + \dfrac{x^2}{2}.$

$$I = \left(\frac{x^3}{3} + \frac{x^2}{2}\right) \ln (x + 1) - \frac{1}{6} \int \frac{2x^3 + 3x^2}{x + 1}\,dx$$

$$= \left(\frac{x^3}{3} + \frac{x^2}{2}\right) \ln (x + 1) - \frac{1}{6} \int \left(2x^2 + x - 1 + \frac{1}{x + 1}\right) dx$$

$$= \left(\frac{x^3}{3} + \frac{x^2}{2}\right) \ln (x + 1) - \frac{1}{9} x^3 - \frac{1}{12} x^2 + \frac{1}{6} x - \frac{1}{6} \ln (x + 1) + C.$$

731. $u = \arcsin x,\ du = \dfrac{dx}{\sqrt{1 - x^2}},\ dv = \dfrac{dx}{\sqrt{1 + x}},\ v = 2\sqrt{1 + x}.$

$$I = 2\sqrt{1 + x} \arcsin x - \int \frac{2\sqrt{1 + x}}{\sqrt{1 - x^2}}\,dx$$

$$= 2\sqrt{1 + x} \arcsin x + 4\sqrt{1 - x} + C.$$

732. $u = x,\ du = dx,\ dv = \tan^2 x\,dx,\ v = \tan x - x.$

$$I = x \tan x - x^2 - \int (\tan x - x)\,dx = x \tan x - x^2 + \ln |\cos x| + \frac{x^2}{2} + C$$

$$= x \tan x - \frac{x^2}{2} + \ln |\cos x| + C.$$

733. $u = \ln \ln x,\ du = \dfrac{1}{\ln x} \cdot \dfrac{1}{x}\,dx,\ dv = \dfrac{1}{x}\,dx,\ v = \ln x.$

$$I = \ln x \cdot \ln \ln x - \int \frac{1}{x}\,dx = \ln x\,(\ln \ln x - 1) + C.$$

Here we could write $\ln x$, instead of $\ln |x|$, because from the given integral it follows that x is positive (moreover, $x > 1$).

734. First we substitute $x^2 = t$, $2x\,dx = dt$.

$$I = \tfrac{1}{2}\int e^{-t}t^2\,dt = \tfrac{1}{2}e^{-t}(-t^2 - 2t - 2) + C$$
$$= -\tfrac{1}{2}e^{-x^2}(x^4 + 2x^2 + 2) + C.$$

We used formula (A).

735. $u = \cos\ln x$, $du = \dfrac{-\sin\ln x}{x}\,dx$, $dv = dx$, $v = x$.

$$I = x\cos\ln x + \int \sin\ln x\,dx$$
$$= x\cos\ln x + x\sin\ln x - \int \cos\ln x\,dx;$$
$$I = \tfrac{1}{2}x(\cos\ln x + \sin\ln x) + C.$$

736. $u = \arcsin x$, $du = \dfrac{dx}{\sqrt{1 - x^2}}$, $dv = \dfrac{x\,dx}{\sqrt{1 - x^2}}$, $v = -\sqrt{1 - x^2}$.

$$I = -\sqrt{1 - x^2}\arcsin x + x + C.$$

737. $u = \arctan x$, $du = \dfrac{dx}{1 + x^2}$,

$$dv = \frac{3x^2 - 1}{2x\sqrt{x}}\,dx = \left(\frac{3}{2}x^{1/2} - \frac{1}{2}x^{-3/2}\right)dx,\ v = x^{3/2} + x^{-1/2} = \frac{x^2 + 1}{\sqrt{x}}.$$

$$I = \frac{x^2 + 1}{\sqrt{x}}\arctan x - 2\sqrt{x} + C.$$

738. We begin by substituting

$$\arcsin x = t,\ \frac{dx}{\sqrt{1 - x^2}} = dt; \qquad x = \sin t,\ 1 - x^2 = \cos^2 t.$$

Then
$$I = \int \frac{t\,dt}{\cos^2 t}.$$

Now integration by parts gives

$$u = t, \qquad du = dt, \qquad dv = \frac{dt}{\cos^2 t}, \qquad v = \tan t.$$

$$I = t\tan t - \int \tan t\,dt = t\tan t + \ln|\cos t| + C$$
$$= \frac{x}{\sqrt{1 - x^2}}\arcsin x + \ln\sqrt{1 - x^2} + C.$$

739. $t = \arctan x$, $dt = \dfrac{dx}{1 + x^2}$, $x = \tan t$.

$$I = \int t \tan^2 t \, dt = t \tan t - \frac{t^2}{2} + \ln|\cos t| + C$$

$$= x \arctan x - \tfrac{1}{2}(\arctan x)^2 - \tfrac{1}{2}\ln(1 + x^2) + C.$$

We used here the solution of Problem 732 and the identity

$$\cos \arctan x = \frac{1}{\sqrt{1 + x^2}}.$$

740. $$I = \int e^x \frac{1 + 2\sin\frac{x}{2}\cos\frac{x}{2}}{2\cos^2\frac{x}{2}} dx = \int e^x \frac{dx}{2\cos^2\frac{x}{2}} + \int e^x \tan\frac{x}{2}\, dx.$$

In the first integral,

$$u = e^x, \qquad du = e^x\, dx, \qquad dv = \frac{1}{2}\frac{dx}{\cos^2(x/2)}, \qquad v = \tan\frac{x}{2}.$$

Now

$$I = e^x \tan\frac{x}{2} - \int e^x \tan\frac{x}{2}\, dx + \int e^x \tan\frac{x}{2}\, dx = e^x \tan\frac{x}{2} + C.$$

743. $I = (Ax^2 + Bx + C)\cos 2x + (Dx^2 + Ex + F)\sin 2x + C_1.$

$$(x^2 + 3x + 5)\cos 2x = (2Ax + B)\cos 2x - 2(Ax^2 + Bx + C)\sin 2x$$
$$+ (2Dx + E)\sin 2x + 2(Dx^2 + Ex + F)\cos 2x.$$

$1 = 2D,\ D = \tfrac{1}{2}; \qquad 3 = 2A + 2E; \qquad 5 = B + 2F;$

$-2A = 0,\ A = 0,\ E = \tfrac{3}{2};$

$-2B + 2D = 0,\ B = \tfrac{1}{2},\ F = \tfrac{9}{4}; \qquad -2C + E = 0,\ C = \tfrac{3}{4}.$

$$I = (\tfrac{1}{2}x + \tfrac{3}{4})\cos 2x + (\tfrac{1}{2}x^2 + \tfrac{3}{2}x + \tfrac{9}{4})\sin 2x + C_1.$$

Here it seems that the general formula (A) for integration by parts would have been more convenient.

745. $x^3 - 3x^2 + 4 = (x + 1)(x - 2)^2.$

Put $\dfrac{x - 5}{(x + 1)(x - 2)^2} = \dfrac{A}{x + 1} + \dfrac{B}{x - 2} + \dfrac{C}{(x - 2)^2}$; then

$$x - 5 = A(x - 2)^2 + B(x + 1)(x - 2) + C(x + 1);$$

$x = -1\colon\ -6 = 9A,\ A = -\tfrac{2}{3}.$

$x = 2\colon\quad -3 = 3C,\ C = -1.$

$x = 0\colon\quad -5 = 4A - 2B + C,\ 2B = -\tfrac{8}{3} - 1 + 5 = \tfrac{4}{3},$

$$B = \tfrac{2}{3}.$$

$$I = \int \left[-\frac{2}{3}\frac{1}{x + 1} + \frac{2}{3}\frac{1}{x - 2} - \frac{1}{(x - 2)^2}\right] dx = \frac{2}{3}\ln\left|\frac{x - 2}{x + 1}\right| + \frac{1}{x - 2} + C.$$

746. Let us substitute $x - 2 = z$; $dx = dz$, $x = 2 + z$.

$$I = \int \frac{4(z+2)+3}{z^3}\,dz = \int \left(\frac{4}{z^2} + \frac{11}{z^3}\right) dz = -\frac{4}{z} - \frac{11}{2z^2} + C$$

$$= -\frac{4}{x-2} - \frac{11}{2(x-2)^2} + C.$$

747. Let $\dfrac{x^3 - 2x^2 + 4}{x^3(x-2)^2} = \dfrac{A}{x} + \dfrac{B}{x^2} + \dfrac{C}{x^3} + \dfrac{D}{x-2} + \dfrac{E}{(x-2)^2}$;

$$x^3 - 2x^2 + 4 = Ax^2(x-2)^2 + Bx(x-2)^2 + C(x-2)^2 + Dx^3(x-2) + Ex^3.$$

$x = 0$: $4 = 4C$, $C = 1$; $\quad x = 2$: $4 = 8E$, $E = \frac{1}{2}$.

Let us subtract the terms in C and E from the given fraction:

$$\frac{x^3 - 2x^2 + 4}{x^3(x-2)^2} - \frac{1}{x^3} - \frac{1}{2(x-2)^2} = \frac{2x^3 - 4x^2 + 8 - 2x^2 + 8x - 8 - x^3}{2x^3(x-2)^2}$$

$$= \frac{x^3 - 6x^2 + 8x}{2x^3(x-2)^2} = \frac{x(x-2)(x-4)}{2x^3(x-2)^2}$$

$$= \frac{x-4}{2x^2(x-2)}.$$

Now we continue the decomposition:

$$\frac{x-4}{2x^2(x-2)} = \frac{A}{x} + \frac{B}{x^2} + \frac{D}{x-2};$$

$$\tfrac{1}{2}(x-4) = Ax(x-2) + B(x-2) + Dx^2.$$

$x = 0$: $-2 = -2B$, $B = 1$; $\quad x = 2$: $-1 = 4D$, $D = -\frac{1}{4}$.

Thus

$$\frac{x-4}{2x^2(x-2)} - \frac{1}{x^2} + \frac{1}{4(x-2)} = \frac{2x - 8 - 4x + 8 + x^2}{4x^2(x-2)}$$

$$= \frac{x^2 - 2x}{4x^2(x-2)} = \frac{x(x-2)}{4x^2(x-2)} = \frac{1}{4x}.$$

Finally,

$$I = \int \left[\frac{1}{4x} + \frac{1}{x^2} + \frac{1}{x^3} - \frac{1}{4(x-2)} + \frac{1}{2(x-2)^2}\right] dx$$

$$= \frac{1}{4}\ln|x| - \frac{1}{x} - \frac{1}{2x^2} - \frac{1}{4}\ln|x-2| - \frac{1}{2(x-2)} + C.$$

We have demonstrated here a slight variation in the method. In this case little was gained.

750. $x^4 + x^3 - x - 1 = (x+1)(x-1)(x^2+x+1)$.

Put $$\frac{12}{x^4 + x^3 - x - 1} = \frac{A}{x+1} + \frac{B}{x-1} + \frac{Cx + D}{x^2 + x + 1},$$

$$12 = A(x-1)(x^2 + x + 1) + B(x+1)(x^2 + x + 1) + (Cx + D)(x+1)(x-1).$$

$x = 1$: $12 = 6B$, $B = 2$; $x = -1$: $12 = -2A$, $A = -6$.

$x = 0$: $12 = -A + B - D$, $D = 6 + 2 - 12 = -4$.

Equating the coefficients of x^3, $A + B + C = 0$, $C = 4$.

$$I = \int \left(-\frac{6}{x+1} + \frac{2}{x-1} + 4\,\frac{x-1}{x^2 + x + 1}\right) dx$$

$$= 2 \ln\left|\frac{x-1}{(x+1)^3}\right| + 4\left[\frac{1}{2}\int \frac{2x+1}{x^2 + x + 1}\,dx - \frac{3}{2}\int \frac{dx}{(x + \frac{1}{2})^2 + \frac{3}{4}}\right]$$

$$= 2 \ln\left|\frac{x-1}{(x+1)^3}\right| + 2 \ln (x^2 + x + 1) - \frac{12}{\sqrt{3}} \arctan \frac{2x+1}{\sqrt{3}} + C$$

$$= 2 \ln\left|\frac{x^3 - 1}{(x+1)^3}\right| - 4\sqrt{3} \arctan \frac{2x+1}{\sqrt{3}} + C.$$

751. $$x^4 + 1 = x^4 + 2x^2 + 1 - 2x^2 = (x^2 + 1)^2 - 2x^2 = (x^2 + \sqrt{2}\,x + 1)(x^2 - \sqrt{2}\,x + 1).$$

Put $$\frac{1}{x^4 + 1} = \frac{Ax + B}{x^2 + \sqrt{2}\,x + 1} + \frac{Cx + D}{x^2 - \sqrt{2}\,x + 1},$$

$$1 = (Ax + \mathrm{B})(x^2 - \sqrt{2}\,x + 1) + (Cx + D)(x^2 + \sqrt{2}\,x + 1).$$

We equate the coefficients of similar terms on both sides:

$$x^3\colon A + C = 0,$$

$$x^2\colon -\sqrt{2}A + B + \sqrt{2}C + D = 0,$$

$$x^1\colon A - \sqrt{2}B + C + \sqrt{2}D = 0,$$

$$x^0\colon B + D = 1.$$

The solutions are $B = D = \frac{1}{2}$, $A = \frac{1}{2\sqrt{2}}$, $C = -\frac{1}{2\sqrt{2}}$.

$$I = \frac{1}{2\sqrt{2}}\left[\int \frac{x + \sqrt{2}}{x^2 + x\sqrt{2} + 1}\,dx - \int \frac{x - \sqrt{2}}{x^2 - x\sqrt{2} + 1}\,dx\right]$$

$$= \frac{1}{2\sqrt{2}}\left[\frac{1}{2}\int \frac{2x + \sqrt{2}}{x^2 + x\sqrt{2} + 1}\,dx + \frac{\sqrt{2}}{2}\int \frac{dx}{\left(x + \frac{\sqrt{2}}{2}\right)^2 + \frac{1}{2}}\right.$$

$$-\frac{1}{2}\int \frac{2x-\sqrt{2}}{x^2 - x\sqrt{2}+1}\,dx + \frac{\sqrt{2}}{2}\int \frac{dx}{\left(x-\frac{\sqrt{2}}{2}\right)^2+\frac{1}{2}}\Bigg]$$

$$= \frac{1}{4\sqrt{2}}\ln\frac{x^2+x\sqrt{2}+1}{x^2-x\sqrt{2}+1} + \frac{\sqrt{2}}{4}\arctan(\sqrt{2}\,x+1)$$

$$+ \frac{\sqrt{2}}{4}\arctan(\sqrt{2}\,x-1) + C.$$

752. We shall use here the substitution $u = x + (1/x)$. It is convenient in some cases, similar to this, in which the denominator is a polynomial with symmetric coefficients while in the numerator the factor $x^2 - 1$ appears. Now

$$\left(1-\frac{1}{x^2}\right)dx = du,$$

$$I = \int \frac{1-\frac{1}{x^2}}{x^2+\frac{1}{x^2}+3\left(x+\frac{1}{x}\right)+5}\,dx = \int \frac{du}{u^2-2+3u+5}$$

$$= \int \frac{du}{u^2+3u+3} = \int \frac{du}{(u+\frac{3}{2})^2+\frac{3}{4}}$$

$$= \frac{2}{\sqrt{3}}\arctan\frac{2u+3}{\sqrt{3}} + C = \frac{2}{\sqrt{3}}\arctan\frac{2x^2+3x+2}{\sqrt{3}x} + C.$$

753. We shall use the substitution

$$u = x - \frac{1}{x}, \qquad du = \left(1+\frac{1}{x^2}\right)dx.$$

$$I = \int \frac{1+\frac{1}{x^2}}{x^2+1+\frac{1}{x^2}}\,dx = \int \frac{du}{u^2+2+1} = \int \frac{du}{u^2+3}$$

$$= \frac{1}{\sqrt{3}}\arctan\frac{u}{\sqrt{3}} + C$$

$$= \frac{1}{\sqrt{3}}\arctan\frac{x^2-1}{\sqrt{3}x} + C.$$

We remark that an equivalent result is obtained by the standard procedure in which we write $x^4 + x^2 + 1 = (x^2 + x + 1)(x^2 - x + 1)$ and employ partial fractions.

754. In integrals of this kind it is sometimes convenient to use the substitution

$$\frac{x-a}{x-b}=t.$$

In our case

$$\frac{x-2}{x+1}=t, \qquad \frac{x+1-x+2}{(x+1)^2}\,dx=\frac{3\,dx}{(x+1)^2}=dt,$$

$$x=\frac{t+2}{1-t}, \qquad x-2=\frac{t+2}{1-t}-2=\frac{t+2-2+2t}{1-t}=\frac{3t}{1-t}.$$

We substitute all these and obtain

$$I=\frac{1}{3}\int\frac{dt}{\left(\frac{3t}{1-t}\right)^3}=\frac{1}{81}\int\left(\frac{1}{t^3}-\frac{3}{t^2}+\frac{3}{t}-1\right)dt$$

$$=\frac{1}{81}\left(-\frac{1}{2t^2}+\frac{3}{t}+3\ln|t|-t\right)+C.$$

It remains to substitute $t=\dfrac{x-2}{x+1}$ and perform the algebraic simplifications.

755. $x^5+2x^3+x=x(x^2+1)^2.$

Put $$\frac{3x^4+x^3+4x^2+1}{x(x^2+1)^2}=\frac{A}{x}+\frac{Bx+C}{(x^2+1)^2}+\frac{Dx+E}{x^2+1},$$

$$3x^4+x^3+4x^2+1$$
$$=A(x^2+1)^2+(Bx+C)x+(Dx+E)(x^2+1)x,$$
$$=x^4(A+D)+Ex^3+x^2(2A+B+D)+x(C+E)+A.$$

$E=1;\quad C+E=0,\ C=-1;\quad A=1;\quad A+D=3,\ D=2;$
$2A+B+D=4,\ B=4-2-2=0.$

$$I=\int\left(\frac{1}{x}-\frac{1}{(x^2+1)^2}+\frac{2x+1}{x^2+1}\right)dx$$

$$=\ln|x|+\int\frac{2x}{x^2+1}\,dx-\int\frac{1-x^2-1}{(x^2+1)^2}\,dx$$

$$=\ln|x|+\ln(x^2+1)+\int\frac{x^2\,dx}{(x^2+1)^2}.$$

In the last integral we substitute $x=\tan t$:

$$\int\frac{x^2\,dx}{(x^2+1)^2}=\int\frac{\tan^2 t\,dt}{\sec^4 t\cos^2 t}=\int\sin^2 t\,dt=\frac{1}{2}\int(1-\cos 2t)\,dt$$

$$=\frac{1}{2}t-\frac{1}{4}\sin 2t+C=\frac{1}{2}\arctan x-\frac{1}{2}\frac{x}{1+x^2}+C.$$

Finally

$$I = \ln |x(x^2 + 1)| + \frac{1}{2} \arctan x - \frac{1}{2} \cdot \frac{x}{1 + x^2} + C.$$

We remark that $\int \frac{dx}{(1 + x^2)^2}$ can also be obtained using the reduction formula of Problem 741.

757. We solve this integral by the method of the last exercise:

$$\int \frac{5x^2 + 6x + 9}{(x - 3)^2(x + 1)^2} dx = \frac{Ax + B}{(x - 3)(x + 1)} + \int \left(\frac{C}{x - 3} + \frac{D}{x + 1}\right) dx.$$

By differentiation,

$$\frac{5x^2 + 6x + 9}{(x - 3)^2(x + 1)^2} = \frac{A(x^2 - 2x - 3) - (Ax + B)(2x - 2)}{(x - 3)^2(x + 1)^2} + \frac{C}{x - 3} + \frac{D}{x + 1}.$$

$$\begin{aligned} x^3 &: 0 = C + D, \\ x^2 &: 5 = -A - C - 5D, \\ x &: 6 = -2B - 5C + 3D, \\ x^0 &: 9 = -3A + 2B - 3C + 9D. \end{aligned}$$

$$A = -5, \qquad B = -3, \qquad C = 0, \qquad D = 0.$$

Consequently,

$$I = -\frac{5x + 3}{(x - 3)(x + 1)} + C.$$

759. First we transform:

$$I = \int \sqrt[4]{\frac{x - 1}{x + 2}} \cdot \frac{dx}{(x - 1)(x + 2)}.$$

Now we substitute $\frac{x - 1}{x + 2} = t^4$, i.e.,

$$x = \frac{1 + 2t^4}{1 - t^4}, \qquad x - 1 = \frac{3t^4}{1 - t^4}, \qquad x + 2 = \frac{3}{1 - t^4}, \qquad dx = \frac{12t^3}{(1 - t^4)^2} dt.$$

$$I = \int \frac{t \cdot 12t^3 \, dt}{(1 - t^4)^2 \frac{3t^4}{1 - t^4} \cdot \frac{3}{1 - t^4}} = \int \frac{4}{3} dt = \frac{4}{3} t + C = \frac{4}{3} \sqrt[4]{\frac{x - 1}{x + 2}} + C.$$

761. $I = \int x^{-1/2}(x^{1/3} + 1)^{-2} \, dx;\ p = 2,\ m = -\frac{1}{2},\ n = \frac{1}{3}.$ This is the first case. We substitute $x = t^6$, $dx = 6t^5 \, dt$.

$$I = \int t^{-3}(t^2 + 1)^{-2} 6t^5 \, dt = 6 \int \frac{t^2}{(t^2 + 1)^2} dt$$

$$= 6\left(\frac{1}{2} \arctan t - \frac{1}{2} \frac{t}{1 + t^2}\right) + C = 3 \arctan \sqrt[6]{x} - \frac{3\sqrt[6]{x}}{1 + \sqrt[3]{x}} + C.$$

We used here the solution of Problem 755.

765. $I = \dfrac{1}{\sqrt{2}} \displaystyle\int \dfrac{dx}{\sqrt{(x+\frac{1}{4})^2 - \frac{25}{16}}}$. We substitute

$$x + \tfrac{1}{4} = \tfrac{5}{4}\cosh t, \qquad dx = \tfrac{5}{4}\sinh t\,dt.$$

$$I = \frac{1}{\sqrt{2}} \int \frac{\frac{5}{4}\sinh t\,dt}{\frac{5}{4}\sinh t} = \frac{t}{\sqrt{2}} + C = \frac{1}{\sqrt{2}}\operatorname{arg\,cosh}\frac{4x+1}{5} + C$$

$$= \frac{1}{\sqrt{2}} \ln\left|\frac{4x+1}{5} + \sqrt{\left(\frac{4x+1}{5}\right)^2 - 1}\right| + C.$$

We remark that the formula of the preceding problem can be used here directly.

766. $I = \dfrac{1}{\sqrt{3}} \displaystyle\int \dfrac{dx}{\sqrt{\frac{1}{3} + \frac{2}{3}x - x^2}} = \dfrac{1}{\sqrt{3}} \displaystyle\int \dfrac{dx}{\sqrt{\frac{4}{9} - (x - \frac{1}{3})^2}}$

$$= \frac{1}{\sqrt{3}} \arcsin\frac{3x-1}{2} + C.$$

The formula of Problem 764 can be used here also.

768. $I = -\dfrac{1}{2} \displaystyle\int \dfrac{-2x+1}{\sqrt{1+x-x^2}}\,dx + \dfrac{1}{2}\displaystyle\int \dfrac{dx}{\sqrt{1+x-x^2}}$

$$= -\sqrt{1+x-x^2} + \frac{1}{2}\int \frac{dx}{\sqrt{\frac{5}{4} - (x-\frac{1}{2})^2}}$$

$$= -\sqrt{1+x-x^2} + \frac{1}{2}\arcsin\frac{2x-1}{\sqrt{5}} + C.$$

770. $I = (Ax^3 + Bx^3 + Cx + D)\sqrt{1-x^2} + p\displaystyle\int \dfrac{dx}{\sqrt{1-x^2}}$.

We differentiate:

$$\frac{x^4}{\sqrt{1-x^2}} = (3Ax^2 + 2Bx + C)\sqrt{1-x^2}$$

$$+ (Ax^3 + Bx^2 + Cx + D)\frac{(-x)}{\sqrt{1-x^2}} + p\frac{1}{\sqrt{1-x^2}};$$

$$x^4 = (3Ax^2 + 2Bx + C)(1-x^2) - (Ax^3 + Bx^2 + Cx + D)x + p.$$

$$-3A - A = 1: \quad A = -\tfrac{1}{4},$$

$$-2B - B = 0: \quad B = 0,$$

$$3A - C - C = 0: C = \tfrac{3}{2}A = -\tfrac{3}{8},$$

$$2B - D = 0: \quad D = 0,$$

$$C + p = 0; \quad p = \tfrac{3}{8}.$$

$$I = -\frac{x}{8}(2x^2+3)\sqrt{1-x^2} + \frac{3}{8}\arcsin x + C.$$

772. $x - 1 = \frac{1}{t}$, $dx = -\frac{1}{t^2}\,dt.$

$$I = \int \frac{\frac{1}{-t^2}\,dt}{\frac{1}{t^2}\sqrt{\left(\frac{1}{t} + 1\right)^2 - 1}} = -\int \frac{t\,dt}{\sqrt{2t + 1}}$$

$$= -\frac{1}{2}\int \frac{2t + 1}{\sqrt{2t + 1}}\,dt + \frac{1}{2}\int \frac{dt}{\sqrt{2t + 1}}$$

$$= -\frac{1}{2}\int \sqrt{2t + 1}\,dt + \frac{1}{2}\int \frac{dt}{\sqrt{2t + 1}}$$

$$= -\frac{1}{4}\cdot\frac{2}{3}(2t + 1)^{3/2} + \frac{1}{2}(2t + 1)^{1/2} + C$$

$$= \sqrt{2t + 1}\left[-\frac{1}{6}(2t + 1) + \frac{1}{2}\right] + C = \frac{\sqrt{2t + 1}}{3}(1 - t) + C.$$

Finally, with the substitutions

$$t = \frac{1}{x - 1}, \qquad 2t + 1 = \frac{x + 1}{x - 1}, \qquad 1 - t = \frac{x - 2}{x - 1},$$

$$I = \frac{1}{3}\sqrt{\frac{x + 1}{x - 1}}\cdot\frac{x - 2}{x - 1} + C.$$

773. First we divide:

$$(x^3 + x - 1) \div (x - 1) = x^2 + x + 2 + \frac{1}{x - 1}.$$

Then we find

$$I_1 = \int \frac{x^2 + x + 2}{\sqrt{x^2 + 2x - 1}}\,dx = (Ax + B)\sqrt{x^2 + 2x - 1} + p\int \frac{dx}{\sqrt{x^2 + 2x - 1}}.$$

Differentiating,

$$\frac{x^2 + x + 2}{\sqrt{x^2 + 2x - 1}} = A\sqrt{x^2 + 2x - 1} + (Ax + B)\frac{x + 1}{\sqrt{x^2 + 2x - 1}} + \frac{p}{\sqrt{x^2 + 2x - 1}}.$$

$$x^2 + x + 2 = A(x^2 + 2x - 1) + (Ax + B)(x + 1) + p.$$

$$\begin{aligned} A + A &= 1, & A &= \tfrac{1}{2}, \\ 2A + A + B &= 1; & B &= -\tfrac{1}{2}, \\ -A + B + p &= 2; & p &= 3. \end{aligned}$$

$$I_1 = (\tfrac{1}{2}x - \tfrac{1}{2})\sqrt{x^2 + 2x - 1} + 3 \ln |x + 1 + \sqrt{x^2 + 2x - 1}| + C.$$

Now we have to integrate

$$I_2 = \int \frac{dx}{(x - 1)\sqrt{x^2 + 2x - 1}}.$$

We substitute $x - 1 = 1/t$, $dx = -1/t^2\, dt$:

$$I_2 = \int \frac{-\frac{1}{t^2}\, dt}{\frac{1}{t}\sqrt{\left(\frac{1}{t} + 1\right)^2 + 2\left(\frac{1}{t} + 1\right) - 1}} = -\int \frac{dt}{\sqrt{2t^2 + 4t + 1}}$$

$$= -\frac{1}{\sqrt{2}} \int \frac{dt}{\sqrt{(t + 1)^2 - \frac{1}{2}}} = -\frac{1}{\sqrt{2}} \ln\left|t + 1 + \sqrt{t^2 + 2t + \frac{1}{2}}\right| + C.$$

$$= -\frac{1}{\sqrt{2}} \ln\left|\frac{\sqrt{2}\,x + \sqrt{x^2 + 2x - 1}}{x - 1}\right| + C_1.$$

The original integral I equals $I_1 + I_2$.

778. We transform

$$I = \int \frac{a - x}{\sqrt{(x - b)(a - x)}}\, dx = \int \frac{-x + a}{\sqrt{-x^2 + (a + b)x - ab}}\, dx.$$

Continuing as in Problem 768, we obtain

$$I = \sqrt{(a - x)(x - b)} + \frac{a - b}{2} \arcsin \frac{2x - a - b}{b - a} + C.$$

779. $$I = \int \frac{x^2 + 2x}{x\sqrt{x^2 + 2x}}\, dx = \int \frac{x + 2}{\sqrt{x^2 + 2x}}\, dx$$

$$= \frac{1}{2}\int \frac{2x + 2}{\sqrt{x^2 + 2x}}\, dx + \int \frac{dx}{\sqrt{(x + 1)^2 - 1}}$$

$$= \sqrt{x^2 + 2x} + \ln |x + 1 + \sqrt{x^2 + 2x}| + C.$$

780. We substitute $x = \sin t$, $dx = \cos t\, dt$.

$$I = \int \frac{\cos t\, dt}{\cos t - 1} = \int \left(1 + \frac{1}{\cos t - 1}\right) dt = t - \int \frac{1}{2 \sin^2 (t/2)}\, dt$$

$$= t + \cot\frac{t}{2} + C = t + \frac{1 + \cos t}{\sin t} + C = \arcsin x + \frac{1 + \sqrt{1 - x^2}}{x} + C.$$

781. $I = \int \frac{dx}{1 + \sqrt{(x+1)^2 + 1}}$. Let us use a hyperbolic substitution,

$$x + 1 = \sinh t, \qquad dx = \cosh t\, dt.$$

$$I = \int \frac{\cosh t}{1 + \cosh t}\, dt = \int \left(1 - \frac{1}{1 + \cosh t}\right) dt = t - \int \frac{dt}{2\cosh^2 (t/2)}$$

$$= t - \tanh \frac{t}{2} + C = t + \frac{1 - \cosh t}{\sinh t} + C$$

$$= \operatorname{argsinh}(x + 1) + \frac{1 - \sqrt{1 + (x+1)^2}}{x + 1} + C$$

$$= \frac{1 - \sqrt{x^2 + 2x + 2}}{x + 1} + \ln (x + 1 + \sqrt{x^2 + 2x + 2}) + C.$$

The same integral can be solved also by the substitution

$$x + 1 = \frac{2z}{1 - z^2}.$$

782. We shall substitute $\sin x = t$, $\cos x\, dx = dt$.

$$I = \int t^6(1 - t^2)^2\, dt = \int (t^6 - 2t^8 + t^{10})\, dt$$

$$= \frac{t^7}{7} - \frac{2t^9}{9} + \frac{t^{11}}{11} + C = \frac{\sin^7 x}{7} - \frac{2\sin^9 x}{9} + \frac{\sin^{11} x}{11} + C.$$

783. We substitute $\cos x = t$, $-\sin x\, dx = dt$.

$$I = -\int \frac{1 - t^2}{t^2}\, dt = \int \left(-\frac{1}{t^2} + 1\right) dt = \frac{1}{t} + t + C = \frac{1}{\cos x} + \cos x + C.$$

784. Put $t = \sin x$, $dt = \cos x\, dx$:

$$I = \int \frac{dt}{(1 - t^2)^2} = \int \frac{dt}{(1 - t)^2(1 + t)^2}.$$

Now we have to decompose the integrand into partial fractions. It seems that here the reduction formula will be more convenient:

$$I = \frac{1}{2}\frac{\sin x}{\cos^2 x} + \frac{1}{2}\int \frac{dx}{\cos x} = \frac{1}{2}\frac{\sin x}{\cos^2 x} + \frac{1}{2}\ln\left|\tan\left(\frac{x}{2} + \frac{\pi}{4}\right)\right| + C.$$

785. Put $\sin x = t$, $\cos x\, dx = dt$:

$$I = \int \frac{dt}{t^3(1 - t^2)} = \int \left(\frac{1}{t^3} + \frac{1}{t} + \frac{1}{2}\cdot\frac{1}{1 - t} - \frac{1}{2}\cdot\frac{1}{1 + t}\right) dt$$

$$= -\frac{1}{2t^2} + \ln \frac{t}{\sqrt{1 - t^2}} + C = -\frac{1}{2\sin^2 x} + \ln |\tan x| + C.$$

787. $$I = \int \left(\frac{1-\cos 2x}{2}\right)^3 dx$$

$$= \tfrac{1}{8}\int (1 - 3\cos 2x + 3\cos^2 2x - \cos^3 2x)\,dx$$

$$= \tfrac{1}{8}x - \tfrac{3}{16}\sin 2x + \tfrac{3}{16}\int (1+\cos 4x)\,dx$$

$$- \tfrac{1}{8}\int (1-\sin^2 2x)\cos 2x\,dx$$

$$= \tfrac{1}{8}x - \tfrac{3}{16}\sin 2x + \tfrac{3}{16}x + \tfrac{3}{64}\sin 4x - \tfrac{1}{16}\sin 2x + \tfrac{1}{48}\sin^3 2x + C$$

$$= \tfrac{5}{16}x - \tfrac{1}{4}\sin 2x + \tfrac{3}{64}\sin 4x + \tfrac{1}{48}\sin^3 2x + C.$$

788. $$I = \int (\sin x\cos x)^4\,dx = \frac{1}{16}\int \sin^4 2x\,dx = \frac{1}{16}\int\left(\frac{1-\cos 4x}{2}\right)^2 dx$$

$$= \frac{1}{64}\int\left(1 - 2\cos 4x + \frac{1+\cos 8x}{2}\right)dx$$

$$= \frac{1}{128}\left(3x - \sin 4x + \frac{1}{8}\sin 8x\right) + C.$$

790. Put $\tan x = t$, $\dfrac{dx}{\cos^2 x} = dt$.

$$I = \int \frac{dt}{\left(\dfrac{t^2}{1+t^2}\right)^2 \dfrac{1}{1+t^2}} = \int \frac{(1+t^2)^3}{t^4}\,dt = -\frac{1}{3t^3} - \frac{3}{t} + 3t + \frac{t^3}{3} + C$$

$$= -\frac{1}{3\tan^3 x} - \frac{3}{\tan x} + 3\tan x$$

$$+ \frac{\tan^3 x}{3} + C.$$

792. $$I = \tfrac{1}{2}\int (\cos 2x - \cos 4x)\sin 2x\,dx$$

$$= \tfrac{1}{8}\sin^2 2x - \tfrac{1}{2}\cdot\tfrac{1}{2}\int (\sin 6x - \sin 2x)\,dx$$

$$= \tfrac{1}{8}\sin^2 2x - \tfrac{1}{8}\cos 2x + \tfrac{1}{24}\cos 6x + C.$$

793. $$I = \int \frac{1-\cos 2x}{2}\cdot\frac{1+\cos 6x}{2}\,dx$$

$$= \tfrac{1}{4}\int [1 - \cos 2x + \cos 6x - \tfrac{1}{2}(\cos 4x + \cos 8x)]\,dx$$

$$= \tfrac{1}{4}x - \tfrac{1}{8}\sin 2x - \tfrac{1}{32}\sin 4x + \tfrac{1}{24}\sin 6x - \tfrac{1}{64}\sin 8x + C.$$

795. Here we shall substitute $\tan x = t$, $dx = \dfrac{dt}{1+t^2}$:

$$I = \int \frac{\frac{dt}{1+t^2}}{1+\frac{3}{1+t^2}} = \int \frac{dt}{4+t^2} = \frac{1}{2}\arctan\frac{t}{2} + C = \frac{1}{2}\arctan\frac{\tan x}{2} + C.$$

796. Put $\tan x = t$, $dx = \frac{dt}{1+t^2}$:

$$I = \int \frac{dt}{(1+t^2)(1+2t)}.$$

$$\frac{1}{(1+t^2)(1+2t)} = \frac{At+B}{1+t^2} + \frac{C}{1+2t},$$

$$(At+B)(1+2t) + C(1+t^2) = 1.$$

$$\begin{aligned} 2A + C &= 0, \\ A + 2B &= 0, \\ B + C &= 1, \end{aligned} \qquad \begin{aligned} C - 4B &= 0, \\ C + B &= 1, \end{aligned} \qquad B = \tfrac{1}{5}, \quad C = \tfrac{4}{5}, \quad A = -\tfrac{2}{5}.$$

$$\begin{aligned} I &= \frac{4}{5}\int \frac{dt}{1+2t} - \frac{1}{5}\int \frac{2t-1}{t^2+1}\,dt \\ &= \frac{2}{5}\ln|1+2t| - \frac{1}{5}\ln(t^2+1) + \frac{1}{5}\arctan t + C \\ &= \frac{1}{5}x + \frac{1}{5}\ln\frac{(1+2\tan x)^2}{1+\tan^2 x} + C = \frac{1}{5}x + \frac{1}{5}\ln(\cos x + 2\sin x)^2 + C \\ &= \frac{1}{5}[x + 2\ln|\cos x + 2\sin x|] + C. \end{aligned}$$

797. $I = -\int \frac{d(\sin x + \cos x)}{\sin x + \cos x} = -\ln|\sin x + \cos x| + C.$

798. $I = \int \frac{dx}{\sqrt{a^2+b^2}\sin(x+\varphi)},$

$$\sin\varphi = \frac{b}{\sqrt{a^2+b^2}}, \qquad \cos\varphi = \frac{a}{\sqrt{a^2+b^2}}.$$

Now
$$I = \frac{1}{\sqrt{a^2+b^2}}\ln\left|\tan\frac{x+\varphi}{2}\right| + C.$$

799. We put $\tan x = t$:

$$I = \int \frac{\sqrt{t}\,dt}{(1+t^2)\frac{t}{\sqrt{1+t^2}}\cdot\frac{1}{\sqrt{1+t^2}}} = \int \frac{dt}{\sqrt{t}} = 2\sqrt{t} + C = 2\sqrt{\tan x} + C.$$

800. $I = \int \frac{\cos^2 x - \sin^2 x}{\sin^4 x}\,dx = \int \frac{\cot^2 x}{\sin^2 x} - \int \frac{dx}{\sin^2 x}$

$= -\frac{1}{3}\cot^3 x + \cot x + C.$

801. $I = \int \frac{\tan^{3/2} x}{\cos^2 x}\,dx = \frac{2}{5}\tan^2 x\sqrt{\tan x} + C.$

805. $I = \int \frac{\dfrac{e^x - e^{-x}}{e^x + e^{-x}}}{\sqrt{1 - \dfrac{e^x - e^{-x}}{e^x + e^{-x}}}}\,dx = \int \frac{e^x - e^{-x}}{\sqrt{e^x + e^{-x}}\sqrt{2e^{-x}}}\,dx.$

Now we put $e^x = t$, $e^x\,dx = dt$:

$$I = \int \frac{t - \dfrac{1}{t}}{\sqrt{t + \dfrac{1}{t}}\sqrt{2 \cdot \dfrac{1}{t}}} \cdot \frac{dt}{t} = \frac{1}{\sqrt{2}} \int \frac{t^2 - 1}{t\sqrt{t^2 + 1}}\,dt$$

$$= \frac{1}{\sqrt{2}} \int \frac{t\,dt}{\sqrt{t^2 + 1}} - \frac{1}{\sqrt{2}} \int \frac{dt}{t\sqrt{t^2 + 1}}$$

$$= \frac{1}{\sqrt{2}}\sqrt{t^2 + 1} + \frac{1}{\sqrt{2}}\ln\left(\frac{1}{t} + \sqrt{\frac{1}{t^2} + 1}\right) + C$$

$$= \frac{1}{\sqrt{2}}\sqrt{e^{2x} + 1} + \frac{1}{\sqrt{2}}\ln\left(e^{-x} + \sqrt{1 + e^{-2x}}\right) + C.$$

We used here the method of Problem 771.

806. $I = \int e^x\sqrt{e^{-x}}\,dx = \int e^{x/2}\,dx = 2e^{x/2} + C.$

808. We substitute $\tanh x = t$, $dt = \dfrac{dx}{\cosh^2 x}$, and

$$\sinh^2 x = \frac{\sinh^2 x}{\cosh^2 x - \sinh^2 x} = \frac{\tanh^2 x}{1 - \tanh^2 x} = \frac{t^2}{1 - t^2}.$$

Now

$$I = \int \frac{dx}{(1 - \sinh^2 x)(1 + \sinh^2 x)} = \int \frac{dx}{(1 - \sinh^2 x)\cosh^2 x}$$

$$= \int \frac{dt}{1 - \dfrac{t^2}{1 - t^2}} = \int \frac{1 - t^2}{1 - 2t^2}\,dt = \int \left[\frac{1}{2} + \frac{1}{2(1 + \sqrt{2}\,t)(1 - \sqrt{2}\,t)}\right]dt$$

$$= \frac{1}{2}t + \frac{1}{4}\int \left(\frac{1}{1 + \sqrt{2}\,t} + \frac{1}{1 - \sqrt{2}\,t}\right)dt = \frac{1}{2}t + \frac{1}{4\sqrt{2}}\ln\frac{1 + \sqrt{2}\,t}{1 - \sqrt{2}\,t} + C$$

$$= \frac{1}{2}\tanh x + \frac{1}{4\sqrt{2}} \ln \frac{1 + \sqrt{2}\tanh x}{1 - \sqrt{2}\tanh x} + C.$$

809. $\cosh x = t$; $\sinh x\,dx = dt$.

$$I = \int \frac{(\cosh^2 x - 1)\sinh x\,dx}{\sqrt{\cosh x}} = \int \frac{t^2 - 1}{\sqrt{t}}\,dt = \int (t^{3/2} - t^{-1/2})\,dt$$

$$= \tfrac{2}{5}t^{5/2} - 2t^{1/2} + C = \tfrac{2}{5}\cosh^2 x\sqrt{\cosh x} - 2\sqrt{\cosh x} + C.$$

810. We shall integrate by parts:

$$u = x,\ du = dx; \qquad dv = \frac{dx}{\cosh^2 x},\ v = \tanh x.$$

$$I = x\tanh x - \int \tanh x\,dx = x\tanh x - \ln(\cosh x) + C.$$

811. We use the hyperbolic formula

$$\cosh a \cosh b = \tfrac{1}{2}[\cosh(a+b) + \cosh(a-b)].$$

$$I = \tfrac{1}{2}\int \cosh 2x(\cosh 2x + \cosh 4x)\,dx$$

$$= \tfrac{1}{2}\int (\cosh^2 2x + \cosh 2x \cosh 4x)\,dx$$

$$= \tfrac{1}{4}\int (\cosh 4x + 1 + \cosh 2x + \cosh 6x)\,dx$$

$$= \tfrac{1}{4}(x + \tfrac{1}{2}\sinh 2x + \tfrac{1}{4}\sinh 4x + \tfrac{1}{6}\sinh 6x) + C.$$

812. $u = \ln(x+1),\ du = \dfrac{dx}{x+1};\ dv = \dfrac{dx}{\sqrt{x+1}},\ v = 2\sqrt{x+1}.$

$$I = 2\sqrt{x+1}\ln(x+1) - 2\int \frac{dx}{\sqrt{x+1}} = 2\sqrt{x+1}\,[\ln(x+1) - 2] + C.$$

813. Let $\tan(x/2) = t$;

$$I = \int \frac{\dfrac{2\,dt}{1+t^2}}{1 + \dfrac{2t}{1+t^2} + \dfrac{1-t^2}{1+t^2}} = \int \frac{dt}{1+t}$$

$$= \ln|1+t| + C = \ln\left|1 + \tan\frac{x}{2}\right| + C.$$

814. $u = \arctan x,\ du = \dfrac{dx}{1+x^2};\ dv = \dfrac{dx}{(1+x)^3},\ v = -\dfrac{1}{2}\cdot\dfrac{1}{(1+x)^2};$

$$I = -\frac{1}{2}\frac{\arctan x}{(1+x)^2} + \frac{1}{2}\int \frac{dx}{(1+x^2)(1+x)^2}.$$

Let $$\frac{Ax+B}{1+x^2} + \frac{C}{1+x} + \frac{D}{(1+x)^2} = \frac{1}{(1+x^2)(1+x)^2},$$

$$(Ax + B)(1 + x)^2 + C(1 + x^2)(1 + x) + D(1 + x^2) = 1;$$

$$A + C = 0, 2A + B + C + D = 0, A + 2B + C = 0, B + C + D = 1;$$

$$A = -\tfrac{1}{2}; \quad C = \tfrac{1}{2}, \quad B = 0, \quad D = \tfrac{1}{2}.$$

$$I = -\frac{\arctan x}{2(x+1)^2} + \frac{1}{2}\int \left[\frac{-x}{2(1+x^2)} + \frac{1}{2(x+1)} + \frac{1}{2(x+1)^2}\right] dx$$

$$= -\frac{\arctan x}{2(x+1)^2} - \frac{1}{8}\ln(1+x^2) + \frac{1}{4}\ln|x+1| - \frac{1}{4(x+1)} + C.$$

815. Use $\tan\frac{x}{2} = t$, $x = 2\arctan t$.

$$I = \int \frac{\left(2\arctan t + \frac{2t}{1+t^2}\right)\frac{2\,dt}{1+t^2}}{1 + \frac{1-t^2}{1+t^2}} = 2\int \left(\arctan t + \frac{t}{1+t^2}\right) dt$$

$$= 2t\arctan t - \ln(1+t^2) + \ln(1+t^2) + C = x\tan\frac{x}{2} + C.$$

We used here the solution of Problem 727.

816. $I = \int \frac{\left(1 - \frac{1}{x^2}\right)dx}{\left(x + \frac{1}{x}\right)\sqrt{\frac{1}{x^2} + x^2}}$. Now we substitute

$$x + \frac{1}{x} = u, \qquad \left(1 - \frac{1}{x^2}\right)dx = du;$$

$$I = \int \frac{du}{u\sqrt{u^2 - 2}}.$$

Now substituting $u^2 - 2 = t^2$, $2u\,du = 2t\,dt$, $\frac{du}{u} = \frac{t\,dt}{t^2 + 2}$,

$$I = \int \frac{t\,dt}{(t^2+2)t} = \frac{1}{\sqrt{2}}\arctan\frac{t}{\sqrt{2}} + C = \frac{1}{\sqrt{2}}\arctan\frac{\sqrt{u^2-2}}{\sqrt{2}} + C$$

$$= \frac{1}{\sqrt{2}}\arctan\sqrt{\frac{1+x^4}{2x^2}} + C.$$

817. $\sin x = t$, $\cos x\,dx = dt$, $x = \arcsin t$.

$$I = \int e^t \frac{\arcsin t\cdot(1-t^2) - \frac{t}{\sqrt{1-t^2}}}{1-t^2}\,dt$$

$$= \int e^t \arcsin t\,dt - \int \frac{e^t\cdot t}{\sqrt{(1-t^2)^3}}\,dt.$$

In the first integral, let

$$u = \arcsin t, \qquad dv = e^t\,dt,$$

$$du = \frac{dt}{\sqrt{1-t^2}}, \qquad v = e^t.$$

In the second, let

$$u = e^t, \qquad dv = \frac{t\,dt}{(1-t^2)^{3/2}},$$

$$du = e^t\,dt, \qquad v = \frac{1}{(1-t^2)^{1/2}}.$$

Now

$$I = e^t \arcsin t - \int \frac{e^t\,dt}{\sqrt{1-t^2}} - \frac{e^t}{\sqrt{1-t^2}} + \int \frac{e^t\,dt}{\sqrt{1-t^2}}$$

$$= e^t\left(\arcsin t - \frac{1}{\sqrt{1-t^2}}\right) + C = e^{\sin x}\left(x - \frac{1}{\cos x}\right) + C.$$

We note that in the following problems different methods may lead to different (but equivalent) answers.

818. $I = \ln(x^2+1) - \arctan x + \dfrac{3}{x+1} + C.$

819. $I = -\dfrac{1}{x^2+4x+5} + \arctan(x+2) + C.$

820. $I = -\ln|x-1| + \dfrac{3}{2}\ln(x^2-2x+5) + \dfrac{1}{2}\arctan\dfrac{x-1}{2} + C.$

821. $I = 2\arctan x + \dfrac{1}{2}\left(-\dfrac{1}{x^2} + \ln\dfrac{x^2+1}{x^2}\right) + C.$

822. $I = \frac{1}{2}(x^2 - \arctan x^2) + C.$

823. $I = \frac{1}{3}(2x - x^2)^{3/2} + C.$

824. $I = \sqrt{2x+4} + \dfrac{3}{2}\ln\left|\dfrac{\sqrt{2x+4}-2}{\sqrt{2x+4}+2}\right| + C.$

825. $I = \sqrt{3x^2-4x+6} + \dfrac{13}{\sqrt{3}}\operatorname{argsinh}\dfrac{3x-2}{\sqrt{14}} + C$

$$= \sqrt{3x^2-4x+6} + \frac{13}{\sqrt{3}}\ln\left[3x - 2 + \sqrt{9x^2-12x+18}\right].$$

826. $I = -\dfrac{1}{3a^2x^3}(a^2-x^2)^{3/2} + C.$

827. $I = \ln\left|\dfrac{x}{a+\sqrt{a^2-x^2}}\right| + \arcsin\dfrac{x}{a} + C.$

828. $I = -\sqrt{4x + 2 - x^2} + 2 \arcsin \dfrac{x-2}{\sqrt{6}} + C.$

829. $I = \dfrac{1}{4}\left(-\dfrac{1}{2\tan^2(x/2)} + 2\ln\left|\tan\dfrac{x}{2}\right| + \dfrac{1}{2}\tan^2\dfrac{x}{2}\right) + C$

$$= \frac{1}{2}\left(\frac{-\cos x}{\sin^2 x} + \ln\left|\tan\frac{x}{2}\right|\right).$$

830. For $b^2 - ac > 0$:

$$I = \frac{1}{a}\cdot\frac{1}{t_1 - t_2}\ln\left|\frac{t-t_1}{t-t_2}\right|, \qquad t_{1,2} = \frac{-b \pm \sqrt{b^2 - ac}}{a}, \qquad t = \tan x.$$

$b^2 - ac = 0$:

$$I = -\frac{1}{a(t - t_1)} + C, \qquad t_1 = -\frac{b}{a}, \qquad t = \tan x.$$

$b^2 - ac < 0$:

$$I = \frac{1}{\sqrt{ac - b^2}}\arctan\frac{at + b}{\sqrt{ac - b^2}} + C.$$

831. $I = \tan x - \frac{3}{2}x + \frac{1}{4}\sin 2x + C.$

832. $I = 2\ln|e^x - 1| - x + C.$

833. $I = \frac{1}{2}x\ln(x^2 + 1) - x + \arctan x + C.$

834. $I = \frac{1}{2}(x^2 - 1)[\ln(x^2 - 1) - 1] + C.$

835. $I = \dfrac{(x^2 - 1)\arctan x + x}{4(1 + x^2)} + C.$

836. $I = -\left(t - \dfrac{2}{3}t^3 + \dfrac{t^5}{5}\right) + C, t = \sqrt{1 - x^2}.$

837. $I = -\frac{1}{2}\cos^2 x + \frac{1}{2}\ln(1 + \cos^2 x) + C.$

838. $I = \frac{1}{4}(x^2 + x\sin 2x + \frac{1}{2}\cos 2x) + C.$

839. $I = (\sin x - \cos x)(x^2 - 2) + 2x(\sin x + \cos x) + C.$

840. $I = \dfrac{3}{8}\arcsin x - \dfrac{x}{8}(3 + 2x^2)\sqrt{1 - x^2} + C.$

841. $I = (x + 1)\arctan\sqrt{x} - \sqrt{x} + C.$

842. $I = 2\ln|u - 1| - \ln(u^2 + u + 1) + \dfrac{2}{\sqrt{3}}\arctan\dfrac{2u + 1}{\sqrt{3}} + C,$

$$u = \sqrt{x + 1}.$$

843. $I = \frac{1}{2}(\ln|\tan x| + \tan x) + C.$

846. $\int_{-2}^{5}\frac{1}{2}(x^2 + 2x + 1)\,dx.$

848. Here we use the same partition as in the solution of Problem 847 and choose the points x_i^* at the left end of each subinterval. We obtain

$$S = \sum_{i=0}^{n-1} e^{i/n} \cdot \frac{1}{n} = \frac{1}{n}\sum_{i=0}^{n-1} e^{i/n}.$$

By the formula for the sum of a geometric progression,

$$S = \frac{1}{n}\frac{(e^{1/n})^n - 1}{e^{1/n} - 1} = \frac{e - 1}{n(e^{1/n} - 1)}.$$

We now have to compute $\lim_{n\to\infty} n(e^{1/n} - 1)$. This limit equals

$$\lim_{x\to\infty} x(e^{1/x} - 1) = \lim_{x\to\infty} \frac{e^{1/x} - 1}{1/x} = \lim_{x\to\infty} \frac{e^{1/x}(-1/x^2)}{-1/x^2} = 1.$$

Consequently,
$$\int_0^1 e^x\,dx = e - 1.$$

The same integral is very easily computed by the Newton-Leibniz formula:

$$\int_0^1 e^x\,dx = e^x\Big|_0^1 = e^1 - e^0 = e - 1.$$

852. $f(x) = \dfrac{x^2+5}{x^2+2}$; $f'(x) = \dfrac{2x(x^2+2) - (x^2+5)\cdot 2x}{(x^2+2)^2} = \dfrac{-6x}{(x^2+2)^2}$.

$f'(x) \leqslant 0$ in $0 \leqslant x \leqslant 2$, i.e., the function decreases in the interval of integration, and consequently

$$m = f(2) = \tfrac{9}{6} = \tfrac{3}{2}, \qquad M = f(0) = \tfrac{5}{2}.$$

We have
$$\frac{3}{2}\cdot 2 = 3 \leqslant \int_0^2 \frac{x^2+5}{x^2+2}\,dx \leqslant 5 = \frac{5}{2}\cdot 2.$$

853. (a) in $[0, 1]$, $x^2 \geqslant x^3$. The equality holds only at the endpoints, and consequently

$$\int_0^1 x^3\,dx < \int_0^1 x^2\,dx.$$

(b) $\displaystyle\int_1^2 x^2\,dx < \int_1^2 x^3\,dx.$

855. Put $f_1(x) = 1$, $f_2(x) = \sqrt{1 + x^3}$. Then

$$\int_0^1 [f_1(x)]^2\,dx = \int_0^1 1\,dx = x\Big|_0^1 = 1;$$

$$\int_0^1 [f^2(x)]^2\,dx = \int_0^1 (1 + x^3)\,dx = x + \frac{x^4}{4}\Big|_0^1 = \frac{5}{4}.$$

And now
$$\int_0^1 \sqrt{1 + x^3}\,dx \leqslant \sqrt{1\cdot\frac{5}{4}} = \frac{\sqrt{5}}{2}.$$

By the previous procedure, $M = \sqrt{2}$, $m = 1$, and

$$1 \leqslant \int_0^1 \sqrt{1 + x^3}\,dx \leqslant \sqrt{2}.$$

$\sqrt{5}/2 < \sqrt{2}$, i.e., the Cauchy-Schwarz estimate is sharper.

857. $$\int_{-a}^{a} \sqrt{a^2 - x^2}\,dx = \frac{1}{2}\left[a^2 \arcsin\frac{x}{a} + x\sqrt{a^2 - x^2}\right]\Bigg|_{-a}^{a}$$

$$= \frac{1}{2}\left[a^2 \cdot \frac{\pi}{2} - a^2\left(-\frac{\pi}{2}\right)\right] = \frac{1}{2}\pi a^2.$$

(Compare Problem 774.) Now

$$\bar{y} = \frac{\frac{1}{2}\pi a^2}{2a} = \frac{\pi a}{4}.$$

The value of the integral can be also obtained using its geometrical interpretation as the area of a semicircle of radius a, which is $\frac{1}{2}\pi a^2$.

859. We use here the fact that

$$A = \int_0^x y\,dx = \int_0^x t^3\,dt = \frac{t^4}{4}\Bigg|_0^x = \frac{x^4}{4}$$

whence

$$\Delta A = \int_0^{x+\Delta x} t^3\,dt - \int_0^x t^3\,dt = \int_x^0 t^3\,dt + \int_0^{x+\Delta x} t^3\,dt$$

$$= \int_x^{x+\Delta x} t^3\,dt = \frac{(x + \Delta x)^4}{4} - \frac{x^4}{4}.$$

ΔA can also be obtained directly from the expression for A:

$$dA = d\left[\int_0^x t^3\,dt\right] = \left[\int_0^x t^3\,dt\right]' dx = x^3\,dx.$$

Now $\alpha = |\Delta A - dA|$. For $x = 4$ and $\Delta x = 1$ we obtain

$$\Delta A = 92.25,\ dA = 64; \qquad \alpha = 28.25,\ \beta = 0.306.$$

For $x = 4$ and $\Delta x = 0.1$:

$$\Delta A = 6.644,\ dA = 6.4; \qquad \alpha = 0.244,\ \beta = 0.0367.$$

861. $$\frac{d}{dx}\int_x^5 \sqrt{1 + t^2}\,dx = -\frac{d}{dx}\int_5^x \sqrt{1 + t^2}\,dt = -\sqrt{1 + x^2}.$$

$$y'(0) = -1, \qquad y'(\tfrac{3}{4}) = -\tfrac{5}{4}.$$

863. (a) $$\frac{d}{dx}\int_{x^3}^2 \ln x\,dx = -\frac{d}{d(x^3)}\int_2^{x^3} \ln dx \cdot \frac{d(x^3)}{dx} = -\ln x^3 \cdot 3x^2$$

$$= -9x^2 \ln x.$$

865. $$\frac{dy}{dt} = -2t \cdot t^4 \ln t^2 = -4t^5 \ln t,$$

$$\frac{dx}{dt} = 2t \cdot t^2 \ln t^2 = 4t^3 \ln t,$$

$$\frac{dy}{dx} = \frac{-4t^5 \ln t}{4t^3 \ln t} = -t^2.$$

866. $I = -\frac{1}{10} \cdot \frac{1}{(11+5x)^2}\Big|_{-2}^{-1} = -\frac{1}{10}\left(\frac{1}{36} - 1\right) = \frac{7}{72}.$

867. $I = \frac{1}{5}(e^x - 1)^5\Big|_0^1 = \frac{1}{5}(e-1)^5.$

868. $I = \int_0^\pi \frac{1+\cos t}{2}\,dt = \left(\frac{t}{2} + \frac{\sin t}{2}\right)\Big|_0^\pi = \frac{\pi}{2}.$

869. $I = \int_0^{a/2}\left(-\frac{dx}{x-a} + \frac{dx}{x-2a}\right) = \ln\left|\frac{x-2a}{x-a}\right|\Bigg|_0^{a/2}$

$= \ln 3 - \ln 2 = \ln 1.5.$

870. $\int_0^\pi \cos^3 x\,dx = \int_0^\pi (1 - \sin^2 x)\cos x\,dx = \sin x - \frac{\sin^3 x}{3}\Big|_0^\pi = 0.$

The average value of $y = \cos^3 x$ in the interval $[0, \pi]$ is 0.

873. $I = x\ln(x+1)\Big|_0^{e-1} - \int_0^{e-1} \frac{x}{x+1}\,dx$

$= e - 1 - \int_0^{e-1} \frac{x+1}{x+1}\,dx + \int_0^{e-1} \frac{dx}{x+1}$

$= e - 1 + [-x + \ln(x+1)]\Big|_0^{e-1}$

$= e - 1 - e + 1 + \ln e = 1.$

874. $I = e^{2x}\sin x\Big|_0^{\pi/2} - 2\int_0^{\pi/2} e^{2x}\sin x\,dx$

$= e^\pi + 2e^{2x}\cos x\Big|_0^{\pi/2} - 4\int_0^{\pi/2} e^{2x}\cos x\,dx = e^\pi - 2 - 4I,$

$$I = \frac{e^\pi - 2}{5}.$$

877. Substitute $1 + x = t^2$, $dx = 2t\,dt$:

$$t_1^2 = 1 + 3 = 4,\ t_1 = 2; \qquad t_2^2 = 1 + 8 = 9,\ t_2 = 3.$$

$$I = \int_2^3 \frac{(t^2-1)\cdot 2t\,dt}{t} = 2\left(\frac{t^3}{3} - t\right)\Big|_2^3 = 2\left(9 - 3 - \frac{8}{3} + 2\right) = \frac{32}{3}.$$

From $t_1^2 = 4$ we can also deduce $t_1 = -2$. But then we cannot choose in addition $t_2 = 3$, because for example, for $t = 0$ $(-2 < 0 < 3)$, $x = -1$ and this value is outside the interval $[3, 8]$. If we choose $t_1 = -2$ we must take $t_2 = -3$, and then we obtain the above result with the opposite sign. This follows from the fact that t is negative in the interval $[-3, -2]$, and $\sqrt{1+x} = -t$ there. In other words, we can choose $t_1 = -2$ and $t_2 = -3$ but we must proceed as follows:

$$I = \int_{-2}^{-3} \frac{(t^2-1)\cdot 2t\,dt}{-t} = -2\left(\frac{t^3}{3} - t\right)\Big|_{-2}^{-3} = -2\left[-9+3+\frac{8}{3}-2\right] = \frac{32}{3}.$$

We could have avoided this discussion by using the substitution $\sqrt{1+x} = t$.

878. $\sqrt{1+\ln x} = t$, $\frac{1}{x}\,dx = 2t\,dt$. $t_1 = \sqrt{1+0} = 1$, $t_2 = \sqrt{1+3} = 2$.

$$I = \int_1^2 \frac{2t\,dt}{t} = 2t\Big|_1^2 = 2.$$

879. $\tan(x/2) = t$, $t_1 = \tan 0 = 0$, $t_2 = \tan(\pi/4) = 1$.

$$I = \int_0^1 \frac{\frac{2\,dt}{1+t^2}}{2\frac{1-t^2}{1+t^2}+3} = \int_0^1 \frac{2\,dt}{2-2t^2+3+3t^2} = 2\int_0^1 \frac{dt}{t^2+5}$$

$$= \frac{2}{\sqrt{5}}\arctan\frac{t}{\sqrt{5}}\Big|_0^1 = \frac{2}{\sqrt{5}}\arctan\frac{1}{\sqrt{5}}.$$

880. We substitute $x = \sinh t$, $dx = \cosh t\,dt$. We denote the new limits by t_1 and t_2. Then

$$I = \int_{t_1}^{t_2} \frac{\cosh t}{\sinh^2 t}\cosh t\,dt = \int_{t_1}^{t_2} \frac{1+\sinh^2 t}{\sinh^2 t}\,dt$$

$$= -\coth t + t\Big|_{t_1}^{t_2} = -\frac{\cosh t}{\sinh t} + t\Big|_{t_1}^{t_2}$$

$$= \left[-\frac{\sqrt{1+x^2}}{x} + \ln(x+\sqrt{1+x^2})\right]\Big|_1^{\sqrt{3}}$$

$$= -\frac{2}{\sqrt{3}} + \sqrt{2} + \ln\frac{\sqrt{3}+2}{1+\sqrt{2}}$$

$$= 2 - \frac{2}{\sqrt{3}} + \ln(\sqrt{6}+2\sqrt{2}-\sqrt{3}-2).$$

882. $t = \frac{x}{2}$, $dt = \frac{dx}{2}$, $t_1 = 0$, $t_2 = \frac{\pi}{2}$;

$$I = \int_0^{\pi/2} \sin^6 t\cdot 2\,dt = 2\cdot\frac{5\cdot 3\cdot 1}{6\cdot 4\cdot 2}\cdot\frac{\pi}{2} = \frac{5}{16}\pi.$$

We used here the formula of Problem 875.

883. $2x = t,\ 2\,dx = dt,\ t_1 = 0,\ t_2 = \pi/2;$

$$I = \int_0^{\pi/2} \cos^7 t \cdot \frac{dt}{2} = \frac{1}{2} \cdot \frac{6\cdot4\cdot2}{7\cdot5\cdot3} = \frac{8}{35}.$$

(Compare Problems 881 and 875.)

885. Substitute $dx = -dt,\ x = a + b - t$:

$$a = a + b - t_1,\ t_1 = b; \qquad b = a + b - t_2,\ t_2 = a.$$

$$I = \int_a^b f(x)\,dx = \int_b^a f(a + b - t)(-dt)$$

$$= \int_a^b f(a + b - t)\,dt = \int_a^b f(a + b - x)\,dx.$$

Two applications of this formula are

$$\int_0^a f(x)\,dx = \int_0^a f(a - x)\,dx;$$

$$\int_0^{\pi/2} f(\cos x)\,dx = \int_0^{\pi/2} f\left[\cos\left(\frac{\pi}{2} - x\right)\right] dx = \int_0^{\pi/2} f(\sin x)\,dx.$$

887. (a) $I = \int_{-a}^{a} f(x)\,dx = \int_{-a}^{0} f(x)\,dx + \int_0^a f(x)\,dx.$

We substitute in the first integral $x = -t$.

$$I = \int_a^0 f(-t)(-dt) + \int_0^a f(x)\,dx = -\int_a^0 f(t)\,dt + \int_0^a f(x)\,dx$$

$$= \int_0^a f(t)\,dt + \int_0^a f(x)\,dx = 2\int_0^a f(x)\,dx.$$

The proof of (b) is analogous.

888. In all three cases the domain of integration is symmetric with respect to 0. In (a) and (b) the integrand is odd and in (c), even. The results follow directly from the formulas of the preceding problem.

892. 0.470.

894. (a) By the solution of Problem 893 we have

$$I = \tfrac{1}{3}\cdot\tfrac{1}{1} = \tfrac{1}{3}.$$

(b) It follows from Problem 893 that this integral does not converge.

896. $$\int_1^M \frac{dx}{x^2(x + 1)} = \int_1^M \left(\frac{1}{x^2} - \frac{1}{x} + \frac{1}{x + 1}\right) dx$$

$$= \left(-\frac{1}{x} + \ln\left|\frac{x + 1}{x}\right|\right)\Bigg|_1^M$$

$$= -\frac{1}{M} + \ln\frac{M + 1}{M} + 1 - \ln 2.$$

$$I = \lim_{M\to\infty}\left(-\frac{1}{M} + \ln\frac{M+1}{M} + 1 - \ln 2\right) = 1 - \ln 2.$$

897. First we compute the indefinite integral $\int x^3 e^{-x^2}\,dx$. We substitute $x^2 = t$, $2x\,dx = dt$:

$$\int x^3 e^{-x^2}\,dx = \tfrac{1}{2}\int te^{-t}\,dt = \tfrac{1}{2}(-te^{-t} - e^{-t}) + C$$

$$= \tfrac{1}{2}(-x^2e^{-x^2} - e^{-x^2}) + C.$$

$$I = \lim_{M\to\infty} \tfrac{1}{2}(-M^2e^{-M^2} - e^{-M^2} + 1) = \tfrac{1}{2}.$$

898. In Problem 721 we found

$$\int e^{-ax}\cos bx\,dx = \frac{b\sin bx - a\cos bx}{a^2+b^2}e^{-ax} + C.$$

Now $$I = \lim_{M\to\infty}\left(\frac{b\sin bM - a\cos bM}{a^2+b^2}e^{-aM} + \frac{a}{a^2+b^2}\right).$$

For $a > 0$ the first term tends to zero, and $I = \dfrac{a}{a^2+b^2}$. For $a \leqslant 0$ the integral does not exist.

899. $\lim\limits_{M\to\infty} (1/b)\sin bM$ does not exist, i.e., the given integral has no meaning.

903 $$\int_0^{a-\alpha}\frac{dx}{\sqrt{a^2-x^2}} = \arcsin\frac{x}{a}\Big|_0^{a-\alpha} = \arcsin\frac{a-\alpha}{a};$$

$$I = \lim_{\alpha\to 0}\arcsin\frac{a-\alpha}{a} = \frac{\pi}{2}.$$

904. $\ln\ln e = \ln 1 = 0$. Consequently, $x = e$ is a point of infinite discontinuity of the function.

$$\int_{e+\alpha}^{10}\frac{dx}{x\ln x\ln\ln x} = \ln\ln\ln x\Big|_{e+\alpha}^{10} = \ln\ln\ln 10 - \ln\ln\ln(e+\alpha).$$

$\lim\limits_{\alpha\to 0}\ln\ln\ln(e+\alpha)$ does not exist and consequently I has no meaning.

905. The point of discontinuity is $x = 1$.

$$\int\frac{dx}{(x-1)(x-3)} = \frac{1}{2}\int\left(-\frac{1}{x-1} + \frac{1}{x-3}\right)dx = \frac{1}{2}\ln\left|\frac{x-3}{x-1}\right| + C;$$

$$I = \lim_{\alpha\to 0}\frac{1}{2}\ln\left|\frac{x-3}{x-1}\right|\Bigg|_0^{1-\alpha} + \lim_{\beta\to 0}\frac{1}{2}\ln\left|\frac{x-3}{x-1}\right|\Bigg|_{1+\beta}^{2}.$$

The limit does not exist, nor does I.

907. $\ln \frac{8}{7}$.

908. 0.

909. Not convergent.

910. $\frac{1}{2}$.

911. $\frac{8}{3}$.

912. π.

913. $\frac{\pi}{4} + \frac{1}{2}\ln 2$.

914. $\pi, r \neq 1\ [0, r = 1]$.

915. $\frac{3e^{\pi} + 2}{13}$.

916. 0.

917. $I_{m,n} = 0,\ m \neq n;\quad I_{m,n} = \pi,\ m = n \neq 0;\quad I_{m,n} = 0,\ m = n = 0.$

$J_{m,n} = 0.$

$K_{m,n} = 0,\ m \neq n;\quad K_{m,n} = \pi,\ m = n \neq 0;\quad K_{m,n} = 2\pi,\ m = n = 0.$

918. $\pi\sqrt{2}/4$.

919. $I \leqslant \sqrt{\frac{6}{5}}$.

920. $\frac{3 + \sqrt{3}}{2}\pi - \frac{3}{2}\ln 2$.

922. $\ln 2$.

923. $\sin x\,(\sin \cos^3 x) - \cos x\,(\sin \sin^3 x)$.

928. $\frac{1}{2}I_0^2$.

929. *Hint.* Divide the domain of integration into two parts $[0,2]$ and $[2,\infty)$ and observe that $e^{-x^2} < e^{-x}$ in the second part.

930. *Hint.* Integrate by parts. $\Gamma(1) = 1$.

933. $A = \int_0^{\pi/4} \frac{\sin^4 x}{\cos^6 x}\,dx = \int_0^{\pi/4} \frac{\tan^4 x}{\cos^2 x}\,dx = \frac{\tan^5 x}{5}\Big|_0^{\pi/4} = \frac{1}{5}.$

934. $y = 0;\ x^2 - x - 2 = 0,\ x_1 = -1,\ x_2 = 2.$

$$A = \int_{-1}^{2} \frac{x^2 - x - 2}{x^3 + 8}\,dx = \frac{1}{3}\int_{-1}^{2} \frac{dx}{x + 2} + \frac{1}{3}\int_{-1}^{2} \frac{2x - 5}{x^2 - 2x + 4}\,dx$$

$$= \frac{1}{3}\ln|x + 2|\Big|_{-1}^{2} + \frac{1}{3}\ln(x^2 - 2x + 4)\Big|_{-1}^{2} - \frac{1}{\sqrt{3}}\arctan\frac{x - 1}{\sqrt{3}}\Big|_{-1}^{2}$$

$$= \frac{1}{3}\ln\frac{16}{7} - \frac{1}{\sqrt{3}}\arctan 3\sqrt{3}.$$

936. $A = \int_0^{2\pi} a(1 - \cos t)\cdot d[a(t - \sin t)]$

$$= a^2 \int_0^{2\pi} \left(1 - 2\cos t + \frac{1 + \cos 2t}{2}\right) dt$$

$$= a^2 \cdot \frac{3}{2}t\Big|_0^{2\pi} = 3\pi a^2.$$

The averages of $\cos t$ and $\cos 2t$ equal zero in the interval $[0,2\pi]$.

937. $$A = 4\int_{\pi/2}^{0} a\sin^3 t\cdot a\cdot 3\cos^2 t(-\sin t)\,dt = 12a^2\int_0^{\pi/2}\sin^4 t\cos^2 t\,dt$$

$$= 3a^2\int_0^{\pi/2}\sin^2 2t\,\frac{1-\cos 2t}{2}\,dt$$

$$= \frac{3a^2}{2}\int_0^{\pi/2}\left(\frac{1-\cos 4t}{2} - \sin^2 2t\cos 2t\right)dt$$

$$= \frac{3}{4}a^2t\Big|_0^{\pi/2} = \frac{3}{8}\pi a^2.$$

The other terms vanish.

938. $y = \pm x\sqrt{1-x^2}$, $-1 \leqslant x \leqslant 1$. By symmetry (with respect to both axes),

$$A = 4\int_0^1 x\sqrt{1-x^2}\,dx = -4\,\frac{2\cdot 1}{3\cdot 2}(1-x^2)^{3/2}\Big|_0^1 = \frac{4}{3}.$$

939. $y = \pm\sqrt{x-x^2} + \arcsin x$. We require $x - x^2 \geqslant 0$, i.e., $0 \leqslant x \leqslant 1$. For $x = 0$, $y = 0$; for $x = 1$, $y = \pi/2$; for $x = \frac{1}{2}$, $y = \pm\frac{1}{2} + \pi/6$ (see Fig. 131).

$$A = \int_0^1 [\sqrt{x-x^2} + \arcsin x - (-\sqrt{x-x^2} + \arcsin x)]\,dx$$

$$= 2\int_0^1 \sqrt{x-x^2}\,dx = 2\int_0^1 \sqrt{\tfrac{1}{4} - (x-\tfrac{1}{2})^2}\,dx.$$

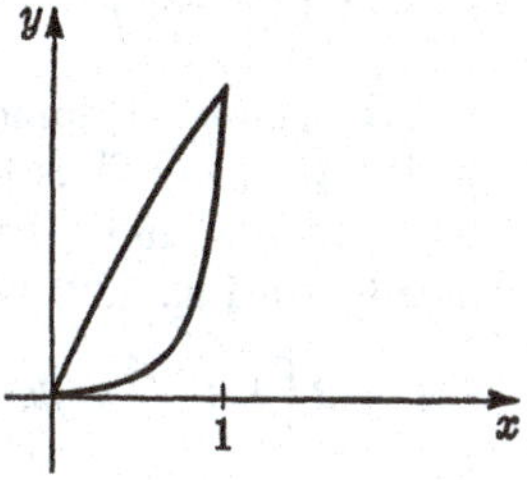

FIGURE 131

We substitute $x - \frac{1}{2} = \frac{1}{2}\sin t$, $dx = \frac{1}{2}\cos t\,dt$. Then

$$x = 0,\ -\tfrac{1}{2} = \tfrac{1}{2}\sin t,\ t = -\pi/2;\quad x = 1,\ t = \pi/2.$$

$$A = 2\int_{-\pi/2}^{\pi/2}\frac{1}{2}\cos t\cdot\frac{1}{2}\cos t\,dt = \frac{1}{2}\int_{-\pi/2}^{\pi/2}\frac{1+\cos 2t}{2}\,dt = \frac{1}{4}\left(\frac{\pi}{2}+\frac{\pi}{2}\right) = \frac{\pi}{4}.$$

940. We have to find two values of t giving the same point of the curve. These are $t = -1$ and $t = 1$. Now for $t = 0$, x attains its minimum value -1. For $0 < t < 1$, y is smaller than for $-1 < t < 0$. Consequently,

$$A = \int_0^{-1}(t^3 - t)\cdot 2t\,dt - \int_0^1 (t^3-t)2t\,dt = 2\int_1^{-1}(t^3-t)t\,dt$$

$$= 2\left[\frac{t^5}{5} - \frac{t^3}{3}\right]_1^{-1} = 2\left(-\frac{1}{5}+\frac{1}{3}-\frac{1}{5}+\frac{1}{3}\right) = \frac{8}{15}.$$

942. By symmetry,

$$A = 2 \cdot \frac{1}{2} \int_0^{\pi} r^2 \, d\theta = 4a^2 \int_0^{\pi} \left(4 + 4\cos\theta + \frac{1 + \cos 2\theta}{2}\right) d\theta$$

$$= 4a^2 \cdot \frac{9}{2}\pi = 18\pi a^2.$$

943. First we shall sketch the given curve. When $0 \leqslant \theta \leqslant 3\pi/2$, r increases from 0 to a. For $\frac{3}{2}\pi \leqslant \theta \leqslant 3\pi$, r decreases back from a to 0. To sketch the curve we shall compute a few points on it.

θ	0	$\frac{\pi}{2}$	π	$\frac{3\pi}{2}$	2π	$\frac{5\pi}{2}$	3π
r	0	$\frac{1}{8}a$	$\frac{3\sqrt{3}}{8}a$	a	$\frac{3\sqrt{3}}{8}a$	$\frac{1}{8}a$	0

For $3\pi < \theta < 6\pi$, r is negative and will be disregarded.

$$\sin\frac{1}{3}\left(\frac{3\pi}{2} - \theta\right) = \sin\frac{1}{3}\left(\frac{3\pi}{2} + \theta\right),$$

i.e., the curve is symmetric with respect to the y axis. Thus the y axis bisects both loops, and the required area (shaded in Fig. 132) is equal to

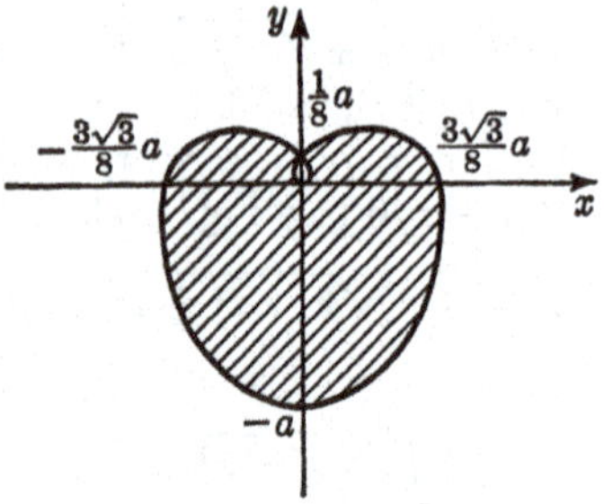

FIGURE 132

$$A = 2\left[\tfrac{1}{2}\int_{\pi/2}^{3\pi/2} r^2 \, d\theta - \tfrac{1}{2}\int_0^{\pi/2} r^2 \, d\theta\right].$$

$$\int r^2 \, d\theta = a^2 \int \sin^6 \frac{\theta}{3} \, d\theta$$

$$= \frac{a^2}{8}\int \left[1 - 3\cos\frac{2\theta}{3} + 3\,\frac{1 + \cos\frac{4\theta}{3}}{2} - \left(1 - \sin^2\frac{2\theta}{3}\right)\cos\frac{2\theta}{3}\right] d\theta$$

$$= \frac{a^2}{8}\left(\frac{5}{2}\theta - 6\sin\frac{2\theta}{3} + \frac{9}{8}\sin\frac{4\theta}{3} + \frac{1}{2}\sin^3\frac{2\theta}{3}\right) + C.$$

By substituting the limits we obtain

$$A = \frac{a^2}{8}\left[\frac{5}{2}\left(\pi - \frac{\pi}{2}\right) - 6\left(-2\sin\frac{\pi}{3}\right) - \frac{9}{8} \cdot 2\sin\frac{2\pi}{3} - \frac{1}{2} \cdot 2\sin^3\frac{\pi}{3}\right]$$

$$= a^2\,\frac{5\pi + 18\sqrt{3}}{32}.$$

944. $-1 \leqslant t \leqslant 1$. $r(t = -1) = r(t = +1) = 0$. θ increases together with t.

$$A = \frac{1}{2}\int_{t=-1}^{t=1} r^2\,d\theta = \frac{1}{2}\int_{-1}^{1}(1-t^2)\left(\frac{1}{\sqrt{1-t^2}} + \frac{t}{\sqrt{1+t^2}}\right)dt$$

$$= \frac{1}{2}\left[\int_{-1}^{1}\sqrt{1-t^2}\,dt + \int_{-1}^{1}\frac{(1-t^2)t}{\sqrt{1+t^2}}\,dt\right].$$

The first integral equals the area of a semicircle of radius one, i.e., $\pi/2$. The second integrand is an odd function of t and as the domain of integration is symmetric with respect to $t = 0$, it equals zero. Consequently,

$$A = \frac{1}{2}\cdot\frac{\pi}{2} = \frac{\pi}{4}.$$

945. The graph of this curve was given in Problem 409. To find the area we pass to polar coordinates, where we have $r^4(\sin^4\theta + \cos^4\theta) = r^2$, i.e.,

$$r^2 = \frac{1}{\sin^4\theta + \cos^4\theta}.$$

By symmetry,

$$A = 8\cdot\frac{1}{2}\int_0^{\pi/4} r^2\,d\theta = 4\int_0^{\pi/4}\frac{d\theta}{\sin^4\theta + \cos^4\theta} = 4\int_0^{\pi/4}\frac{d\theta}{1 - \frac{1}{2}\sin^2 2\theta}.$$

We substitute $\tan 2\theta = z$. Then

$$\sin^2 2\theta = \frac{z^2}{1+z^2}, \qquad d\theta = \frac{1}{2}\frac{dz}{1+z^2}.$$

We obtain the improper integral

$$A = 4\int_0^{\infty}\frac{\frac{1}{2}\frac{dz}{1+z^2}}{1 - \frac{1}{2}\frac{z^2}{1+z^2}} = 4\int_0^{\infty}\frac{dz}{2+z^2}$$

$$= 4\lim_{M\to\infty}\frac{1}{\sqrt{2}}\arctan\frac{z}{\sqrt{2}}\bigg|_0^M = 2\sqrt{2}\lim_{M\to\infty}\arctan\frac{M}{\sqrt{2}} = \pi\sqrt{2}.$$

946. The parametric equations of this curve are

$$x = \frac{3at}{1+t^3}, \qquad y = \frac{3at^2}{1+t^3}.$$

When $0 \leqslant t < \infty$, the corresponding point describes the loop from the origin and back to it. Let t_0 correspond to maximum x. Then

$$A = \int_{\infty}^{t=t_0} y\,dx - \int_{t=0}^{t_0} y\,dx = -\int_0^{\infty} y\,dx$$

$$= -\int_0^{\infty}\frac{3at^2}{1+t^3}\cdot\frac{3a(1-2t^3)}{(1+t^3)^2}\,dt = -3a^2\int_0^{\infty}\frac{3t^2(1-2t^3)}{(1+t^3)^3}\,dt.$$

We substitute $1 + t^3 = u$, $3t^2\,dt = du$.

$$A = -3a^2 \int_1^\infty \frac{3 - 2u}{u^3}\,du = 3a^2 \int_1^\infty \left(\frac{2}{u^2} - \frac{3}{u^3}\right) du$$

$$= 3a^2 \lim_{M\to\infty} \left(-\frac{2}{u} + \frac{3}{2u^2}\right)\Bigg|_1^M = 3a^2\left(2 - \frac{3}{2}\right) = \frac{3}{2}\,a^2.$$

948. $y^2 = 4\,\dfrac{2 - x}{x}$. We require $\dfrac{2 - x}{x} \geqslant 0$, i.e., $0 < x \leqslant 2$; hence there is no horizontal asymptote. Now

$$x = \frac{8}{4 + y^2}, \qquad \lim_{y\to\infty} x = 0,$$

i.e., the y axis is an asymptote. By symmetry,

$$A = 2\int_0^\infty \frac{8}{4 + y^2}\,dy = 4\pi.$$

949. The asymptote is the x axis. The required area is given by

$$A = 2\int_0^\infty x^2 e^{-x^2}\,dx.$$

$u = x$, $du = dx$, $dv = xe^{-x^2}\,dx$, $v = -\frac{1}{2}e^{-x^2}$.

$$A = \lim_{M\to\infty} (-xe^{-x^2})\Big|_0^M + \int_0^\infty e^{-x^2}\,dx = \frac{\sqrt{\pi}}{2}.$$

The first expression is zero and the second is known to equal $\sqrt{\pi}/2$. See note following Problem 899.

951. $\dfrac{3\pi + 2}{9\pi - 2}$.

952. a^2.

953. $\dfrac{\pi}{2} - \dfrac{1}{3}$.

954. $2 - \dfrac{2}{e}$.

955. $\frac{23}{4}\pi a^2$.

956. $\frac{4}{3}a^2$.

958. $y' = \dfrac{1}{2}\left[\sqrt{x^2 - 1} + \dfrac{x^2}{\sqrt{x^2 - 1}} - \dfrac{1}{\sqrt{x^2 - 1}}\right] = \sqrt{x^2 - 1}$.

$$L = \int_1^{a+1} \sqrt{1 + x^2 - 1}\,dx = \frac{x^2}{2}\Bigg|_1^{a+1} = \frac{1}{2}\,(a^2 + 2a).$$

959. $$L = \int_0^{1/2} \sqrt{1 + \left(\frac{-2x}{1 - x^2}\right)^2}\,dx = \int_0^{1/2} \sqrt{\frac{1 - 2x^2 + x^4 + 4x^2}{(1 - x^2)^2}}\,dx$$

$$= \int_0^{1/2} \frac{1 + x^2}{1 - x^2}\,dx = \int_0^{1/2} \left(-1 + \frac{1}{1 + x} + \frac{1}{1 - x}\right) dx$$

$$= -x + \ln\left|\frac{1 + x}{1 - x}\right|\Bigg|_0^{1/2} = \ln 3 - \frac{1}{2}.$$

960. $y' = x^4 - \frac{1}{4x^4}$.

$$L = \int_{1/2}^{1} \sqrt{1 + x^8 - \frac{1}{2} + \frac{1}{16x^8}}\, dx = \int_{1/2}^{1} \sqrt{x^8 + \frac{1}{2} + \frac{1}{16x^8}}\, dx$$

$$= \int_{1/2}^{1} \left(x^4 + \frac{1}{4x^4}\right) dx = \frac{x^5}{5} - \frac{1}{12x^3}\Big|_{1/2}^{1} = \frac{1}{5} - \frac{1}{12} - \frac{1}{160} + \frac{8}{12} = \frac{373}{480}.$$

961. $L = \int_0^x \sqrt{1 + \frac{x^2}{p^2}}\, dx = \frac{1}{p}\left[\frac{1}{2} x\sqrt{x^2 + p^2} + \frac{p^2}{2} \ln\left(x + \sqrt{x^2 + p^2}\right)\right]\Big|_0^x$

$$= \frac{x}{2p}\sqrt{x^2 + p^2} + \frac{p}{2} \ln \frac{x + \sqrt{x^2 + p^2}}{p}.$$

963. $x_t' = (f''' + f') \cos t, \quad y' = -(f''' + f') \sin t,$

$$(x_t')^2 + (y_t')^2 = (f''' + f')^2.$$

$$L = \int_{t_1}^{t_2} (f''' + f')\, dt = f''(t) + f(t)\Big|_{t_1}^{t_2}$$

$$= f''(t_2) + f(t_2) - f''(t_1) - f(t_1).$$

964. The curve is an astroid with unequal axes. We use a parametric representation,

$$x = a \cos^3 t, \qquad y = b \sin^3 t,$$

$$dx = -3a \cos^2 t \sin t\, dt, \qquad dy = 3b \sin^2 t \cos t\, dt.$$

By symmetry,

$$L = 4 \int_0^{\pi/2} \sqrt{9a^2 \cos^4 t \sin^2 t + 9b^2 \sin^4 t \cos^2 t}\, dt$$

$$= 12 \int_0^{\pi/2} \sin t \cos t \sqrt{a^2 \cos^2 t + b^2 \sin^2 t}\, dt.$$

Put $\sin t = z$, $\cos t\, dt = dz$, $z_1 = 0$, $z_2 = 1$:

$$L = 12 \int_0^1 z\sqrt{a^2(1 - z^2) + b^2 z^2}\, dz = 12 \int_0^1 z\sqrt{a^2 + (b^2 - a^2)z^2}\, dz$$

$$= 12 \cdot \frac{2}{3} \cdot \frac{1}{2} [a^2 + (b^2 - a^2)z^2]^{3/2} \cdot \frac{1}{b^2 - a^2}\Big|_0^1$$

$$= \frac{4}{b^2 - a^2} [(a^2 + b^2 - a^2)^{3/2} - (a^2)^{3/2}]$$

$$= \frac{4(b^3 - a^3)}{(b + a)(b - a)} = 4 \frac{a^2 + ab + b^2}{a + b}.$$

For an astroid with $a = b$ we have

$$L = 4 \cdot \frac{3a^2}{2a} = 6a.$$

965. $$dx = a\left(-\sin t + \frac{1}{\tan (t/2)} \cdot \frac{1}{\cos^2 (t/2)} \cdot \frac{1}{2}\right) dt$$

$$= a\left(-\sin t + \frac{1}{\sin t}\right) dt$$

$$= a\,\frac{1 - \sin^2 t}{\sin t}\, dt = a \cos t \cot t\, dt;$$

$$dy = a \cos t\, dt; \qquad dL = a\,|\cos t|\,\sqrt{1 + \cot^2 t}\, dt = a\,|\cot t|\, dt.$$

The value of t corresponding to $(0,a)$ is $\pi/2$. The value of t corresponding to (x,y) for $x > 0$ satisfies $\pi/2 < t < \pi$. In this interval $|\cot t| = -\cot t$ and $\sin t > 0$, i.e.,

$$L = -a \int_{\pi/2}^{t} \cot t\, dt = -a \ln \sin t\Big|_{\pi/2}^{t}$$

$$= -a \ln \sin t = -a \ln \frac{y}{a} = a \ln \frac{a}{y}.$$

967. $$L = \int_0^{2\pi} \sqrt{a^2\theta^2 + a^2}\, d\theta = a \int_0^{2\pi} \sqrt{1 + \theta^2}\, d\theta$$

$$= \frac{a}{2}\,[\theta\sqrt{1 + \theta^2} + \ln (\theta + \sqrt{1 + \theta^2})]\Big|_0^{2\pi}$$

$$= \frac{a}{2}\,[2\pi\sqrt{1 + 4\pi^2} + \ln (2\pi + \sqrt{1 + 4\pi^2})].$$

968. $y' = \sqrt{\cos x}$;

$$L = \int_{-\pi/2}^{\pi/2} \sqrt{1 + \cos x}\, dx = 2 \int_0^{\pi/2} \sqrt{2} \cos \frac{x}{2}\, dx = 4\sqrt{2} \sin \frac{x}{2}\Big|_0^{\pi/2}$$

$$= 4\sqrt{2} \cdot \frac{\sqrt{2}}{2} = 4.$$

970. Let us pass to the parametric equations

$$x = a \cos t, \qquad y = b \sin t;$$

$$dL = \sqrt{a^2 \sin^2 t + b^2 \cos^2 t}\, dt = \sqrt{a^2 - (a^2 - b^2) \cos^2 t}\, dt$$

$$= a\sqrt{1 - \epsilon^2 \cos^2 t}\, dt.$$

Here $\epsilon = \sqrt{a^2 - b^2}/a$ is the eccentricity of the ellipse.

$$L = 4a \int_0^{\pi/2} \sqrt{1 - \epsilon^2 \cos^2 t}\, dt.$$

This indefinite integral cannot be computed in terms of elementary functions. For $a = \sqrt{2}$ and $b = 1$, we obtain $\epsilon^2 = (2 - 1)/2 = \frac{1}{2}$; then

$$L = 4\sqrt{2}\int_0^{\pi/2}\sqrt{1 - \tfrac{1}{2}\cos^2 t}\,dt = 4\int_0^{\pi/2}\sqrt{2 - \cos^2 t}\,dt$$

$$= 4\int_0^{\pi/2}\sqrt{1 + \sin^2 t}\,dt$$

and, comparing with Problem 969, we see that this is the length of one wave of the cosine curve, therefore also of the sine curve.

971. $$dL = \sqrt{(dx)^2 + (dy)^2} = \sqrt{(x')^2 + 1}\,dy = \sqrt{\left(\frac{y}{2} - \frac{1}{2y}\right)^2 + 1}\,dy$$

$$= \sqrt{\frac{y^2}{4} + \frac{1}{4y^2} + \frac{1}{2}}\,dy = \left(\frac{y}{2} + \frac{1}{2y}\right)dy.$$

$$L = \int_1^e \left(\frac{y}{2} + \frac{1}{2y}\right)dy = \left(\frac{y^2}{4} + \frac{1}{2}\ln y\right)\Bigg|_1^e = \frac{e^2}{4} + \frac{1}{2} - \frac{1}{4} = \frac{e^2 + 1}{4}.$$

972. $$L = \int_{\theta_1}^{\theta_2}\sqrt{r^2 + (r')^2}\,d\theta = \int_{r=1}^{r=3}\sqrt{r^2 + \left(\frac{1}{d\theta/dr}\right)^2}\left|\frac{d\theta}{dr}\right|dr$$

$$= \int_1^3 \sqrt{r^2\left(\frac{d\theta}{dr}\right)^2 + 1}\,dr = \int_1^3 \sqrt{\frac{r^2}{4}\left(1 - \frac{1}{r^2}\right)^2 + 1}\,dr$$

$$= \frac{1}{2}\int_1^3 \sqrt{r^2 - 2 + \frac{1}{r^2} + 4}\,dr = \frac{1}{2}\int_1^3 \left(r + \frac{1}{r}\right)dr$$

$$= \frac{1}{2}\left(\frac{r^2}{2} + \ln r\right)\Bigg|_1^3 = \frac{1}{2}\left(\frac{9}{2} + \ln 3 - \frac{1}{2}\right) = 2 + \frac{1}{2}\ln 3.$$

973. $\frac{14}{3}$.

974. $2\pi^2 a$.

975. $a(2\pi - \tanh \pi)$.

979. $$v = \pi\int_0^{\pi/2}(x^2 + \sin x)^2\,dx = \pi\int_0^{\pi/2}(x^4 + 2x^2\sin x + \sin^2 x)\,dx.$$

$$\int_0^{\pi/2} x^2 \sin dx = -x^2\cos x\Big|_0^{\pi/2} + 2\int_0^{\pi/2} x\cos x\,dx$$

$$= 2x\sin x\Big|_0^{\pi/2} - 2\int_0^{\pi/2}\sin x\,dx$$

$$= 2\cdot\frac{\pi}{2} + 2\cos x\Big|_0^{\pi/2} = \pi - 2.$$

$$v = \pi\left[\frac{x^5}{5}\Bigg|_0^{\pi/2} + 2(\pi - 2) + \left(\tfrac{1}{2}x - \tfrac{1}{4}\sin 2x\right)\Bigg|_0^{\pi/2}\right]$$

$$= \pi\left(\frac{\pi^5}{160} + 2\pi - 4 + \frac{\pi}{4}\right) = \pi\left(\frac{\pi^5}{160} + \frac{9\pi}{4} - 4\right).$$

981. $$v = \pi\int_0^{2\pi} a^2(1 - \cos t)^2 a(1 - \cos t)\,dt$$

$$= \pi a^3 \int_0^{2\pi} \left[1 - 3\cos t + \frac{3(1 + \cos 2t)}{2} - \cos^3 t\right] dt$$

$$= \pi a^3 \cdot \frac{5}{2} t \Big|_0^{2\pi} = 5\pi^2 a^3.$$

The average in $[0,2\pi]$ of all other functions in the integrand is zero.

983. See Figure 133.

$$y = b \pm \sqrt{a^2 - x^2};$$

$$v = 2\pi \int_0^a [(b + \sqrt{a^2 - x^2})^2 - (b - \sqrt{a^2 - x^2})^2]\, dx$$

$$= 2\pi \int_0^a 4b\sqrt{a^2 - x^2}\, dx = 2\pi \cdot 4b \cdot \frac{\pi a^2}{4} = 2\pi^2 a^2 b.$$

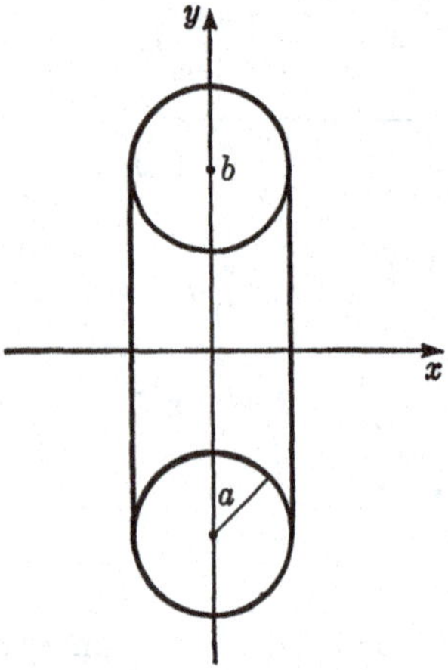

FIGURE 133

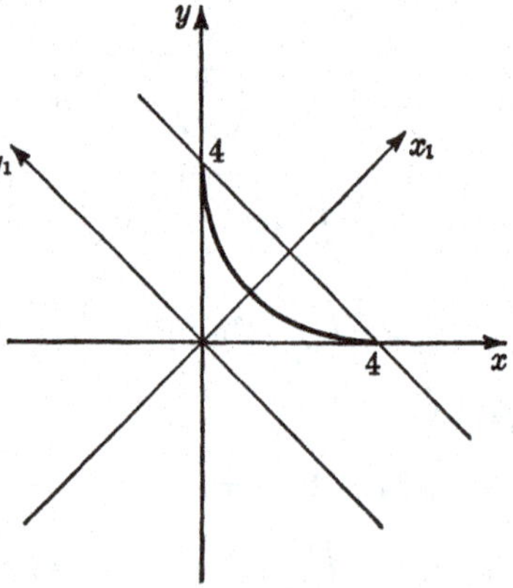

FIGURE 134

984. See Figure 134. We shall rotate the axes through the angle $\pi/4$. Then

$$x = (x_1 - y_1)\frac{\sqrt{2}}{2}, \qquad y = (x_1 + y_1)\frac{\sqrt{2}}{2}.$$

The equation of $x + y = 4$ is now $x_1 = 2\sqrt{2}$. The equation of the parabola is $x + y + 2\sqrt{xy} = 4$, i.e.,

$$\sqrt{2}x_1 + \sqrt{2}\sqrt{x_1^2 - y_1^2} = 4, \qquad x_1 + \sqrt{x_1^2 - y_1^2} = 2\sqrt{2},$$

$$x_1^2 - y_1^2 = 8 - 4\sqrt{2}x_1 + x_1^2, \qquad y_1^2 = 4\sqrt{2}x_1 - 8.$$

$$y_1 = 0, \qquad x_1 = \sqrt{2}.$$

Now $$v = \pi \int_{\sqrt{2}}^{\sqrt{8}} (4\sqrt{2}\, x_1 - 8)\, dx_1 = 4\sqrt{2}\, \pi.$$

985. By the formula used in Problem 980 we have

$$v = \frac{2}{3}\pi a^3 \int_0^{\pi} (1 + \cos\theta)^3 \sin\theta\, d\theta = -\frac{2}{3}\pi a^3 \frac{(1 + \cos\theta)^4}{4}\Big|_0^{\pi} = \frac{8}{3}\pi a^3.$$

987. $$v = 2\pi \int_0^{\infty} e^{-2x^2}\, dx$$

$$= 2\pi \cdot \frac{1}{\sqrt{2}} \int_0^{\infty} e^{-u^2}\, du = \frac{2\pi}{\sqrt{2}} \cdot \frac{\sqrt{\pi}}{2} = \pi\sqrt{\frac{\pi}{2}}.$$

We used the known result

$$\int_0^{\infty} e^{-u^2}\, du = \frac{\sqrt{\pi}}{2}.$$

989. First we shall find the area of a symmetric parabolic segment with base b and altitude H. By Simpson's formula we obtain

$$A = \frac{b}{6}(0 + 4H + 0) = \frac{2bH}{3}$$

(this result is exact). Now we shall find the length of the chord b of a circle as a function of its distance from the center.

$$b = 2\sqrt{R^2 - x^2};$$

the required volume is now

$$v = 2\int_0^R \frac{2}{3} H \cdot 2\sqrt{R^2 - x^2}\, dx = \frac{8}{3} H \cdot \frac{\pi R^2}{4} = \frac{2\pi R^2 H}{3}.$$

990. $\frac{\pi}{2}\left(\arctan 2 - \frac{\pi}{4} + \frac{1}{10}\right)$. **991.** $\pi ab(h/2)$.

992. $\frac{2}{3}abH$. **993.** $\frac{1}{4}$. **994.** $4\pi abc/3$

995. $\pi abH/3$. **996.** $\frac{4}{21}\pi a^3$. **997.** $\frac{32}{105}\pi a^3$.

999. $$P = 2\pi \int_{x=0}^{3a} y\sqrt{1 + (y')^2}\, dx = 2\pi \int_0^{3a} \sqrt{y^2 + (yy')^2}\, dx.$$

We have $y^2 = 4ax$, thus $2yy' = 4a$ or $yy' = 2a$. Now

$$P = 2\pi \int_0^{3a} \sqrt{4ax + 4a^2}\, dx = 2\pi \cdot \frac{4}{3}(ax + a^2)^{3/2} \cdot \frac{1}{a}\Big|_0^{3a}$$

$$= \frac{8\pi}{3a}(8a^3 - a^3) = \frac{56}{3}\pi a^2.$$

1000. $y^2 = R^2 - x^2$, $yy' = -x$. Now

$$P = 2\pi \int_a^b \sqrt{y^2 + (yy')^2}\, dx = 2\pi \int_a^b \sqrt{R^2 - x^2 + x^2}\, dx = 2\pi R(b - a).$$

We see that the circular area of a spherical zone depends only on its altitude $h = b - a$, and not on its position in the sphere. For $h = 2R$, we obtain the area of the whole spherical surface:

$$2\pi R \cdot 2R = 4\pi R^2.$$

1001. The curve intersects the x axis at $x = 0$ and $x = 3a$. The loop is symmetric with respect to this axis. We have

$$18ayy' = 9a^2 - 12ax + 3x^2.$$

Now $P = 2\pi \int_0^{3a} \sqrt{\frac{9a^2x - 6ax^2 + x^3}{9a} + \frac{(3a^2 - 4ax + x^2)^2}{36a^2}}\, dx$

$$= \frac{2\pi}{6a} \int_0^{3a} \sqrt{9a^4 + x^4 + 12a^3x - 2a^2x^2 - 4ax^3}\, dx$$

$$= \frac{\pi}{3a} \int_0^{3a} (3a^2 + 2ax - x^2)\, dx = \frac{\pi}{3a}(9a^3 + 9a^3 - 9a^3) = 3\pi a^2.$$

1002. $P = 2\pi \cdot 2 \int_0^{\pi/2} a \sin^3 t \sqrt{9a^2 \cos^4 t \sin^2 t + 9a^2 \sin^4 t \cos^2 t}\, dt$

$$= 4\pi \int_0^{\pi/2} 3a^2 \sin^3 t \sin t \cos t\, dt = 12\pi a^2 \left. \frac{\sin^5 t}{5} \right|_0^{\pi/2} = \frac{12}{5}\pi a^2.$$

1003. $P = 2\pi \int_0^{\pi} r \sin\theta \sqrt{r^2 + (r')^2}\, d\theta$

$$= 2\pi \int_0^{\pi} a(1 + \cos\theta) \sin\theta \cdot a \cdot 2 \cos\frac{\theta}{2}\, d\theta$$

$$= 4\pi a^2 \int_0^{\pi} 2\cos^2\frac{\theta}{2} \cdot 2 \sin\frac{\theta}{2} \cos\frac{\theta}{2} \cos\frac{\theta}{2}\, d\theta$$

$$= 16\pi a^2 \int_0^{\pi} \cos^4\frac{\theta}{2} \sin\frac{\theta}{2}\, d\theta$$

$$= -16\pi a^2 \cdot \frac{2}{5} \left. \cos^5\frac{\theta}{2} \right|_0^{\pi} = \frac{32}{5}\pi a^2.$$

1004. $y' = -e^{-x}$, $P = 2\pi \int_0^{\infty} e^{-x} \sqrt{1 + e^{-2x}}\, dx.$

We substitute $e^{-x} = t$, $-e^{-x}\, dx = dt$:

$$\int e^{-x} \sqrt{1 + e^{-2x}}\, dx = -\int \sqrt{1 + t^2}\, dt$$

$$= -\tfrac{1}{2}[t\sqrt{t^2 + 1} + \ln(t + \sqrt{t^2 + 1})] + C$$

$$= -\tfrac{1}{2}[e^{-x}\sqrt{1 + e^{-2x}} + \ln(e^{-x} + \sqrt{1 + e^{-2x}})] + C.$$

$$P = -\pi \lim_{M\to\infty} [e^{-M}\sqrt{1 + e^{-2M}} + \ln(e^{-M} + \sqrt{1 + e^{-2M}})$$

$$- \sqrt{2} - \ln(1 + \sqrt{2})] = \pi[\sqrt{2} + \ln(1 + \sqrt{2})].$$

1005. $\dfrac{2\pi\sqrt{2}}{5}(e^{\pi} - 2)$. **1006.** $\dfrac{\pi}{27}(10\sqrt{10} - 1)$.

1007. $\dfrac{64}{3}\pi a^2$. **1008.** $\dfrac{\pi}{6}(5\sqrt{5} - 1)$.

1009. $2\pi a^2(2 - \sqrt{2})$.

1012. By symmetry, $\bar{x} = \pi/2$.

$$M_x = \tfrac{1}{2}\int_0^{\pi} y^2\,dx = \tfrac{1}{2}\int_0^{\pi} \sin^2 x\,dx = \tfrac{1}{4}\pi,$$

$$A = \int_0^{\pi} \sin x\,dx = 2;$$

$$\bar{y} = \frac{\frac{1}{4}\pi}{2} = \frac{\pi}{8}.$$

1013. By symmetry, $\bar{x} = \pi a$.

$$M_x = \frac{1}{2}\int_0^{2\pi} a^2(1 - \cos t)^2 \cdot a(1 - \cos t)\,dt$$

$$= \frac{1}{2}a^3\int_0^{2\pi} (1 - \cos t)^3\,dt = \frac{1}{2}a^3 \cdot 5\pi = \frac{5\pi a^3}{2}$$

(cf. Prob. 981). By Problem 936,

$$A = 3\pi a^2, \qquad \text{i.e., } \bar{y} = \frac{5\pi a^3/2}{3\pi a^2} = \frac{5}{6}a.$$

1014. By symmetry, $\bar{x} = \pi a$. $dL = 2a \sin\dfrac{t}{2}\,dt$ (see Prob. 962).

$$M_x = \int_{x=0}^{x=2\pi a} y\,dL = \int_{t=0}^{2\pi} 2a\sin^2\frac{t}{2}\,2a\sin\frac{t}{2}\,dt$$

$$= 4a^2\int_0^{2\pi} \sin^3\frac{t}{2}\,dt = 4a^2\int_0^{2\pi}\left(1 - \cos^2\frac{t}{2}\right)\sin\frac{t}{2}\,dt$$

$$= 4a^2\left(-2\cos\frac{t}{2} + \frac{2}{3}\cos^3\frac{t}{2}\right)\Bigg|_0^{2\pi} = 4a^2\left(2 - \frac{2}{3} + 2 - \frac{2}{3}\right) = \frac{32}{3}a^2.$$

By Problem 962, $L = 8a$, i.e.,

$$\bar{y} = \tfrac{32}{3}a^2 \div 8a = \tfrac{4}{3}a.$$

1016. By symmetry $\bar{y} = 0$. We use the formula of Problem 1015.

$$\int_0^{2\pi} r^3\cos\theta\,d\theta$$

$$= a^3\int_0^{2\pi}(1 + \cos\theta)^3\cos\theta\,d\theta$$

$$= a^3\int_0^{2\pi}\left[\cos\theta + 3\,\frac{1 + \cos 2\theta}{2} + 3(1 - \sin^2\theta)\cos\theta + \frac{(1 + \cos 2\theta)^2}{4}\right]d\theta$$

$$= a^3\left(\frac{3}{2}\theta + \frac{1}{4}\theta + \frac{1}{8}\theta\right)\Big|_0^{2\pi} = \frac{15}{4}\pi a^3;$$

$$\int_0^{2\pi} r^2\,d\theta = a^2\int_0^{2\pi}(1+\cos\theta)^2\,d\theta = a^2\int_0^{2\pi}\left(1 + 2\cos\theta + \frac{1+\cos 2\theta}{2}\right)d\theta$$

$$= \frac{3}{2}a^2\cdot 2\pi = 3\pi a^2.$$

Therefore

$$\bar{x} = \frac{2}{3}\cdot\frac{\frac{15}{4}\pi a^3}{3\pi a^2} = \frac{5}{6}a.$$

1018. $L = \int_{\pi/2}^{\pi}\sqrt{r^2 + (r')^2}\,d\theta = \int_{\pi/2}^{\pi}\sqrt{a^2e^{2\theta} + a^2e^{2\theta}}\,d\theta$

$$= a\sqrt{2}\int_{\pi/2}^{\pi} e^{\theta}\,d\theta = a\sqrt{2}(e^{\pi} - e^{\pi/2});$$

$$M_y = \int_{\pi/2}^{\pi} r\cos\theta\sqrt{r^2 + (r')^2}\,d\theta = \int_{\pi/2}^{\pi} ae^{\theta}\cos\theta\, a\sqrt{2}\,e^{\theta}\,d\theta$$

$$= a^2\sqrt{2}\int_{\pi/2}^{\pi} e^{2\theta}\cos\theta\,d\theta$$

$$= a^2\sqrt{2}\,\frac{e^{2\theta}}{5}(2\cos\theta + \sin\theta)\Big|_{\pi/2}^{\pi} = \frac{a^2\sqrt{2}}{5}(-2e^{2\pi} - e^{\pi}).$$

$$\bar{x} = \frac{M_y}{L} = -\frac{1}{5}a\,\frac{2e^{2\pi} + e^{\pi}}{e^{\pi} - e^{\pi/2}}.$$

By the same procedure we find

$$\bar{y} = \frac{a}{5}\,\frac{e^{2\pi} - 2e^{\pi}}{e^{\pi} - e^{\pi/2}}.$$

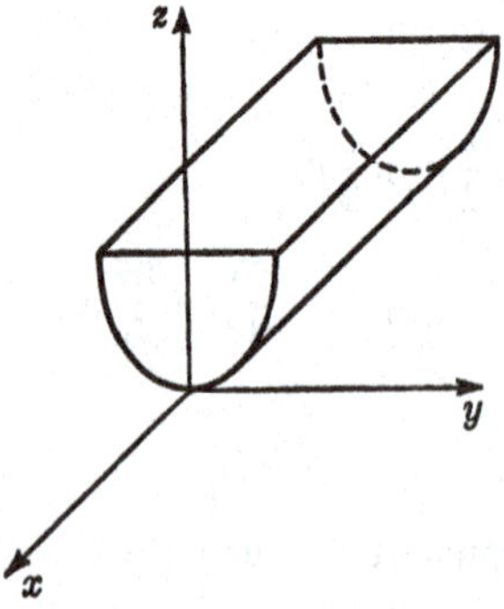

FIGURE 135

1020. We choose a coordinate system as in Figure 135. By symmetry, $\bar{y} = 0$, $\bar{x} = -b/2$. The equation of the parabola in the yz plane will be $z = ky^2$. To find k we substitute

$$H = k\left(\frac{a}{2}\right)^2, \qquad \text{i.e., } k = \frac{4H}{a^2}.$$

$\bar{z}$ of the solid will clearly be equal to $\bar{z}$ of the area cut from it by the yz plane, i.e.,

$$\bar{z} = \frac{2\int_0^{a/2}\frac{1}{2}(H^2 - z^2)\,dy}{2\int_0^{a/2}(H - z)\,dy} = \frac{1}{2}\,\frac{\int_0^{a/2}\left(H^2 - \frac{16H^2}{a^4}y^4\right)dy}{\int_0^{a/2}\left(H - \frac{4H}{a^2}y^2\right)dy}$$

$$= \frac{1}{2} \frac{H^2 \frac{a}{2} - \frac{16}{5} \frac{H^2}{a^4} \cdot \frac{a^5}{32}}{H \frac{a}{2} - \frac{4H}{3a^2} \cdot \frac{a^3}{8}} = \frac{3}{5} H.$$

1021. $(\frac{1}{6}, \frac{11}{30})$.

1022. $\left(0, 0, \frac{15\pi - 32}{8(3\pi - 4)}\right)$.

1023. $\frac{1}{3}H$.

1024. $(\frac{7}{3}, 0)$.

1025. $\left(0, 0, \frac{3}{8} \cdot \frac{(r + d)^4 - r^4}{(r + d)^3 - r^3}\right)$.

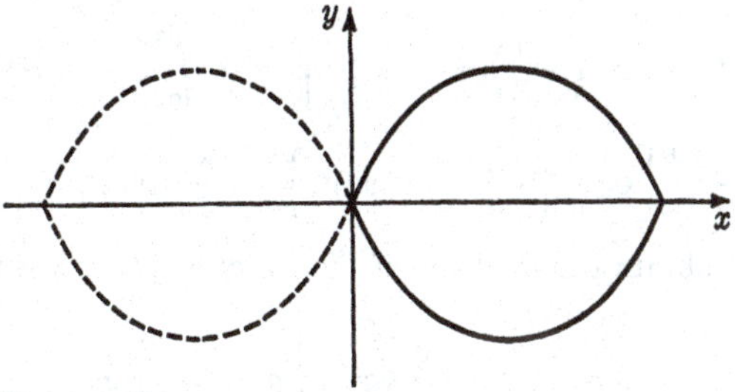

FIGURE 136

1028. See Figure 136. By the Pappus theorems we obtain, using the results of Problems 962 and 936,

$$P = 2\pi \cdot \pi a \cdot 2 \cdot 8a = 32\pi^2 a^2,$$
$$v = 2\pi \cdot \pi a \cdot 2 \cdot 3\pi a^2 = 12\pi^3 a^3.$$

1030. See Figure 137. The parametric equations of the ellipse are

$$x = 2 \cos t, \qquad y = \sin t.$$

$$\begin{aligned} I_x &= 4 \int_0^1 y^2 x \, dy \\ &= 4 \int_0^{\pi/2} \sin^2 t \cdot 2 \cos t \cdot \cos t \, dt \\ &= 8 \int_0^{\pi/2} \frac{\sin^2 2t}{4} \, dt = \int_0^{\pi/2} (1 - \cos 4t) \, dt = \frac{\pi}{2}. \end{aligned}$$

1032. $\frac{1}{12} a^4$.

1034. See Figure 138. Let the y axis be the axis of revolution. The equation of the ellipse is

$$\frac{x^2}{a^2} + \frac{y^2}{b^2} = 1.$$

$$I_y = 2 \int_0^a 2\pi x y \, dx \cdot x^2 = 4\pi \int_0^a x^3 y \, dx = 4\pi \frac{b}{a} \int_0^a x^3 \sqrt{a^2 - x^2} \, dx;$$

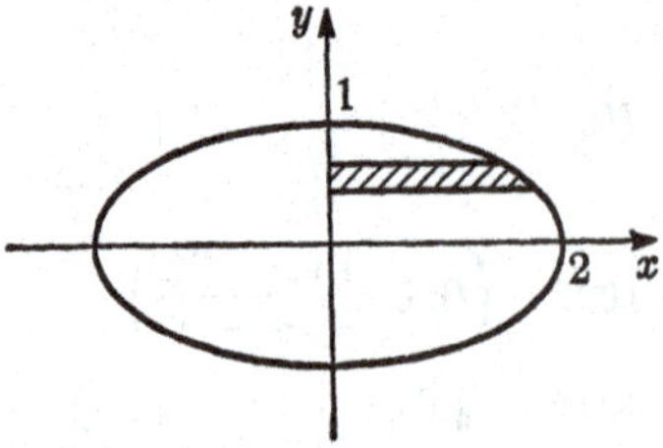

FIGURE 137

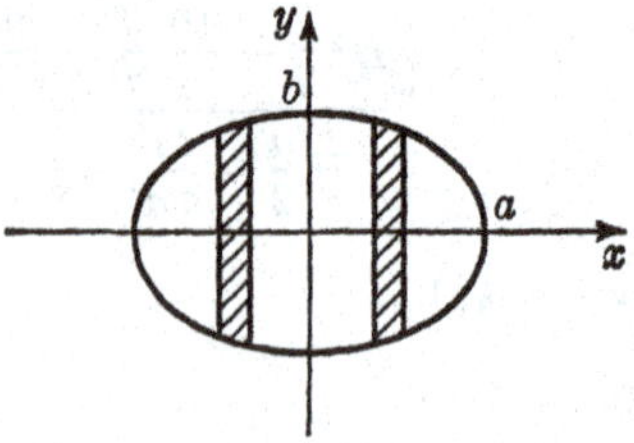

FIGURE 138

this can be integrated by parts (or by substitution), and yields

$$I_y = 4\pi \frac{b}{a}\left[x^2 \cdot \left(-\frac{1}{3}\right)(a^2 - x^2)^{3/2}\right]_0^a + \frac{1}{3}\int_0^a (a^2 - x^2)^{3/2} 2x\,dx$$

$$= 4\pi \frac{b}{a}\left[0 + \frac{1}{3} \cdot \frac{2}{5}(a^2 - x^2)^{5/2}\right]_0^a = \frac{8\pi}{15} a^4 b.$$

For $a = b$ we obtain the moment of inertia of a sphere with respect to its diameter:

$$I = \frac{8\pi}{15} a^5 = \frac{4}{3}\pi a^3 \cdot \frac{2}{5} a^2 = v \cdot \frac{2}{5} a^2.$$

1035. $\frac{64}{3}\pi a^5$.

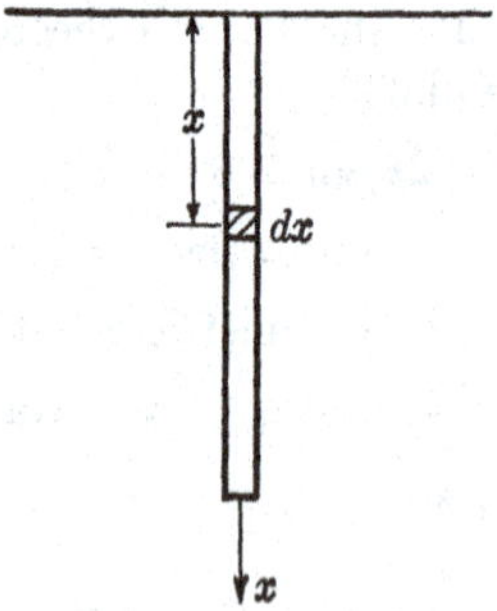

FIGURE 139

1037. See Figure 139. The section dx of the wire is subject to the force $(l - x)a\gamma$, and it becomes longer by

$$\frac{dx}{aE}(l - x)a\gamma = \frac{\gamma}{E}(l - x)\,dx.$$

For the whole wire we have

$$\Delta l = \int_0^l \frac{\gamma}{E}(l - x)\,dx = \frac{\gamma}{E}\left(l^2 - \frac{1}{2}l^2\right) = \frac{\gamma l^2}{2E}.$$

1038. We assume that e_2 moves along a straight line passing through e_1.

(It can be shown that the work does not depend on the path followed by e_2.) The work is

$$A = -\int_\infty^a \frac{e_1 \cdot 1}{x^2}\,dx = \lim_{M\to\infty} \frac{e_1}{x}\Big|_M^a = \frac{e_1}{a}.$$

(The minus sign accounts for the fact that the direction of the force acting on e_2 is opposite to that of the motion.)

1040. The main difference from Problem 1039 is that now A is a function of z:

$$A = \pi\left(\frac{z}{H}R\right)^2 = \frac{\pi R^2}{H^2}z^2.$$

We obtain, as in Problem 1039,

$$-\frac{\pi R^2}{H^2}z^2\,dz = k\sqrt{2gz}\,a\,dt.$$

$$\int_0^T dt = -\int_H^0 \frac{\pi R^2}{H^2ka\sqrt{2g}}z^{3/2}\,dz,$$

$$T = \frac{\pi R^2}{H^2ka\sqrt{2g}}\cdot\frac{2}{5}H^{5/2} = \frac{\pi R^2}{5ka}\sqrt{\frac{2H}{g}}.$$

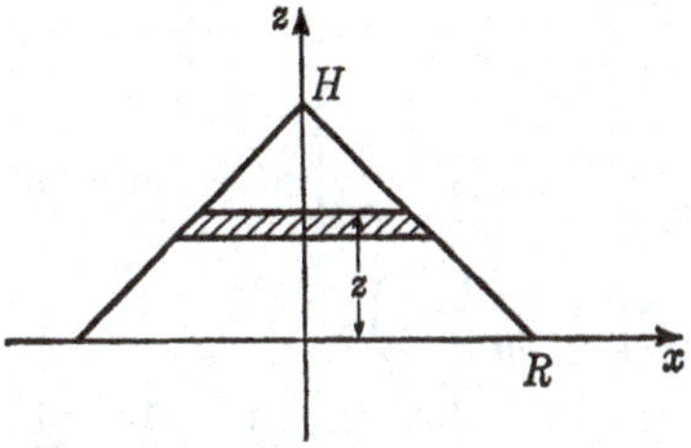

FIGURE 140

1042. See Figure 140. The weight of the shaded layer is

$$\pi\left[\frac{R(H-z)}{H}\right]^2 dz\cdot\gamma.$$

The work corresponding to pouring this layer of sand is

$$\pi\gamma\frac{R^2}{H^2}(H-z)^2\,dz\cdot z;$$

hence the total work is

$$W = \pi\gamma\frac{R^2}{H^2}\int_0^H (H-z)^2z\,dz = \pi\gamma\frac{R^2}{H^2}\cdot H^4\left(\frac{1}{2}-\frac{2}{3}+\frac{1}{4}\right) = \frac{1}{12}\pi\gamma R^2H^2$$

$$= \frac{1}{3}\pi R^2H\gamma\cdot\frac{1}{4}H = v\gamma\cdot\frac{1}{4}H.$$

Note that $\frac{1}{4}H$ is the height of the centroid of the embankment, and $v\gamma$ is the weight of the sand.

1044. It is known from thermodynamics that $dW = p\,dv$, and if the temperature is constant, then $pv = C$ (C a constant). Now

$$W = \int_v^{\frac{1}{2}v} \frac{C}{v}\,dv = C \ln v\Big|_v^{\frac{1}{2}v} = C \ln \frac{1}{2} = -C \ln 2 = -pv \ln 2.$$

The minus sign is due to the fact that the work is performed on the gas, not obtained from it.

1045. $K = \frac{1}{2}\cdot 2 \int_0^R 2\pi x\sqrt{R^2 - x^2}\,dx\;\gamma\cdot x^2\omega^2 = 2\pi\gamma\omega^2 \int_0^R x^3\sqrt{R^2 - x^2}\,dx.$

Substituting

$$\sqrt{R^2 - x^2} = u, \qquad R^2 - x^2 = u^2; \qquad -2x\,dx = 2u\,du,$$

$$K = 2\pi\gamma\omega^2 \int_R^0 -(R^2 - u^2)u\cdot u\,du = 2\pi\gamma\omega^2\left(R^2\frac{u^3}{3} - \frac{u^5}{5}\right)\Big|_0^R$$

$$= \frac{4\pi\omega^2\gamma R^5}{15} = \frac{1}{2} I\omega^2.$$

(I is the moment of inertia of the sphere with respect to its diameter. Compare Problem 1034.)

1047. $\pi\gamma R^3 L.$

1048. $v = \frac{c}{n}\ln\frac{m}{m - nt},\ h = h_0 + \frac{c}{n^2}\left[nt - (m - nt)\ln\frac{m}{m - nt}\right].$

1050. $u_n = \frac{1}{6}\left(\frac{1}{2n - 1} - \frac{1}{2n + 5}\right).$

$$S_n = \frac{1}{6}\left(1 - \frac{1}{7} + \frac{1}{3} - \frac{1}{9} + \frac{1}{5} - \frac{1}{11} + \frac{1}{7} - \frac{1}{13} + \cdots\right.$$

$$+ \frac{1}{2n - 7} - \frac{1}{2n - 1} + \frac{1}{2n - 5} - \frac{1}{2n + 1}$$

$$\left. + \frac{1}{2n - 3} - \frac{1}{2n + 3} + \frac{1}{2n - 1} - \frac{1}{2n + 5}\right)$$

$$= \frac{1}{6}\left(1 + \frac{1}{3} + \frac{1}{5} - \frac{1}{2n + 1} - \frac{1}{2n + 3} - \frac{1}{2n + 5}\right).$$

$$S = \lim_{n\to\infty} S_n = \tfrac{1}{6}(1 + \tfrac{1}{3} + \tfrac{1}{5}) = \tfrac{23}{90}.$$

1051. $u_n = \frac{1}{n(n + 1)(n + 2)} = \frac{A}{n} + \frac{B}{n + 1} + \frac{C}{n + 2};$

$A(n + 1)(n + 2) + Bn(n + 2) + Cn(n + 1) = 1;$

$n = 0,\ A = \frac{1}{2};\ n = -1,\ B = -1;\ n = -2,\ C = \frac{1}{2};$

$$u_n = \frac{1}{2}\left(\frac{1}{n} - \frac{2}{n+1} + \frac{1}{n+2}\right).$$

$$S_n = \frac{1}{2}\left(1 - \frac{2}{2} + \frac{1}{3} + \frac{1}{2} - \frac{2}{3} + \frac{1}{4} + \frac{1}{3} - \frac{2}{4} + \frac{1}{5} + \frac{1}{4} - \frac{2}{5}\right.$$

$$+ \frac{1}{6} + \ldots + \frac{1}{n-3} - \frac{2}{n-2} + \frac{1}{n-1} + \frac{1}{n-2} - \frac{2}{n-1}$$

$$\left. + \frac{1}{n} + \frac{1}{n-1} - \frac{2}{n} + \frac{1}{n+1} + \frac{1}{n} - \frac{2}{n+1} + \frac{1}{n+2}\right)$$

$$= \frac{1}{2}\left(\frac{1}{2} - \frac{1}{n+1} + \frac{1}{n+2}\right).$$

$$S = \lim_{n\to\infty} S_n = \tfrac{1}{4}.$$

1052. $u_n = (\tfrac{3}{6})^n + (\tfrac{2}{6})^n = (\tfrac{1}{2})^n + (\tfrac{1}{3})^n;$

$$S_n = \frac{1}{2}\,\frac{1 - (\frac{1}{2})^n}{1 - \frac{1}{2}} + \frac{1}{3}\cdot\frac{1 - (\frac{1}{3})^n}{1 - \frac{1}{3}} = 1 - \left(\frac{1}{2}\right)^n + \frac{1}{2}\left[1 - \left(\frac{1}{3}\right)^n\right].$$

$$S = \lim_{n\to\infty} S_n = 1 + \tfrac{1}{2} = \tfrac{3}{2}.$$

1053. $$u_n = \frac{2n+1}{n^2(n+1)^2} = \frac{n^2 + 2n + 1 - n^2}{n^2(n+1)^2}$$

$$= \frac{(n+1)^2}{n^2(n+1)^2} - \frac{n^2}{n^2(n+1)^2} = \frac{1}{n^2} - \frac{1}{(n+1)^2};$$

$$S_n = 1 - \frac{1}{4} + \frac{1}{4} - \frac{1}{9} + \frac{1}{9} - \frac{1}{16} + \ldots + \frac{1}{(n-1)^2} - \frac{1}{n^2}$$

$$+ \frac{1}{n^2} - \frac{1}{(n+1)^2} = 1 - \frac{1}{(n+1)^2};$$

$$S = \lim_{n\to\infty} S_n = 1.$$

1055. The known inequality $\sin \pi/2^n < \pi/2^n$ ensures the convergence of the given series. ($\pi/2^n$ is a term of a convergent geometric series.)

1056. $\dfrac{1}{(n+1)(n+4)} < \dfrac{1}{n^2}.$ The series $\dfrac{1}{n^2}$ is known to converge ($p = 2 > 1$), thus the given series converges also. By the method of Problem 1050, we have $S_n = \tfrac{13}{16}$.

1058. $\dfrac{1}{3n-1} > \dfrac{1}{3n}.$ But the series

$$\frac{1}{3}\cdot 1 + \frac{1}{3}\cdot\frac{1}{2} + \frac{1}{3}\cdot\frac{1}{3} + \ldots + \frac{1}{3}\cdot\frac{1}{n} + \ldots$$

diverges, and therefore so does the given series.

1059. Diverges.

1061. By the theorem of Problem 1060,

$$\lim_{n\to\infty} \frac{\dfrac{1}{\sqrt{n^2+2n}}}{\dfrac{1}{n}} = \lim_{n\to\infty} \frac{n}{\sqrt{n^2+2n}} = \lim_{n\to\infty} \frac{1}{\sqrt{1+\dfrac{2}{n}}} = 1.$$

The given series diverges, since $\sum_{n=1}^{\infty} \frac{1}{n}$ is known to be divergent.

1063. $$r = \lim_{n\to\infty} \frac{(n+1)\tan(\pi/2^{n+1})}{n\tan(\pi/2^n)}$$

$$= \lim_{n\to\infty} \frac{n+1}{n} \cdot \frac{\tan(\pi/2^{n+1})}{2^{\pi/(n+1)}} \frac{(\pi/2^n)}{\tan(\pi/2^n)} \cdot \frac{1}{2}$$

$$= 1\cdot 1\cdot 1 \cdot \frac{1}{2} = \frac{1}{2}.$$

The series converges.

1064. $r = \lim_{n\to\infty} \frac{(n+1)^2\cdot 3^n}{3^{n+1}\cdot n^2} = \frac{1}{3} \lim_{n\to\infty} \frac{(n+1)^2}{n^2} = \frac{1}{3} \lim_{n\to\infty} \left(1+\frac{1}{n}\right)^2 = \frac{1}{3}.$

The series converges.

1065. Converges. **1066.** Converges.

1068. See Problem 1067. $3/e > 1$, i.e., the series diverges.

1071. $t = \lim_{n\to\infty} \frac{n}{2n+1} = \frac{1}{2}.$ The series converges.

1072. Converges.

1073. $$t = \lim_{n\to\infty} \sqrt[n]{\frac{\left(\dfrac{n+1}{n}\right)^{n^2}}{3^n}} = \lim_{n\to\infty} \frac{\left(\dfrac{n+1}{n}\right)^n}{3} = \frac{e}{3} < 1.$$

The series converges.

1077. $f(x) = \frac{1}{x \ln x};$

$$\int_2^{\infty} f(x)\,dx = \lim_{M\to\infty} \int_2^M \frac{dx}{x\ln x} = \lim_{M\to\infty} (\ln\ln M - \ln\ln 2) = \infty.$$

The series does not converge.

1078. Converges.

1079. $\frac{1}{(n+1)^{3/2}} < \frac{1}{n^{3/2}}.$ The series converges $(p = \frac{3}{2} > 1)$.

1081. $\lim_{n\to\infty} \frac{u_{n+1}}{u_n} = \lim_{n\to\infty} \frac{(n+1)!10^{5(n-1)}}{10^{5n}\cdot n!} = \lim_{n\to\infty} \frac{n+1}{10^5} = \infty.$

The series diverges.

1082. $\lim_{n\to\infty} u_n = \lim_{n\to\infty} \frac{n}{1000(n+1)} = \frac{1}{1000}.$

The series diverges.

1083. Converges. (Use d'Alembert's test.)

1084. $\lim_{n\to\infty} \sqrt[n]{u_n} = \lim_{n\to\infty} \arctan\frac{1}{n} = 0.$

The series converges.

1085. $$\lim_{n\to\infty} \frac{\sin\dfrac{\pi}{2n+1}}{\sin\dfrac{\pi}{2n}} = \lim_{n\to\infty} \frac{\sin\dfrac{\pi}{2n+1}}{\dfrac{\pi}{2n+1}} \cdot \frac{\dfrac{\pi}{2n}}{\sin\dfrac{\pi}{2n}} \cdot \frac{2n}{2n+1} = 1\cdot 1\cdot 1 = 1.$$

No decision is arrived at, so we try the comparison test. We shall show that beginning with a certain x, $\sin(\pi/2x) > 1/x$.

$$\lim_{x\to\infty} \frac{\sin(\pi/2x)}{1/x} = \lim_{x\to\infty} \frac{\cos(\pi/2x)\cdot(-\pi/2x^2)}{-1/x^2} = \lim_{x\to\infty} \frac{\pi}{2}\cos\frac{\pi}{2x} = \frac{\pi}{2}.$$

The function $\dfrac{\sin(\pi/2x)}{1/x}$ $(x > 0)$ is continuous, and consequently there exists an x_0 such that for $x > x_0$ the value of the function is greater than 1. (Its limit for $x \to \infty$ is $\pi/2 = 1.57\ldots.$) This proves that the given series diverges.

1087. We shall investigate the series $\sum_{n=1}^{\infty} \frac{n^n}{(n!)^2}$:

$$\lim_{n\to\infty} \frac{u_{n+1}}{u_n} = \lim_{n\to\infty} \frac{(n+1)^{n+1}(n!)^2}{[(n+1)!]^2 n^n} = \lim_{n\to\infty} \frac{(n+1)(n+1)^n}{(n+1)^2 n^n}$$
$$= \lim_{n\to\infty} \frac{1}{n+1}\left(1+\frac{1}{n}\right)^n = \lim_{n\to\infty} \frac{e}{n+1} = 0.$$

The series converges, i.e., the truth of the given limit is established.

1088. We consider the series $\sum_{n=1}^{\infty} \frac{(n!)^n}{n^{n^2}}$. By Cauchy's test,

$$\lim_{n\to\infty} \sqrt[n]{u_n} = \lim_{n\to\infty} \frac{n!}{n^n}.$$

The latter limit can be computed also by the same method. We consider the series

$$\sum_{n=1}^{\infty} v_n = \sum_{n=1}^{\infty} \frac{n!}{n^n};$$

$$\lim_{n\to\infty} \frac{v_{n+1}}{v_n} = \lim_{n\to\infty} \frac{(n+1)!n^n}{(n+1)^{n+1}n!} = \lim_{n\to\infty} \frac{n^n}{(n+1)^n} = \lim_{n\to\infty} \frac{1}{\left(1+\frac{1}{n}\right)^n} = \frac{1}{e}.$$

This series converges, i.e., $\lim\limits_{n\to\infty} n!/n^n = 0$. Consequently the former series converges too, and the given limit is established.

1089. Converges.

1090. Diverges.

1091. Diverges.

1092. Diverges.

1093. Converges for $p > 1$.

1094. Converges.

1095. Converges.

1096. Converges.

1098. $|u_n| = \dfrac{1}{(2n-1)^3} \leqslant \dfrac{1}{n^3}$,

This comparison shows that the given series converges absolutely.

1099. By d'Alembert's test,

$$\lim_{n\to\infty} \frac{|u_{n+1}|}{|u_n|} = \lim_{n\to\infty} \frac{100n^2 \cdot 2^{n-1}}{2^n \cdot 100(n-1)^2} = \lim_{n\to\infty} \left(\frac{n}{n-1}\right)^2 \cdot \frac{1}{2} = \frac{1}{2}.$$

The series converges absolutely.

1100. $\lim\limits_{n\to\infty} a_n = \lim\limits_{n\to\infty} (-1)^{n+1} \dfrac{n+1}{4n}$.

This limit is not zero and the series does not converge.

1101. By comparison with $1/n$ we prove that there is no absolute convergence. By Leibniz's test,

$$\frac{1}{\ln(n+1)} < \frac{1}{\ln n} \quad \text{and} \quad \lim_{n\to\infty} \frac{1}{\ln n} = 0.$$

The series converges conditionally.

1102. Converges conditionally.

1103. Converges absolutely.

1104. $\dfrac{1}{n - \ln n} > \dfrac{1}{n}$, and comparison with the harmonic series shows that there is no absolute convergence. By Leibniz's test,

$$\frac{1}{n+1-\ln(n+1)} - \frac{1}{n-\ln n} = \frac{n - \ln n - n - 1 + \ln(n+1)}{[(n+1) - \ln(n+1)](n - \ln n)}$$

$$= \frac{\ln\left(1+\frac{1}{n}\right) - 1}{[(n+1) - \ln(n+1)](n - \ln n)}$$

The denominator is positive ($n > \ln n$) and the numerator clearly negative, because for $n \geqslant 1$, $1 + 1/n < e$. Consequently $b_{n+1} < b_n$. Now

$$\lim_{n\to\infty} b_n = \lim_{n\to\infty} \frac{1}{n - \ln n} = 0,$$

because
$$\lim_{x\to\infty} \frac{\ln x}{x} = \lim_{x\to\infty} \frac{1}{x} = 0,$$

and consequently

$$\lim_{x\to\infty} (x - \ln x) = \lim_{x\to\infty} x\left(1 - \frac{\ln x}{x}\right) = \infty.$$

The conditional convergence is established.

1106. $\left|\frac{1}{n^2}\cos n\alpha\right| \leqslant \frac{1}{n^2}$. The series converges absolutely.

1107. Diverges.

1108. Converges conditionally.

1109. Converges conditionally.

1115. (a) converges; (b) diverges.

1116. Converges.

1119. $\left|\frac{u_{n+1}}{u_n}\right| = \frac{|x|^{n+1}n^2}{(n+1)^2|x|^n} = |x|\frac{1}{\left(1+\frac{1}{n}\right)^2}$. We require

$$\frac{|x|}{\left(1+\frac{1}{n}\right)^2} \leqslant q < 1$$

for any $n > n_0$. This will be true for any $-1 < x < 1$, and there the series converges. For $|x| > 1$ the above ratio becomes, for large n, greater than 1 and the series diverges. For $x = \pm 1$ we obtain the convergent series

$$\pm 1 + \frac{1}{2^2} \pm \frac{1}{3^2} + \ldots \pm \frac{1}{(2n-1)^2} + \frac{1}{(2n)^2} \pm \ldots.$$

As a result, the domain of convergence is $-1 \leqslant x \leqslant 1$.

1120. $r_n = \left|\frac{x^{n+1}\sqrt{n}}{\sqrt{n+1}\,x^n}\right| = |x|\sqrt{\frac{n}{n+1}}.$

For $|x| < 1$ the series clearly converges. For $|x| > 1$ it diverges. For $x = 1$, we have the series

$$1 + \frac{1}{\sqrt{2}} + \ldots + \frac{1}{\sqrt{n}} + \ldots,$$

which diverges. For $x = -1$, we have

$$-1 + \frac{1}{\sqrt{2}} - \frac{1}{\sqrt{3}} + \ldots + \frac{(-1)^n}{\sqrt{n}} + \ldots,$$

which converges by the Leibniz test. The domain of convergence is thus $-1 \leqslant x < 1$.

1122. $\left|\frac{x^n}{1+x^{2n}}\right| \leqslant |x^n|$, and for $|x| < 1$ the given series converges. For $|x| > 1$,

$$\left|\frac{x^n}{1+x^{2n}}\right| < \left|\frac{x^n}{x^{2n}}\right| = \left|\frac{1}{x^n}\right|.$$

But in this case $\sum_{n=1}^{\infty} |1/x^n|$ converges, and the same is true for the given series. If $x = 1$, we have $\frac{1}{2} + \frac{1}{2} + \ldots$; if $x = -1$, we have $-\frac{1}{2} + \frac{1}{2} - \frac{1}{2} + \frac{1}{2} - \ldots$. There is no convergence in either case. The domain of convergence is $x \neq \pm 1$.

1123. For any x, $|\sin x| \leqslant |x|$. The series

$$\frac{|x|}{2} + \frac{|x|}{4} + \ldots + \frac{|x|}{2^n} + \ldots$$

converges for every x, and thus the given series also converges.

1124. $t_n = \left|\frac{(n+1)xe^{nx}}{e^{(n+1)x}nx}\right| = \frac{n+1}{ne^x}$, $\lim_{n\to\infty} t_n = \frac{1}{e^x}$.

For $x > 0$, $1/e^x < 1$ and the series converges. For $x < 0$, it diverges. For $x = 0$ we have $0 + 0 + \ldots$ and this series clearly converges. The domain of convergence is $x \geqslant 0$.

1125. Converges for any x.

1126. Converges for $|x - \pi k| \leqslant \pi/6$ ($k = 0, \pm 1, \pm 2, \ldots$).

1127. $x \neq 0$.

1128. Diverges everywhere.

1129. $|x| \geqslant 1$.

1131. $\frac{1}{n^2(1+n^2x^2)} < \frac{1}{n^2}$ for any x.

$1/n^2$ converges, hence the given series converges uniformly everywhere.

1132. $\left|\frac{\sin nx}{2^n}\right| < \frac{2}{2^n}$; the answer as in Problem 1131.

1133. Converges uniformly in the interval $(-\infty, \infty)$.

1136. $\lim_{n\to\infty}\left|\frac{u_{n+1}}{u_n}\right| = \lim_{n\to\infty}\frac{|x^{n+1}|}{(n+1)!}\cdot\frac{n!}{|x^n|} = \lim_{n\to\infty}\frac{|x|}{n+1} = 0$, for any x.

The series converges at every point of the given interval $0 \leqslant x < \infty$. Now for any x in this interval,

$$R_{n+1}(x) = \frac{x^{n+1}}{(n+1)!} + \frac{x^{n+2}}{(n+2)!} + \ldots > \frac{x^{n+1}}{(n+1)!}.$$

For a fixed ϵ and an arbitrarily large n_0, an x can always be found such that $x^{n_0+1}/(n_0+1)! > \epsilon$; i.e., there is no uniform convergence in the domain $0 \leqslant x < \infty$. It is easy to show, however, that in every bounded subinterval $0 \leqslant x \leqslant b$ such convergence does exist.

1137. No.

1143. For $|x| < 1$ we have

$$f(x) = 1 + x + x^2 + \ldots + x^n + \ldots = \frac{1}{1-x}.$$

The sum in (a) is built from derivatives of the terms of $f(x)$. If $1 + 2x + \ldots + nx^{n-1} + \ldots$ converges uniformly in a certain subinterval of $-1 < x < 1$, then it will be equal there to $f'(x) = 1/(1-x)^2$. But for any $|x| \leqslant q < 1$, $|nx^{n-1}| \leqslant nq^{n-1}$, and since $\sum_{n=1}^{\infty} nq^{n-1}$ converges

$$(\lim_{n\to\infty} (n+1)q^n/nq^{n-1} = q < 1),$$

the assertion about the uniform convergence of $\sum_{n=1}^{\infty} nx^{n-1}$ is true. For any $|x| < 1$, a q can be found such that $|x| \leqslant q < 1$ and consequently the formula

$$\sum_{n=1}^{\infty} nx^{n-1} = \frac{1}{(1-x)^2}$$

holds for $|x| < 1$. In a similar way we obtain

$$1 + 3x + \ldots + \frac{n(n+1)}{2} x^{n-1} + \ldots = \frac{1}{(1-x)^3}.$$

1144. It is easy to show that the series of the first and second derivatives of the terms of the given series converge uniformly in any bounded interval. We thus have, for any x,

$$J_0'(x) = \sum_{n=1}^{\infty} (-1)^n \frac{2nx^{2n-1}}{(n!)^2 \cdot 2^{2n}}, \qquad J_0''(x) = \sum_{n=1}^{\infty} (-1)^n \frac{2n(2n-1)x^{2n-2}}{(n!)^2 \cdot 2^{2n}}.$$

The coefficient of x^{2n-2} in the sum $J_0''(x) + \dfrac{J_0'(x)}{x} + J_0(x)$ equals

$$\frac{(-1)^n 2n(2n-1)}{(n!)^2 \cdot 2^{2n}} + \frac{(-1)^n 2n}{(n!)^2 \cdot 2^{2n}} + (-1)^{n-1} \frac{1}{[(n-1)!]^2 \cdot 2^{2n-2}}$$
$$= \frac{(-1)^n}{(n!)^2 \cdot 2^{2n}} [2n(2n-1) + 2n - n^2 \cdot 4] = 0.$$

Consequently every term in the above sum equals zero and $J_0(x)$ satisfies the given differential equation.

1146. $r = \lim\limits_{n\to\infty} \dfrac{n}{n+1} = 1$, $R = 1$.

At $x = 1$ the series converges, at $x = -1$ it diverges. The domain of convergence is $-1 < x \leqslant 1$.

1147. $r = \lim\limits_{n\to\infty} \dfrac{(n+1)!}{n!} = \infty,\ R = 0;$

i.e., the series converges only at $x = 0$.

1149. $\lim\limits_{n\to\infty} \dfrac{(2n-1)(2n-1)!}{(2n+1)(2n+1)!} = \lim\limits_{n\to\infty} \dfrac{2n-1}{(2n+1)^2\cdot 2n} = 0.$

Consequently R equals ∞.

1150. $r = \lim\limits_{n\to\infty} \dfrac{(n+1)^{n+1}n!}{(n+1)!\,n^n} = \lim\limits_{n\to\infty} \left(1 + \dfrac{1}{n}\right)^n = e,\ R = \dfrac{1}{e}.$

By methods not introduced in this book it can be shown that for $x = 1/e$ this series diverges and for $x = -1/e$ it converges.

1151. By Cauchy's test,

$$t = \lim_{n\to\infty} \sqrt[n]{|a_n|} = \lim_{n\to\infty} n = \infty, \qquad R = 0.$$

1152. $r = \lim\limits_{n\to\infty} \dfrac{[\ln(n+2)](n+1)}{(n+2)\ln(n+1)} = 1,\ R = 1.$

At $x = 1$ the terms of the series are greater than the corresponding terms of the harmonic series and there is no convergence. At $x = -1$ the series converges by Leibniz's test. The domain of convergence is $-1 \leqslant x < 1$.

1153. $t = \lim\limits_{n\to\infty} \sqrt[n]{|a_n|} = \lim\limits_{n\to\infty} \left(\dfrac{n+1}{n}\right)^n = e,\ R = \dfrac{1}{e}.$

At $x = 1/e$ we obtain

$$u_n = \left(\frac{n+1}{n}\right)^{n^2} \left(\frac{1}{e}\right)^n$$

and this term does not tend to zero when $n \to \infty$. The same occurs at $x = -1/e$. Consequently the domain of convergence is $-1/e < x < 1/e$.

1155. $-6 \leqslant x \leqslant -4$.

1156. $-3 < x < 3$.

1157. $-1 \leq x \leq 1$.

1158. $-\frac{11}{5} \leqslant x - \frac{9}{5}$.

1164. Compare Problem 1162.

$$y = \sqrt{x^3} = x^{3/2} = [1 + (x-1)]^{3/2}$$

$$= 1 + \frac{3}{2}(x-1) + \frac{3}{2}\cdot\frac{1}{2}\cdot\frac{1}{2}(x-1)^2 + \frac{3}{2}\cdot\frac{1}{2}\left(-\frac{1}{2}\right)\cdot\frac{1}{3!}(x-1)^3 + \cdots.$$

The series represents $\sqrt{x^3}$ for $-1 \leqslant x - 1 \leqslant 1$, i.e., $0 \leqslant x \leqslant 2$.

1170. $y = x\cos x - \sin x = x \sum\limits_{k=0}^{\infty} \dfrac{(-1)^k x^{2k}}{(2k)!} - \sum\limits_{k=0}^{\infty} (-1)^k \dfrac{x^{2k+1}}{(2k+1)!}$

$$= \sum_{k=0}^{\infty} (-1)^k x^{2k+1} \left[\frac{1}{(2k)!} - \frac{1}{(2k+1)!}\right]$$

$$= \sum_{k=0}^{\infty} (-1)^k x^{2k+1} \frac{2k}{(2k+1)!}.$$

1171. $y = \ln(10 + x) = \ln\left[10\left(1 + \frac{x}{10}\right)\right] = \ln 10 + \ln\left(1 + \frac{x}{10}\right)$

$$= \ln 10 + \frac{x}{10} - \frac{1}{2}\left(\frac{x}{10}\right)^2 + \frac{1}{3}\left(\frac{x}{10}\right)^3$$

$$- \ldots + (-1)^{n-1}\frac{1}{n}\left(\frac{x}{10}\right)^n + \ldots, \quad -10 < x \leqslant 10.$$

1172. $y = (1 + x^2)^{1/2}$

$$= 1 + \frac{1}{2}x^2 + \frac{1}{2}\left(-\frac{1}{2}\right)\cdot\frac{1}{2}x^4 + \frac{1}{2}\left(-\frac{1}{2}\right)\left(-\frac{3}{2}\right)\cdot\frac{1}{3!}x^6 + \ldots,$$

$$-1 \leqslant x \leqslant 1.$$

1175. $y' = \frac{1}{\sqrt{1 - x^2}} = (1 - x^2)^{-1/2}$

$$= 1 + \frac{1}{2}x^2 + \frac{1\cdot 3}{2^2}\cdot\frac{1}{2}x^4 + \frac{1\cdot 3\cdot 5}{2^3}\cdot\frac{1}{3!}x^6 + \ldots$$

$$+ \frac{1\cdot 3\cdots(2n-1)}{2^n}\cdot\frac{1}{n!}x^{2n} + \ldots.$$

$$y = x + \frac{1}{2}\cdot\frac{x^3}{3} + \frac{1\cdot 3}{2^2}\cdot\frac{1}{2!}\cdot\frac{x^5}{5} + \frac{1\cdot 3\cdot 5}{2^3}\cdot\frac{1}{3!}\cdot\frac{x^7}{7} + \ldots$$

$$+ \frac{1\cdot 3\cdots(2n-1)}{2^n}\cdot\frac{1}{n!}\cdot\frac{x^{2n+1}}{2n+1} + \ldots.$$

As in Problem 1174, this series represents $y = \arcsin x$ in the interval $-1 \leqslant x \leqslant 1$.

1176. $y' = \frac{1}{2}\left(\frac{1}{1+x} + \frac{1}{1-x}\right) = \frac{1}{1-x^2},$

$$y' = 1 + x^2 + x^4 + x^6 + \ldots + x^{2n} + \ldots.$$

$$\frac{1}{2}\ln\frac{1+x}{1-x} = x + \frac{x^3}{3} + \frac{x^5}{5} + \frac{x^7}{7} + \ldots + \frac{x^{2n+1}}{2n+1} + \ldots.$$

This series represents the function for $-1 < x < 1$. At $x = \pm 1$ it does not converge.

1179. $\sum_{n=0}^{\infty} \frac{2^{n/2}\cos(n\pi/4)}{n!} x^n.$

The series represents the given function everywhere.

1180. $(x-1)+\frac{5}{2}(x-1)^2+\frac{11}{6}(x-1)^3$

$$+6\sum_{n=4}^{\infty}\frac{(-1)^n}{n(n-1)(n-2)(n-3)}\cdot(x-1)^n,\ 0<x\leqslant 2.$$

1181. $\sum_{n=0}^{\infty}\left(1+\frac{1}{1!}+\frac{1}{2!}+\frac{1}{3!}+\ldots+\frac{1}{n!}\right)x^n,\ -1<x<1.$

1182. $\operatorname{argsinh} x=\ln(x+\sqrt{1+x^2})=x-\frac{x^3}{2\cdot 3}+\frac{1\cdot 3}{2^2}\cdot\frac{1}{2!}\frac{x^5}{5}+\ldots$

$$+(-1)^n\frac{1\cdot 3\cdot 5\ldots(2n-1)}{2^n\cdot n!}\cdot\frac{x^{2n+1}}{2n+1}+\ldots,|x|\leqslant 1.$$

1184. $f(x)=x^6e^x=x^6\left(1+\frac{x}{1!}+\frac{x^2}{2!}+\frac{x^3}{3!}+\frac{x^4}{4!}+\frac{x^5}{5!}+\ldots\right),$

$\frac{f^{(11)}(0)}{11!}=\frac{1}{5!}$, i.e., $f^{(11)}(0)=11\cdot 10\cdot 9\cdot 8\cdot 7\cdot 6=332{,}640.$

1186. $\lim_{x\to 0}\left(\frac{1}{x^2}-\cot^2 x\right)$

$$=\lim_{x\to 0}\frac{\sin^2 x-x^2\cos^2 x}{x^2\sin^2 x}$$

$$=\lim_{x\to 0}\frac{\left(x-\frac{x^3}{3!}+\ldots\right)^2-x^2\left(1-\frac{x^2}{2!}+\frac{x^4}{4!}-\ldots\right)^2}{x^2\left(x-\frac{x^3}{3!}+\ldots\right)^2}$$

$$=\lim_{x\to 0}\frac{x^2-2\frac{x^4}{3!}-x^2+2\frac{x^4}{2!}+\text{higher terms}}{x^4+\text{higher terms}}$$

$$=\lim_{x\to 0}\frac{\frac{2}{3}x^4+\ldots}{x^4+\ldots}=\frac{2}{3}.$$

Note that the power series in the numerator and denominator are continuous functions of x.

1189. $=\sqrt[3]{70}=\sqrt[3]{64+6}=\sqrt[3]{64}\sqrt[3]{1+\frac{6}{64}}=4(1+0.0937)^{1/3}$

$$=4\left(1+\frac{1}{3}\cdot 0.0937-\frac{1}{3}\cdot\frac{2}{3}\cdot\frac{1}{2}\cdot 0.0937^2\right.$$

$$\left.+\frac{1}{3}\cdot\frac{2}{3}\cdot\frac{5}{3}\cdot\frac{1}{2\cdot 3}0.0937^3-\ldots\right)$$

$$=4(1+0.0312-0.00098+\ldots)=4\cdot 1.0302=4.121.$$

Estimation of the remainder shows that the number of terms computed here suffices.

1193. $$\int_0^x e^{-x^2}\,dx = \int_0^x \left[1 - \frac{x^2}{1} + \frac{x^4}{2!} - \frac{x^6}{3!} + \ldots + (-1)^n \frac{x^{2n}}{n!} + \ldots\right] dx$$

$$= x - \frac{x^3}{3} + \frac{x^5}{5 \cdot 2!} - \ldots + (-1)^n \frac{x^{2n+1}}{(2n+1) \cdot n!} + \ldots.$$

This series represents the integral everywhere.

1194. $$\int_0^{1/2} \frac{dx}{\sqrt{1+x^4}} = \int_0^{1/2} (1+x^4)^{-1/2}\,dx = \int_0^{1/2} (1 - \tfrac{1}{2}x^4 + \ldots)\,dx$$

$$\approx \left(x - \tfrac{1}{10}x^5\right)\Big|_0^{1/2} = \tfrac{1}{2} - \tfrac{1}{320} = \tfrac{159}{320} = 0.4969.$$

To estimate the error we can use the fact that the signs of the terms in the expansion alternate and the terms decrease in absolute value. Thus the error is not greater than the first omitted term:

$$-\frac{1}{2} \cdot \left(-\frac{3}{2}\right) \cdot \frac{1}{2!} \cdot \frac{1}{9} x^9 \Big|_0^{1/2} = \frac{1}{24} \cdot \left(\frac{1}{2}\right)^9 \approx \frac{1}{24 \cdot 512} < 0.0001.$$

1195. This integral can be computed exactly (see Prob. 751), but the result contains logarithms, the calculation of which requires tables and thus leads to approximate results after all. It is more convenient to use a Maclaurin expansion:

$$\int_0^{1/2} \frac{dx}{1+x^4} = \int_0^{1/2} (1 - x^4 + x^8 - x^{12} + \ldots)\,dx$$

$$= \left(x - \frac{x^5}{5} + \frac{x^9}{9} - \frac{x^{13}}{13} + \ldots\right)\Big|_0^{1/2}.$$

We calculate the third term:

$$\frac{(\frac{1}{2})^9}{9} = \frac{1}{9 \cdot 512} = \frac{1}{4608} < 0.0005.$$

Consequently (since the signs alternate) it is enough to compute the first two terms:

$$\left(x - \frac{x^5}{5}\right)\Big|_0^{1/2} = \frac{1}{2} - \frac{1}{5} \cdot \frac{1}{32} = 0.494.$$

1196. By symmetry,

$$A = 2\int_0^{1/2} \sqrt{x^3+1}\,dx$$

$$= 2\int_0^{1/2} (1 + \tfrac{1}{2}x^3 + \tfrac{1}{2}(-\tfrac{1}{2}) \cdot \tfrac{1}{2}x^6 + \ldots)\,dx$$

$$= 2(x + \tfrac{1}{8}x^4 - \tfrac{1}{8} \cdot \tfrac{1}{7}x^7 + \ldots)\Big|_0^{1/2}.$$

The signs of the terms alternate and their absolute values decrease. The third term is approximately 0.0002, i.e., with the required accuracy

$$A = 2(\tfrac{1}{2} + \tfrac{1}{8}\cdot\tfrac{1}{16}) = 1 + \tfrac{1}{64} = 1.016.$$

1197. 343.4.

1198. 0.105361.

1199. 0.10033.

1200. 0.785.

1201. $\dfrac{x^3}{3} - \dfrac{x^7}{7\cdot 3!} + \dfrac{x^{11}}{11\cdot 5!} - \ldots + (-1)^n \dfrac{x^{4n+3}}{(4n+3)(2n+1)!} + \ldots.$

1203. $\sum_{n=0}^{\infty}\left[2\left(\dfrac{x}{2}\right)^n - \dfrac{5}{3}\left(\dfrac{x}{3}\right)^n\right]$, $-2 < x < 2$.

1204. $\frac{1}{2}\arctan\frac{1}{2}$.

1206. $\dfrac{1}{18}\ln\dfrac{(x-1)^2}{x^2+2} - \dfrac{1}{3(x-1)} - \dfrac{2\sqrt{2}}{9}\arctan\dfrac{x}{\sqrt{2}} + C.$

1208. (a) $A_{\max} = \frac{1}{2}ab(1+\sqrt{2})$; (b) $A_{\min} = 0$.

1209. $\sqrt[5]{h^5 + \dfrac{20p^2V}{\pi}} - h.$

1211. $e^{3/17}$.

1212. $\dfrac{d^3y}{dt^3} + y = 0.$

1213. $-\dfrac{1}{3}\arcsin\dfrac{3}{x+2} + C.$

1215. Min $(\ln 3, -15 + 6\ln 3)$; Max $(0, -7)$.

1217. sbt $= \frac{5}{8}$; sbn $= 160$.

1218. $\arctan\frac{87}{451}$.

1219. $\dfrac{x^2}{2}\arcsin x - \dfrac{1}{4}\arcsin x + \dfrac{1}{4}x\sqrt{1-x^2} + C.$

1220. (a) $r = \sqrt{\frac{2}{3}}R$, $H = (2/\sqrt{3})R$.

(b) $r = R\cos(\frac{1}{2}\arctan 2)$, $H = 2R\sin(\frac{1}{2}\arctan 2)$.

1221. $y' = \dfrac{x^2 - ay}{ax - y^2}$, $y'' = \dfrac{2a^3xy}{(ax - y^2)^3}$.

1222. $\frac{32}{81}\pi R^3$.

1223. Diverges.

1224. 1.

1225. $\ln\dfrac{1+x^2}{|1-x^2|} + C.$

1227. 0.530.

1228. $1 - \dfrac{1}{24}\ln 7 - \dfrac{\sqrt{3}}{12}\arctan\dfrac{\sqrt{3}}{5}.$

1229. (a) $\frac{1}{3}\%$; (b) $\frac{1}{3}\%$.

1230. $p\sqrt{2}$, $q\sqrt{2}$.

1232. $\frac{1}{4}\arcsin\sqrt{x} - \frac{1}{4}\sqrt{x - x^2}(1 - 2x) + C.$

1233. 1.63 hours after the moment of leaving.

1235. $x - \frac{3}{2}x^2 + \frac{11}{6}x^3 - \frac{23}{12}x^4 + \ldots.$

1238. $\frac{1}{2}e^x(x^2\sin x - x^3\cos x + 2x\cos x - \sin x - \cos x)$.

1239. The condition is $P^2 \geqslant 24S$.

1240. $y' = 1 + \frac{1}{x^2} - \frac{1}{(x+1)^2} - \frac{1}{(x-1)^2}$.

$y^{(k)} = (-1)^k k! \left[\frac{1}{(x+1)^{k+1}} + \frac{1}{(x-1)^{k+1}} - \frac{1}{x^{k+1}}\right]$, $k > 1$.

1241. $(-1)^{n-1} \frac{2(n-3)!(2x+n)}{(x+1)^{n-1}}$.

1244. $\frac{1}{\sqrt{b^2 - a^2}} \ln \left| \frac{\sqrt{b+a} + \sqrt{b-a} \cdot \tan \frac{x}{2}}{\sqrt{b+a} - \sqrt{b-a} \cdot \tan \frac{x}{2}} \right| + C$.

1245. $p = \sinh \frac{ax}{H}$.

1246. $\frac{1}{4} \ln 2 + \frac{3\sqrt{7}}{14} \arctan \frac{\sqrt{7}}{7}$.

1249. $(a - p, \sqrt{2p(a-p)}, \sqrt{p(2a-p)}$, if $a \geqslant p$;

$(0, 0)$, $|a|$, if $a < p$.

1250. $\frac{1}{16}(x + \frac{1}{8} \sin 4x - \frac{1}{8} \sin 8x - \frac{1}{24} \sin 12x) + C$.

1251. $(-4, -6.4)$.

1253. $\pi ab \cos (\alpha - \beta)$.

1254. $2\alpha + \beta = 2\pi n + \frac{\pi}{2}$. For $\beta = 10°$, $\alpha = 40° + 180°n$; $y = 0.704$.

1255. 103°, 77°, 1.03.

1256. *Hint.* Write y as a sum of two functions.

1258. (a) $\left(\ln \frac{b}{a}, \frac{b}{a}\right)$; (b) $\frac{b}{a}(e^{a-1} - a)$.

1259. $\frac{\pi^3 + 6\pi^2 - 24\pi - 48}{384}$.

1262. $t = 0.67$, $s = 1.85$.

1263. $\frac{13}{60}$.

1264. $(\frac{14}{3}, 72)$.

1266. The error is less than 0.0002.

1267. $\frac{1}{8}(\cos 2x - \frac{1}{2} \cos 4x - \frac{1}{3} \cos 6x) + C$.

1268. $\frac{ab}{2}\left(\arctan \frac{a \tan \beta}{b} - \arctan \frac{a \tan \alpha}{b}\right)$, $\frac{\pi}{2} > \beta > \alpha > \frac{-\pi}{2}$.

1271. Converges for $p < 1$.

1272. $-\frac{1}{2(x-1)} + \frac{1}{2} \ln |x - 1| - \frac{1}{4} \ln (x^2 + 1) + \arctan x + C$.

1273. $1/\sqrt{t}$.

1275. $\pm\frac{4}{3}$.

1276. The angle 2α at the vertex satisfies $\sin\alpha = \frac{1}{3}$.

1277. $\frac{27}{e^2}$.

1278. $\frac{1}{2}$.

1281. $\sum_{n=0}^{\infty} (-1)^{n+1} \frac{3^{2n+1}}{(2n+1)!}\left(x - \frac{\pi}{3}\right)^{2n+1}$.

INDEX

Absolute convergence, 235
Absolute error, 129, 131, 181
Acceleration, 37, 266
Accumulation point, 1
Addition of series, 238
Algebraic function, 13
Alternating series, 236
Analytic function, 252
Angle between curves, 107, 112
Antiderivative, 134
Applications
 of derivatives, 37
 of extrema, 80–84
 of integration, 199
 of series, 259–264
Approximation
 by differentials, 129
 of integrals, 187–191
 of roots (Newton's method), 120
 by Taylor's formula, 59–64
Arc length, 204
Archimedes' spiral, 110, 206
Area, plane, 200
Area, surface, 214, 219, 220
Argument, 12
Arithmetic-geometric mean, 7
Arithmetic-harmonic mean, 7, 275
Arithmetic operations on series, 237
Astroid, 111, 116, 119, 202, 213, 425
Asymptotes, 88–93, 203
Average value of function, 180

Base of natural logarithms, 14, 25
Basic formulas
 for definite integrals, 174
 for derivatives, 38, 39
 for differentials, 126
 for indefinite integrals, 135
Binomial formula, 57, 158, 254, 316
Binomial integrals, 158
Binomial series, 254, 316
Bound, greatest lower, 2
Bound, least upper, 2
Bounded function, 174
Bounded sequence, 2

Cardioid, 112, 113, 206, 212, 214, 217
Cauchy's criterion, 235
Cauchy's form of remainder, 253
Cauchy's test, 229, 233
Cauchy's theorem, 57
Cauchy-Schwarz inequality, 179
Center of curvature, 114, 118
Center of gravity, 215–219
Centroid, 215–219
Chain rule for derivatives, 38
Change of variable in integrals, 184
Circle of curvature, 114
Circle, osculating, 114
Cissoid of Diocles, 347
Coefficients, undetermined, 149, 152, 157, 161, 162, 256, 262
Comparison test, 228
Composite function, 15, 21, 251
 derivative of, 41
Concavity, 85, 94
Conditional convergence, 235
Cone, 83, 84, 213, 218, 224
Constant of integration, 134
Contact, order of, 113
Continuity, 30, 36, 47, 93
 of elementary functions, 32
 of functions defined by series, 242
 one-sided, 31
 uniform, 32
Convergence
 absolute, 235
 Cauchy's criterion, 235
 Cauchy's integral test, 229, 233
 Cauchy's test, 229
 comparison test, 228
 conditional, 235
 criteria for, 228, 229, 235, 236, 241, 249
 D'Alembert's test, 229, 249
 Dirichlet's test, 236
 domain of, 241–248
 of integrals, 191
 Leibniz's test, 236
 necessary condition for, 228
 radius of, 248
 ratio test, 229
 of sequences, 2

Convergence (*cont'd*):
  of series, 228
  test by definition, 228
  uniform, 241, 242, 249, 250, 251
Criteria for
  convergence of integrals, 191
  convergence of series, 228, 229, 235, 236, 241, 249
  inflection point, 85
  maxima and minima, 71, 72
Criterion, Cauchy's, 235
Curvature, 116
  center of, 114, 118
  circle of, 114
  radius of, 114
Curve tracing, 93–104
Curve, vertex of, 117
Cycloid, 110, 113, 114, 117, 119, 202, 205, 206, 211, 217
Cylinders, 12, 33, 83, 84, 105, 199, 208, 212, 213

D'Alembert's test, 229, 231, 232, 233, 234, 249
Decreasing function, 69
Decreasing sequence, 1
Definite integral, 172
  applications, 199
  basic properties, 174
  change of variable in, 184
  estimation, 175, 179
  evaluation, 176
  with variable limits, 181
Definition, extension of, 31, 33
Dependent variable, 12
Derivative
  applications, 37, 68
  basic formulas, 38, 39
  definition, 36
  evaluation of, 40–56
  four-step rule, 37
  higher order, 49–51
  of implicit functions, 45, 50, 78
  infinite, 47
  of integrals, 175, 181
  notation, 36
  one-sided, 46
Differential, 126–131
  approximation by, 129
  of arc length, 204
  of area, 181
  of higher order, 131
  invariance of form, 128, 132
Differential equation, 44, 45, 50, 51, 56, 248
Differentiation, 37
Differentiation (*cont'd*):
  by definition, 37, 48, 77
  graphical, 53
  of implicit functions, 45, 50, 78, 182
  logarithmic, 44, 75
  parametric, 46, 50, 80, 87, 92
  of series (term by term), 242, 245, 246, 251
Dirichlet's theorem, 236
Discontinuity, points of, 32, 33, 34, 93, 174, 193, 194
Divergence, of series, 228
  of improper integrals, 194, 195
Domain, of definition, 12, 23
  of convergence, 241–248
  of function, 12, 23, 93
Dummy variable, 176

e = base of natural logarithms, 14, 25
Electric current, 37, 131, 197, 222
Elementary functions, 14–16
  algebraic, 13
  composite, 15, 21
  exponential, 14
  hyperbolic, 14, 19
  inverse, 14, 19
  inverse hyperbolic, 15, 21
  inverse trigonometric, 15, 20
  irrational, 14, 97
  logarithmic, 14, 18
  polynomial, 13
  power, 13
  rational, 13, 17, 94
  trigonometric, 14, 19
Ellipsoid, 208, 222
Epsilon-delta (and ϵ-N) statements, 2–5, 7, 23, 25, 26, 30, 32, 33
Equations
  approximate solution, 120
  parametric, 46, 78, 100, 105, 115, 202, 205
Error (*see* Approximation)
Error, absolute, 129, 181
  relative, 129, 181
Estimation of definite integrals, 175, 179
Euler substitution, 159–160
Even function, 16
Evolute, 110, 118
Expansion, in infinite series, 252
  in partial fractions, 152, 307–309
Explicit form of function, 45
Exponential base, e, 25
Exponential function, 14
  integration of, 169
Extension of domain of definition, 31, 33
Extrema (*See* Maxima, Minima)

Flex points, 85
Flow of liquid, 223
Folium of Descartes, 107, 363

Infinite series
  multiplication of, 237, 238
  numerical, 227
  power, 248–264
  rearrangement of terms, 238
  sum of, 227
  Taylor's, 250, 259–264
  uniform convergence, 241, 242
Inflection points, 85, 94
Integrability, 174, 175, 191
Integral rational function, 13
Integral test (convergence), 228
Integrals, 134
  basic properties, 135
  binomial, 158
  definite, 172
  derivatives of, 175, 181
  formulas for, 135
  improper, 191, 229
  indefinite, 134
  irrational, 157
  trigonometric, 165
Integrand, 143
Integration
  approximate, 187, 260–262, 263, 264
  constant of, 134
  of exponential functions, 169
  of hyperbolic functions, 169
  or irrational functions, 157–160
  of miscellaneous functions, 170
  by parts, 144–145, 168, 169
  of rational functions, 151–152, 307–309
  of series (term by term), 242, 251, 258, 263
  special methods, 149, 150, 156, 159, 161–164, 166–168
  by substitution, 40–143, 157–160, 163, 164, 166, 168, 169, 185
  of trigonometric functions, 165
  by undertermined coefficients, 149
Intercepts on axes, 93
Intermediate value theorem, 31
Interval of convergence, 241–250
Invariance of differential form, 128, 132
Inverse functions, 14, 19
  derivatives of, 40
Inverse hyperbolic functions, 14, 19
Inverse trigonometric functions, 14, 19
Involute, 112, 118, 119
Irrational functions, 14, 97
  integration of, 157–160
Irrational integrals, 157

Lagrange's form of remainder, 60, 253, 254
Lagrange's theorem, 57
Law of the mean, 57–59
Least squares principle, 84
Least upper bound, 1
Least value of function, 72, 78, 82
Left-hand continuity, 31
  derivative, 46
  limit, 24
Leibniz' formula, 51, 52
Leibniz' test, 236, 243
Lemniscate, 107, 117, 203, 209
Length of arc, 204
L'Hôpital's rule, 65, 296
Limits
  basic, 25
  defining improper integral, 191
  evaluation of, 9, 26–30
  of a function, 23, 24, 26
  of integration, 172, 173
  lower, 1, 172
  one-sided, 24
  of a sequence, 2
  theorems on, 2, 3, 24
  upper, 1, 172
  variable upper, 175, 181
Liquid flow, 223
Ln x, 14
Local maxima, minima, 71–72
Logarithmic differentiation, 44, 75
Logarithmic function, 14, 18
Logarithmic spiral, 110, 119, 218
Lower bound, greatest, 2
Lower limit of sequence, 1

Maclaurin's formula, 60–64
Maclaurin's series, 250, 259–264
Mass, moment of, 215
Mathematical induction, 6, 49, 51, 56, 77
Maxima, minima, 71–84, 94
  existence of, 31
  necessary and sufficient conditions, 71–72
Mean value of a function, 180
Mean value theorem
  for derivatives, 57–59
  for integrals, 175–180
Minima (*See* Maxima)
Moment of inertia, 220
Moment of mass, 215
Monotone function, 31, 70, 93
Monotone sequence, 1
Multiplication of series, 238

n-th derivative of product, 51
Natural logarithm, 14

Necessary condition (convergence), 228
Newton's binomial formula, 51, 158, 254, 316
Newton-Leibniz formula, 176
Newton's method, 120
Newton's serpentine, 342
Normal, 107
Notation for derivative, 36
 for integrals, 134
Number defined by sequence, 5
Numerical integration, 187

Odd function, 6
One-sided continuity, 31
One-sided derivative, 46
One-sided limit, 24
Order of contact, 113
Orthogonal curves, 112
Osculating circle, 114

Pappas' theorems, 219
Paraboloid, 212, 218
Parametric differentiation, 46, 50, 80, 87, 92, 118
Parametric equations, 46, 50, 78, 80, 87, 92, 100, 105, 107, 115, 202, 204
Partial fractions, 64, 152, 258, 307–309
Partition, 172
Parts, integration by, 144–145, 168, 169
 generalized (formula A), 145
Period, 15
Periodic functions, 15, 22, 93
Physics problems, 37, 83, 84, 131, 174, 220, 222
Plane areas, 173, 200
Point of accumulation, 1
Point of discontinuity, 32, 33, 34, 93, 174, 193–194
Point of inflection, 85, 94
Polar coordinates, 68, 101, 105, 109, 115, 200, 202, 204, 210, 217
Polar subnormal, 109
Polar subtangent, 109
Polynomial function, 13
Power function, 13
Power series, 248–264
Principal part of infinitesimal, 129, 199

Radius of convergence, 248
Radius of curvature, 114
Range of a function, 12
Rate of change, 68
Rational functions, 13, 17, 94
 integration of, 151, 152
Rationalization, 157, 165
Ratio test (convergence), 229, 231–233, 249
Rearrangement of series, 238
Rectangle formula (integration), 187
Reduction formulas, 56, 148, 149, 166, 167, 183, 197
Relative error, 129–131, 181
Relative maxima, minima, 71
Remainder term, 60
 Cauchy's form, 60, 253
 Lagrange's form, 60, 253, 254
Riemann's theorem, 238
Right-hand continuity, 31
Right-hand derivative, 46
Right-hand limit, 24
Rolle's theorem, 57

Second derivative, 49
Sequence, 1
 bounded, 2
 convergent, 2
 limit of, 2
 monotone, 1
 number defined by, 5
 subsequence, 4
Series (see Infinite Series)
Serpentine, 342
Short methods (integration) 156
Simpson's approximation to an integral, 189, 206
Slope of tangent, 37
Solid of revolution, 208, 209, 212
Sphere, 68, 83, 84, 123, 214, 218
Spiral of Archimedes, 110, 206
Spiral, logarithmic, 110, 119, 218
Steiner's theorem, 221
Subnormal, 107
Subsequence, 4
Substitution, integration by, 140–143, 157–160, 163, 164, 166, 168, 169
Subtangent, 107
Sum of series, 227
Surface area, 214, 219, 220
Surface of revolution, 199, 214
Symmetry, 93, 94

Tangent, 37, 107
 slope of, 37
Taylor's formula, 60–64
Taylor's series, 252, 259–264
Term by term differentiation, 240
Term by term integration, 240
Tests
 for convergence of integrals, 191

Tests (*cont'd*):
for convergence of series, 228, 229, 235, 236, 241, 249
for inflection points, 85
for maxima and minima, 71, 72
Theorem of the mean
for derivatives, 57–59
for integrals, 175–180
Toricelli's law, 223
Torus, 212, 219
Tractrix, 206
Transcendental functions, 14, 97
Transformation of coordinates, 101–102
Trapezoidal formula, 188
Trigonometric functions, 14, 19
integration of, 165
inverse trigonometric functions, 15, 20
Trigonometric integrals, 165
Trigonometric substitution (integration by), 143

Unbounded function, integral of, 191
Undetermined coefficients, 149, 152, 161–162, 256
Uniform continuity, 32
Uniform convergence, 241, 242
Upper bound, least, 2
Upper limit of integral, 172
Upper limit of sequence, 1

Value, intermediate, 31
Variable, dependent, 12
independent, 12
Variable limits (integration), 175, 181
Velocity, 37, 68, 174
Velocity curve, 53, 54
Vertex of curve, 117
Volumes, 199, 207, 219, 220

Wallis' formula, 183
Weierstrass' test, 241, 244–245
Work, 174, 222, 224, 225

Zero exponent, 32

LIST OF GREEK LETTERS

α	alpha
β	beta
γ	gamma
δ	delta
ϵ	epsilon
ζ	zeta
η	eta
θ	theta
ι	iota
κ	kappa
λ	lambda
μ	mu
ν	nu
ξ	xi
o	omicron
π	pi
ρ	rho
σ	sigma
τ	tau
υ	upsilon
ϕ	phi
χ	chi
ψ	psi
ω	omega

A CATALOG OF SELECTED

DOVER BOOKS

IN SCIENCE AND MATHEMATICS

Astronomy

CHARIOTS FOR APOLLO: The NASA History of Manned Lunar Spacecraft to 1969, Courtney G. Brooks, James M. Grimwood, and Loyd S. Swenson, Jr. This illustrated history by a trio of experts is the definitive reference on the Apollo spacecraft and lunar modules. It traces the vehicles' design, development, and operation in space. More than 100 photographs and illustrations. 576pp. 6 3/4 x 9 1/4. 0-486-46756-2

EXPLORING THE MOON THROUGH BINOCULARS AND SMALL TELESCOPES, Ernest H. Cherrington, Jr. Informative, profusely illustrated guide to locating and identifying craters, rills, seas, mountains, other lunar features. Newly revised and updated with special section of new photos. Over 100 photos and diagrams. 240pp. 8 1/4 x 11. 0-486-24491-1

WHERE NO MAN HAS GONE BEFORE: A History of NASA's Apollo Lunar Expeditions, William David Compton. Introduction by Paul Dickson. This official NASA history traces behind-the-scenes conflicts and cooperation between scientists and engineers. The first half concerns preparations for the Moon landings, and the second half documents the flights that followed Apollo 11. 1989 edition. 432pp. 7 x 10. 0-486-47888-2

APOLLO EXPEDITIONS TO THE MOON: The NASA History, Edited by Edgar M. Cortright. Official NASA publication marks the 40th anniversary of the first lunar landing and features essays by project participants recalling engineering and administrative challenges. Accessible, jargon-free accounts, highlighted by numerous illustrations. 336pp. 8 3/8 x 10 7/8. 0-486-47175-6

ON MARS: Exploration of the Red Planet, 1958-1978--The NASA History, Edward Clinton Ezell and Linda Neuman Ezell. NASA's official history chronicles the start of our explorations of our planetary neighbor. It recounts cooperation among government, industry, and academia, and it features dozens of photos from Viking cameras. 560pp. 6 3/4 x 9 1/4. 0-486-46757-0

ARISTARCHUS OF SAMOS: The Ancient Copernicus, Sir Thomas Heath. Heath's history of astronomy ranges from Homer and Hesiod to Aristarchus and includes quotes from numerous thinkers, compilers, and scholasticists from Thales and Anaximander through Pythagoras, Plato, Aristotle, and Heraclides. 34 figures. 448pp. 5 3/8 x 8 1/2. 0-486-43886-4

AN INTRODUCTION TO CELESTIAL MECHANICS, Forest Ray Moulton. Classic text still unsurpassed in presentation of fundamental principles. Covers rectilinear motion, central forces, problems of two and three bodies, much more. Includes over 200 problems, some with answers. 437pp. 5 3/8 x 8 1/2. 0-486-64687-4

BEYOND THE ATMOSPHERE: Early Years of Space Science, Homer E. Newell. This exciting survey is the work of a top NASA administrator who chronicles technological advances, the relationship of space science to general science, and the space program's social, political, and economic contexts. 528pp. 6 3/4 x 9 1/4. 0-486-47464-X

STAR LORE: Myths, Legends, and Facts, William Tyler Olcott. Captivating retellings of the origins and histories of ancient star groups include Pegasus, Ursa Major, Pleiades, signs of the zodiac, and other constellations. "Classic." – *Sky & Telescope.* 58 illustrations. 544pp. 5 3/8 x 8 1/2. 0-486-43581-4

A COMPLETE MANUAL OF AMATEUR ASTRONOMY: Tools and Techniques for Astronomical Observations, P. Clay Sherrod with Thomas L. Koed. Concise, highly readable book discusses the selection, set-up, and maintenance of a telescope; amateur studies of the sun; lunar topography and occultations; and more. 124 figures. 26 halftones. 37 tables. 335pp. 6 1/2 x 9 1/4. 0-486-42820-6

Chemistry

MOLECULAR COLLISION THEORY, M. S. Child. This high-level monograph offers an analytical treatment of classical scattering by a central force, quantum scattering by a central force, elastic scattering phase shifts, and semi-classical elastic scattering. 1974 edition. 310pp. 5 3/8 x 8 1/2. 0-486-69437-2

HANDBOOK OF COMPUTATIONAL QUANTUM CHEMISTRY, David B. Cook. This comprehensive text provides upper-level undergraduates and graduate students with an accessible introduction to the implementation of quantum ideas in molecular modeling, exploring practical applications alongside theoretical explanations. 1998 edition. 832pp. 5 3/8 x 8 1/2. 0-486-44307-8

RADIOACTIVE SUBSTANCES, Marie Curie. The celebrated scientist's thesis, which directly preceded her 1903 Nobel Prize, discusses establishing atomic character of radioactivity; extraction from pitchblende of polonium and radium; isolation of pure radium chloride; more. 96pp. 5 3/8 x 8 1/2. 0-486-42550-9

CHEMICAL MAGIC, Leonard A. Ford. Classic guide provides intriguing entertainment while elucidating sound scientific principles, with more than 100 unusual stunts: cold fire, dust explosions, a nylon rope trick, a disappearing beaker, much more. 128pp. 5 3/8 x 8 1/2. 0-486-67628-5

ALCHEMY, E. J. Holmyard. Classic study by noted authority covers 2,000 years of alchemical history: religious, mystical overtones; apparatus; signs, symbols, and secret terms; advent of scientific method, much more. Illustrated. 320pp. 5 3/8 x 8 1/2. 0-486-26298-7

CHEMICAL KINETICS AND REACTION DYNAMICS, Paul L. Houston. This text teaches the principles underlying modern chemical kinetics in a clear, direct fashion, using several examples to enhance basic understanding. Solutions to selected problems. 2001 edition. 352pp. 8 3/8 x 11. 0-486-45334-0

PROBLEMS AND SOLUTIONS IN QUANTUM CHEMISTRY AND PHYSICS, Charles S. Johnson and Lee G. Pedersen. Unusually varied problems, with detailed solutions, cover of quantum mechanics, wave mechanics, angular momentum, molecular spectroscopy, scattering theory, more. 280 problems, plus 139 supplementary exercises. 430pp. 6 1/2 x 9 1/4. 0-486-65236-X

ELEMENTS OF CHEMISTRY, Antoine Lavoisier. Monumental classic by the founder of modern chemistry features first explicit statement of law of conservation of matter in chemical change, and more. Facsimile reprint of original (1790) Kerr translation. 539pp. 5 3/8 x 8 1/2. 0-486-64624-6

MAGNETISM AND TRANSITION METAL COMPLEXES, F. E. Mabbs and D. J. Machin. A detailed view of the calculation methods involved in the magnetic properties of transition metal complexes, this volume offers sufficient background for original work in the field. 1973 edition. 240pp. 5 3/8 x 8 1/2. 0-486-46284-6

GENERAL CHEMISTRY, Linus Pauling. Revised third edition of classic first-year text by Nobel laureate. Atomic and molecular structure, quantum mechanics, statistical mechanics, thermodynamics correlated with descriptive chemistry. Problems. 992pp. 5 3/8 x 8 1/2. 0-486-65622-5

ELECTROLYTE SOLUTIONS: Second Revised Edition, R. A. Robinson and R. H. Stokes. Classic text deals primarily with measurement, interpretation of conductance, chemical potential, and diffusion in electrolyte solutions. Detailed theoretical interpretations, plus extensive tables of thermodynamic and transport properties. 1970 edition. 590pp. 5 3/8 x 8 1/2. 0-486-42225-9

Engineering

FUNDAMENTALS OF ASTRODYNAMICS, Roger R. Bate, Donald D. Mueller, and Jerry E. White. Teaching text developed by U.S. Air Force Academy develops the basic two-body and n-body equations of motion; orbit determination; classical orbital elements, coordinate transformations; differential correction; more. 1971 edition. 455pp. 5 3/8 x 8 1/2. 0-486-60061-0

INTRODUCTION TO CONTINUUM MECHANICS FOR ENGINEERS: Revised Edition, Ray M. Bowen. This self-contained text introduces classical continuum models within a modern framework. Its numerous exercises illustrate the governing principles, linearizations, and other approximations that constitute classical continuum models. 2007 edition. 320pp. 6 1/8 x 9 1/4. 0-486-47460-7

ENGINEERING MECHANICS FOR STRUCTURES, Louis L. Bucciarelli. This text explores the mechanics of solids and statics as well as the strength of materials and elasticity theory. Its many design exercises encourage creative initiative and systems thinking. 2009 edition. 320pp. 6 1/8 x 9 1/4. 0-486-46855-0

FEEDBACK CONTROL THEORY, John C. Doyle, Bruce A. Francis and Allen R. Tannenbaum. This excellent introduction to feedback control system design offers a theoretical approach that captures the essential issues and can be applied to a wide range of practical problems. 1992 edition. 224pp. 6 1/2 x 9 1/4. 0-486-46933-6

THE FORCES OF MATTER, Michael Faraday. These lectures by a famous inventor offer an easy-to-understand introduction to the interactions of the universe's physical forces. Six essays explore gravitation, cohesion, chemical affinity, heat, magnetism, and electricity. 1993 edition. 96pp. 5 3/8 x 8 1/2. 0-486-47482-8

DYNAMICS, Lawrence E. Goodman and William H. Warner. Beginning engineering text introduces calculus of vectors, particle motion, dynamics of particle systems and plane rigid bodies, technical applications in plane motions, and more. Exercises and answers in every chapter. 619pp. 5 3/8 x 8 1/2. 0-486-42006-X

ADAPTIVE FILTERING PREDICTION AND CONTROL, Graham C. Goodwin and Kwai Sang Sin. This unified survey focuses on linear discrete-time systems and explores natural extensions to nonlinear systems. It emphasizes discrete-time systems, summarizing theoretical and practical aspects of a large class of adaptive algorithms. 1984 edition. 560pp. 6 1/2 x 9 1/4. 0-486-46932-8

INDUCTANCE CALCULATIONS, Frederick W. Grover. This authoritative reference enables the design of virtually every type of inductor. It features a single simple formula for each type of inductor, together with tables containing essential numerical factors. 1946 edition. 304pp. 5 3/8 x 8 1/2. 0-486-47440-2

THERMODYNAMICS: Foundations and Applications, Elias P. Gyftopoulos and Gian Paolo Beretta. Designed by two MIT professors, this authoritative text discusses basic concepts and applications in detail, emphasizing generality, definitions, and logical consistency. More than 300 solved problems cover realistic energy systems and processes. 800pp. 6 1/8 x 9 1/4. 0-486-43932-1

THE FINITE ELEMENT METHOD: Linear Static and Dynamic Finite Element Analysis, Thomas J. R. Hughes. Text for students without in-depth mathematical training, this text includes a comprehensive presentation and analysis of algorithms of time-dependent phenomena plus beam, plate, and shell theories. Solution guide available upon request. 672pp. 6 1/2 x 9 1/4. 0-486-41181-8